Updates in Hypertension and Cardiovascular Protection

Series Editors
Giuseppe Mancia, Milan, Italy
Enrico Agabiti-Rosei, Brescia, Brescia, Italy

The aim of this series is to provide informative updates on both the knowledge and the clinical management of a disease that, if uncontrolled, can very seriously damage the human body and is still among the leading causes of death worldwide. Although hypertension is associated mainly with cardiovascular, endocrine, and renal disorders, it is highly relevant to a wide range of medical specialties and fields – from family medicine to physiology, genetics, and pharmacology. The topics addressed by volumes in the series *Updates in Hypertension and Cardiovascular Protection* have been selected for their broad significance and will be of interest to all who are involved with this disease, whether residents, fellows, practitioners, or researchers.

More information about this series at https://link.springer.com/bookseries/15049

C. Venkata S. Ram
Boon Wee Jimmy Teo • Gurpreet S. Wander
Editors

Hypertension and Cardiovascular Disease in Asia

Editors
C. Venkata S. Ram
Apollo Hospitals
Apollo Institute for Blood Pressure
Management
Hyderabad, India

University of Texas
Southwestern Medical Center
Dallas, USA

Macquarie University
Faculty of Medical and Health Sciences
Sydney, Australia

Gurpreet S. Wander
Department of Cardiology
Hero DMC Heart Institute, Dayanand
Medical College
Ludhiana, Punjab, India

Boon Wee Jimmy Teo
Head, Division of Nephrology
Department of Medicine
Yong Loo Lin Sch of Med Natl Univ of Sin
Singapore, Singapore

ISSN 2366-4606   ISSN 2366-4614   (electronic)
Updates in Hypertension and Cardiovascular Protection
ISBN 978-3-030-95736-0   ISBN 978-3-030-95734-6   (eBook)
https://doi.org/10.1007/978-3-030-95734-6

This Springer imprint is published by the registered company Springer Nature Switzerland AG
The registered company address is: Gewerbestrasse 11, 6330 Cham, Switzerland

Foreword

Understanding and addressing cardiovascular disease (CVD) and hypertension in Asia is like completing a challenging 1000-piece jigsaw puzzle. The problem is so grand that it accounts for the 10.8 million annual deaths in Asia, comprising 58% of the 18.6 million reported CVD deaths worldwide in 2019. Close to 40% of these deaths were deemed premature, defined as deaths before the age of 70 years. Uncontrolled hypertension remains the real-world "Freddy Krueger" which kills its victims, not with razors and gloved hands, but with heart attacks, strokes, and other hypertension-related complications.

Describing hypertension as a serial killer is actually an understatement. It has killed in pandemic proportions, and the excess deaths due to it have apparently increased in the past 2 years, as the whole world grapples from the deadly strangle of another pandemic—COVID-19. A double whammy indeed!

And just like the complexity of a multi-piece puzzle, the CVD and hypertension enigma in Asia is the consequence of the complex effects of varied factors among the countries in the world's most populated continent. These include the diverse socioeconomic, cultural and living environments, demography, and Westernized lifestyles. With different healthcare systems and strategies being employed, Asian countries also differ in their capacity to achieve CVD prevention and treatment goals.

With the multifactorial diversity and heterogeneity, it is understandable that the CVD spectrum may also differ in Asia as a region. It is well known that the cardiovascular phenotype is quite different in Asia, with stroke being associated more with hypertension than ischemic heart disease; in contrast to the Western world wherein the opposite is true. Within some of the countries comprising Asia, the diversity in clinical disease presentation may also be apparent, since some countries may be characterized by a multiethnic composition. The CVD and hypertension pandemic may also pan out differently among the different Asian countries, depending on the socioeconomic stage or development a particular Asian country is in.

It is humbling to admit that we still find some odd pieces that don't fit in the CVD and hypertension jigsaw puzzle in Asia. And just like when we solve challenging puzzles, we need to sort out the pieces first, so we would know where they might fit. Gathering all the updated information one could gather on all aspects of hypertension and CVD in Asia—from epidemiology to disease presentation to management—is a big step in helping clinicians, researchers, and policy makers understand the complex picture which hypertension and CVD in Asia present.

This book, *Hypertension and Cardiovascular Disease in Asia*, is definitely a big accomplishment that fills in a major gap in the understanding and management of hypertension and CVD in a continent that is considered the disease epicenter in the world. Our hats off to the editors—Doctors C. Venkata S. Ram, Boon Wee Jimmy Teo, and Gurpreet S. Wander; and all the world-renowned contributing authors for this remarkable feat.

This book will definitely occupy a prominent place in my library; I hope it does in yours, too.

Rafael R. Castillo
(Retired), Adventist University of the Philippines, College of Medicine
Silang, Cavite, Philippines

Preface

We are privileged to introduce this book on *Hypertension and Cardiovascular Disease in Asia* to readers interested in preserving public health by reducing the disease burden in the community. As Asia is going through a cardiovascular disease endemic, it is timely to push back the wave through medical education and preventive strategies. The grim statistics show a high incidence of cardiovascular disease in Asia compelling policy makers and medical practitioners to action. Advances in prediction technology will enable medical practitioners to take an active stance in disease prevention, changing the paradigm from disease management to "anticipatory" medicine. While individual patient-level disease prevention is important, a greater impact is achieved if the majority of the population can be prevented from ever having hypertension and cardiovascular disease.

The current epidemiology, clinical profile, and consequences of hypertension and cardiovascular disease in Asia are discussed extensively. From population science to individual therapeutic implications, clinicians will gain additional knowledge. We should discard the status quo and embark on a firm footing to safeguard public health in Asia. We will live a longer and a healthier life. One unique feature of this compendium is that it does not have the typical textbook structure where a sequence of reading is required; instead you can start the journey from any chapter based on your preference or interest.

We would like to thank the European Society of Hypertension and its leadership (Drs. Giuseppe Mancia, Enrico Agabiti-Rosei, and Reinhold Kreutz) for giving us the special opportunity to assemble this book devoted to Asian countries. It is timely that the ESH asked us to look at the pattern of hypertension and cardiovascular disease in Asia. The term "Eurasia" indicates the contiguous landmass of Europe and Asia. There is no geographic demarcation between the two continents. It is an ethnic and cultural demarcation at best. Asia has 49 countries with 5 being transcontinental (partly in Asia and Europe) and is the largest continent of the world. It has 30% of the land area of the world and has 60% of the world population. The pattern of these diseases in Asia is important also to Europeans. With increasing international migration, public health becomes an important matter of great economic significance. Asia has very diverse climates and cultures. The United Nations statistics department divides Asia into 6 regions. The body build, genetic ancestry, disease prevalence, and phenotypes are different due to varying climates and diverse lifestyles.

The economic situation of different regions is also variable. These challenges should be kept in mind while going through this book.

We are grateful to all the authors for their valuable and expert contributions to this book. They have covered the various aspects of hypertension and cardiovascular disease with an Asian perspective. They have a vast experience in the field and have done original work in cardio-metabolic diseases in the Asian countries. We are grateful to the staff of Springer Nature publications for working with us closely in completing this book. We acknowledge the excellent managerial help provided by Ms. N.Madhavi Latha and meticulous proofing and corrections done by Ms. Nicola Ryan.

Hyderabad, India   C. Venkata S. Ram
Singapore, Singapore   Boon Wee Jimmy Teo
Ludhiana, India   Gurpreet S. Wander

Contents

Epidemiology of Hypertension in Asia

Rody G. Sy, Elmer Jasper B. Llanes,
Felix Eduardo R. Punzalan, Jaime Alfonso M. Aherrera,
and Paula Victoria Catherine Y. Cheng

1.1 Geographical Definition of Asia

Asia is the largest and most populous continent, with roughly 60% of the total population in the world. It is divided into five regions, namely Middle East, East Asia, South Asia, Southeast Asia, and Central Asia. Asian countries of interest in this chapter would include Afghanistan, Armenia, Azerbaijan, Bahrain, Bangladesh, Bhutan, Brunei, Cambodia, China, Georgia, India, Indonesia, Iran, Iraq, Israel, Japan, Jordan, Kazakhstan, Kuwait, Kyrgyzstan, Laos, Lebanon, Malaysia, Maldives, Mongolia, Myanmar, Nepal, North/South Korea, Oman, Pakistan, Philippines, Qatar, Russia, Saudi Arabia, Singapore, Sri Lanka, Syria, Tajikistan, Thailand, Timor-Leste, Turkey, Turkmenistan, United Arab Emirates, Uzbekistan, Vietnam, and Yemen. Much of the literature reviewed would involve the countries in Southeast Asia (SEA) (e.g., Malaysia, Indonesia, Thailand, Singapore, Philippines, Vietnam, Laos, Cambodia, Myanmar, Brunei, East Timor), Japan, and countries in the middle East. This chapter discusses the prevalence, incidence, and risk factors of hypertension across the Asian countries. It is important to highlight that the various literature reviewed here would have different definitions of hypertension and blood pressure control, as well as different methods in determining blood pressure measurements. Period of evaluation also varies from study to study. Caution must be exercised when interpreting the presented numbers, specifically for the prevalence and incidence data.

R. G. Sy (✉) · E. J. B. Llanes · F. E. R. Punzalan
J. A. M. Aherrera · P. V. C. Y. Cheng
Division of Cardiovascular Medicine, Department of Medicine,
Philippine General Hospital, University of the Philippines, Manila, Philippines
e-mail: ebllanes@up.edu.ph; frpunzalan@up.edu.ph;
jmaherrera@up.edu.ph; pycheng@up.edu.ph

© The Author(s), under exclusive license to Springer Nature Switzerland AG 2022
C. V. S. Ram et al. (eds.), *Hypertension and Cardiovascular Disease in Asia*,
Updates in Hypertension and Cardiovascular Protection,
https://doi.org/10.1007/978-3-030-95734-6_1

1.2 Prevalence of CVD and Hypertension as a Risk Factor Worldwide

Hypertension is a major cause of premature death worldwide and remains as the most common, readily recognizable, and modifiable risk factor for myocardial infarction, stroke, heart failure, atrial fibrillation, and peripheral arterial disease. It is estimated that 1.13 billion people worldwide have hypertension, most (two-thirds) of whom live in low- and middle-income countries [1]. Although hypertension has often been regarded as a "silent killer," it is still a major modifiable risk factor for cardiovascular morbidity and mortality. In 2013, death from complications of hypertension (e.g., coronary artery disease, stroke) was estimated by the WHO to be responsible for 9.4 million deaths worldwide every year [2]. The global burden of hypertension is rising due to escalating obesity and population aging, and is projected to affect 1.5 billion persons—one third of the world's population—by 2025 [3–5]. High blood pressure (BP) continues to be the largest single contributor to the global burden of disease, causing two-thirds of all cerebrovascular accidents (strokes) and half of all ischemic heart disease worldwide, and thus 9.4 million deaths each year [3].

1.2.1 Definitions of Hypertension

Historically, hypertension has long been defined as a systolic blood pressure (SBP) of more than or equal to 140 mmHg and/or a diastolic blood pressure (DBP) of 90 mmHg or more. Over the past decade, guidelines have been revised to adapt to the evolving data. In 2017, the American College of Cardiology/American Heart Association (ACC/AHA) taskforce has lowered the cutoffs for the definition of hypertension—130–139 mmHg SBP/80–89 mmHg DBP [6]; whereas the European Society of Cardiology retains the traditional definition of more than or equal to 140/90 mmHg [7]. In Asia, certain countries have developed their own definitions of hypertension, summarized in Table 1.1.

1.3 Prevalence of Hypertension in Asia

Blood pressure levels and the prevalence of hypertension vary among countries and among subpopulations within a country. The overall prevalence of hypertension in Asia in 2015 has been similar to the global prevalence [18]. During the past four decades, the highest worldwide BP levels have shifted from high-income countries to low-income countries in south Asia, while BP remained persistently high in central and eastern Europe [19]. In Southeast Asia (SEA), about one-third of the adult population have hypertension and an estimated 1.5 million deaths are associated with hypertension annually [20]. A comprehensive review for SEA reported a prevalence of hypertension at 35% [21]. In the Urban population, a systematic review was performed in 2021 and found that out of 37,630 individuals, 12,842 had hypertension [PPE of 32.14%, (95% CI: 30.2–3.8)] [4]. Table 1.2 summarizes the available data on the prevalence of hypertension across Asian countries.

Table 1.1 Definitions of hypertension across Asian countries (compared to AHA and ESH) [8]

	SBP <120 and DBP <80	SBP 120–129 and DBP <80	SBP 130–139 and/or DBP 80–89	SBP 140–159 and/or DBP 90–99	SBP 160–179 and/or DBP 100–109	SBP ≥180 and/or DBP ≥110	SBP ≥140 and/or DBP <90
China 2018 [9]	Normal	High normal	High normal	Grade 1 HPN	Grade 2 HPN	Grade 3 HPN	ISH
Korea 2018 [10]	Normal	Elevated	Pre-HPN	Grade 1 HPN	Grade 2 HPN	Grade 2 HPN	ISH
Japan 2019 [11]	Normal	High normal	Elevated	Grade 1 HPN	Grade 2 HPN	Grade 3 HPN	ISH
Singapore 2017 [12]	Normal	Normal[a]	High normal[b]	Grade 1 HPN	Grade 2 HPN	Grade 3 HPN	ISH
Malaysia 2006 [13]	Optimal	Normal[a]	High normal[b]	Stage I HPN	Stage II HPN	Stage III HPN	ISH
Thailand 2015 [14]	Optimal	Normal[a]	High normal[b]	Grade 1 HPN	Grade 2 HPN	Grade 3 HPN	ISH
India 2019 [15]	Optimal	Normal[a]	High normal[b]	Stage 1 HPN	Stage 2 HPN	Stage 3 HPN	ISH[c]
Saudi Arabia 2018 [16]	Normal	Pre-HPN	Pre-HPN	Grade I HPN	Grade II HPN	Grade III HPN	ISH
AHA 2017 [6]	Normal	Elevated	Grade 1 HPN	Grade 2 HPN	Grade 2 HPN	Grade 2 HPN	N/A
ESH 2018 [7]	Optimal	Normal[a]	High normal[b]	Grade 1 HPN	Grade 2 HPN	Grade 3 HPN	ISH
Philippines 2020 [17]	Normal	Borderline	Borderline	HPN	HPN	HPN	N/A

[a] DBP: 80–84 mmHg
[b] DBP: 85–89 mmHg
[c] Grade 1 ISH is defined as SBP 140–159 mmHg and DBP <90 mmHg. Grade 2 ISH is defined as an SBP >160 mmHg and DBP <90 mmHg

Table 1.2 Prevalence of hypertension in Asia

Country	Author	N	Prevalence
Central Asia			
Tajikistan	*Demographic and Health Survey 2017* [22]		6% in women aged 15–49 y/o
Kyrgyzstan	*Polupanov, 2020* [23]	1330	34.1% (36.7% in men, 30.5% in women)
Kazakhstan	*Supiyev, 2015* [24] *Aringazina, 2018* [25]	497	70% (65% in men, 75% in women) 24.3%
Uzbekistan	*Aringazina, 2018* [25]	5462	8.3%
Russia	*Erina, 2019* [26]	20,652	50.2%
East Asia			
China	*Wang, 2020* [27] *Li, 2020* [28]	797 983,476	24.2%[a] 41.3%
Japan	*Ministry of Health, Labor and Welfare, Japan, 2017* [29]		Age 40–74: 60% in men; 41% in women Age > 75: 74% in men; 77% in women
Mongolia	*Potts, 2020* [30] *Li, 2016* [31]	4515 3251	25.6% (31.9% in men, 25.5% in women) 28.6% (13.9% in men, 14.7% in women)
South Korea	*Kim, 2021* [32]		23.5% (28% in men, 18.6% in women)
Hong Kong	*Population Health Survey* [33]	2347	27.7% (30.1% in men, 25.5% in women)
Taiwan	*Cheng, 2020* [34]		24.1%
Middle East			
Jordan	*Khader, 2019* [35]	4056	41.4% in men, 28.3% in women
Syria	*Ratnayake, 2020* [36]	915	17.2% ≥ 18y/o, 39.2% ≥ 30y/o
Egypt	*Ibrahim, 1995* [37]	6733	26.3%
Middle East*	*Yusufali, 2017* [38]	10,516	33%
Saudi Arabia	*Saeed, 2013* [39] *Alhabib, 2020* [40]	10,735 2047	15.2% 30.3% (32.8% in men, 27% in women)
Qatar	*Bener, 2004* [41]	1208	32.1%
Turkey	*Sengul, 2016* [42]	5437	30.3% (28.4% in men, 32.3% in men)
Israel	*Abu-Saad, 2014* [43]	763	32.5% (36.1% in men, 28.9% in women)
Iran	*Eghbali, 2018* [44] *Yusufali, 2017* [38]	2107 6013	17.3% (18.9% in men, 15.5% in women) 28%
Iraq	*Saka, 2020* [45]	1480	54.7% (63.4% in men, 51.8% in women)[b]
Kuwait	El-Reshaid, 1999 [46]	2836	26.3% (28.3% in men, 22.9% in women)

Table 1.2 (continued)

Country	Author	N	Prevalence
Oman	Al Riyami, 2003 [47]	7011	25.2%
Palestine	Yusufali 2017 [38]	1545	38%
UAE	Yusufali 2017 [38]	917	52%
South Asia			
India	Gupta, 2002 [48]	1673	36.4% in men, 37.5% in women
	Mohan, 2001 [49]	1175	
	Gupta, 1999 [50]	99,589	14%
	Anchala 2014 [51]	326,644	43.8% in men, 44.5% in women
			29.8%
Sri Lanka	Katulanda, 2014 [52]	4485	23.7% (23.4% in men, Women 23.8%)[b]
Bangladesh	Islam, 2018 [53]	1036	6.9% (4.5% in men, 8.9% in women)[b]
	Hasan, 2021 [54]	4856	31% (23.6% in men, 38.1% in women)[b]
Pakistan	Jafar, 2003 [55]	9442	19% (20.2% in men, 18% in women)
Nepal	Prajapati, 2020 [56]	617	23% (27.6% in men, 22.4% in women)
	Karmacharya, 2017 [57]	1073	27.8% (25.6% in men, 12.2% in women)
Maldives	Aboobakur, 2010 [58]	2028	31.5%
Southeast Asia			
Brunei	Lupat 2016 [59]	5063	48.3%
Cambodia	Gupta 2013 [60]	5017	15.3%
Indonesia	Peltzer 2018 [61]	29,965	33.4% (31% in men, 35.4% in women)
	Christiani 2016 [62]	15,802	32.1%
Laos	Lao Peoples 2010 [63]	4180	24.9% in men; 20% in women
Malaysia	Ab Majid 2018 [64]	11,288	35.3%
	Abdul-Razak 2016 [65]		47.9% (50.9% in men, 45.6% in women)
Myanmar	Bjertness 2016 [66]	7319	29.9%
Singapore	NPHS 2017 [67]	4100	21.5% (23.6% in men, 19.6% in women)
	Liew SJ 2019 [68]	10,215	31.1%[c]
Thailand	Tiptaradol 2012 [69]	36,877	19.1%
Vietnam	Son 2012 [70]	9823	25.1%
	Meiqiari 2019 [71]		21.1%
Philippines	NNHES 2019 [72]	3334	19%
	Sison 2013 [73]		28%

[a] No difference between sexes but prevalence increasing with age
[b] Increasing with age
[c] Chinese 1519/4817 = 31.5%; Malay 879/2659 = 33.1%; Indian 780/2739 = 28.5%
[*] Includes Iran, Palestinian Territory, Saudi Arabia, UAE

1.3.1   Central Asia

Prevalence of hypertension in this region of Asia is highly varied. The countries with the low overall prevalence are Tajikistan at 6% and Uzbekistan at 8.3%. These data came from their national surveys which mostly included women and relatively young individuals aged 15–50 years old which could explain the low prevalence of hypertension in these areas. However, in Russia, Kyrgyzstan, and Kazakhstan the prevalence rates were 50.2%, 34.1%, and 70%, respectively since it included a wider age group. No significant differences were seen in both sexes in this region.

1.3.2   East Asia

The prevalence in this region was somewhat similar in every country approximately at 25% with the exception of China having a higher rate at 41.3%. Men and increasing age showed a higher prevalence rate.

1.3.3   Middle East Asia

Hypertension prevalence in this region ranges from 25 to 54.7%. Some countries have no recent data such as in Kuwait and Egypt wherein published dated back to the 1990s. Others have very small population and even are refugees brought about by civil unrest. These factors are factors to the widely varied prevalence rate in this region. Men and increasing age were more prevalent for hypertension.

1.3.4   South Asia

All countries in this region have a prevalence rate of hypertension of approximately 30% with the exception of Pakistan at 19%. However, the Pakistani data was taken nearly two decades ago. More hypertensive women than men were seen in this region. Older individuals showed an increasing prevalence rate as well.

1.3.5   Southeast Asia

Prevalence of hypertension was low in Thailand and Cambodia at 19.1% and 15.3%, respectively while higher in Brunei and Malaysia at 48%. There were no significant differences in prevalence between sexes. Older age remains to be a factor for increasing prevalence of hypertension which was similar to other regions in Asia.

The prevalence of hypertension is increasing fastest in developing countries, where poor hypertension treatment and control contribute to the growing epidemic of cardiovascular disease. For example, in India, the prevalence of raised BP rose from 5% in the 1960s to 12% in the 1990s, to >30% in 2008 [74]. The prevalence increases with age, from 13.7% in the third decade to 64% in the sixth decade. The reasons for the recent rural–urban convergence in hypertension are not well understood but could be due to the recent rapid changes in the lifestyle of those living in rural areas including increase in salt intake [75]. In Indonesia, the prevalence rose from 8% in the year 1995 to 32% in the year 2008 and 33.4% in the year 2015 [61, 74]. In Myanmar, the Ministry of Health reported an increase in hypertension prevalence, from 18% to 31% in males, and from 16 to 29% in females during 2004–2009 [74]. In the Philippines, the prevalence of hypertension has been increasing. Several cross-sectional studies have shown that the numbers are steadily increasing; from 11% in 1992 to 25% in 2008. The National Nutrition and Health Survey (NNHES) of the Food and Nutrition Research Institute conducted in 2012 indicated a small decline in the prevalence of individuals with hypertension, about 22.3% [76, 77]. Later data gathered in 2018 from FNRI (2019) showed the prevalence of elevated blood pressure at 19%. It was higher among males (20.7%) compared to females (15.7%) [72]. In China, in 2008, the sample included 131,520 individuals with a mean age of 54.1 years (SD, 9.7), of which 64,430 (48.9%) were men. In 2018, 10,926 participants accomplished the study, with 4390 men (40.2%) and an average age of 60 years. The crude prevalence of hypertension was 44.3% in 2008 and 60.6% in 2018. The age-standardized prevalence increased from 44.7 to 53.6% from 2008 to 2018 [78].

The prevalence of hypertension in Singapore fell from 27.3% in 1998 to 24.9% in 2004 and 23.5% in 2010 [67]. In Malaysia, prevalence remained largely stable at 34.6% in 2006, 33.6% in 2011, and 35.5% in 2015 [64]. In Korea, age-standardized prevalence of hypertension modestly decreased from 26% in 1998 to 23.5% in 2018. However, the absolute number of people with hypertension, has exceeded 12 million [32]. In Jordan from 2009 to 2017, there was nonsignificant decrease in hypertension prevalence of 2.7% among men and 1.1% among women [35]. The trend in the shifting of hypertension burden from higher income countries to lower income countries has also been suggested in a recent analysis. A pooled global age-standardized prevalence of raised BP found that, over the past four decades, mean BP decreased from 1975 to 2015 in high-income Western and Asia Pacific countries. Highest worldwide BP levels have shifted from high-income countries to low-income countries in South Asia and Sub-Saharan Africa [18, 19].

1.4 Incidence of Hypertension in Asia

There were several cohort studies that investigated the development of hypertension over a period termed as incidence rate [79–87]. Difference in follow-up time frame, different number of population size studied and expressed rates (in percentage or in person-years) makes it difficult to compare between countries. Published data on incidence rates are quite sparse but from here we can identify factors that have contributed to the development of hypertension.

Obesity, overweight, and increasing age are common factors that lead to hypertension [79–87]. Hypertriglyceridemia was a unique predictive factor for Iran as

Table 1.3 Incidence rate of hypertension in different Asian countries

Thailand	*Thawornchaisit, 2013* [79]	57,588 Thai open university open 4-year follow-up	4% (Men 5.2%, Women 2.1%)
Tehran, Iran	*Asgari, 2020* [80]	6533 Tehran Lipid and Glucose study 13.1-year follow-up	All 36.1 per 1000 person-years Men 34.9 per 1000 person-years Women 38.7 per 1000 person-years
Central Iran	*Talaei, 2013* [81]	3283 Isfahan Cohort 7-year follow-up	22.1%
Chennai, India	*Mohan, 2020* [82]	1650 CURES cohort 9-year follow-up	28.7 per 1000 person-years
Chennai/ Karachi, India	*Prabhakaran, 2017* [83]	12,504 CARRS cohort 2-year follow-up	16.2%
China	*Luo, 2021* [84]	12,952 China Health and Nutrition Survey (CHNS)	40.8 per 1000 person-years (95%CI 38.3–43.4) between 1993 and 1997 48.6 per 1000 person-years (95%CI 46.1–51.0) between 2011 and 2015
Korea	*Yu, 2021* [85]	21,172 National Health Insurance Service-Health Screening 12-year follow-up	76.4%
Japan	*Oka, 2018* [86]	22.45 million who underwent med check-up last 2010	Wakayama prefecture 28.3% for men 20.9% for women
Philippines	*LIFECARE Sy, unpublished* [87]	2089 Filipinos 4-year follow-up	15%

seen in its two cohorts [80, 81]. Glucose intolerance was a predictive factor for both Iran and India [80–83] while alcohol consumption was both common factors for India and China [83, 84]. High normal blood pressure levels were a factor for Iran and the Philippines [80, 87]. In the cohort of Japan, an increase in level of activity measured by a pedometer and usage of public transport negatively correlated with development of hypertension [86] (Table 1.3).

In Thailand, hypertension was associated with increasing age, BMI, and the presence of comorbidities. Income and education were not predictive factors. In men, the risk for hypertension is with physical inactivity, smoking, alcohol, and fast food intake. In women, hypertension was correlated with having a partner [79].

A study in Iran, the multivariate-adjusted model controlled for all study covariates showed that significant contributors to hypertension include age, male sex, general and central obesity, hypertriglyceridemia, impaired fasting glucose, diabetes mellitus, baseline BP at least 120/80 mmHg (nonoptimal BP), and parental history of hypertension. Higher education level and more than 10% decrease in waist circumference over 7-year follow-up represented protective effects [81].

In the Chennai and Karachi regions in India, higher body weight, BMI, age, glucose intolerance, low socioeconomic status, and current heavy alcohol use were significantly associated with an increased risk of incident hypertension [83].

Incident hypertension in China increased with age, BMI, and increased alcohol consumption, while lower risk was noted with higher educational attainment and the female sex [84].

In Japan, unique associations between living conditions and risk of hypertension were seen. Incidence of hypertension showed a significant negative correlation with walking and medical check-ups, and a significant positive correlation with light-vehicle usage and slope of habitable land. Between the number of steps and variables related to the living environment, number of rail stations showed a significant positive correlation, while, standard and light-vehicle usage showed significant negative correlation [84].

An unpublished study in the Philippines by Sy et al. in 2015 (LIFECARE study) determined that predictive factors for hypertension include age, male sex, abdominal obesity, family history of hypertension, and pre-hypertension BP levels [87].

1.5 Awareness and Control of Hypertension in Asian Countries

There is an abundance of literature for hypertension control in the Western population, but data in Asia has been limited. In 2015, 1 in 4 men and 1 in 5 women had hypertension [2]. Fewer than 1 in 5 people with hypertension have the problem under control [2]. Awareness of a diagnosis of hypertension, treatment rate, and BP control vary from country to country; and available data are summarized in Table 1.4. Note that the numbers presented are less favorable compared to Western data such as in the United States showing hypertension awareness at 81%, treatment rates at 74%, and hypertension control at 53% [88].

Hypertension awareness is an important initial step toward BP control. In several countries in SEA, awareness level of hypertension is less than 50%, but 56–70% in the more affluent countries in the region. Of those who are aware, about half are on treatment, but control rates to BP levels below 140/90 remains dismally low [77]. Generally, BP control is low in the Asian countries, compared to Western nations with treatment rates ranging from 51 to 80% and control rates ranging from 27 to 66% [18, 88].

The term "control" has been variably defined in various guidelines. The ACC/AHA guidelines in 2017 suggest a BP target <140/90 mmHg for those with an atherosclerotic cardiovascular disease (ASCVD) risk of less than 10% and a target of <130/80 mmHg for those with a 10-year ASCVD risk of >10%. The Singaporean guidelines in 2017 [12] and Malaysian guidelines in 2013 [13] have a similar recommended target of <140/90 for most patients, <150/90 mmHg for those 80 years and above, <140/80 mmHg for those with diabetes mellitus, and <130/80 mmHg for those with significant albuminuria. A systematic review evaluating medication adherence factors for hypertension in developing countries cited other factors such

Table 1.4 Awareness of diagnosis, treatment rate, and overall blood pressure control for patients with hypertension in different Asian countries

Country	Author	Awareness	Treatment rate[a]	Overall BP control[b]
Singapore	*Epidemiology 2010* [90]	73.7%		67.4%
	Liew, 2019 [68]	51%	74.8%	37.6%
Thailand	*Tiptaradol, 2012* [69]	48.4%	42%	14.9%
Vietnam	*Son, 2012* [70]	48.4%	61%	36.3%
	Meiqiari, 2019 [71]	9.3%	4.7%	
Myanmar	*Bjertness, 2016* [66]	27.8%	8.7%	
Malaysia	*Ab Majid, 2018* [64]		83.2%	37.4%
Laos	*Lao People 2010* [63]		<20%	
Indonesia	*Peltzer, 2018* [61]	42.9%	11.5%	14.3%
Philippines	*Sison, 2007* [91]	16%		20%
	Sison, 2013 [73]	19%	75%	27%
India	*Wander, 2018* [92]	42%	38%	20%
Middle East[a]	*Yusufali, 2017* [38]	49%	47%	19%
Jordan	*Khader, 2019* [35]	57.7% men		30.7% men
		62.5% women		30.7% women
Saudi Arabia	*Saeed, 2013* [39]	57.8%	78.9%	45%
	Alhabib, 2020 [40]	61.1%	58.9%	30.6%
China	*Xing, 2019* [93]	48.5%	38%	14.9%
	Xing, 2020 [78]	54.2%—2008	42%—2008	3.7%—2008
		47.5%—2018	35.4%—2018	3.6%—2018
Japan	*JSH 2019* [11]	67%	56%	40% men
				45% women
South Korea	*Kim, 2021* [94]	25%—1998	22%—1998	5%—1998
		65%—2016	61%—2016	44%—2016
Taiwan	*Cheng, 2019* [34]	72.8%	57.2%	63.4%
Turkey	*Sengul, 2016* [42]	40.7%—2003	31.1%—2003	8.1%—2003
		54.7%—2012	47.4%—2012	28.7%—2012
Kuwait	*El-Reshaid, 1999* [46]	23%		
Iran	*Eghbali, 2018* [44]	69.2%	92.4%	59.2%
Nepal	*Karmacharya, 2017* [57]	43.6%	76.1%	35.3%
Mongolia	*Li, 2016* [31]	66.34%	87.03%	43.58%
	Potts, 2020 [30]	69.7%	46.8%	24%
Kazakhstan	*Supiyev, 2015* [24]	91%	77%	34%

[a] Number of hypertensive patients on treatment
[b] Defined as BP < 140/90

as cost barriers, irregular follow-ups, and competing availability of traditional herbal remedies [18, 89].

1.6 Risk Factors of Hypertension

The risk factors for developing hypertension are divided into modifiable and non-modifiable risk factors. The modifiable risk factors include diet, physical activity, alcohol consumption and tobacco smoking, and obesity or being overweight. In contrast, the nonmodifiable risk factors include family history of hypertension, age >65 years, and the presence of other comorbidities, including diabetes and chronic kidney diseases [4, 95]. In a 2021 review, strong risk factors associated with hypertension are male sex, low education status, low socioeconomic level, higher BMI, increased waist circumference, dyslipidemia, and smoking [4].

1.6.1 Salt Intake

Dietary salt intake, a strong risk factor for hypertension, also contributes significantly to the poor control of hypertension in SEA. Salt intake in SEA is high, ranging 10–17 g a day, which is 2–3 times as high as the recommended daily salt consumption of <5 g/day by the World Health Organization [18, 96]. In countries such as Japan, high salt intake is one of the risk factors purported to be the cause of high prevalence of hypertension and stroke in the country, as salt intake in the 1950s using 24 h urine collection was estimated to be as high as 25 g/day. In the more recent National Health and Nutrition survey in Japan in 2016, the mean daily intake has decreased to 9.9 g with a goal of further reducing it to <8 g for men and <7 g for women by the year 2022 [11]. This can be one factor as to the declining prevalence of hypertension in the country. In Korea, estimated daily intake of salt is around 10 g. Several studies have shown an independent association between salt intake and blood pressure among patients with metabolic syndrome [10]. In Ningbo, China, the mean dietary intake was noted to be 13 g, and this was significantly associated with the risk of developing hypertension and pre-hypertension [97]. The Taiwan Society of Cardiology Hypertension guidelines recommend limiting sodium to 2–4 g/day as part of the lifestyle interventions to control hypertension [98].

1.6.2 Low Education Status and Low Socioeconomic Level

In the present study, low education and low socioeconomic level are associated with hypertension among the urban population [4, 62, 70, 99]. This is probably because higher education may aid better health literacy of the importance of a healthy lifestyle (e.g., diet, physical activity, regular check-ups) [4]. Likewise, individuals with a low socioeconomic status might be unemployed or may have jobs with fewer health benefit packages [100, 101]. In India for example, about 70% of patients meet treatment expenses "out of pocket" since they have no insurance coverage. Treatment cost has important bearing on drug compliance in India [15]. In a published study from China looking into the effect of socioeconomic and geographical factors affecting hypertension and the presence of usual comorbidities (dyslipidemia, diabetes, and coronary artery disease), it was shown that hypertensive patients with higher incomes had higher levels of cholesterol but hypertensive patients with lower incomes had higher prevalence of coronary artery disease. This was postulated to be due to increased intake of fatty food among those with higher income, but with increased opportunities to intervene earlier to prevent coronary artery disease as compared to their lower income counterparts [102].

1.6.3 Overweight/Obesity

A high body mass index has also been linked to hypertension. Being obese or over-weight has a twofold likelihood to be associated with hypertension compared to being nonobese [103, 104]. A high waist circumference is also associated with two- to three-fold increased risk of being hypertensive, compared to a normal waist cir-cumference [103, 105]. The combined presence of high waist circumference and high BMI is superior to individual indices for predicting hypertension [106, 107]. The average prevalence of obesity in SEA is 7.4%, with the highest prevalence in Brunei and Malaysia at 15%, followed by Thailand with 11% [20]. Latest data from Indonesia in 2019 showed prevalence of obesity at 21.8% [108].

On the other hand, in countries like Japan, the phenotype of patients who are hypertensive used to be lean patients with high salt intake. However, in recent years, there was a two-fold increase of the proportion of obese persons, corre-sponding to the increase in metabolic syndrome prevalence in the country. However, overall numbers of obesity prevalence is still very low compared to Caucasian counterparts [11]. In a recent study published in Korea which followed up a total of 115,456 individuals who were prehypertensive over a course of 12 years and they identified risk factors associated with progression to hypertension. A higher BMI ($\geq$30 kg/m^2) was invariably associated with the development of hypertension and is the strongest predictor among the risk factors, while a lower BMI appeared to confer protective effects [109]. Similar findings were seen in South Asia where a meta-analysis published in 2014 showed that general obesity and central obesity were both associated with hypertension [110]. In a systematic review conducted for the Middle Eastern countries, it was found that the overall prevalence of obesity in this region was 24.5%, with more women who were obese compared to men [111].

1.6.4 Alcohol Consumption

Alcohol intake and hypertension have long been associated but its correlation with other factors such as sex or race have been unclear. In a recent study, the pooled analysis of 22 articles with over 400,000 total participants, it was seen that the type of alcohol and sex affects the risk for hypertension at low levels of ethanol intake. In terms of race, black people may have increased risk of hypertension compared to Asians at the same level of alcohol consumption [112]. In another meta-analysis of 36 trials, it was seen that alcohol intake reduction also lowered BP in a dose-dependent manner with a possible threshold effect. This was especially true for patients who had more than two drinks per day [113]. In a meta-analysis in Myanmar, a pooled analysis of three studies showed an associated higher risk to have hypertension in patients with alcohol intake, though no specific amount of alcohol intake was mentioned [114]. The Taiwanese Society of Hypertension rec-ommends limitation of alcohol intake to <20 g of alcohol per day in women and <30 g in men [98].

1.6.5 Smoking

In the 2009 Global Health Risk report of the WHO, tobacco smoking ranks second as the risk factor with the highest attributable deaths worldwide, with smoking accounting for as much as 8.7% of deaths worldwide [115]. In South Asia, the pooled odds ratio of a meta-analysis showed that smokers had a higher likelihood of having hypertension [110]. In SEA, the average prevalence of smoking is around 21.1%, with the highest prevalence seen in Timor-Leste at 42%, followed by Laos at 27%, and the Philippines at 24%. There was a notable higher frequency of smokers among males compared to females. The average prevalence of male smokers is at 38.6%, and of female smokers, 3.7% [20]. In the Middle East, a similar distribution of smokers exists, with 2.9% prevalence in women compared to 28.8% in men. Overall prevalence was estimated at 15.6% [111]. Smoking cessation is not only recommended to control blood pressure but also to reduce the risk of other cardiovascular disease [98].

1.6.6 Physical Inactivity

The association of physical activity and hypertension has been established in so much so as increased physical activity (moderate intensity, at least 30 min per day for 5–7 days a week) decreases BP in patients with established hypertension. Endurance training, more than other types of exercise, can reduce BP the most. This also translates to mortality benefit in patients with hypertension [7]. On the other hand, a recent meta-analysis was not able to establish an association between physical activity and incident hypertension [116]. Data regarding the level of physical activity in the general population in different Asian countries are difficult to ascertain due to lack of published studies on the topic as well as differences in definitions and methods of determining the level of physical activity. In South Asia, a meta-analysis showed that physical inactivity was associated with hypertension [110]. In a different systematic review for South Asians, they found that females, skilled workers, professionals, and those with higher education tended to have more inactive lifestyles [117]. Among the Hui population in China, physical exercise was seen to be a protective factor against having incident hypertension [118]. Regular exercise should therefore be part of the lifestyle interventions prescribed for the control of hypertension [98].

1.6.7 Dyslipidemia

Prevalence of the coexistence of hypertension and dyslipidemia is 15–31%. The coexistence of the two risk factors has more than an additive adverse impact on the vascular endothelium, resulting in enhanced atherosclerosis. The odds of developing hypertension is not as high as compared to the previously mentioned risk factors; nonetheless, it still has a significant association with hypertension and other cardiovascular diseases [62, 99]. Hypercholesterolemia is similarly highly prevalent in SEA. The average prevalence in the region is 41.2%; with the highest prevalence of 57.50% in Singapore, 55.50% in Brunei and Thailand, 52.10% in Malaysia, and 43.4% in the Philippines [119].

1.6.8   Diabetes Mellitus

Hypertension is also a common comorbid condition with type 2 diabetes mellitus (T2DM) in the region, with high BP coexisting in 40–60% of individuals with T2DM, and vice-versa. This dual problem likely accounts for the increased deaths due to cardiovascular disease (CVD), which remains the leading cause of mortality in the continent [77]. In a Korean study of pre-hypertensives, a personal history of T2DM was also shown to be associated with increased risk for developing hypertension [109]. Similar findings were seen in a meta-analysis conducted in South Asia which showed an association between diabetes and dyslipidemia [110]. Diabetes was also prevalent in the Middle Eastern countries with an estimated 10.5% of the population having the disease, with a significant association with obesity [111]. In SEA, T2DM has an average prevalence of 7.18%. In Malaysia and Thailand, the prevalence is 10%, followed by Brunei with 9% [20]. In the Philippines, prevalence is increasing with more than a two-fold increase over 15 years [72]. The double burden of diabetes and hypertension increases the risk for cardiovascular morbidity and mortality and control of both is necessary.

1.7   Burden of Hypertension: Hypertension Related Morbidity and Mortality

South Asians (e.g., India, Pakistan, Bangladesh, Nepal, Sri Lanka) have been shown to experience their first myocardial infarction almost 10 years earlier compared with people from other countries. This increase is largely due to a high prevalence of risk factors, including hypertension [120, 121].

In a project to pool data from major cohort studies in Japan, Evidence for Cardiovascular Prevention from Observational Cohorts in Japan (EPOCH-JAPAN), a meta-analysis of ten cohort studies, with approximately 70,000 included individuals, showed that the association between blood pressure level and cardiovascular mortality risk was almost linear in middle-aged (40–64 years) and early-phase older (65–74 years) people. The slope was stronger in younger people, and the risk was lowest in those with blood pressure levels of <120/80 mmHg. Furthermore, the EPOCH-JAPAN revealed a significant association also between blood pressure level and mortality due to heart failure [122].

Hypertension is known to be associated with cardiovascular morbidity and mortality and is a strong risk factor for cardiovascular deaths. In the recent Japanese Society of Hypertension guidelines published in 2019, hypertension ranks as the number one risk factor with the highest attributable cardiovascular deaths, followed by low physical activity and smoking [11]. A similar proclamation was made by the World Health Organization in 2009 where hypertension was ranked as the leading risk factor with the most attributable deaths globally, estimated to be 12.8% of all deaths worldwide. In the same report, 51% of strokes and 45% of ischemic heart disease deaths worldwide are attributable to hypertension [115].

In a systematic review published in 2020 that investigated the costs of hypertension globally, majority (51.5%) of the money spent is still ascribed to direct costs of the disease such as those used for medications, diagnostics, hospitalization etc. as compared to indirect costs (48.5%) such as the economic losses brought about by absenteeism or premature death due to the disease or its complications. The estimated total cost per year of hypertension can reach as high as US Dollars (USD) 316 trillion in the United States. In the same study, some Asian countries such as Indonesia have also been included with estimated total cost of USD 1.32 trillion for the treatment of hypertension. Estimated cost per person in Indonesia is 30, which was lowest, compared to the USD 6250 that is utilized in the United States, which was highest among the countries included. In terms of costs of complications of hypertension, the amount used for treatment of stroke is one of the most utilized means to measure this value. In the same study, cost of stroke per person can range from USD 380 (Indonesia) to over USD 40,000 (Spain) [123].

1.8   Summary

Hypertension remains to be prevalent in Asia with a shift of the burden from developed to developing countries. The burden for risk factors for hypertension is also high. If these factors are not controlled alongside hypertension, it will lead to significant socioeconomic and health impacts on the Asian population.

References

1. WHO Website. https://www.who.int/news-room/fact-sheets/detail/h.
2. World Health Organization. A global brief on hypertension. Geneva: World Health Organization; 2013.
3. Poulter NR, Prabhakaran D, Caulfield M. Hypertension. Lancet. 2015;386:801–12.
4. Mohammed Nawi A, Mohammad Z, Jetly K, Abd Razak MA, Ramli NS, Wan Ibadullah WAH, Ahmad N. The prevalence and risk factors of hypertension among the urban population in Southeast Asian countries: a systematic review and meta-analysis. Int J Hypertens. 2021;2021:6657003. https://doi.org/10.1155/2021/6657003.
5. Kingue S, Ngoe CN, Menanga AP, Jingi AM, Noubiap JJN, Fesuh B, Nouedoui C, Andze G, Muna WFT. Prevalence and risk factors of hypertension in urban areas of cameroon: a nationwide population-based cross-sectional study. J Clin Hypertens. 2015;17:819–24.
6. Whelton PK, Carey RM, Aronow WS, et al. 2017 ACC/AHA/AAPA/ABC/ACPM/AGS/APhA/ASH/ASPC/NMA/PCNA guideline for the prevention, detection, evaluation, and management of high blood pressure in adults: executive summary: a report of the American college of cardiology/American Heart Association task. Hypertension. 2018;71(6):1269–324. https://doi.org/10.1161/HYP.0000000000000066.

7. Mancia G, De Backer G, Dominiczak A, et al. 2018 ESC/ESH guidelines for themanagement of arterial hypertension. Eur Heart J. 2018;39(33):3021–104. https://doi.org/10.1097/HJH.0b013e3281fc975a.
8. Liu J. Highlights of the 2018 Chinese hypertension guidelines; 2020. p. 4–9.
9. Liu LS, Wu ZS, Wang JG, et al. 2018 Chinese guidelines for prevention and treatment of hypertension—a report of the revision committee of Chinese guidelines for prevention and treatment of hypertension. J Geriatr Cardiol. 2019;16:182–245.
10. Lee H-Y, Shin J, Kim G-H, et al. 2018 Korean Society of Hypertension guidelines for the management of hypertension: part II-diagnosis and treatment of hypertension. Clin Hypertens. 2019;25:4–9.
11. Umemura S, Arima H, Arima S, et al. The Japanese Society of Hypertension guidelines for the management of hypertension (JSH 2019). Hypertens Res. 2019;42:1235–481.
12. Ministry of Health Singapore. MOH clinical practice guidelines 1/2017: hypertension; 2017.
13. Ministry of Health Malaysia. Clinical practice guidelines management of hypertension; 2014.
14. Buranakitjaroen P, Sitthisook S, Wataganara T, et al. 2015 Thai hypertension guideline. 3; 2015.
15. Shah SN, Munjal YP, Kamath SA, et al. Indian guidelines on hypertension-IV (2019). J Hum Hypertens. 2020;34:745–58.
16. Saudi Hypertension Management Society. Saudi hypertension guidelines 2018. Saudi Com Heal Spec. 2018:47–8.
17. Ona DID, Jimeno CA, Jasul GV, et al. Executive summary of the 2020 clinical practice guidelines for the management of hypertension in the Philippines. J Clin Hypertens. 2021:1–14.
18. Chua YT, Wong WK, Gollamudi SPK, Cheang C, Leo H. Hypertension trends in Asia. Hypertens J. 2018;4
19. Zhou B, Bentham J, Di Cesare M, et al. Worldwide trends in blood pressure from 1975 to 2015: a pooled analysis of 1479 population-based measurement studies with 19·1 million participants. Lancet. 2017;389:37–55.
20. World Health Organization. Noncommunicable diseases country profiles. 2018. https://www.who.int/nmh/publications/ncd-profiles-2018/en.
21. WHO south east asia region. Blood pressure, take control. Reg Heal Forum, WHO South-East Asia Reg. 2013;17:1–83.
22. Tajikistan M of H and SP of P. Tajikistan Demographic and Health Survey; 2017.
23. Polupanov AG, Khalmatov AN, Altymysheva AT, Lunegova OS, Mirrakhimov AE, Sabirov IS, Kontsevaya AV, Dzhumagulova AS, Mirrakhimov EM. The prevalence of major cardiovascular risk factors in a rural population of the Chui region of Kyrgyzstan: the results of an epidemiological study. Anatol J Cardiol. 2020;24:183–91.
24. Supiyev A, Kossumov A, Utepova L, Nurgozhin T. Prevalence, awareness, treatment and control of arterial hypertension in Astana, Kazakhstan. A cross-sectional study. Public Health. 2015;129(7):948–53.
25. Aringazina A, Arkhipov V, Aringazina A. Burden of the cardiovascular diseases in Central Asia. Cent Asian J Glob Heal. 2018;7(1):321. https://doi.org/10.5195/cajgh.2018.321.
26. Erina AM, Rotar OP, Solntsev VN, Shalnova SA, Deev AD, Baranova EI, Konradi AO, Boytsov SA, Shlyakhto EV. Epidemiology of arterial hypertension in Russian Federation—importance of choice of criteria of diagnosis. Kardiologiia. 2019;59:5–11.
27. Wang L, Li N, Heizhati M, et al. Prevalence, awareness, treatment, control, and related factors of hypertension among Tajik nomads living in Pamirs at high altitude. Int J Hypertens. 2020;2020:1–10. https://doi.org/10.1155/2020/5406485.
28. Li X, Wu C, Lu J, Chen B, Li Y, Yang Y, Hu S, Li J. Cardiovascular risk factors in China: a nationwide population-based cohort study. Lancet Public Heal. 2020;5:e672–81.
29. Ministry of Health Labour and Welfare Japan. The National Health and Nutrition Survey in Japan; 2016.
30. Potts H, Baatarsuren U, Myanganbayar M, et al. Hypertension prevalence and control in Ulaanbaatar, Mongolia. J Clin Hypertens. 2020;22:103–10.

31. Li G, Wang H, Wang K, et al. Prevalence, awareness, treatment, control and risk factors related to hypertension among urban adults in Inner Mongolia 2014: differences between Mongolian and Han populations. BMC Public Health. 2016;16:1–10.

32. Lee HH, Cho SMJ, Lee H, Baek J, Bae JH, Chung WJ, Kim HC. Korea heart disease fact sheet 2020: analysis of nationwide data. Korean Circ J. 2021;51:2–5.

33. Report of Population Health Survey, Hong Kong. https://www.chp.gov.hk/files/pdf/dh_phs_2014_15_full_report_eng.pdf.

34. Cheng HM, Lin HJ, Wang TD, Chen CH. Asian management of hypertension: current status, home blood pressure, and specific concerns in Taiwan. J Clin Hypertens. 2020;22: 511–4.

35. Khader Y, Batieha A, Jaddou H, Rawashdeh SI, El-Khateeb M, Hyassat D, Khader A, Ajlouni K. Hypertension in Jordan: prevalence, awareness, control, and its associated factors. Int J Hypertens. 2019;2019:3210617. https://doi.org/10.1155/2019/3210617.

36. Ratnayake R, Rawashdeh F, Abualrub R, et al. Access to care and prevalence of hypertension and diabetes among Syrian refugees in Northern Jordan. JAMA Netw Open. 2020;3:1–13.

37. Ibrahim M, Rizk H. Hypertension prevalence, awareness, treatment, and control in Egypt: results from the Egyptian National Hypertension Project (NHP). Hypertension. 1995;6:886–90.

38. Yusufali AM, Khatib R, Islam S, Alhabib KF, Bahonar A, Swidan HM, Khammash U, Alshamiri MQ, Rangarajan SYS. Prevalence, awareness, treatment and control of hypertension in four Middle East countries. J Hypertens. 2017;35:1457–64.

39. Saeed AA, Al-hamdan NA, Bahnassy AA, Abdalla AM, Abbas MAF, Abuzaid LZ. Prevalence, awareness, treatment, and control of hypertension among Saudi adult population: a national survey. Int J Hypertens. 2011;2011:1–8.

40. Alhabib KF, Batais MA, Almigbal TH, Alshamiri MQ, Altaradi H, Rangarajan S, Yusuf S. cardiovascular disease risk factors in the Saudi population: results from the Prospective Urban Rural Epidemiology study (PURE-Saudi). BMC Public Health. 2020;20:1–14.

41. Bener A, Al-Suwaidi J, Al-Jaber K, Al-Marri S, Elbagi IE. Epidemiology of hypertension and its associated risk factors in the Qatari population. J Hum Hypertens. 2004;18:529–30.

42. Sengul S, Akpolat T, Erdem Y, Derici U, Arici M, Sindel S. Changes in hypertension prevalence, awareness, treatment, and control rates in Turkey from 2003 to 2012. J Hypertens. 2016;34:1207–17.

43. Abu-saad K, Chetrit A, Eilat-adar S, Alpert G, Atamna A, Gillon-keren M, Rogowski O, Ziv A, Kalter-leibovici O. Blood pressure level and hypertension awareness and control differ by marital status, sex, and ethnicity: a population-based study. Am J Hypertens. 2014;27:1511–20.

44. Eghbali M, Khosravi A, Feizi A, Mansouri A, Mahaki B. Prevalence, awareness, treatment, control, and risk factors of hypertension among adults: a cross-sectional study in Iran. Epidemiol Heal. 2018;40:1–9.

45. Saka M, Shabu S, Shabila N. Prevalence of hypertension and associated risk factors in older adults in Kurdistan, Iraq. East Mediterr Heal J. 2020;26:265–72.

46. El-reshaid K, Al-owaish R, Diab A. Hypertension in Kuwait: the past, present and future. Saudi J Kidney Dis Transpl. 1999;10:357–64.

47. Al Riyami AA, Afifi M. Clustering of cardiovascular risk factors among Omani adults. East Mediterr Heal J. 2003;9:893–903.

48. Gupta R, Gupta V, Sarna M, Bhatnagar S, Thanvi J, Sharma V, Singh A, Gupta J, Kaul V. Prevalence of coronary heart disease and risk factors in an urban Indian population: Jaipur Heart Watch-2. Indian Heart J. 2002;54:59–66.

49. Mohan V, Deepa R, Shanthi Rani S, Premalatha G. Prevalence of coronary artery disease and its relationship to lipids in a selected population in South India: the Chennai Urban Population Study (CUPS no. 5). J Am Coll Cardiol. 2001;38:682–7.

50. Gupta PC, Gupta R. Hypertension prevalence and blood pressure trends among 99,589 subjects in Mumbai, India. Indian Heart J. 1999;51:691.

51. Anchala R, Kannuri NK, Pant H, Khan H, Franco OH, Di Angelantonio E, Prabhakaran D. Hypertension in India: a systematic review and meta-analysis of prevalence, awareness, and control of hypertension. J Hypertens. 2014;32:1170–7.
52. Katulanda P, Ranasinghe P, Jayawardena R, Constantine GR, Rezvi Sheriff MH, Matthews DR. The prevalence, predictors and associations of hypertension in Sri Lanka: a cross-sectional population based national survey. Clin Exp Hypertens. 2014;36:484–91.
53. Islam JY, Zaman MM, Haq SA, Ahmed S, Al- Quadir Z. Epidemiology of hypertension among Bangladeshi adults using the 2017 ACC/AHA Hypertension Clinical Practice Guidelines and Joint National Committee 7 Guidelines. J Hum Hypertens. 2018;32:668–80.
54. Hasan M, Khan MSA, Sutradhar I, Hossain MM, Hossaine M, Yoshimura Y, Choudhury SR, Sarker M, Mridha MK. Prevalence and associated factors of hypertension in selected urban and rural areas of Dhaka, Bangladesh: findings from SHASTO baseline survey. BMJ Open. 2021;11:1–12.
55. Jafar TH, Levey AS, Jafary FH, White F, Gul A, Rahbar MH, Khan AQ, Hattersley A, Schmid CH, Chaturvedi N. Ethnic subgroup differences in hypertension in Pakistan. J Hypertens. 2003;21:905–12.
56. Prajapati D, Poudel P, Hirachan A, et al. Prevalence and determinants of systemic hypertension in inhabitants of high altitude of Nepal. Asian J Med Sci. 2020;11:12–6.
57. Karmacharya BM, Koju RP, LoGerfo JP, Chan KCG, Mokdad AH, Shrestha A, Sotoodehnia N, Fitzpatrick AL. Awareness, treatment and control of hypertension in Nepal: findings from the Dhulikhel Heart Study. Heart Asia. 2017;9:1–8.
58. Aboobakur M, Latheef A, Mohamed AJ, Moosa S, Pandey RM, Krishnan A, Prabhakaran D. Surveillance for non-communicable disease risk factors in Maldives: results from the first STEPS survey in Male. Int J Public Health. 2010;55:489–96.
59. Lupat A, Hengelbrock J, Luissin M, Fix M, Bassa B, Craemer EM, Becher H, Meyding-Lamadé U. Brunei epidemiological stroke study: patterns of hypertension and stroke risk. J Hypertens. 2016;34:1416–22.
60. Gupta V, LoGerfo JP, Raingsey PP, Fitzpatrick AL. The prevalence and associated factors for prehypertension and hypertension in Cambodia. Heart Asia. 2013;5:253–8.
61. Peltzer K, Pengpid S. The prevalence and social determinants of hypertension among adults in Indonesia: a cross-sectional population-based national survey. Int J Hypertens. 2018;2018:5610725. https://doi.org/10.1155/2018/5610725.
62. Christiani Y, Byles JE, Tavener M, Dugdale P. Gender inequalities in noncommunicable disease risk factors among Indonesian urban population. Asia-Pacific J Public Heal. 2016;28:134–45.
63. Committee S, D CPM, D IBM, Investigator P, Officer T. Report on STEPS survey on non communicable diseases risk factors in Vientiane Capital city, Lao PDR; 2010.
64. Ab Majid NL, Omar MA, Khoo YY, Mahadir Naidu B, Ling Miaw Yn J, Rodzlan Hasani WS, Mat Rifin H, Abd Hamid HA, Robert Lourdes TG, Mohd Yusoff MF. Prevalence, awareness, treatment and control of hypertension in the Malaysian population: findings from the National Health and Morbidity Survey 2006–2015. J Hum Hypertens. 2018;32:617–24.
65. Abdul-Razak S, Daher AM, Ramli AS, et al. Prevalence, awareness, treatment, control and socio demographic determinants of hypertension in Malaysian adults. BMC Public Health. 2016;16:1–10.
66. Bjertness MB, Htet AS, Meyer HE, Htike MMT, Zaw KK, Oo WM, Latt TS, Sherpa LY, Bjertness E. Prevalence and determinants of hypertension in Myanmar—a nationwide cross-sectional study. BMC Public Health. 2016;16:590. https://doi.org/10.1186/s12889-016-3275-7.
67. Ministry of Health. Executive summary on national population health survey 2016/17; 2017. p. 1–4.
68. Liew SJ, Lee JT, Tan CS, Koh CHG, Van Dam R, Müller-Riemenschneider F. Sociodemographic factors in relation to hypertension prevalence, awareness, treatment and control in a multi-ethnic Asian population: a cross-sectional study. BMJ Open. 2019;9:1–10.

69. Tiptaradol S, Aekplakorn W. Prevalence, awareness, treatment and control of coexistence of diabetes and hypertension in Thai population. Int J Hypertens. 2012;2012:386453. https://doi.org/10.1155/2012/386453.
70. Son PT, Quang NN, Viet NL, Khai PG, Wall S, Weinehall L, Bonita R, Byass P. Prevalence, awareness, treatment and control of hypertension in Vietnamresults from a national survey. J Hum Hypertens. 2012;26:268–80.
71. Meiqari L, Essink D, Wright P, Scheele F. Prevalence of hypertension in Vietnam: a systematic review and meta-analysis. Asia-Pacific J Public Heal. 2019;31:101–12.
72. Patalen C. Expanded National Nutrition Survey: 2019 results health and nutritional status of Filipino adults, 20–59 years old; 2019. p. 20–59.
73. Sison J. Presyon 3; 2013.
74. World Health Organization (WHO). High blood pressure—country experiences and effective interventions utilized across the European Region. World Health Organization; 2013. p. 1–30.
75. Gupta R, Gaur K, Ram CVS. Emerging trends in hypertension epidemiology in India. J Hum Hypertens. 2019;33(8):575–87.
76. Oliva RV. A review on the status of hypertension in six Southeast Asian countries. Hypertens J. 2019;5:5–8.
77. Castillo R. Prevalence and management of hypertension in Southeast Asia; 2016:2016.
78. Xing L, Liu S, Jing L, et al. Trends in prevalence, awareness, treatment, and control of hypertension in rural Northeast China: 2008 to 2018. Biomed Res Int. 2020;2020:1456720. https://doi.org/10.1155/2020/1456720.
79. Thawornchaisit P, De Looze F, Reid CM, Seubsman S, Sleigh AC, Cohort T, Team S. Health risk factors and the incidence of hypertension: 4-year prospective findings from a national cohort of 60 569 Thai Open University students. BMJ Open. 2013;3(6):e002826. https://doi.org/10.1136/bmjopen-2013-002826.
80. Asgari S, Moazzeni SS, Azizi F, Abdi H, Khalili D, Hakemi MS, Hadaegh F. Sex-specific incidence rates and risk factors for hypertension during 13 years of follow-up: the Tehran Lipid and Glucose Study. Glob Heart. 2020;15:1–13.
81. Talaei M, Sadeghi M, Mohammadifard N, Shokouh P, Oveisgharan S, Sarrafzadegan N. Incident hypertension and its predictors: the Isfahan Cohort Study. J Hypertens. 2014;32(1):30–38.
82. Mohan V, Anjana RM, Unnikrishnan R. Incidence of hypertension among Asian Indians: 10 year follow up of the Chennai Urban Rural Epidemiology Study (CURES-153). J Diabetes Complicat. 2020;34(10):107652.
83. Prabhakaran D, Jeemon P, Ghosh S, Shivashankar R, Ajay VS, Kondal D, Gupta R, Ali MK, Mohan D, Mohan V, Kadir MM, Tandon N, Reddy KS, Venkat Narayan KM. Prevalence and incidence of hypertension: results from a representative cohort of over 16,000 adults in three cities of South Asia. Indian Heart J. 2017;69(4):434–41. https://doi.org/10.1016/j.ihj.2017.05.021.
84. Luo Y, Xia F, Yu X, Li P, Huang W. Long-term trends and regional variations of hypertension incidence in China: a prospective cohort study from the China Health and Nutrition Survey, 1991–2015. BMJ Open. 2021;11:1–9.
85. Yu ES, Hong K, Chun BC. A longitudinal analysis of the progression from normal blood pressure to stage 2 hypertension: a 12-year Korean cohort. BMC Public Health. 2021;21(1):61.
86. Oka M, Yamamoto M, Mure K, Takeshita T, Arita M. Relationships between lifestyle, living environments, and incidence of hypertension in Japan (in men): based on participant's data from the nationwide medical check-up. PLoS One. 2016;11:1–9.
87. Sy RG. Hypertension among apparently healthy adult Filipinos in the LIFECARE Philippine cohort.
88. Joffres M, Falaschetti E, Gillespie C, Robitaille C, Loustalot F, Poulter N, Mcalister FA, Johansen H, Baclic O, Campbell N. Hypertension prevalence, awareness, treatment and control in national surveys from England, the USA and Canada, and correlation with stroke and ischaemic heart disease mortality: a cross-sectional study. BMJ Open. 2013;3(8):e003423.

89. Dhar L, Dantas J, Ali M. A systematic review of factors influencing medication adherence to hypertension treatment in developing countries. Open J Epidemiol. 2017;7:211–50.
90. Division E and DC. National Health Survey 2010; 2010. Singapore.
91. Sison J, Arceo LP, Buan N, Reynaldo E, Torres DJ, Trinidad E, Chua P, Punzalan A, Yape IM, Bautista A, Lapitan R. Philippine Heart Association-Council on Hypertension Report on Survey of Hypertension and Target Organ Damage (PRESYON 2-TOD*): a report on prevalence of hypertension, awareness, treatment profile and control rate. Philipp J Cardiol. 2007;35:1–9.
92. Wander GS, Ram CVS. Blood pressure—methods to record & numbers that are significant: lets make a tailored suit to suit us. Indian J Med Res. 2018:435–8.
93. Xing L, Jing L, Tian Y, et al. Urban-Rural disparities in status of hypertension in northeast China: a populationbased study, 2017-2019. Clin Epidemiol. 2019;11:801–820. Published 2019 Sep 3. https://doi.org/10.2147/CLEP.S218110.
94. Kim K, Ji E, Choi J, Kim S, Ahn S. Ten-year trends of hypertension treatment and control rate in Korea. Sci Rep. 2021;11:6966.
95. Kaddumukasa M, Kayima J, Nakibuuka J, Blixen C, Welter E, Katabira E, Sajatovic M. Modifiable lifestyle risk factors for stroke among a high risk hypertensive population in Greater Kampala , Uganda; a cross-sectional study. BMC Res Notes. 2017;10:675.
96. World Health Organization (WHO). Sodium intake and iodized salt in the South-East Asia Region; 2014. p. 29–30.
97. Lin Y, Mei Q, Qian X, He T. Salt consumption and the risk of chronic diseases among Chinese adults in Ningbo city. Nutr J. 2020;19:1–10.
98. Chiang C, Wang T, Ueng K, et al. Guidelines of the Taiwan Society of Cardiology and the Taiwan Hypertension Society for the management of hypertension. J Chinese Med Assoc. 2015;78:1–47.
99. Visanuyothin S, Plianbangchang S, Somrongthong R. Appearance and potential predictors of poorly controlled hypertension at the primary care level in an urban community. J Multidiscip Healthc. 2017;11:131–8.
100. Wang Y, Chen J, Wang K, Edwards C. Education as an important risk factor for the prevalence of hypertension and elevated blood pressure in Chinese men and women. J Hum Hypertens. 2006;20:898–900.
101. Olack B, Wabwire-mangen F, Smeeth L, Montgomery JM, Kiwanuka N, Breiman RF. Risk factors of hypertension among adults aged 35–64 years living in an urban slum Nairobi, Kenya. BMC Public Health. 2015;15:1251.
102. Wang J, Ma JJ, Liu J, Zeng DD, Song C, Cao Z. Prevalence and risk factors of comorbidities among hypertensive patients in China. Int J Med Sci. 2017;14:201–12.
103. Tian S, Dong G, Wang D, Liu M, Lin Q, Meng X, Xu L, Hou H, Ren Y. Factors associated with prevalence, awareness, treatment and control of hypertension in urban adults from 33 communities in China: the CHPSNE Study. Hypertens Res. 2011;34:1087–92.
104. Shen Y, Chang C, Zhang J, Jiang Y, Ni B, Wang Y. Prevalence and risk factors associated with hypertension and prehypertension in a working population at high altitude in China: a cross-sectional study. Environ Health Prev Med. 2017;22:1–8.
105. Leung LCK, Sung RYT, So H, et al. Prevalence and risk factors for hypertension in Hong Kong Chinese adolescents: waist circumference predicts hypertension, exercise decreases risk. Arch Dis Child. 2011;96:804–9.
106. Hu L, Huang X, You C, Li J, Hong K, Li P, Wu Y. Prevalence and risk factors of prehypertension and hypertension in Southern China. PLoS One. 2017;12:1–15.
107. Kalani Z, Salimi T, Rafiei M. Comparison of obesity indexes BMI, WHR and WC in association with hypertension: results from a Blood Pressure Status Survey in Iran. J Cardiovasc Dis Res. 2015;6:72–7.
108. Ministry of Health Indonesia. Indonesia Health Profile 2019; 2019.
109. Yu ES, Hong K, Chun BC. Incidence and risk factors for progression from prehypertension to hypertension: a 12-year Korean Cohort Study. J Hypertens. 2020;38:1755–62.

110. Neupane D, McLachlan CS, Sharma R, Gyawali B, Khanal V, Mishra SR, Christensen B, Kallestrup P. Prevalence of hypertension in member countries of South Asian Association for Regional Cooperation (SAARC): systematic review and meta-analysis. Med (United States). 2014;93:1–10.
111. Motlagh B, O'Donnell M, Yusuf S. Prevalence of cardiovascular risk factors in the middle east: a systematic review. Eur J Cardiovasc Prev Rehabil. 2009;16:268–80.
112. Liu F, Liu Y, Sun X, et al. Nutrition, metabolism & cardiovascular diseases race- and sex-specific association between alcohol consumption and hypertension in 22 cohort studies: a systematic review and meta-analysis. Nutr Metab Cardiovasc Dis. 2020;30(8):1249–59. https://doi.org/10.1016/j.numecd.2020.03.018.
113. Roerecke M, Kaczorowski J, Tobe SW, Gmel G, Hasan OSM, Rehm J. The effect of a reduction in alcohol consumption on blood pressure: a systematic review and meta-analysis. Lancet Public Heal. 2017;2:e108–20.
114. Naing C, Aung K. Prevalence and risk factors of hypertension in Myanmar: a systematic review and meta-analysis. Med (United States). 2014;93:1–9.
115. World Health Organization (WHO). Global health risks: mortality and burden of disease attributable to selected major risks; 2009.
116. Cleven L, Krell-roesch J, Nigg CR, Woll A. The association between physical activity with incident obesity, coronary heart disease, diabetes and hypertension in adults: a systematic review of longitudinal studies published after 2012. BMC Public Health. 2020;20:1–15.
117. Ranasinghe CD, Ranasinghe P, Jayawardena R, Misra A. Physical activity patterns among South-Asian adults: a systematic review. Int J Behav Nutr Phys Act. 2013;10:1–11.
118. Zhang Y, Fan X, Li S, Wang Y, Shi S, Lu H, Yan F. Prevalence and risk factors of hypertension among Hui population in China. Medicine (Baltimore). 2021;100(18):e25192.
119. Farzadfar F, Finucane MM, Danaei G, Pelizzari PM, Cowan MJ, Paciorek CJ, Singh GM. National, regional, and global trends in serum total cholesterol since 1980: systematic analysis of health examination surveys and epidemiological studies with 321 country-years and 30 million participants. Lancet. 2011;377:578–86.
120. Joshi P. Risk factors for early myocardial infarction in South Asians compared with individuals. J Am Med Assoc. 2015;297:286–94.
121. Rehman H, Samad Z, Raj S, Merchant AT, Narula JP, Mishra S, Virani SS, Asia S. Epidemiologic studies targeting primary cardiovascular disease prevention in South Asia. Indian Heart J. 2018;70(5):721–30. https://doi.org/10.1016/j.ihj.2018.01.029.
122. Fujiyoshi A, Ohkubo T, Miura K, Murakami Y, Nagasawa S. Blood pressure categories and long-term risk of cardiovascular disease according to age group in Japanese men and women. Hypertens Res. 2012;35:947–53.
123. Wierzejska E, Giernaś B, Lipiak A, Karasiewicz M, Cofta M. Systematic review/meta-analysis. A global perspective on the costs of hypertension: a systematic review. Arch Med Sci. 2020;16:1078–91.

Pathophysiology and Mechanisms of Hypertension (Asian Context)

2

Leilani B. Mercado-Asis

2.1 Genetics of Hypertension

Essential hypertension is a heritable disease with a complex genetic trait caused by multiple susceptibility genes the effects of which are modulated by gene–environment and gene–gene interactions [1]. These genetic determinants involve multiple genes that make it challenging to study blood pressure BP variations in general population [2]. BP may be dependent on a genetic pattern of many loci with influence at variance according to race [3], gender [4], age, or lifestyle [5].

Table 2.1 summarizes the linkage of genes studies that cause Mendelian forms of hypertension. Almost all are geared toward increasing BP through genetic mutations involving net renal salt reabsorption [6], mutations in enzymes like aldosterone synthase (glucocorticoid-remediable aldosteronism), enzymes synthesizing steroids that activate the mineralocorticoid (MC) receptor like 11-β hydroxylase (apparent mineralocorticoid excess), beta and gamma subunits of the renal epithelial sodium channel (Liddle's syndrome), the serine-threonine kinases (WNK1 and 4 in pseudohypoaldosteronism type 2), and mutation in the MC receptor (hypertension exacerbated by pregnancy) [7].

There are candidate genes analyzed based on their known biological or physiological function. Conspicuous of these are genes from the renin-angiotensin-aldosterone system which polymorphisms have affected the regulation of blood pressure: angiotensinogen [8], renin, angiotensin-converting enzyme [9], angiotensin (AT) II receptor type 1 [10], and aldosterone synthase [11, 12]. These genetic cascade abnormalities have resulted into enhanced renal tubular sodium reabsorption that leads to hypertension. G protein-coupled receptors (GPCRs) for

L. B. Mercado-Asis (✉)
Section of Endocrinology and Metabolism, Department of Medicine, Faculty of Medicine and Surgery, University of Santo Tomas, Manila, Philippines
e-mail: lmasis@ust.edu.ph

© The Author(s), under exclusive license to Springer Nature Switzerland AG 2022
C. V. S. Ram et al. (eds.), *Hypertension and Cardiovascular Disease in Asia*,
Updates in Hypertension and Cardiovascular Protection,
https://doi.org/10.1007/978-3-030-95734-6_2

Table 2.1 Mendelian forms of human hypertension, disease entities, and gene mutation loci

Disease entity	Gene mutation locus
Glucocorticoid-remediable aldosteronism	Aldosterone synthase and 11 β-hydroxylase
Apparent mineralocorticoid excess	11 β-hydroxylase
Liddle's syndrome	ENaCβ- or γ subunit
Pseudohypoaldosteronism type 2 (Gordon's syndrome)	WNK kinase 1 and 4
Hypertension exacerbated by pregnancy	Ligand-binding domain of the mineralocorticoid receptor

endothelins, α and β adrenoceptors, angiotensin II (ANG II), and vasopressin have also been implicated in the development of hypertension. Similarly, regulators of G proteins signaling (RGS2) have been reported to influence vasoconstrictors including ANG II, endothelin I, noradrenalin, thromboxane, thrombin, and vasopressin [13]. Augmented and chronic vasoconstriction lead to severe elevation of BP and vascular hypertrophy in knockout mice [14].

Although different mutations in the same gene may cause hypertension, monogenic disorders of BP regulations are uncommon and may not explain BP variability in the general population [15]. With the advent of the genomic era, it is now generally accepted that the most common form of hypertension is a complex trait with a polygenic basis and environmental influences that may also exert effects through epigenetic changes which could even be transmitted across generations [16–19].

Genome-wide association studies (GWAS) permit the investigation of most variability due to common traits in the human genome. This approach examines unrelated individuals in a population using genotypes of a large number of polymorphic markers in subjects with marked BP elevation compared with healthy controls. A single nucleotide polymorphisms (SNPs) array is used as marker in GWAS mapping studies to identify hypertension susceptibility loci. In a recent report of one million people, 535 new loci associated with BP traits were identified [20]. A GWAS of BP traits: systolic blood pressure (SBP, diastolic BP (DBP) and pulse pressure (PP) in people of European ancestry drawn from UK Biobank [21] and the International Consortium of Blood Pressure Genome-Wide Association Studies (ICBP) [22, 23] was reported [24]. The estimated SNP-wide heritability of BP traits for SBP, DBP, and PP with a gain in percentage of BP variance increased from 2.8% for the 274 previously published loci to 5.7%. When association analysis was extended to unrelated individuals from Africa and South Asia, blood pressure variants in combination were associated with 6.1 mmHg and 7.4 mm Hg higher, sex-adjusted mean SBP, respectively, in the two ethnic groups compared to European descent.

In the assessment of tissue enrichment of blood pressure loci, enrichment was greatest in the cardiovascular system; blood vessels, heart, adrenal tissues, and adipose tissues. The enrichment of BP gene expression in the adrenal tissue gives light to autonomous aldosterone production by the adrenal glands thought to be responsible for 5–10% of all hypertension rising to 22% amongst people with resistant hypertension [25, 26]. Other new novel loci include genes involved in vascular remodeling like vascular endothelial growth factor A (VEGFA) and fibroblast growth factor 5 (FGF5) which are linked to enhanced angiogenesis, atherosclerosis, and vascular smooth muscle cell differentiation [27, 28].

Lastly, novel loci analyzed provided insight on blood pressure variants and lifestyle exposures [20]. There are genetic associations with daily food intake, urinary sodium, and creatinine concentration, body mass index (BMI), weight, waist circumference, and intake of water, coffee, and tea. SNP rs13107325 novel locus is associated with frequency of drinking alcohol, time spent watching television, and association of BMI with weight and waist circumference. Most importantly, regulator analysis identified several BP therapeutic targets such as calcium channels, angiotensinogen, natriuretic peptide receptor, angiotensin-converting enzyme, angiotensin receptors and endothelin receptors [20].

The human genetic linkage of essential hypertension has been demonstrated in the study of Ciolac and his group [29]. The hemodynamic, metabolic, and neurohumoral biomarkers for risk of hypertension among normotensive offsprings with hypertensive parents have shown to be significantly different versus those offsprings with normotensive or only one parent with hypertension. Nonhypertensive young individuals with hypertensive parents have higher pulse wave velocity, fasting insulin, insulin-to-glucose ratio, DBP response to exercise, and altered catecholamine, endothelin-1, and nitric oxide response to exercise [29]. The most likely explanation to the increased BP response to exercise commonly found among normotensive offspring of hypertensive parents appears to be a failure to reduce total peripheral resistance [30].

Indeed, genetic studies of a human genome of an individual at risk for essential hypertension is a very promising approach where precision medicine takes its role in decreasing morbidity and mortality of the disease.

2.2 Neurohumoral Regulation of Blood Pressure

2.2.1 The Sympathetic Nervous System

Neurogenic form of hypertension is characterized by increase in BP with sympathetic overactivity, loss of parasympathetic mediated cardiac function and complicated by the renin-aldosterone-angiotensin circuit with influence on the cardiovascular system [31, 32]. Studies have shown that the adrenergic overdrive that characterizes hypertension is not stable but instead follows the BP increase and the progression from uncomplicated to complicated stages that may occur in the course of the disease [33].

It is well-established that sympathetic nervous system regulates blood pressure via the modulation of peripheral vascular tone and cardiac output [31]. Both the surgical sympathectomy [34] and use of antihypertensive medications that lower sympathetic activity [35] provide an indication of the clinical evidence for a significant neurogenic component to hypertension. As elaborated by the groups of Fisher and Platon [32] and Mancia and Grassi [36] sympathetic neural drive has been demonstrated in various strata of the population, such as in the young, middle-aged, and elderly hypertensives; in pregnancy- induced hypertension; and in systolic-diastolic hypertension or an isolated elevation of systolic BP [37]. The same has been documented in patients with both elevated BP and metabolic risk factors, such as obesity, metabolic syndrome, or diabetes mellitus [38]. These observations have led to the conclusion that in hypertension, sympathetic hyperactivity is a generalized occurrence, irrespective of the heterogeneous clinical aspects that accompany a high BP condition.

In the study of the relationship between adrenergic outflow and BP values [39], the number of neural bursts and burst amplitude were progressively greater in individuals with elevated BP compared with the normotensive individuals. Other studies revealed that sympathetic activation is normally more pronounced in complicated than in noncomplicated stages of hypertension. Similarly, sympathetic activation were found to be pronounced in various disease conditions such as, hypertensive patients with impaired renal function [40, 41] and those with adverse cardiac outcomes like left ventricular hypertrophy, impaired left ventricular diastolic function, systolic heart failure, and ventricular arrhythmias [42]. Finally, in resistant hypertensives who do not respond despite treatment with multiple antihypertensive drugs [43] muscle sympathetic nerve firing is more pronounced than in patients with hypertension who respond to the usual antihypertensive drug regimen [44]. There data suggest that activation of the adrenergic nervous system evolves from less to more severe hypertensive states.

The clinical model of excess catecholamine secretion in adrenal chromaffin tumors so-called pheochromocytoma (PHEO) clearly demonstrates the influence of sympathetic stimulation on the cardiovascular system leading to marked elevation of BP [45, 46]. The marked stimulation of the adrenoceptors $\alpha1$, $\alpha2$, $\beta1$, $\beta2$, $\beta3$, and dopaminergic receptors D1 and D2 from the excessive circulating catecholamines from the adrenal chromaffin tumors cause hemodynamic changes that alters the cardiovascular system milieu and leads to elevation of BP [46]. In general, $\alpha1$ receptors, mostly found in smooth muscle, peripheral arteries and veins cause vasoconstriction upon stimulation and increasing systemic pressure. Stimulation of $\alpha2$-adrenergic receptors located on smooth muscles will result in arterial vasodilation and coronary vasoconstriction; in PHEO typical manifestations may include diaphoresis and orthostatic hypotension. Stimulation of $\beta1$-adrenergic receptors has a positive chronotropic and inotropic effect in the heart and will also result in release of renin. In PHEO this can contribute to hypertension, palpitations, and tachycardia [46–48]. It has been repeatedly demonstrated the causative relationship of hypercatecholaminemia with severe elevation in BP [49, 50]. Removal of the unilateral adrenal chromaffin tumor [51] or merely performing unilateral adrenalectomy in bilateral syndromic PHEO [52] or removal of the dominant adrenal tumor in bilateral PHEO [53] normalizes or decreases BP level with no required antihypertensive medication or decrease in the number of antihypertensive drugs postoperatively [53–55].

2.2.2 The Autonomic Nervous System

The modulation of the autonomic nervous system on the cardiovascular system and BP regulation has long been demonstrated with various physiological and functional investigations. With environmental manipulation and at rest, there is a gradual reduction of the bradycardic and tachycardic responses to baroreceptor stimulation and deactivation in mild to more severe degrees of BP elevation, respectively [39]. Experimental and clinical investigations have tested the hypothesis that the origin, progression, and outcome of human hypertension are related to dysfunctional autonomic cardiovascular control with much influence from disturbed activation of the sympathetic division [36]. In animal studies, hypertension adverse outcomes have demonstrated to be attributed to both an increase in sympathetic nerve activity and a reduction of vagal cardiac tone [56, 57]. Subsequent investigations in human showed similar autonomic alterations that may have a causative or co-causative role in the pathophysiology of human hypertension [58–60].

Sympathetic overdrive seen in individuals with essential hypertension have been linked to reduced inhibitory influence of the arterial baroreceptors. In various animal studies, arterial baroreceptor denervation is followed by a marked increase of BP variability, with little or no change in long-term mean blood pressure level [60, 61]. In humans, with environmental manipulation, Sinski and his group demonstrated that deactivation of carotid body chemoreceptors reduces sympathetic activity in hypertensive subjects and decrease BP [61, 62]. This reflex mechanism is thought to be more involved in BP stabilization than in the regulation of its average values [36]. Furthermore, the arterial baroreflex loses much of its ability to control heart rate, but it continues to effectively modulate BP and sympathetic activity [39]. These observations point to the role of the autonomic nervous system in the maintenance and progression of sympathetic drive as BP severity increases.

The participation of the autonomic nervous system in the regulation of BP have also been demonstrated in several studies with invasive approaches like deep brain stimulation (DBS) of the periaqueductal gray matter (PAG), vagus nerve stimulation, baroreflex activation therapy, and renal denervation or carotid body ablation, whereby effective lowering of BP and heart rate [63, 64] has been attained. High and low-frequency peripheral nerve stimulation has been reported to produce systolic and diastolic BP changes and decrease heart rate both in hypertensive patients [65] and healthy subjects [66]. Furthermore, in animal studies, stimulation of the median nerve inhibits sympatho-excitatory cardiovascular responses and thereby decreases cardiac sympathetic drive and reverses hypertension [67]. Recently, Bang and his colleague demonstrated the BP-lowering effect of transcutaneous electrical activation of C-fiber in the median nerve among hypertensive individuals with slight reduction in heart rate [64].

2.2.3 The Renin-Angiotensin-Aldosterone System

While the baroreceptor reflex responds in a short-term manner to decreased arterial pressure, the renin-angiotensin-aldosterone system or RAAS is responsible for more chronic alterations to elevate the BP in a prolonged manner [68, 69].

The RAAS regulates BP and fluid balance through three (3) stimuli: (1) decrease in blood volume, (2) decrease in blood sodium level, and (3) increase in blood potassium [70, 71]. Renin from the juxtaglomerular cells in the kidney is secreted to stimulate angiotensinogen, produced in the liver, to be converted to the hormone angiotensin I (ANG I). The conversion of angiotensin I to angiotensin II is catalyzed by an enzyme called angiotensin-converting enzyme (ACE). ACE is found primarily in the vascular endothelium of the lungs and kidneys. After ANG I is converted to ANG II, it has effects on the kidney, adrenal cortex, arterioles, and brain by binding to angiotensin II type I (AT1) and type II (AT2) receptors. The role of AT receptors has been shown to cause vasodilation by nitric oxide generation. In the plasma, angiotensin II has a half-life of 1–2 min, at which point peptidases degrade it into angiotensin III and IV. Angiotensin III (ANG III) has been shown to have 100% of the aldosterone stimulating effect of ANG II, but 40% of the pressor effects, while angiotensin IV (ANG IV) has further decreased the systemic effect. In the proximal convoluted tubule of the kidney, angiotensin II acts to increase Na-H exchange, increasing sodium reabsorption. Increased levels of Na in the body acts to increase the osmolarity of the blood, leading to a shift of fluid into the blood volume and extracellular space (ECF). This increases the arterial pressure of the patient. ANG II also acts on the adrenal cortex, specifically the zona glomerulosa. Here, it stimulates the release of aldosterone, a hormone that causes an increase in sodium reabsorption and potassium excretion at the distal tubule and collecting duct of the nephron. The increased total body sodium leads to an increase in osmolarity and subsequent increase in blood and ECF volume. In contrast to ANG II, aldosterone is a steroid hormone. As a result, it produces change by binding to nuclear receptors and altering gene transcription. Thus, the effects of aldosterone may take hours to days to begin, while the effects of ANG II are rapid. The effect of ANG II on vasoconstriction takes place in systemic arterioles. Here, ANG II binds to G protein-coupled receptors, leading to a secondary messenger cascade that results in potent arteriolar vasoconstriction. This acts to increase total peripheral resistance, causing an increase in BP. Finally, ANG II acts on the brain. Here, it has three effects. First, it binds to the hypothalamus, stimulating thirst and increased water intake. Second, it stimulates the release of antidiuretic hormone (ADH) by the posterior pituitary. ADH or vasopressin, acts to increase water reabsorption in the kidney by inserting aquaporin channels at the collecting duct. Finally, ANG II decreases the sensitivity of the baroreceptor reflex. This diminishes baroreceptor response to an increase in BP, which would be counterproductive to the goal of the RAAS. The net effect of these interactions is an increase in total body sodium, total body water, and vascular tone [70–73].

2.2.3.1 The Central Nervous System and Angiotensin II and Angiotensin III

Several investigations have demonstrated the participation of the central nervous system (CNS) in the pathophysiology of essential hypertension [74]. As mentioned, the CNS influences BP thru the sympatho-humoral mechanisms which involve the sympathetic nerve activity (SNA), adrenocorticotropic hormone (ACTH), growth hormone, ANG II) and vasopressin. Increased circulating ANG II generated within the brain increases sympathetic nerve activity leading to elevation of the plasma levels of vasopressin and ACTH and thereby causing increased BP [75]. The major mechanisms of actions for these RAS inhibitors or receptor blockers are mediated primarily by blocking the detrimental effects of the classic angiotensinogen/renin/ACE/ANG II/AT1/aldosterone axis [74–76].

The long-term pressor effect of central ANG II has been shown to be associated with the progressive activation of a slow neuromodulatory pathway [77]. This pathway is mediated by a sequence that involves local aldosterone synthesis, activation of mineralocorticoid receptor (MR), and benzamil-sensitive epithelial sodium channels (ENaCs), increased ENaC activity and local synthesis of brain digitalis (ouabain)-like compounds [77–79]. Hamlyn and his group elaborated a novel neuroendocrine humoral and vascular pressor pathway for brain ANG II. The proximal components of this axis are neuronal pathways activated by brain ANG II that depend upon central aldosterone and MRs. The distal components of the axis include up-regulated circulating levels of endogenous ouabain (EO) and related steroids, and functional reprogramming of arterial function due to increased expression of arterial myocyte proteins that raise arterial myocyte Ca^{2+} and myogenic tone and that enhance sympathetic responses [75, 80]. The long-term increases in central ANG II and circulating EO sustain BP via the combined effects of heightened sympathetic activity and the functional reprogramming of arterial function. Alteration in the axis by various stimuli like, stress, hormones, and environmental factors lead to hypertension.

Li and colleagues elaborated the current understanding on the entire RAS superfamily, including the classic angiotensinogen/renin/ACE/ANG II/AT1 receptor axis, the prorenin/renin/prorenin receptor (PRR or Atp6ap2)/MAP kinases ERK1/2/V-ATPase axis, the ANG II/APA/ANG III/AT2/NO/cGMP axis, the ANG I/ANG II/ACE2/ANG (1–7)/Mas receptor axis, and the ANG III/APN/ANG IV/IRAP/AT4 receptor axis [68]. The first two axes represent the powerful vasopressor systems, which are physiologically required to maintain normal cardiovascular, BP, and renal homeostasis. However, overactivation of these two axes of the RAS plays a critical role in the development of cardiovascular and kidney diseases and hypertension [68, 75, 76].

ACE is the primary enzyme responsible for ANG II formation [81, 82]. ANG II is metabolized to form ANG III, primarily by aminopeptidase A (APA) [83, 84]. In spontaneously hypertensive rats (SHR) and ANG II-infused rats, high levels of APA and high APA activity [78, 85] are expressed. In the extrarenal tissues, the pituitary express high levels of APA or APA activity, followed by the median eminence, and paraventricular nuclei, which are linked to the central regulation of cardiovascular function and BP [86, 87]. Since APA is the key enzyme for the degradation of ANG II, its implications in cardiovascular diseases have been studied in APA-KO mice [88]. Mitsui et al. showed that basal systolic blood pressure was significantly elevated in APA-KO mice, and with infusion of ANG II BP further increased signifying an enhanced response due to the lack of APA to metabolize ANG II [88]. Interestingly, consumption in APA-KO mice was significantly elevated, suggesting that ANG II rather than ANG III regulates water drinking behavior or thirst [89].

ANG III is the key metabolite of the effector peptide ANG II [78]. The metabolism of ANG II and generation of ANG III and their structure and activity relationship have been extensively investigated [75, 83]. The presence of ANG III has been demonstrated in various tissues. Wysocki and colleagues reported that the level of plasma ANG III was remarkably higher than that of ANG II in wild-type mice on the C57BL6 genetic background, whereas the kidney ANG III level was much lower than that of plasma [90]. In normal rat plasma, the ANG II level is about 50% of that for ANG I, whereas the ANG III is about 50% of ANG II. Based on the available data, ANG III levels are most likely lower in the plasma and kidney than ANG II [78].

The physiological roles of ANG III as a vasopressor or a vasodepressor remains to be further elucidated. Carey and colleagues demonstrated that ANG III is less efficacious than ANG II in increasing BP and stimulating aldosterone production [91, 92]. A novel role of ANG III in the kidney in inducing natriuresis via activation of AT2 receptors has been demonstrated [93]. Interestingly, ANG III significantly increased urinary sodium excretion in the presence of an AT1 receptor blocker, suggesting a possible role of AT2 receptors in mediating ANG III-induced natriuretic response unmasked by AT1 receptor blockade [94]. Furthermore, a recent report by Kopf et al. demonstrated there is an obligatory metabolism of ANG II to ANG III in the adrenal zona glomerulosa cells not for adrenal secretion but for relaxation of adrenal arteries [95].

Lastly, the AT2 receptor plays a key role in mediating the vasodepressor responses, and cardiorenal protective effects to the activation of the APA/ANG III/AT2/cGMP axis [96, 97]. Angiotensinogen/renin/ACE/ANG II/AT1 receptor axis is an important factor to induce vasoconstriction, increase BP, and promote growth and fibrotic and proinflammatory responses [98, 99]. The most commonly described signaling pathways for the AT2 receptor include ANG II-induced inhibition of the protein phosphotyrosine phosphatase (PTP) activity, phospholipase A(2), nitric oxide, and cyclic guanosine monophosphate (cGMP) [96, 97]. Angiotensin receptor blockers (ARBs) selectively blocks the actions of the angiotensinogen/ renin/ACE/ ANG II/AT1 receptor axis by binding and occupying AT1 receptors in target tissues, which leads to several-fold increases in the circulating ANG II levels. ANG II then may bind and activate the unopposed AT2 receptors in cardiovascular and kidney tissues to induce vasodepressor effects [96, 97].

2.2.4   Isolated Systolic Hypertension and Its Pathophysiology

Isolated systolic hypertension (ISH) is defined as systolic BP (SBP) $\geq$130 mmHg and DBP <80 mmHg in the American guidelines and SBP $\geq$140 mmHg and DBP <90 mmHg in the European and most hypertension guidelines [100, 101]. ISH predominates among the elderly and is now becoming a common form of hypertension among the young adults and adolescents [102–104]. With aging, SBP continues a linear rise but DBP starts to fall after mid-50 years of age [102, 105]. This phenomenon contributes to the increasing pulse pressure with increasing age. The predominant cause of ISH is vascular aging-associated large artery stiffening. In the young, ISH is often associated with several secondary causes. These illnesses involving physico-hormonal and vascular abnormalities include accelerated atherosclerosis from chronic kidney disease, repaired coarctation of the aorta, peripheral vascular disease, insulin resistance, osteoporosis with vascular calcifications, thyrotoxicosis, and altered elastin formation during intrauterine fetal growth retardation [106, 107].

The major pathophysiology of ISH is a reduction in the compliance of the large elastic arteries. This is in contrast to the long-term vasoconstriction of arterioles causing systolic-diastolic hypertension (SDH) [108]. The hemodynamics of ISH patients are characterized by a marked increase in arterial stiffness and aortic characteristic impedance, a marked increase in central SBP and central pulse pressure, a mild increase in total peripheral resistance, and a moderate increase in wave reflections [109–112]. ISH is often classified into the burned-out type, which is a result of aging-related decrease in DBP in the established SDH, and the de novo type caused by a novel increase of SBP [108]. In the Framingham study population with ISH, as many as 40% did not go through a period of SDH [113].

Longstanding ISH leads to adverse effects including increased inflammatory cytokines levels, activation of the renin angiotensin-aldosterone system and sympathetic tone, endothelial dysfunction, and end-organ hypoperfusion [114]. These untoward effects in turn further enhance arterial stiffness through influence from reduced vasodilatation, accelerated atherosclerosis, and left ventricle remodeling, leading to worsened ISH and higher cardiovascular risk if ISH remains uncontrolled [114–116].

2.3   Summary

Essential, primary, or idiopathic hypertension accounts for 95% of all cases of hypertension. It is a heterogenous disorder with various causal elements that accounts for the elevation in BP. There are two primary elements in the etiopathogenesis of essential hypertension: genetic and neurohumoral factors. These factors have been demonstrated to intertwine with environmental and behavioral influences like lifestyle, obesity, alcohol, and salt intake that brought about various phenotypes of the disease. The recent report on the genome-wide association studies (GWAS) on the human genome with newly discovered novel loci sheds light on the association of blood pressure with environmental variants and lifestyle exposure. The contribution of peripheral and central neurohumoral factors in the development of essential hypertension has lately gained significant research interest and advancement. It is well-established the sympathetic nervous system regulates blood pressure via the modulation of peripheral vascular tone and cardiac output through the adrenergic receptors with increased in the number of sympathetic neural bursts and amplitude of burst. The sympathetic overdrive in individuals with essential hypertension has also been linked to reduced inhibitory control of the autonomic nervous system. While the baroreceptor reflex responds in a short-term manner to decreased arterial pressure, the renin-angiotensin-aldosterone system or RAAS has been shown to be responsible for more chronic alterations to elevate the blood pressure affecting both the vascular and cardiac remodeling. Lastly, ISH might dominate the cardiovascular problem involving elevation of BP owing to a growing global aging population and young adults developing metabolic problems that will lead to vascular stiffness, activation of RAAS and eventual endothelial dysfunction. Overall, further research to understand other gray areas in the pathophysiology of hypertension are warranted to specifically define genotypes and phenotypes of individuals at risk for timely intervention and thus prevent long-term cardiovascular complications.

References

1. Singh M, Singh AK, Pandey S, Chandra S, Singh KA, Gambhir IS. Molecular genetics of essential hypertension. Clin Exp Hypertens. 2016;38(3):268–77.
2. Doris PA. Hypertension genetics, single nucleotide polymorphisms, and the common disease: common variant hypothesis. Hypertension. 2002;39:323–31.
3. Luft FC, Miller JZ, Grim CE, Fineberg NS, Christian JC, Daugherty SA, et al. Salt sensitivity and resistance of blood pressure. Age and race as factors in physiological responses. Hypertension. 1991;17:1102–8.
4. Higaki J, Baba S, Katsuya T, Sato N, Ishikawa K, Mannami T, et al. Deletion allele of angiotensin-converting enzyme gene increases risk of essential hypertension in Japanese men: the Suita Study. Circulation. 2000;101:2060–5.
5. Staessen JA, Wang J, Bianchi G, Birkenhager WH. Essential hypertension. Lancet. 2003;361:1629–41.
6. Lifton RP, Gharavi AG, Geller DS. Molecular mechanisms of human hypertension. Cell. 2001;104:545–56.
7. Geller DS, Farhi A, Pinkerton N, Fradley M, Moritz M, Spitzer A, et al. Activating mineralocorticoid receptor mutation in hypertension exacerbated by pregnancy. Science. 2000;289:119–23.
8. Inoue I, Nakajima T, Williams CS, Quackenbush J, Puryear R, Powers M, et al. A nucleotide substitution in the promoter of human angiotensinogen is associated with essential hypertension and affects basal transcription in vitro. J Clin Invest. 1997;99:1786–97.
9. O'Donnell CJ, Lindpaintner K, Larson MG, Rao VS, Ordovas JM, Schaefer EJ, et al. Evidence for association and genetic linkage of the angiotensin-converting enzyme locus with hypertension and blood pressure in men but not women in the Framingham Heart Study. Circulation. 1998;97:1766–72.
10. Brand E, Chatelain N, Mulatero P, Fery I, Curnow K, Jeunemaitre X, et al. Structural analysis and evaluation of the aldosterone synthase gene in hypertension. Hypertension. 1998;32:198–204.
11. Lifton RP, Dluhy RG, Powers M, Rich GM, Cook S, Ulick S, et al. A chimaeric 11 beta-hydroxylase/aldosterone synthase gene causes glucocorticoid-remediable aldosteronism and human hypertension. Nature. 1992;355:262–5.
12. Rigat B, Hubert C, Alhenc-Gelas F, Cambien F, Corvol P, Soubrier F. An insertion/deletion polymorphism in the angiotensin I-converting enzyme gene accounting for half the variance of serum enzyme levels. J Clin Invest. 1990;86:1343–6.
13. Haximer SP, Watson N, Linder ME, Blumer KJ, Hepler JR. RGS2/G0S8 is a selective inhibitor of Gqalpha function. Proc Natl Acad Sci U S A. 1997;94:14389–93.
14. Heximer SP, Knutsen RH, Sun X, Kaltenbronn KM, Rheee M, Peng N, et al. Hypertension and prolonged vasoconstrictor signaling in RGS2-deeficient mice. J Clin Invest. 2003;111:445–52.
15. Lander E, Kruglyak L. Genetic dissection of complex traits. Nat Genet. 1995;11:241–7.
16. Patel RS, Masi S, Taddei S. Understanding the role of genetics in hypertension. Euro Heart J. 2017;38:2309–12.
17. Ehret GB, Caulfield MJ. Genes for blood pressure: an opportunity to understand hypertension. Eur Heart J. 2013;34:951–61.

18. Dominiczak A, Delles C, Padmanabhan S. Genomics and precision medicine for clinicians and scientists in hypertension. Hypertension. 2017;69:e10–3.
19. Wang J, Gong L, Tan Y, Hui R, Wang Y. Hypertensive epigenetics: from DNA methylation to microRNAs. J Hum Hypertens. 2015;29:575–82.
20. Evangelou E, Warren HR, Mosen-Ansorena D, Mifsud B, Pazoki R, Gao H, et al. Genetic analysis of over 1 million people identifies 535 new loci associated with blood pressure traits. Nat Genet. 2018;50:1412–25.
21. Sudlow C, Gallacher J, Allen N, Beral V, Burton P, Danesh J, et al. UK Biobank: an open access resource for identifying the causes of a wide range of complex diseases of middle and old age. PLoS Med. 2015;12:e1001779.
22. Wain LV, Vaez A, Jansen R, Joehanes R, van der Most PJ, Mesut Erzurumluoglu A, et al. Novel blood pressure locus and gene discovery using genome-wide association study and expression data sets from blood and the kidney. Hypertension. 2017;70:e4–e19.
23. Ehret GB, Munroe PB, Rice KM, Bochud M, Johnson AD, Chasman DI, et al. Genetic variants in novel pathways influence blood pressure and cardiovascular disease risk. Nature. 2011;478:103–9.
24. Evangelou E, Warren H R, Mosen-Ansorena D, Mifsud B, Pazoki R, Gao H, et al. Genetic analysis of over one million people identifies 535 new loci associated with blood pressure traits. Nat. Genet. 2018;50:1412–25.
25. Douma S, Petidis K, Doumas M, Papaefthimiou P, Triantafyllou A, Kartali N, et al. Prevalence of primary hyperaldosteronism in resistant hypertension: a retrospective observational study. Lancet. 2008;371(9628):1921–6.
26. Brown JM, Siddiqui M, Calhoun DA, Carey RM, Hopkins PN, Williams GH, et al. The unrecognized prevalence of primary aldosteronism: a cross-sectional study. Ann Intern Med. 2020;173:10–20.
27. Warren HR, Evangelou E, Cabrera CP, Gao H, Ren M, Mifsud B, et al. Genome-wide association analysis identifies novel blood pressure loci and offers biological insights into cardiovascular risk. Nat Genet. 2017;49:403–15.
28. Oliveira-Paula GH. Polymorphisms in VEGFA gene affect the antihypertensive responses to enalapril. Eur J Clin Pharmacol. 2015;71:949–57.
29. Ciolac EG, Bocchi EA, Bortolotto LA, Carvalho VO, Greve JMD, Guimaraes GV. Haemodynamic, metabolic and neuro-humoral abnormalities in young normotensive women at high familial risk for hypertension. J Hum Hypertens. 2010;24:814–22.
30. Bond V Jr, Franks BD, Tearney RJ, Wood B, Melendez MA, Johnson L, et al. Exercise blood pressure response and skeletal muscle vasodilator capacity in normotensives with positive and negative family history of hypertension. J Hypertens. 1994;12:285–90.
31. Esler M. The 2009 Carl Ludwig Lecture: pathophysiology of the human sympathetic nervous system in cardiovascular diseases: the transition from mechanisms to medical management. J Appl Physiol. 2010;108(2):227–37.
32. Fisher JP, Paton JFR. The sympathetic nervous system and blood pressure in humans: implications for hypertension. J Hum Hypertens. 2012;26:463–75.
33. Smith PA, Graham LN, Mackintosh AF, Stoker JB, Mary DA. Relationship between central sympathetic activity and stages of human hypertension. Am J Hypertens. 2004;17:217–22.
34. Morrissey DM, Brookes VS, Cooke WT. Sympathectomy in the treatment of hypertension: review of 122 cases. Lancet. 1953;1:403–8.
35. Sica DA. Centrally acting antihypertensive agents: an update. J Clin Hypertens. 2007;9:399–405.
36. Mancia G, Grassi G. The autonomic nervous system and hypertension. Circ Res. 2014;114:1804–14.
37. Grassi G, Seravalle G, Bertinieri G, Turri C, Dell'Oro R, Stella M, et al. Sympathetic and reflex alterations in systo-diastolic and systolic hypertension of the elderly. J Hypertens. 2000;18:587–93.
38. Thorp AA, Schlaich MP. Relevance of sympathetic nervous system activation in obesity and metabolic syndrome. J Diabetes Res. 2015;2015:341583.

39. Grassi G, Cattaneo BM, Lanfranchi SG, Mancia G. Baroflex control of sympathetic nerve activity in essential and secondary hypertension. Hypertension. 1998;31:68–72.
40. Grassi G, Quarti-Trevano F, Seravalle G, Arenare F, Volpe M, Furiani S, Dell-Oro R, Mancia G. Early sympathetic activation in the initial clinical stages of chronic renal failure. Hypertension. 2011;57:846–51.
41. Kaur J, Young BE, Fadel PJ. Sympathetic overactivity in chronic kidney disease: consequences and mechanisms. Int J Mol Sci. 2017;18:1682–94.
42. Grassi G. Sympathetic neural activity in hypertension and related diseases. Am J Hypertens. 2010;23:1052–60.
43. Calhoun DA, Booth JN, Oparil S, Irvin MR, Shimbo D, Lackland DT, Howard G, et al. Refractory hypertension: determination of prevalence, risk factors, and comorbidities in a large, population-based cohort. Hypertension. 2014;63:451–8.
44. Prevalence and clinical characteristics of patients with true resistant hypertension in central and Eastern Europe: data from the BP-CARE study. J Hypertens. 2013;10:2018–24.
45. Mercado-Asis LB, Castillo RR. Clinical presentation, diagnosis, and management of primary aldosteronism and pheochromocytoma. Hypertens J. 2019;5:5.
46. Mercado-Asis LB, Siao RM, Amba NF. Evolving clinical presentation and assessment of pheochromocytoma: a review. J Med Univ Santo Tomas. 2017;1:5–23.
47. Eisenhofer G, Lenders JW, Timmers HJLM, Mannelli M, Grebe SK, Hofbauer LC, et al. Measurements of plasma methoxytyramine, normetanephrine, and metanephrine as discriminators of different hereditary forms of pheochromocytoma. Clin Chem. 2011;57:411–20.
48. Pacak K. Pheochromocytoma: a catecholamine and oxidative stress disorder. Endocr Regul. 2011;45:65–90.
49. van Berkel A, Lenders J, Timmers HJLM. Biochemical diagnosis of phaeochromocytoma and paraganglioma. Eur J Endocrinol. 2014;170:R109–19.
50. Pussard E, Chaouch A, Said T. Radioimmunoassay of free plasma metanephrines for the diagnosis of catecholamine producing tumors. Clin Chem Lab Med. 2014;52:437–44.
51. Osinga TE, van den Eijnden HA, Kema IP, Kerstens MN, Dullaart RPF, de Jong WHA, et al. Unilateral and bilateral adrenalectomy for pheochromocytoma requires adjustment of urinary and plasma meetanephrine reference ranges. J Clin Eendocrinol Metab. 2013;98:1976–083.
52. Zhou GW, Wei Y, Chen X, Jiang XH, Li XY, Ning G, et al. Diagnosis and surgical treatment of multiple endocrine neoplasia. Chin Med J. 2009;122:1495–500.
53. Mercado-Asis LB, Tingcungco AG, Bolong DT, Lopez RA, Caguioa EV, Yamamoto ME, et al. Diagnosis of small adrenal pheochromocytoma by adrenal venous sampling with glucagon stimulation test. Int J Endocrinol Metab. 2011;9:323–9.
54. Malong CLP, Tanchee-Ngo MJ, Torres-Salvador P, Pacak K, Mercado-Asis LB. Removal of dominant adrenal lateralized by glucagon-stimulated adrenal venous sampling alleviates hypertension in bilateral pheochromocytoma. J Life Sci. 2013;7:586–91.
55. Gomez MF, Gan FR, Mendoza E, Mercado LB. Systemic hormonal unloading in unilateral adrenalectomy in a patient with bilateral adrenal hyperplasia: a case report. J Med Univ Santo Tomas. 2019;3:303–8.
56. Oparil S. The sympathetic nervous system in clinical and experimental hypertension. Kidney Int. 1986;30:437–52.
57. Mark AL. The sympathetic nervous system in hypertension: a potential long-term regulator of arterial pressure. J Hypertens Suppl. 1996;14:S159–65.
58. Mancia G, Grassi G, Giannattasio C, Seravalle G. Sympathetic activation in the pathogenesis of hypertension and progression of organ damage. Hypertension. 1999;34:724–8.
59. Esler M. Sympathetic nervous system moves toward center stage in cardiovascular medicine: from Thomas Willis to resistant hypertension. Hypertension. 2014;63:e25–32.
60. Sinski M, Lewandowski J, Przybylski J, Zalewski P, Symonides B, Abramczyk P, et al. Deactivation of carotid body chemoreceptors by hyperoxia decreases blood pressure in hypertensive patients. Hypertens Res. 2014;37:858–62.
61. Cowleey AW Jr. Long-teerm control of arterial blood pressure. Physiol Rev. 1992;72:231–300.

62. Sinski M, Lewandowski J, Przybylski J, Bidiuk J, Abramczyk P, Ciarka A, Gaciong Z. Tonic activity of carotid body chemoreceptors contributes to the increased sympathetic drive in essential hypertension. Hypertens Res. 2012;35:487–91.
63. Grassi G. Assessment of sympathetic cardiovascular drive in human hypertension: achievements and perspectives. Hyperrtension. 2009;54:690–7.
64. Bang SK, Ryu Y, Chang S, Im CK, Bae JH, Gwak YS, et al. Attenuation of hypertension by C-fiber stimulation of the human median nerve and the concept-based novel device. Sci Rep. 2018;8:14967.
65. de Jong MR, Adiyaman A, Gal P, Smit JJJ, Delnoy PPHM, Heeg J, van Hasselt BAAM, et al. Renal nerve stimulation-induced blood pressure changes predict ambulatory blood pressure response after renal denervation. Hypertension. 2016;68:707–14.
66. Maver J, Struci M, Accetto R. Autonomic nervous system in normotensive subjects with a family history of hypertension. Clin Auton Res. 2004;14:369–75.
67. Elayan HH, Sun P, Milic M, Liu F, Bao X, Ziegler MG. Cardiovascular responses to electrical stimulation of sympathetic nerves in the pithed mouse. Auton Neurosci. 2008;30:49–52.
68. Li XC, Zhang J, Zhuo JL. The vasoprotective axes of the renin-angiotensin system: physiological relevance and therapeutic implications in cardiovascular, hypertensive and kidney diseases. Pharmacol Res. 2017;125:21–38.
69. Munoz-Durango N, Fuentes CA, Castillo AE, Gonzalees-Gomeez LM, Veecchiola A, Fardella CE, Kaleergis AM. Role of the renin-angiotensin-aldosterone system beyond blood pressure regulation: molecular and cellular mechanisms involved in end-organ damage during arterial hypertension. Int J Mol. 2016;17:797–814.
70. Ren L, Lu X, Danser AHJ. Revisiting the brain renin-angiotensin system—focus on novel therapies. Curr Hypertens Rep. 2019;21:28–35.
71. Nehme A, Zouein FA, Zayeri ZD, Zibara K. An update on the tissue renin angiotensin system and its role in physiology and pathology. J Cardiovasc Dev Dis. 2019;6:14–31.
72. Sztechman D, Czarzasta K, Cudnoch-Jedrzejewska A, Szczepanska-Sadowska E, Zera T. Aldosterone and mineralocorticoid receptors in regulation of the cardiovascular system and pathological remodelling of the heart and arteries. J Physiol Pharmacol. 2018;69.
73. Santos RAS, Oudit GY, Verano-Braga T, Canta G, Steckelings UM, Bader M. The renin-angiotensin system: going beyond the classical paradigms. Am J Physiol Heart Circ Physiol. 2019;316:H958–70.
74. Reen L, Lu X, Danser AHJ. Revisiting the brain renin-angiotensin system—focus on novel therapies. Curr Hypertens Rep. 2019;21:28–35.
75. Hamlyn JM, Linde CI, Gao J, Huang BS, Golovina VA, Balustein MP, et al. Neuroendocrine humoral and vascular components in the pressor pathway for brain angiotensin II: a new axis in long term blood pressure control. PLoS One. 2014;9:e108916.
76. Lu J, Wang H, Ahmad M, Keshtkar-Jahromi M, Blaustein MP, Hamlyn JM, Leeneen FHH. Central and peripheral slow-pressor mechanisms contributing to angiotensin II-salt hypertension in rats. Cardiovasc Res. 2018;114:233–46.
77. Leenen FH. The central role of the brain aldosterone "ouabain" pathway in salt-sensitive hypertension. Biochem Biophys Acta. 2010;1802:1132–9.
78. Huang BS, Ahmadi M, White RA, Leenen FH. Central neuronal activation and pressor responses induced by circulating ANG II: role of the brain aldosterone "ouabain" pathway. Am J Physiol Heart Circ Physiol. 2010;299:H422–30.
79. Osborn JW, Olson DM, Guzman P, Toney GM, Fink GD. The neurogenic phase of angiotensin II-salt hypertension is prevented by chronic intracerebroventricular administration of benzamil. Physiol Rep. 2014;2:e00245.
80. Wang Y, Chen L, Wier WG, Zhang J. Intravital Forster resonance energy transfer imaging reveals elevated [Ca^{2+}], and enhanced sympathetic tone in femoral arteries of angiotensin II-infused hypertensive biosensor mice. J Physiol. 2013;591:5321–36.
81. Gonzales-Villalobos RA, Billet S, Kim C, Satou R, Fuchs S, Berrnstein KE, et al. Intrarenal angiotensin-converting enzyme induces hypertension in response to angiotensin I infusion. J A Soc Nephrol. 2011;22:449–59.

82. Bernstein KEE. Two ACEs and a heart. Nature. 2002;417:799–802.
83. Wilson WL, Roques BP, Llorens-Cortes C, Speth RC, Harding JW, Wright JW. Roles of brain angiotensins II and III in thirst and sodium appetite. Brain Res. 2005;1060:108–17.
84. Bodineau L, Frugiere A, Marc Y, Inguimbert N, Fassot C, Balavoine F, Roques B, Llorens-Cortes C. Orally active aminopeptidase A inhibitors reduce blood pressure: a new strategy for treating hypertension. Hypertension. 2008;51:1318–25.
85. Raina H, Zhang Q, Rhee AY, Pallone TL, Wier WG. Sympathetic nerves and the endothelium influence the vasoconstrictor effect of low concentrations of ouabain in pressurized small arteries. Am J Physiol Heart Circ Physiol. 2010;298:H2093–101.
86. Hamlyn JM, Laredo J, Shah JR, Lu ZR, Hamilton BP. 11-hydroxylation in the biosynthesis of endogenous ouabain: multiple implications. Ann N Y Acad Sci. 2003;986:685–93.
87. Mellon SH, Griffin LD, Compagnone NA. Biosynthesis and action of neurosteroids. Brain Res Rev. 2001;37:3–12.
88. Asano N, Ogura T, Mimura Y, Kishida M, Kataoka H, Otsuka F, et al. Renal AT1 receptor: autographic localization and quantification in rat. Res Commun Mol Pathol Pharmacol. 1998;100(2):161–70.
89. Vinson GP. Glomerulosa function and aldosterone synthesis in the rat. Mol Cell Endocrinol. 2004;217:59–65.
90. Wysocki J, Ye M, Batlle D. Plasma and kidney angiotensin peptides: importance of the aminopeptidase A/angiotensin III axis. Am J Hypertens. 2015;28:1418–28.
91. Ye P, Kenyon CJ, MacKenzie SM, Seckl JR, Fraser R, et al. Regulation of aldosterone synthase gene expression in the rat adrenal gland and central nervous system by sodium and angiotensin II. Endocrinology. 2003;144:3321–8.
92. Carey RM, Vaughan ED Jr, Peach MJ, Ayers CR. Activity of (des-Aspartyl[1])-angiotensin II and angiotensin II in man. Differences in blood pressure and adrenocortical response during normal and low sodium intake. J Clin Invest. 1978;61:20–31.
93. Kemp BA, Bell JF, Rottkamp DM, Howell NL, Shao W, Navar LG, et al. Intrarenal angiotensin III is the predominant agonist for proximal tubule AT2 receptors. Hypertension. 2012;60:387–95.
94. Padia SH, Howell NL, Siragy HM, Carey RM. Renal angiotensin type 2 receptors mediate natriuresis via angiotensin III in the angiotensin II type 1 receptor-blocked rat. Hypertension. 2006;47:537–44.
95. Kopf PG, Park S, Herrnreiter A, Krause C, Roques BP, Campbell WB. Obligatory metabolism of angiotensin II to angiotensin III for zona glomerulosa cell-mediated relaxations of bovine adrenal cortical arteries. Endocrinology. 2018;159:238–47.
96. de Gasparo M, Catt KJ, Inagami T, Wright JW, Unger T. International union of pharmacology, XXIII. The angiotensin II receptors. Pharmacol Rev. 2000;52:415–72.
97. Forrester SJ, Booz GW, Sigmund CD, Coffman TM, Kawai T, Rizzo V, et al. Angiotensin II signal transduction: an update on mechanisms of physiology and pathophysiology. Physiol Rev. 2018;98(3):1627–738.
98. Carey RM. Update on angiotensin AT2 receptors. Curr Opin Nephrol Hypertens. 2017;26:91–6.
99. Siragy HM, Carey RM. Angiotensin type 2 receptors: potential importance in the regulation of blood pressure. Curr Opin Nephrol Hypertens. 2001;10:99–103.
100. Whelton PK, Carey RM, Aronow WS, Casey DE Jr, Collins KJ, Himmelfarb CD, et al. 2017 ACC/AHA/AAPA/ABC/ACPM/AGS/APhA/ASH/ASPC/NMA/PCNA guideline for the prevention, detection, evaluation, and management of high blood pressure in adults: a report of the American College of Cardiology/American Heart Association Task Force on clinical practice guidelines. J Am Coll Cardiol. 2018;71(19):e127–248.
101. Williams B, Mancia G, Spiering W, Rosei EA, Azizi M, Burner M, Clement DL, et al. 2018 ESC/ESH guidelines for the management of arterial hypertension. Eur Heart J. 2018;39(33):3021–104.

102. Grebla RC, Rodriguez CJ, Borrell LN, Pickering TG. Prevalence and determinants of isolated systolic hypertension among young adults: the 1999–2004 US National Health and Nutrition Examination Survey. J Hypertens. 2010;28(1):15–23.
103. Yano Y, Stamler J, Garside DB, Daviglus ML, Franklin SS, Carnethon MR, et al. Isolated systolic hypertension in young and middle-aged adults and 31-year risk for cardiovascular mortality: the Chicago Heart Association Detection Project in Industry Study. J Am Coll Cardiol. 2015;65(4):327–35.
104. McEniery CM, Franklin SS, Cockcroft JR, Wilkinson IB. Isolated systolic hypertension in young people is not spurious and should be treated: pro side of the argument. Hypertension. 2016;68(2):269–75.
105. Burt VL, Whelton P, Roccella EJ, Brown C, Cutler JA, Higgins M, et al. Prevalence of hypertension in the US adult population. Results from the Third National Health and Nutrition Examination Survey, 1988–1991. Hypertension. 1995;25(3):305–13.
106. Franklin SS. Elderly hypertensives: how are they different? J Clin Hypertens. 2012;14(11):779–86.
107. Palatini P, Rosei EA, Avolio A, Bilo G, Casiglia E, Ghiadoni L, et al. Isolated systolic hypertension in the young: a position paper endorsed by the European Society of Hypertension. J Hypertens. 2018;36(6):1222–36.
108. Umemura S, Arima H, Arima S, Asayama K, Dohi Y, Hirooka Y, et al. The Japanese Society of Hypertension guidelines for the management of hypertension (JSH 2019). Hypertens Res. 2019;42(9):1235–481.
109. Avolio AP, Deng FQ, Li WQ, Luo YF, Huang ZD, Xing LF, et al. Effects of aging on arterial distensibility in populations with high and low prevalence of hypertension: comparison between urban and rural communities in China. Circulation. 1985;71(2):202–10.
110. Avolio A. Ageing and wave reflection. J Hypertens Suppl. 1992;10(6):S83–6.
111. Mitchell GF, Lacourciere Y, Ouellet JP, Izzo JL, Neutel J, Kwerwin LJ, et al. Determinants of elevated pulse pressure in middle-aged and older subjects with uncomplicated systolic hypertension: the role of proximal aortic diameter and the aortic pressure-flow relationship. Circulation. 2003;108(13):1592–8.
112. Wallace SM, Yasmin, McEniery CM, Maki-Petaja KM, Booth AD, Cockcroft JR, et al. Isolated systolic hypertension is characterized by increased aortic stiffness and endothelial dysfunction. Hypertension. 2007;50(1):228–33.
113. Franklin SS, Barboza MG, Pio JR, Wong ND. Blood pressure categories, hypertensive subtypes, and the metabolic syndrome. J Hypertens. 2006;24(10):2009–16.
114. Mph CB, Goel S, Messerli FH, Bavishi C, Goel S, Messerli FH. Isolated systolic hypertension: an update after SPRINT. Am J Med. 2010;129(12):1251–8.
115. Kocemba J, Kawecka-Jaszcz K, Gryglewska B, Grodzicki T. Isolated systolic hypertension: pathophysiology, consequences and therapeutic benefits. J Hum Hypertens. 1998;12:621–6.
116. Kario K, Chen CH, Park S, Park CG, Hoshide S, Cheng HM, et al. Consensus document on improving hypertension management in Asian patients, taking into account Asian characteristics. Hypertension. 2018;71(3):375–82.

Summary of Recent Guidelines on Hypertension

3

Nihar Mehta and Sachna Shetty

3.1   Introduction

About 20–35% of the worldwide adult population has hypertension. It is the commonest noncommunicable disease and contributes to target organ damage in the heart, kidneys, and brain. Therefore, hypertension is a major contributor to cardiovascular disease (CVD) morbidity and mortality worldwide and in India. Taking into account the special geographical and climatic background, dietary habits, literacy levels and socioeconomic variables, there are significant regional variations in the manifestation and management of hypertension in different regions.

Substantial progress has been made in understanding the epidemiology, pathophysiology, and risk associated with hypertension, and there is a wealth of evidence showing that lowering blood pressure (BP) can substantially reduce premature morbidity and mortality. A number of proven, highly effective, and well-tolerated lifestyle and drug treatment strategies can facilitate this reduction in BP. Despite this, BP control rates remain abysmal worldwide. Therefore, hypertension remains the major preventable cause of CVD and all-cause death globally and in Asia.

Hypertension is a lifelong condition and therefore often requires long-term pharmacological treatment. Good hypertension management program is important if we want to reduce the overall impact of noncommunicable disease. However, economic factors have a significant impact on hypertension management protocols and feasibility. A comparison of some of these important economic and healthcare delivery system factors is shown in Table 3.1.

N. Mehta (✉) · S. Shetty
Jaslok Hospital & Research Centre, Mumbai, Maharashtra, India

© The Author(s), under exclusive license to Springer Nature Switzerland AG 2022   39
C. V. S. Ram et al. (eds.), *Hypertension and Cardiovascular Disease in Asia*,
Updates in Hypertension and Cardiovascular Protection,
https://doi.org/10.1007/978-3-030-95734-6_3

Table 3.1 Differences in economic and healthcare delivery system factors between India, the US, and Europe

Parameters	USA	European Union	India
Population (2017) (million)	325.7	512.4	1339.1
GDP (2017) (trillion USD)	19.3	17.2	2.5
Per capita income (USD)	59,531	33,715	1939
Health expenditure (% of GDP)	16.8	7.9	3.9
Health expenditure per capita (USD)	9536	2192	63

GDP gross domestic product, *USD* United States dollar

Recently, various societies have published guidelines on management of hypertension: the American College of Cardiology/American Hypertension Association (ACC/AHA) in 2017 [1], the European Society of Cardiology/European Society of Hypertension (ESC/ESH) in 2018 [2], Indian Guidelines on Hypertension (IGH) IV in 2019 [3], the International Society of Hypertension in 2020 [4], and Canadian guidelines 2020 [5]. These have made some significant changes, including a change in definition of hypertension, changes in the target BP (based on the findings of the Systolic Blood Pressure Intervention Trial [SPRINT]), greater use of home blood pressure monitoring (HBPM) and ambulatory blood pressure monitoring (ABPM), reduced interest in renal angioplasty and renal denervation therapy due to recent data, and increased use of mineralocorticoid receptor antagonists (e.g., spironolactone) for resistant hypertension. In addition, there is new epidemiological data from different regions on hypertension and hypertension-mediated organ damage (HMOD).

This article provides a summary of the salient features of several current hypertension guidelines to help practicing physicians to streamline management of this important public health problem [6–9].

3.2   Definition and Classification

All currently used definitions of hypertension are arbitrary. There is some evidence that the risk of cardiovascular events in Asian Indians is higher at relatively lower levels of BP [10]. There is a continuous relationship between BP level and the risk of complications. Hazard ratio values for coronary heart disease (CHD) and stroke were 1.1–1.5 for a systolic BP (SBP)/diastolic BP (DBP) of 120–129/80–84 mmHg compared with <120/80 mmHg, and 1.5–2.0 for SBP/DBP 130–139/85–89 mmHg versus <120/80 mmHg [1].

The BP cutoff values used to define hypertension, has been a topic of intense debate over the last few years. Major guidelines recommend that hypertension is diagnosed when a person's office SBP is ≥140 mmHg and/or their DBP is ≥90 mmHg, or any BP level in patients taking antihypertensive medication (Table 3.2). Corresponding home and ambulatory BP values have also been defined (Table 3.3). The latest ACC/AHA guidelines changed the BP threshold for defining hypertension to ≥130/80 mmHg (Table 3.4). However, Indian [3], European [2], and ISH [4] guidelines, and many others, maintain the earlier definition of 140/90 mmHg. The lower BP cutoff value for defining hypertension in the ACC/AHA guideline was associated with an increase in prevalence of hypertension from

30 to 42% [6]. Definitions for each stage of hypertension also differ between the US and European guidelines (Table 3.5). Staging cutoff values in the IGH IV and ISH guidelines are similar to those in the ESC/ESH guidelines. However, all these guidelines do not differ markedly in terms of the target BP to be achieved.

Table 3.2 Classification of BP for adults aged ≥18 years (IGH IV [3] and ESC/ESH [2])

Category	Systolic (mmHg)		Diastolic (mmHg)
Optimal	<120	and	<80
Normal	<130	and	<85
High normal	130–139	or	85–89
Hypertension			
Stage 1	140–159	or	90–99
Stage 2	160–179	or	100–109
Stage 3	≥180	or	>110
Isolated systolic hypertension			
Grade 1	140–159	and	<90
Grade 2	>160	and	<90

Table 3.3 Diagnosis of hypertension based on office, home, and ambulatory BP values (ESC/ESH 2018 [2], IGH IV [3], and ISH 2020 [4])

	SBP (mmHg)	DBP (mmHg)
Office BP	≥140	≥90
Mean home BP	≥135	≥85
Ambulatory BP		
Mean daytime	≥135	≥85
Mean night-time	≥120	≥70
Mean 24-h	≥130	≥80

Table 3.4 Classification of BP in adults (ACC/AHA 2017) [1]

BP category	SBP (mmHg)		DBP (mmHg)
Normal	<120	and	<80
Elevated	120–129	and	<80
Hypertension			
Stage 1	130–139	or	80–89
Stage 2	≥140	or	≥90

Table 3.5 Comparison of ACC/AHA [1] and ESC/ESH [2] hypertension guidelines regarding definition and grading/staging of hypertension

Parameter	ACC/AHA	ESC/ESH
Definition (mmHg)	>130/80	>140/90
Grading of normal pressure (mmHg)	Normal: <120/80	Optimal: <120/80
	Elevated: 120–129/<80	Normal: 120–129/80–84
		High normal: 130–139/85–89
Grading of hypertension (mmHg)	Grade 1: 130–139/80–89	Grade 1: 140–159/90–99
	Grade 2: ≥140/90	Grade 2: 160–179/100–109
		Grade 3: ≥180/110
Target BP	≤65 years: <130/80	<65 years: <130/80
	≥ 65 years: <130/80	≥65 years: <140/80

3.3 Measurement of Blood Pressure

BP can naturally vary widely over different time periods and the diagnosis of high BP should not be made on a single office measurement. Usually, 2–3 office readings taken at 1- to 4-week intervals (depending on the BP) should be taken to confirm the diagnosis of hypertension. The exception would be if BP is ≥180/110 mmHg or there is evidence of HMOD. Recommended methods for BP recording are similar in all the guidelines [1–5]. One important feature of more recent guidelines, including those from Asia, is the greater emphasis on out-of-office BP monitoring [11, 12]. The latest Canadian guidelines include HBPM as a part of the criteria to diagnose hypertension. Most other guidelines suggest that HBPM is used for diagnosis and follow up [1–4].

3.4 Epidemiology

The prevalence of hypertension is variable in different parts of the world. India and China are the two most populated countries, and between them they include 42% of the world's population. However, there are significant differences in the pattern and prevalence of hypertension between different Asian countries [10]. As a complication of hypertension, stroke is more common in China and Korea while South Asians have a higher prevalence of CVD [10]. However, some features are common to patients with hypertension across Asia (Box 3.1) [10].

In addition to the differences from other continents in Asia, there are also significant differences in the epidemiology of hypertension within various Asian countries. For example, in India there are rural versus urban differences in prevalence of hypertension [3]. Also, the prevalence of hypertension has increased over the last two to three decades in some developing countries [13]. The level of control of BP also varies between regions. In some developed countries such as the USA, BP control has been achieved in 50% of individuals [1]. However, in India, the BP control rate is only 20% in urban areas and 10% in rural areas [3]. The need for public awareness and better compliance is one of the reasons behind the development of country-specific guidelines.

Box 3.1: Features of Hypertension in Asians
- Masked hypertension is more common
- BP variability is more common
- Early morning surge and nocturnal rise are more common
- Region-wide differences in prevalence
- HBPM practices are variable
- Higher BP at lower BMI than in Europe

Table 3.6 Multiplication factors for calculating cardiovascular disease risk in immigrants according to the ESC/ESH guidelines [2]

Region of origin	Multiplication factor
Southern Asia	1.4
Sub-Saharan Africa	1.3
Caribbean	1.3
Western Asia	1.2
Northern Africa	0.9
Eastern Asia	0.7
Southern America	0.7

3.5 Hypertension and CVD Risk

Hypertension often coexists with other CVD risk factors such as metabolic syndrome, diabetes mellitus and dyslipidemia. This multiplies CVD risk, meaning that risk stratification is an important part of hypertension management.

Consideration of CVD risk has been suggested in nearly all current guidelines [1–4]. The ESC/ESH 2018 guidelines use the Systemic Coronary Risk Evaluation System (SCORE) for CVD risk, while the ACC/AHA guidelines use atherosclerotic CVD (ASCVD) risk for assessment of an individual's overall risk. Most scoring systems include five major risk factors including age, sex, dyslipidemia, diabetes mellitus and smoking. Risk modifiers that increase CVD risk estimation by the SCORE system of the ESC/ESH guidelines include obesity, physical inactivity, psychosocial stress, family history of premature CVD, autoimmune and inflammatory disorders, major psychiatric disorders, human immunodeficiency virus infection, atrial fibrillation, left ventricular (LV) hypertrophy, chronic kidney disease, obstructive sleep apnea and social deprivation. South Asians, especially Indians, are genetically at higher risk for CVD events. A multiple of 1.4 is recommended for calculating CVD risk in South Asians [2] (Table 3.6).

In addition to CVD risk factors, the presence of HMOD and associated clinical conditions (ACC) reduce the threshold for initiating pharmacotherapy. The prognosis of the patient, the urgency to initiate therapy, and BP targets during therapy should be based on the overall risk stratification depending on risk factors, HMOD and ACC.

3.6 Nonpharmacological Therapy

The primary goal of therapy is to effectively control BP to prevent, delay, or reverse the complications of hypertension and thus reduce overall CVD risk. Nonpharmacological approaches and lifestyle modifications are usually lifelong. Dietary patterns and other lifestyle factors vary significantly by region (Fig. 3.1). For example, salt intake is high in some regions of Asia, including India and China. Different guidelines have advocated salt restriction to a variable extent due to the prevalent pattern of diet [1, 4] (Table 3.7). Even the ESC/ESH guidelines suggest restriction of salt to less than 5 g/day [2]. Most guidelines recommend the DASH

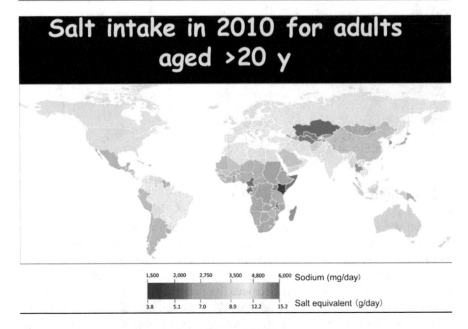

Fig. 3.1 Salt intake in different regions of the world [10]

Table 3.7 Daily salt intake recommended in different guidelines

	ACC/AHA guidelines	IGH–IV guidelines
Year	2017	2019
Daily salt intake recommended	3.75 g/day	6 g/day

diet (IGH IV [3] and ACC/AHA [1]). Potassium supplementation has also been recommendation in some guidelines (ACC/AHA) [1] but is less emphasized in others (ESC/ESH) [2]. Alcohol restriction is important and is equally recommended by all guidelines [14–16].

3.7   Threshold and Target of Drug Therapy

After assessing the patient and performing individualized risk stratification, the next step is to implement lifestyle modifications, then drug therapy can be considered. The threshold for starting drug therapy is somewhat lower in the ACC/AHA [1] guidelines than in the ESC/ESH [2] and IGH IV [3] guidelines due to the different definitions in these guidelines. However, the BP goal during treatment is same in all these guidelines (Fig. 3.2). The targets in the latest version of the ESC/ESH guidelines (2018) [2] are different to those in the earlier (2013) version (ESC 2013) [17] based on the availability of new data to support these recommendations (Fig. 3.3).

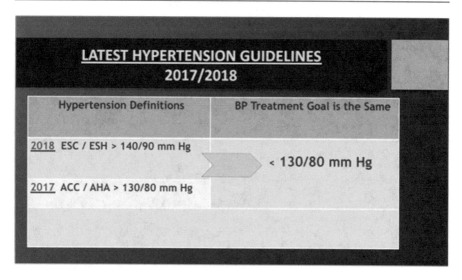

Fig. 3.2 Differences in definition and goal of treatment in the ESC/ESH and ACC/AHA guidelines [8]

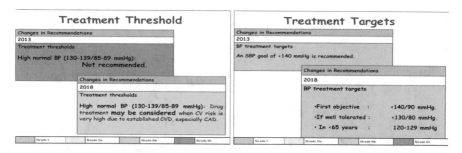

Fig. 3.3 Changes in treatment thresholds and targets between the 2013 and 2018 ESC/ESH guidelines

This change in the threshold and target is similar in other guidelines also [3, 4]. Thus, although the guidelines differ in their definition of hypertension, the approach to treatment is quite similar.

The ACC/AHA guidelines provide the same target of 130/80 mm Hg for all patients with hypertension, irrespective of age and comorbidities [1]. A significant addition in the ESC/ESH guidelines is the recommendation relating to the target DBP level, which should not be below 70 mmHg (Table 3.8) [2]. The table also shows the range in which systolic and diastolic blood pressure should be kept. This is different from the ACC/AHA guidelines, which only give a single BP value [1]. The Indian guidelines (IGH IV) provide a range of SBP and DBP values for different individuals depending on age and comorbidities (Table 3.9) [3].

Table 3.8 The desirable range of diastolic blood pressure in the ESC/ESH guidelines [2]

| Age group | Office SBP treatment target ranges (mmHg) | | | | | DBP target range (mmHg) |
	Hypertension	+ Diabetes	+ CKD	+ CAD	+ Stroke/ TIA	
18–65 years	Target to 130 or lower if tolerated Not <120	Target to 130 or lower if tolerated Not <120	Target to <140–130 if tolerated	Target to 130 or lower if tolerated Not <120	Target to 130 or lower if tolerated Not <120	<80–70
65–79 years	Target to <140–130 if tolerated	Target to <140–130 if tolerated	Target to <140–130 if tolerated	Target to <140–130 if tolerated	Target to <140–130 if tolerated	<80–70
≥80 years	Target to <140–130 if tolerated	Target to <140–130 if tolerated	Target to <140–130 if tolerated	Target to <140–130 if tolerated	Target to <140–130 if tolerated	<80–70
DBP target range	<80–70	<80–70	<80–70	<80–70	<80–70	

CKD chronic kidney disease, *TIA* transient ischemic attack

Table 3.9 BP thresholds for treatment initiation and target BP according to the IGH IV 2019 guidelines [3]

Subjects	Threshold to start treatment (≥), mmHg	Target BP range, mmHg
Age < 65 years		
High ASCVD risk	140/90	120–130/70–80
Low ASCVD risk	140/90	130–140/70–80
Age 65–80 years	140/90	130–140/70–80
Age > 80 years	140–150/90	130–140/70–80
With other risk factors		
Diabetes	140/90	130–140/70–80
History of stroke, TIA	140/90	130–140/70–80
Chronic kidney disease	140/90	130–140/70–80
Coronary artery disease	130/80	120–130/70–80
Heart failure	130/80	120–130/70–80

3.8 Pharmacotherapy

Four classes of antihypertensive agents are recommended by the ACC/AHA guidelines for the first-line treatment of hypertension [1]. These are angiotensin-converting enzyme inhibitors (ACEIs), angiotensin receptor blockers (ARBs), calcium channel blockers (CCBs), and diuretics. Beta-blockers are not recommended as first-line therapy by the ACC/AHA guidelines [1]. However, the ESC/ESH, IGH IV and ISH 2020 guidelines recommend that beta-blockers can be used as first-line agents in special situations, such as young individuals and those with angina, heart failure and after myocardial infarction [2–4]. Specific agents are mentioned within each guideline based on the local availability of specific agents within each class [18–20].

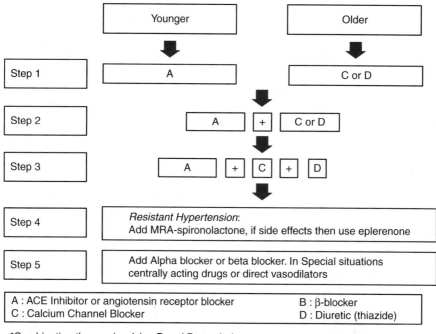

Fig. 3.4 Algorithm for approach to antihypertensive drug therapy in the IGH IV guidelines [3]

Most guidelines recommend the use of ARBs or ACEIs as first-line therapy in individuals aged <55–60 years [2, 3]. In elderly patients, CCBs and diuretics are recommended as first-line antihypertensive therapy [2, 3]. The IGH IV guidelines provide an algorithm for starting therapy and progressing to combination therapy for different individuals (Fig. 3.4).

Combination therapy is emphasized in all the recent guidelines [1–4]. The ESC/ESH guidelines recommend the use of a single pill combination (SPC) to improve compliance with therapy [2]. The ACC/AHA guidelines recommend specific combinations that are approved in that country [1]. Some guidelines even recommend triple combination SPCs because these are available in many countries.

3.9 Conclusion

The first guidelines on hypertension, were the JNC guidelines which were followed across the globe. However, after the JNC stopped issuing guidelines the ACC/AHA guidelines are followed in America. Various medical bodies have issued guidelines which have more relevance to the people of that country. There are some differences in the definition and management approach of various guidelines. However, since they are all based on the recent evidence, they are similar on most issues. In all these guidelines there is emphasis on better control of risk factors and blood pressure levels to reduce the TOD. Also, combination therapy is emphasized by all guidelines from across the globe. Asians are genetically and socially different from people in other regions and hence physicians need to follow guidelines from their country while heavy a knowledge of the other guidelines as well, since hypertension is the commonest non communicable disease.

References

1. Whelton PK, Carey RM, Aronow WS, et al. 2017ACC/AHA/AAPA/ABC/ACPM/AGS/APhA/vASH/ASPC/NMA/PCNA guideline for the prevention, detection, evaluation, and management of high blood pressure in adults. Hypertension. 2018;71:e13–e115.
2. Williams B, Mancia G, Spiering W, et al. 2018 ESC/ESH guidelines for the management of arterial hypertension. The Task Force for the management of arterial hypertension of the European Society of Cardiology (ESC) and the European Society of Hypertension (ESH). Eur Heart J. 2018;39:3021–104.
3. Shah SN, Core Committee Members: Billimoria AB, Mukherjee S, Kamath S, Munjal YP, Maiya M, Wander GS, Mehta N. Indian guidelines on management of hypertension (I.G.H) – IV 2019. Suppl J Assoc Phys India (JAPI). 2019;67(9):1–48.
4. Unger T, Borghi C, Charchar F, et al. 2020 International Society of Hypertension global hypertension practice guidelines. Hypertension. 2020;75:1334–57.
5. Rabi DM, McBrien KA, Sapir-Pichhadze R, Nakhla M, Ahmed SB, Dumanski SM, et al. Hypertension Canada's 2020 comprehensive guidelines for the prevention, diagnosis, risk assessment, and treatment of hypertension in adults and children. Can J Cardiol. 2020;36(5):596–624.
6. Wander GS, Ram CVS. Global impact of 2017 American Heart Association/American College of Cardiology hypertension guidelines—a prospective from India. Circulation. 2018;137:549–50.
7. Wander GS, Ram CVS. Blood pressure—methods to record & numbers that are significant: lets make a tailored suit to suit us. Indian J Med Res. 2018;147(5):435–8.
8. Wander GS, Ram CVS. Optimal blood pressure goals recommended by the latest hypertension guidelines: India may benefit the most. Eur Heart J. 2018;39(33):3012–6.
9. Wander GS, Gupta R, Ram CVS. Western guidelines bring in cardiovascular risk prediction along with blood pressure levels for initiation of antihypertensive drugs: is the pitch ready for Indians. J Hum Hypertens. 2019;33(8):566–7.
10. Kario K, Chia YC, Sukonthasarn A, Turana Y, Shin J, Chen CH, et al. Diversity of and initiatives for hypertension management in Asia—why we need the HOPE Asia Network. J Clin Hypertens (Greenwich). 2020;22(3):331–43.
11. Bobrie G, Genès N, Vaur L, Clerson P, Vaisse B, Mallion JM, et al. Is "isolated home" hypertension as opposed to "isolated office" hypertension a sign of greater cardiovascular risk? Arch Intern Med. 2001;161(18):2205–11.
12. Staessen JA, Den Hond E, Celis H, Fagard R, Keary L, Vandenhoven G, et al.; Treatment of Hypertension Based on Home or Office Blood Pressure (THOP) Trial Investigators. Antihypertensive treatment based on blood pressure measurement at home or in the physician's office: a randomized controlled trial. JAMA. 2004;291(8):955–64.
13. Gupta R, Ram CVS. Hypertension epidemiology in India: emerging aspects. Curr Opin Cardiol. 2019;34(4):331–41.

14. The effects of nonpharmacologic interventions on blood pressure of persons with high normal levels. Results of the Trials of Hypertension Prevention, Phase I. JAMA. 1992;267(9):1213–20.
15. Mozaffarian D, Fahimi S, Singh GM, Micha R, Khatibzadeh S, Engell RE, et al.; Global Burden of Diseases Nutrition and Chronic Diseases Expert Group. Global sodium consumption and death from cardiovascular causes. N Engl J Med. 2014;371(7):624–34.
16. Whelton PK, He J, Cutler JA, Brancati FL, Appel LJ, Follmann D, et al. Effects of oral potassium on blood pressure. Meta-analysis of randomized controlled clinical trials. JAMA. 1997;277(20):1624–32.
17. Mancia G, Fagard R, Narkiewicz K, Redón J, Zanchetti A, Böhm M, et al.; Task Force Members. 2013 ESH/ESC guidelines for the management of arterial hypertension: the Task Force for the management of arterial hypertension of the European Society of Hypertension (ESH) and of the European Society of Cardiology (ESC). J Hypertens. 2013;31(7):1281–357.
18. Takashima N, Ohkubo T, Miura K, Okamura T, Murakami Y, Fujiyoshi A, et al.; NIPPON DATA80 Research Group. Long-term risk of BP values above normal for cardiovascular mortality: a 24-year observation of Japanese aged 30 to 92 years. J Hypertens. 2012;30(12):2299–306.
19. Czernichow S, Ninomiya T, Huxley R, Kengne AP, Batty GD, Grobbee DE, et al. Impact of blood pressure lowering on cardiovascular outcomes in normal weight, overweight, and obese individuals: the Perindopril Protection Against Recurrent Stroke Study trial. Hypertension. 2010;55(5):1193–8.
20. Ogden LG, He J, Lydick E, Whelton PK. Long-term absolute benefit of lowering blood pressure in hypertensive patients according to the JNC VI risk stratification. Hypertension. 2000;35(2):539–43.

Cardiovascular Disease and Hypertension in Migrant Asian Populations in the West

4

Avneet Singh, Rajiv Jauhar, and Joseph Diamond

4.1 Introduction

Cardiovascular disease (CVD) remains the leading cause of mortality and morbidity worldwide, with variable distribution between developed and low- and middle-income nations. In Western nations, improvements in lifestyle and healthcare interventions have been associated with a reduction in the overall incidence of CVD. However, the beneficial impact of these changes and interventions is not uniformly distributed among the diverse population groups that live in these countries. The starkest example of this disparity is the disproportionately high incidence of CVD and its associated risk factors among immigrants from South Asia to the Western nations of Europe and the USA [1]. South Asians in Western countries have a much higher incidence of both CVD risk factors and CVD itself. This may be explained by a higher incidence of traditional risk factors including diabetes [2], but also disparities in lifestyle, diet, socioeconomic status, literacy levels, and access to healthcare. This has a significant social and economic impact on the countries where South Asians live. Contemporary trials in North America and Europe have not specifically addressed the high CVD burden in Asian immigrants. A scientific statement from the American Heart

A. Singh (✉)
Department of Cardiology, Donald and Barbara Zucker School of Medicine/Northwell, Hempstead, NY, USA
e-mail: asingh1@northwell.edu

R. Jauhar
Department of Cardiology, Donald and Barbara Zucker School of Medicine/Northwell, North Shore University Hospital, Manhasset, NY, USA
e-mail: Rjauhar@northwell.edu

J. Diamond
Department of Cardiology, Donald and Barbara Zucker School of Medicine/Northwell, Long Island Jewish Medical Center, Queens, NY, USA
e-mail: Jdiamond@northwell.edu

© The Author(s), under exclusive license to Springer Nature Switzerland AG 2022
C. V. S. Ram et al. (eds.), *Hypertension and Cardiovascular Disease in Asia*,
Updates in Hypertension and Cardiovascular Protection,
https://doi.org/10.1007/978-3-030-95734-6_4

Association (AHA) highlights the importance of recognising the growing epidemic of heart disease among this group and the need for specific research to help understand and treat these high-risk populations [3]. This section discusses the various aspects of CVD among Asian migrants living in Western regions.

4.2   Epidemiology

Asian Americans represent the fastest growing group of immigrants to the USA, UK, Australia, and other Western nations around the world. US census data from 2019 estimated that there were 18.9 million Asian Americans living in the United States, representing 5.7% of the total population [3]. Asian Americans are broadly categorised into three groups based on geographical origin: East Asia, Southeast Asia, or South Asia (i.e., the Indian subcontinent). South Asians are a mixed group of immigrants from countries including India, Pakistan, Bangladesh, Bhutan, Nepal, Sri Lanka, Maldives, and the diaspora from Guyana, Trinidad, Fiji, Kenya, and other nations [4, 5]. Indians account for 80% of US immigrants from South Asia, followed by Pakistani, Bangladeshi, Nepali, Sri Lankan, and Bhutanese [6]. These communities are extremely heterogeneous with respect to linguistic and cultural beliefs, education levels, diet, and access to healthcare. This diversity makes it challenging to apply standard methods of healthcare delivery and risk assessment tools to predict and treat CVD. The variable CVD mortality rates in various Asian American subgroups is shown in Fig. 4.1 [3].

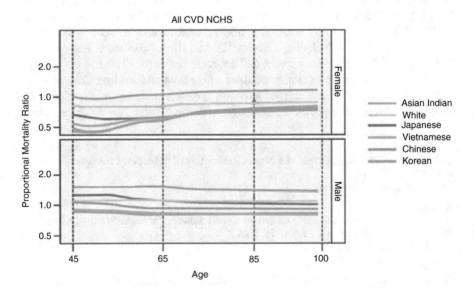

Fig. 4.1 Proportional mortality rates for cardiovascular diseases in Asian American subgroups. (Modified from AHA scientific statement article) [3]

The earliest epidemiological evidence of coronary artery disease (CAD) among South Asians in the Western region is from UK. Based on data from observational studies, it was seen that South Asians not only had premature CAD but also had more extensive atherosclerosis with multivessel involvement [7]. The Southall and Brent Revisited (SABRE) study examined a community-based cohort from North and West London and found that rates of CAD and stroke over a 20-year follow-up were significantly elevated in South Asian migrants compared with British Europeans and this difference was only partially attributed to traditional risk factors [8].

In the USA, the self-reported prevalence of CAD, diabetes mellitus and dyslipidaemia was higher among Indian physicians and their families compared with a matched Caucasian population [9]. In addition, a population-based study from California showed that during the period from 1985 to 1990, Asian Indians had disproportionately higher cardiovascular mortality compared with other ethnic groups [1]. While other ethnic groups showed a reduction in mortality over time, Asian Indians showed a 16% increase in all-cause mortality and 5% increase in CAD mortality [1]. For reasons not clearly understood, the overall incidence of CAD is lower among Chinese populations but these individuals have higher mortality after myocardial infarction compared with other groups [10]. An epidemiological survey from Spain also showed significantly higher prevalence of CV risk factors and CVD among South Asians, especially Bangladeshis and Pakistanis, and a lower incidence among Chinese immigrants [11].

The survey by EA Enas et al. described above [1] assessed the prevalence of CAD and its risk factors in selected first-generation immigrants to the US. These immigrants were Asian Indian physicians and their family members, a total of 1688 individuals (1131 men and 557 women aged ≥ 20 years). These individuals were compared with a native White American population, based on the Framingham Offspring Study. The age-adjusted prevalence of myocardial infarction and/or angina was approximately three times higher in Asian Indian versus White American men (7.2% vs 2.5%; $p < 0.0001$), but not in Asian Indian versus White American women (0.3% vs 1%; $p = 0.64$). Asian Indians also had higher prevalence of type 2 diabetes mellitus (7.6% vs 1%; $p < 0.0001$). The prevalence of hypertension was lower in Asian Indian compared with White American men (14.2% vs 19.1%, $p < 0.008$) but similar in the two populations of women (11.3% vs 11.4%). Asian Indians were significantly less likely to smoke than White Americans (1.3% vs 27%; $p < 0.0001$), and less likely to be obese (4.2% vs 22%; $p < 0.0001$). The latter findings may reflect an overall healthier population given that the Asian Indian study population consisted of the families of healthcare professionals. Overall, CVD risk factors in individuals of South Asian origin differ from those in individuals of other ethnicities, with a higher prevalence of central obesity, metabolic syndrome and diabetes [12]. Each of these risk factors, plus hypertension, is discussed below.

4.3 Risk Factors in Migrant Asians

4.3.1 Impaired Glucose Tolerance and Type 2 Diabetes Mellitus

Diabetes is one of the strongest predictors of CVD in the general population. There has been a significant increase in the number of people with diabetes mellitus in the Indian subcontinent. Government data estimated that the prevalence of diabetes mellitus in India is as high as 14.2%. There are many genetic and environmental factors responsible for development of diabetes mellitus [13, 14]. Insulin resistance may be the cause of a high proportion of type 2 diabetes in South Asia. Both healthy individuals and those with diabetes from South Asia have been shown to have elevated insulin levels in response to an oral glucose load compared with Europeans, suggesting baseline insulin resistance [15].

Although the exact aetiology of insulin resistance in this population is not well established, Asian Indians have greater amounts of visceral fat as measured by CT scan compared with matched Caucasians [16]. There are over 60 genes that have been linked to the development of diabetes mellitus, but to date no specific gene has been identified to explain the higher incidence of diabetes mellitus in South Asian populations. MASALA study investigators found that South Asians had a significantly higher incidence of diabetes mellitus (23%) compared with other ethnic groups, and suggested that this may be related to lower beta-cell function [17]. The high prevalence of diabetes mellitus is likely to be the result of interplay between genetic predisposition and environmental factors, and this will remain the focus of future research to target at-risk individuals at an early age.

4.3.2 Dyslipidaemia

Dyslipidaemia is a well-established predictor of atherosclerotic cardiovascular disease (ASCVD), with elevated low-density lipoprotein (LDL) and low high-density lipoprotein (HDL) levels being the strongest risk factors. There are marked differences in the lipid disorder subtypes between different racial and ethnic subgroups in the US. South Asians have a much higher incidence of lipid disorders, typically with elevated triglycerides, low HDL, and only mild elevation of LDL, compared with African Americans, Hispanics, and Caucasians [18]. Furthermore, observational studies show a unique pattern of elevated apolipoprotein B100/apolipoprotein A-I ratio, lipoprotein-a (Lp-a), and cholesteryl ester transfer protein [19, 20]. These findings are likely related to a combination of factors including genetic predisposition, dietary/lifestyle choices, and metabolic syndrome.

4.3.3 Obesity

Obesity rates are increasing across the globe due to changes in lifestyles and diet. Population-based studies have shown a correlation between obesity and underlying cardiometabolic diseases. Asians may not have the same level of obesity based on traditional indices such as body mass index (BMI), but the risk of developing diabetes mellitus and CVD even at lower BMI values is higher among South Asian populations [21]. For this reason, the World Health Organization defines obesity at a BMI of 23 kg/m^2 in Asians rather than the 25 kg/m^2 value used in other populations. As noted above, there is higher distribution of central and visceral fat in South Asians.

In a pooled data analysis from MASALA and MESA, South Asians had an unfavourable distribution of body fat, with high levels of intermuscular and hepatic fat and lower lean muscle mass [22]. They also had higher levels of resistin and lower levels of adiponectin, which may play a role in insulin resistance and the development of truncal obesity [14]. Based on these observations, it is recommended that abdominal diameter index or waist-to-hip ratio (WHR) may provide a better indication of abnormal adipose tissue than BMI, which accounts for overall body fat [23].

4.3.4 Hypertension

There is a strong association between hypertension and ASCVD including CAD, peripheral arterial disease, cerebrovascular disease, and chronic kidney disease. The MESA study investigators did not find a significant difference in the prevalence of hypertension between Asians and non-Hispanic Whites [23]. US Centers for Disease Control (CDC) data, obtained using interviews and physical examinations of a sample of the civilian noninstitutionalised population, found that the overall incidence of hypertension in Asians living in the USA to be high, but relatively stable. In the MASALA study, 43% of men and 35% of women had hypertension [24]. Furthermore, it appears that individuals from South Asia develop elevated blood pressure (BP) early in life, along with multiple metabolic derangements. There are complex biological, social, dietary, and environmental factors that may contribute to these differences. Specific details relating to hypertension in individuals from different parts of Asia living in the West are provided below.

Prevalence in India: In 2018, the estimated prevalence of hypertension in India was estimated to be 24.5% (95% confidence interval [CI] 24.2–24.9%) in men and 20.0% (95% CI 19.7–20.3%) in women [25]. Prevalence varied considerably by state, being highest in the northern states of Punjab and Himachal Pradesh, the southern state of Kerala, and the north-eastern states of Sikkim and Nagaland. Increasing age was a more important risk factor for elevated BP than any socioeconomic factors. However, important socioeconomic factors associated with a higher prevalence of hypertension included residence in urban versus rural region, and higher household wealth. The impact of level of education on prevalence of hypertension was relatively small [26]. Surveys of Asian immigrants to the United States showed that cardiovascular risk factors varied, depending on the origin of the individuals. Having some understanding of their original culture may help healthcare providers to understand health attitudes, behaviours and cultural beliefs that influence health outcomes.

Hypertension in migrant Asians: Koirala et al. reviewed health data for more than 500,000 U.S. adults from 2010 to 2018, including nearly 34,000 Asian immigrants [18]. Of those immigrants, 26% came from the Indian subcontinent, 45% from Southeast Asia and 29% from the rest of Asia. They found substantial differences in heart disease risk factors, particularly when it came to obesity, hypertension, and diabetes, the so-called cardiometabolic risk (CMR) factors. Compared to Whites, immigrants from the Indian subcontinent had a higher prevalence of diabetes (16.1% for those immigrating from the Indian subcontinent vs 9.5% in white American adults). The prevalence of being overweight or obese was also higher in immigrants from the Indian subcontinent, with prevalence of 79.2% vs 65% respectively. The prevalence of hypertension, however, was lower in the immigrants from the Indian subcontinent, 27.2% vs 34% respectively. Immigrants from the Indian subcontinent also had lower rates of active cigarette smoking.

Hypertension in migrant South Asians: In addition to region of origin, other variables that need to be considered are the eventual destination, level of education, baseline and subsequent income, ownership of land, ownership of consumer durables (e.g., car, television, tractor, animals, etc.), type of employment and religious/cultural beliefs as they pertain to health. In one study of individuals from Kerala, India who immigrated to the Gulf countries, age-adjusted hypertension prevalence was 57.6% in these migrants compared with 31.7% in non-migrants ($p = 0.05$). Unlike migrants to the US, migrants to the Gulf States were more likely to have hypertension than non-migrants after adjusting for age (odds ratio 3.00, 95% CI 1.83–4.94) [27]. It is not clear why migrants to the Gulf States had higher rates of hypertension than migrants to the US because there are so many variables that contribute to hypertension. These include the cutoff BP values used to define hypertension, genetic predisposition, socioeconomic and cultural factors of migratory populations (as outlined above), and the socioeconomic changes that occur with migration.

In the Indian Social Class and Heart Survey, individuals living in rural north India who had higher overall socioeconomic status had a higher prevalence of hypertension that was associated with increased body weight and a sedentary lifestyle [28]. This contrasts with what is seen in developed countries (e.g., US, Great Britain) where there is a strong inverse association between social status and CVD, and the prevalence of coronary risk factors and high BP were higher in lower social classes [29]. Studies in Indian migrants to Britain reported a higher prevalence of hypertension, stroke mortality, and BMI than in their compatriots in India [12]. This suggests that socioeconomic factors and lifestyle changes that occur with immigration to Western countries may be important factors in the development of hypertension [30, 31]. By 2015, there were over 43.3 million immigrants in the US, comprising 13.5% of the population [32]. The top five countries of origin in 2015 were India, China, Mexico, Philippines, and Canada [33, 34].

Data from the National Health Interview Survey (NHIS), a population-based survey of civilian, noninstitutionalised US adults aged ≥18 years was conducted by the National Center for Health Statistics (NCHS) [35]. Of the 41,717 immigrants included in this study, 2712 (6.5%) were from the Indian subcontinent [35]. Immigrants from the Indian subcontinent, Mexico, Central America, and the Caribbean had the highest prevalence of obesity and diabetes, while those from Russia and Southeast Asia had the highest prevalence of hypertension [27]. The overall prevalence of hypertension in individuals from the Indian subcontinent was 23.3% (95% CI 21.1–25.5%), with prevalence rates of 21.8% (95% CI 18.9–24.6) in men and 24.3% (95% CI 20.9–27.7) in women [35]. This was a relatively young group with mean age 40.7 ± 0.5 years, and the prevalence of hypertension does increase significantly with increasing age. The prevalence of hypertension in Russian immigrants was 26% (95% 23.1–28.8), and these patients were older (mean age 53.5 ± 0.4 years) [27]. The prevalence of obesity in immigrants from the Indian subcontinent was 77.6%, (80.4% in men and 76.0% in women), and the prevalence of diabetes mellitus was 14.3%, (16.3% in men and 11.4% in women) [35].

The findings from these trials confirm that the prevalence of hypertension in migrating populations is influenced by numerous factors. Genetic predisposition is modified by environmental factors, which undergo great change in migrating populations. The impact of migration from rural to urban environments, and from one nation to another, will create changes that have a noticeable impact on the prevalence of hypertension and other CVD risk factors. Some studies have reported higher rates of diabetes, obesity, hypertension, and CAD among immigrants from the Indian subcontinent worldwide (including Indians, Pakistanis, Bangladeshis, and Sri Lankans), though not uniformly, particularly with respect to hypertension. Post-migration socioeconomic challenges, poor education, dietary changes, behavioural changes (e.g., level of physical activity), and lack of health insurance are among some of the most important factors that are associated with an increased risk of hypertension and cardiometabolic disease among immigrants. As such, it behoves health care providers to better understand the cultural and socioeconomic challenges faced by immigrants in their new homelands and take into consideration the health behaviours and cultural beliefs that influence health outcomes.

4.3.5 Diet, Physical Activity, and Smoking

Diet, physical activity, smoking, and other psychosocial factors correlate with the development of lipid disorders, diabetes, and hypertension, and ultimately contributing to ASCVD. The dietary pattern of Asian Americans, especially those of Indian origin, often includes an extremely high proportion of carbohydrates and saturated fats, while lacking dietary fibre and fruit. This is likely related to the inherent dietary habits from native counties, compounded by socialisation and the adoption of more processed foods in Western societies. Lack of physical activity has a direct association with most important CVD risk factors. Although there may not be a reliable way to determine physical activity among various ethnic groups, there seems to be less self-awareness and lack of availability of fitness resources among Asians.

4.4 Screening and Risk Reduction

Early detection, screening of high-risk individuals and appropriate early interventions remains the cornerstone of preventing clinical events in any community. South Asian communities have some unique biological and nonbiological links to future risk of ASCVD, and these need to be used as screening tools. Traditional risk assessment tools do not account for heterogeneity among the Asian populations and may underestimate the risk of CAD. The QRISK2 calculator [36], derived from a large data set in the UK, has country-based ethnicity as an independent data point, and may therefore provide more accurate prediction of risk and future CVD events.

We recommend an individualised approach using thorough assessment of traditional risk factors, the Framingham risk score, ASCVD risk scores and the QRISK2 calculator, complemented by diet/lifestyle questionnaires and assessment of visceral fat distribution using WHR to identify at-risk patients. Numerous biomarkers may become useful as screening tools to identify high-risk subjects. These include homocysteine, plasminogen activator inhibitor-1, Lp-a, C-reactive protein (CRP), leptin, interleukin-6, tumour necrosis factor-α, leptin, adiponectin, and resistin. However, most of these currently remain under investigation. Levels of high sensitivity CRP are elevated in South Asians, even after adjusting for other CVD risk factors [37, 38]. In our opinion, this is an easily tested biomarker that may help identify at-risk patients. Genetic studies are still under investigation, but genetic testing may have a role to play as a screening tool in the future.

Several pharmacological therapies that target lipid and metabolic disorders include statins, ezetimibe-statin combinations, PCSK-9 inhibitors, sodium-glucose cotransporter-2 inhibitors and glucagon-like peptide analogues. However, most of the clinical outcome data obtained in patients using these therapies comes from trials that did not have a significant representation of Asian communities. In smaller studies, it appears that these agents have a similar impact on biomarkers in Asian patients, but their correlation with hard clinical endpoints in Asian subgroups has not been well studied [39, 40]. It is recommended that patients be treated as per standard guidelines of the respective professional societies. However, further research focusing on the impact of various interventions is needed to guide the best evidence-based therapies for Asian populations.

Revascularisation strategies and interventional cardiology techniques have evolved over the last three decades. It was historically seen that there was a significant ethnic disparity in the healthcare distribution between Asians versus non-Hispanic Whites. Several initiatives have helped narrow this gap. Most notably, "Get with The Guidelines-Coronary Artery Disease Program (GWTG-CAD)" facilitated better and timely care for patients presenting to hospital with CAD. However, despite this, Asian Americans are at higher risk of mortality compared with non-Hispanic White patients [41]. Furthermore, Indians and Chinese have a disproportionately higher incidence of myocardial infarction within 1 year of index coronary angiography [42]. Similarly, South Asians have higher mortality and morbidity after coronary artery bypass graft surgery, despite adjustment for other risk factors [43].

The comparatively worse outcome may be explained by a higher prevalence of CV risk factors, especially diabetes mellitus, more multivessel CAD, and a smaller vessel calibre. However, there are other barriers to care that negatively impact outcomes. Asians may not seek preventative care, routine healthcare screenings, and ultimately may have a delay in presentation to healthcare facilities. Medication noncompliance and lower utilisation of cardiac rehabilitation programmes may further explain poor outcomes. Significant cultural, linguistic, and economic heterogeneity within Asian communities poses a challenge for equitable and uniform quality healthcare delivery. Community approaches that consider these barriers must be instituted to improve awareness, education, and utilisation of available resources.

4.5   Future Research

In this chapter, we have addressed the high burden of CVD and hypertension in Asians living in modern Western countries. There are complex genetic, environmental, dietary, lifestyle, and cultural factors that need to be investigated further to better understand and guide interventions to improve the health of this growing population. We need more ethnicity-specific data to formulate risk calculation tools to stratify the minority groups living in the US, the UK, and other Western nations. Large registries like the MASALA study will provide valuable data to better understand these heterogeneous groups. Importantly, large population-based clinical trials must be designed with appropriate representation of various ethnic groups to determine the efficacy of new therapies. Some of the ongoing and recently updated studies on the prevalence of hypertension and CVD in migrant populations from Asia are shown in Fig. 4.2.

Fig. 4.2 Recent studies (ongoing and complete) on atherosclerotic cardiovascular disease risk factors in South Asian populations and prevalence in the UK, the USA and Canada

References

1. Enas EA, Garg A, Davidson MA, Nair VM, Huet BA, Yusuf S. Coronary heart disease and its risk factors in first-generation immigrant Asian Indians to the United States of America. Indian Heart J. 1996;48(4):343–53.
2. Bhatnagar D, Anand IS, Darrington PN, Patel DJ, Wander GS, Mackners MI, Creed F, Son BT, Chandrashekhar Y, Winterbotham M, Britt RP, e Keil J, Sutton GC. Coronary risk factors in people from the Indian subcontinent living in West London and their siblings in India. Lancet. 1995;345:405–9.
3. Volgman AS, Palaniappan LS, Aggarwal NT, Gupta M, Khandelwal A, Krishnan AV, et al.; American Heart Association Council on Epidemiology and Prevention; Cardiovascular Disease and Stroke in Women and Special Populations Committee of the Council on Clinical Cardiology; Council on Cardiovascular and Stroke Nursing; Council on Quality of Care and Outcomes Research; and Stroke Council. Atherosclerotic cardiovascular disease in South Asians in the United States: epidemiology, risk factors, and treatments: a scientific statement from the American Heart Association. Circulation. 2018;138(1):e1–34.
4. Asian American Federation. A demographic snapshot of South Asians in the United States. 2012. http://saalt.org/wp-content/uploads/2012/09/Demographic-Snapshot-Asian-American-Foundation-2012.pdf. Accessed 9 Nov 2021.
5. US Census Bureau profile of general population and housing characteristics, 2010. https://www.census.gov/history/pdf/2010angelscamp.pdf. Accessed 9 Nov 2021.
6. Immigration Policy Center. The passage from India: a brief history of Indian immigration to the U.S. 2002. http://www.issuelab.org/resource/the_passage_from_india_a_brief_history_of_indian_immigration_to_the_u_s. Accessed 9 Nov 2021.
7. Donaldson LJ, Taylor JB. Patterns of Asian and non-Asian morbidity in hospitals. Br Med J (Clin Res Ed). 1983;286(6369):949–51.
8. Hughes LO, Raval U, Raftery EB. First myocardial infarctions in Asian and white men. BMJ. 1989;298(6684):1345–50.
9. Tillin T, Hughes AD, Mayet J, Whincup P, Sattar N, Forouhi NG, et al. The relationship between metabolic risk factors and incident cardiovascular disease in Europeans, South Asians, and African Caribbeans: SABRE (Southall and Brent Revisited)—a prospective population-based study. J Am Coll Cardiol. 2013;61(17):1777–86.
10. Palaniappan L, Wang Y, Fortmann SP. Coronary heart disease mortality for six ethnic groups in California, 1990-2000. Ann Epidemiol. 2004;14(7):499–506.
11. Jin K, Ding D, Gullick J, Koo F, Neubeck L. A Chinese immigrant paradox? Low coronary heart disease incidence but higher short-term mortality in Western-dwelling Chinese immigrants: a systematic review and meta-analysis. J Am Heart Assoc. 2015;4(12):e002568.
12. Chambers JC, Eda S, Bassett P, Karim Y, Thompson SG, Gallimore JR, et al. C-reactive protein, insulin resistance, central obesity, and coronary heart disease risk in Indian Asians from the United Kingdom compared with European whites. Circulation. 2001;104:145–50.
13. Sanghera DK, Been L, Ortega L, Wander GS, Mehra NK, Aston CE, Mulvihill JJ, Ralhan S. Testing the association of novel meta analysis derived diabetes risk genes with type II diabetes and related metabolic traits in Asian Indian Sikhs. J Hum Genet. 2009;54:162–8.
14. Sanghera DK, Demirci Y, Been L, Ortega L, Ralhan SK, Wander GS, Mehra NK, Singh JR, Mulvihill JJ, Kamboh MI. (2008): Novel polymorphisms in *PPARG1, PPARG2,* and *ADIPOQ* genes increase type 2 diabetes risk in Asian Indian Sikhs. Metabolism. 2010;59:492–501.
15. Satish P, Vela E, Bilal U, Cleries M, Kanaya AM, Kandula N, et al. Burden of cardiovascular risk factors and disease in five Asian groups in Catalonia: a disaggregated, population-based analysis of 121 000 first-generation Asian immigrants. Eur J Prev Cardiol. 2021;zwab074.
16. Mohan V, Sharp PS, Cloke HR, Burrin JM, Schumer B, Kohner EM. Serum immunoreactive insulin responses to a glucose load in Asian Indian and European type 2 (non-insulin-dependent) diabetic patients and control subjects. Diabetologia. 1986;29(4):235–7.

17. Raji A, Seely EW, Arky RA, Simonson DC. Body fat distribution and insulin resistance in healthy Asian Indians and Caucasians. J Clin Endocrinol Metab. 2001;86(11):5366–71.
18. Gujral UP, Narayan KM, Kahn SE, Kanaya AM. The relative associations of β-cell function and insulin sensitivity with glycemic status and incident glycemic progression in migrant Asian Indians in the United States: the MASALA study. J Diabetes Complicat. 2014;28(1):45–50.
19. Frank AT, Zhao B, Jose PO, Azar KM, Fortmann SP, Palaniappan LP. Racial/ethnic differences in dyslipidemia patterns. Circulation. 2014;129(5):570–9.
20. Joshi P, Islam S, Pais P, Reddy S, Dorairaj P, Kazmi K, et al. Risk factors for early myocardial infarction in South Asians compared with individuals in other countries. JAMA. 2007;297(3):286–94.
21. Sharobeem KM, Patel JV, Ritch AE, Lip GY, Gill PS, Hughes EA. Elevated lipoprotein (a) and apolipoprotein B to AI ratio in South Asian patients with ischaemic stroke. Int J Clin Pract. 2007;61(11):1824–8.
22. WHO Expert Consultation. Appropriate body-mass index for Asian populations and its implications for policy and intervention strategies. Lancet. 2004;363(9403):157–63.
23. Shah AD, Kandula NR, Lin F, Allison MA, Carr J, Herrington D, et al. Less favorable body composition and adipokines in South Asians compared with other US ethnic groups: results from the MASALA and MESA studies. Int J Obes. 2016;40(4):639–45.
24. Smith DA, Ness EM, Herbert R, Schechter CB, Phillips RA, Diamond JA, et al. Abdominal diameter index: a more powerful anthropometric measure for prevalent coronary heart disease risk in adult males. Diabetes Obes Metab. 2005;7(4):370–80.
25. Brister SJ, Hamdulay Z, Verma S, Maganti M, Buchanan MR. Ethnic diversity: South Asian ethnicity is associated with increased coronary artery bypass grafting mortality. J Thorac Cardiovasc Surg. 2007;133(1):150–4.
26. Geldsetzer P, Manne-Goehler J, Theilmann M, Davies JI, Awasthi A, Vollmer S, et al. Diabetes and hypertension in India: a nationally representative study of 1.3 million adults. JAMA Intern Med. 2018;178(3):363–72.
27. Koirala B, Commodore-mensah Y, Turkson-ocran, RA, Baptiste D, Davidson P, Dennison Himmelfarb CR. Heterogeneity of cardiovascular disease risk factors among asian immigrants: insights from the 2010–2018 National Health Interview Survey. Abstract. November 13, 2020. American Heart Association.
28. Begam NS, Srinivasan K, Mini GK. Is migration affecting prevalence, awareness, treatment and control of hypertension of men in Kerala, India? J Immigr Minor Health. 2016;18(6):1365–70.
29. Singh RB, Sharma JP, Rastogi V, Niaz MA, Singh NK. Prevalence and determinants of hypertension in the Indian social class and heart survey. J Hum Hypertens. 1997;11(1):51–6.
30. Rogot E, Sorlie PD, Johnson N. A mortality study of 1.3 million persons by demographic, social and economic 1994; 344: 101–106. factors: 1979 to 1985 follow-up. Bethesda, MD: National Heart, Lung and Blood Institute, NIH Publi-Appendix cation No. 92-3297; 1992.
31. Keil JE, Britt RP, Weinrich MC, Hollis Y, Keil BW. Hypertension in Punjabi females: comparison between migrants to London and natives in India. Hum Biol. 1980;52(3):423–33.
32. Bhatnagar D, Anand IS, Durrington PN, Patel DJ, Wander GS, Mackness MI, et al. Coronary risk factors in people from the Indian subcontinent living in West London and their siblings in India. Lancet. 1995;345(8947):405–9.
33. Zong J, Batalova J, Hallock J. Frequently requested statistics on immigrants and immigration in the United States.
34. Passel J, Rohal M. Modern immigration wave brings 59 million to US, driving population growth and change through 2065. Pew Research Center; 2015.
35. Commodore-Mensah Y, Selvin E, Aboagye J, Turkson-Ocran RA, Li X, Himmelfarb CD, et al. Hypertension, overweight/obesity, and diabetes among immigrants in the United States: an analysis of the 2010-2016 National Health Interview Survey. BMC Public Health. 2018;18(1):773.
36. Hippisley-Cox J, Coupland C, Vinogradova Y, Robson J, Minhas R, Sheikh A, et al. Predicting cardiovascular risk in England and Wales: prospective derivation and validation of QRISK2. BMJ. 2008;336:1475–82.

37. Pradhan AD, Manson JE, Rifai N, Buring JE, Ridker PM. C-reactive protein, interleukin 6, and risk of developing type 2 diabetes mellitus. JAMA. 2001;286:327–34.
38. Sattar N, Gaw A, Scherbakova O, Ford I, O'Reilly DS, Haffner SM, et al. Metabolic syndrome with and without C-reactive protein as a predictor of coronary heart disease and diabetes in the West of Scotland Coronary Prevention Study. Circulation. 2003;108:414–9.
39. Kandula NR, Kanaya AM, Liu K, Lee JY, Herrington D, Hulley SB, et al. Association of 10-year and lifetime predicted cardiovascular disease risk with subclinical atherosclerosis in South Asians: findings from the Mediators of Atherosclerosis in South Asians Living in America (MASALA) study. J Am Heart Assoc. 2014;3(5):e001117.
40. Yusuf S, Bosch J, Dagenais G, Zhu J, Xavier D, Liu L, et al. Cholesterol lowering in intermediate-risk persons without cardiovascular disease. N Engl J Med. 2016;374(21):2021–31.
41. Madan M, Vira T, Rampakakis E, Gupta A, Khithani A, Balleza L, et al. A randomized trial assessing the effectiveness of ezetimibe in South Asian Canadians with coronary artery disease or diabetes: the INFINITY study. Adv Prev Med. 2012;2012:103728.
42. Qian F, Ling FS, Deedwania P, Hernandez AF, Fonarow GC, Cannon CP, et al. Care and outcomes of Asian-American acute myocardial infarction patients: findings from the American Heart Association Get With The Guidelines-Coronary Artery Disease program. Circ Cardiovasc Qual Outcomes. 2012;5(1):126–33.
43. Manjunath L, Chung S, Li J, Shah H, Palaniappan L, Yong CM. Heterogeneity of treatment and outcomes among Asians with coronary artery disease in the United States. J Am Heart Assoc. 2020;9(10):e014362.

Effect of Genetic Ancestry (Racial Factors) on Hypertension in Asian Countries

5

Rajeev Gupta

5.1 Introduction

Hypertension or high blood pressure (BP) is the most important contributor to disease and disability worldwide [1]. According to Global Burden of Diseases study, in 2019 there were 828 million adults (95% uncertainty interval 768–888 million) with systolic blood pressure (SBP) ≥140 mmHg [2]. On the other hand, the Noncommunicable Disease Risk Factor Collaboration (NCDRiSC) estimated the number of individuals with hypertension (SBP ≥140 mmHg and/or diastolic blood pressure [DBP] ≥90 mmHg or known hypertension) as more than one billion in 2015 [3].

There are large global variations in the prevalence of hypertension and hypertension-related mortality. The global prevalence of hypertension has been determined by NCDRiSC and it was calculated that more than half of all individuals with hypertension reside in the countries of East, South-East, and South Asia (Fig. 5.1). The highest number of individuals with hypertension are in East and South-East Asian countries (332 million), followed by South Asia (258 million), and Central Asia, Middle East, and North Africa (79 million). These values are much higher than in high-income countries in Europe and North America (141 million), Sub-Saharan Africa (107 million), Latin America (87 million), and Central and Eastern Europe (87 million) [4].

R. Gupta (✉)
Department of Preventive Cardiology and Internal Medicine, Eternal Heart Care Centre and Research Institute, Mount Sinai New York Affiliate, Jaipur, India

Academic Research Development Unit, Rajasthan University of Health Sciences, Jaipur, India
e-mail: drrajeev.gupta@eternalheart.org

© The Author(s), under exclusive license to Springer Nature Switzerland AG 2022
C. V. S. Ram et al. (eds.), *Hypertension and Cardiovascular Disease in Asia*,
Updates in Hypertension and Cardiovascular Protection,
https://doi.org/10.1007/978-3-030-95734-6_5

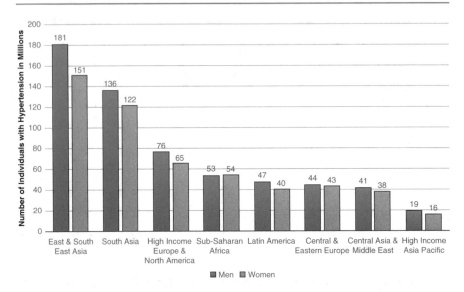

Fig. 5.1 Number of individuals with hypertension in different regions of the world (NCDRiSC 2015) [3]

The Global Burden of Diseases Study has also reported that high SBP (above the ideal of 110–115 mmHg) is the most important cause of disease burden and disability (disability-adjusted life-years [DALYs]) globally [1], and identified hypertension as the most important cause of death not only in high-income countries but also in upper-middle and lower middle-income countries [2, 4]. It has also been reported that while the number of high SBP-related deaths and DALYs is declining in high- and upper middle-income countries, rates are either increasing or not declining in most lower middle-income and low-income countries [4, 5]. Age-standardized DALYs due to high SBP in the year 2019 are shown in Fig. 5.2. Globally, hypertension resulted in the loss of 3042 DALYs per 100,000 person-years in 2019 [5]. The number of DALYs was highest in upper-middle sociodemographic index (SDI) countries and lowest in low SDI countries [5]. An important observation is that the number of DALYs is significantly lower in high SDI countries, suggesting a lower hypertension prevalence, higher age of onset, better treatment and control, and lower mortality related to high blood pressure (BP) compared with mid and low SDI countries [6].

Figure 5.2 also highlights the geographic variation in hypertension-related DALYs. These are highest in Central Asia, Europe, East and South-East Asia, the Middle East, and North Africa [5]. Country-level hypertension-related disease burden shows that Asian countries have the highest burden of high BP-related premature disease and deaths (Fig. 5.3). A low number of DALYs are seen in most of Western Europe, North America, Japan, and Australasia, suggesting an inverse association between hypertension-related disease burden and socioeconomic development [7].

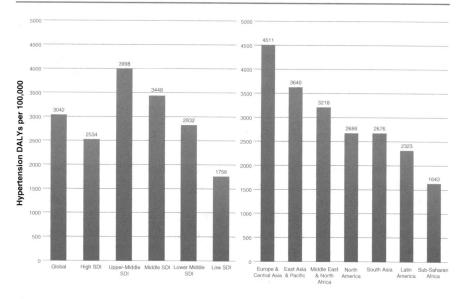

Fig. 5.2 Age-standardized DALYs/100,000 due to high SBP in countries with different sociode-mographic index levels (left) and different geographic regions (right). (Data source: Global Burden of Diseases Study 2019) [4]

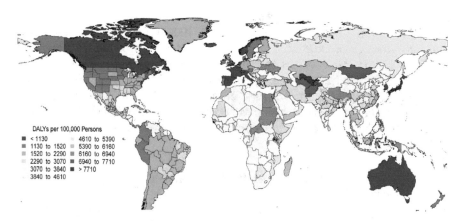

Fig. 5.3 Map of age-standardized DALYs due to high SBP. (Global Burden of Disease Study 2019) [2]

Although hypertension is highly prevalent in Asian countries, there is significant between-country and within-country variation in rates of hypertension. The rates of hypertension awareness, treatment and control are low in most Asian countries [7, 8]. The age-adjusted prevalence of hypertension is declining in most of the high-income countries of Europe, North America, Asia, and Latin America but is increasing in most of the countries of South Asia, South-East Asia, and the Middle East [5]. Multiple factors have been suggested to account for this increase, including race and ethnicity, genes, unhealthy diet (high intake of salt, trans-fats and alcohol, and low intake of fruits and vegetables), physical inactivity, obesity and abdominal obesity, social determinants, and environmental factors [8]. This chapter focusses on genetic ancestry, racial groups and ethnicity and highlights their importance in hypertension in different countries of the Asian continent.

5.2   Asia: Hypertension Epidemiology

Asia is the largest continent in the world and has five separate and well-demarcated zones—East Asia, South Asia, South-East Asia, Central Asia, and West Asia (Middle East), each with a population of 0.5–1.5 billion. North Asia is mainly composed of Russia and not included in the present review. Within Asia there are a few high-income (e.g., Japan, Taiwan, Saudi Arabia and Gulf countries, Singapore), many middle-income (e.g., China, Iran, West and Central Asian countries, Malaysia), lower middle-income (e.g., India, Pakistan, Indonesia, Philippines), and many low-income countries. The epidemiology of hypertension is different in these countries [4, 5].

Global Burden of Disease (2019) has reported substantial differences in the burden of hypertension (DALYs) and deaths due to raised SBP in various Asian regions (Table 5.1) [5]. The highest burden per 100,000 population is in Central Asian countries while the lowest is in South Asian countries. Hypertension-related deaths are also the highest in Central Asian countries (208/100,000) and the lowest in South Asian countries (105/100,000) [5]. However, in terms of absolute numbers, East Asian and South Asian countries have the highest burden (Fig. 5.1) [4]. In the large countries of East and South Asia, hypertension epidemiology is characterized by significant geographic and urban–rural differences, premature onset of high BP, varying levels of hypertension awareness, treatment and control, and disparate outcomes [8]. These differences are also observed between various countries in Asia [7].

Table 5.1 High SBP-related disease burden (DALYs and deaths) in various Asian regions (Global Burden of Disease Study 2019) [1]

Region	Population (million)	DALYs (95% UI) per 100,000	Deaths (95% UI) per 100,000
Asia	4678.2[a]	3202 (2834–3567)	139 (113–152)
South Asia	1805.2	2672 (2304–3083)	105 (90–121)
East Asia	1472.2	3815 (3197–4449)	182 (150–214)
Central Asia	935.3	4888 (4311–5498)	208 (180–235)
South-East Asia	673.7	3764 (3324–4215)	151 (132–170)
Middle East and North Africa	608.7	3126 (2682–3602)	132 (113–152)
High-income Asia-Pacific	187.3	2080 (1753–2390)	127 (98–151)

DALYs disability-adjusted life-years, *UI* uncertainty interval
Data source: http://ghdx.healthdata.org/gbd-results-tool
[a] Source: https://www.worldometers.info/world-population/asia-population/

5.3   Race and Hypertension in Asia

At the beginning of twentieth century, most people including many scientists believed that race was an established fact and that divisions between races were fundamentally biological [9]. Race as a determinant of chronic diseases has received significant attention and evaluation of racial differences in various socioeconomic parameters, disease incidence and health outcomes has been studied for decades [10–12]. Human beings have been wrongly classified into racial categories based on differences in skin-color, superficial features and bony characteristics [13, 14]. This racial discrimination reached its peak with the development of eugenics, which is the study of or belief in the possibility of improving the qualities of the human species or a human population, especially by such means as discouraging reproduction by persons having genetic defects or presumed to have inheritable undesirable traits (negative eugenics) or encouraging reproduction by persons presumed to have inheritable desirable traits (positive eugenics) [13]. There is also an attempt to abandon fixed racial typologies and to reimagine race as populations defined by difference in the frequencies of certain genes or traits [11]. Population genetics remains at the core of scientific understanding of race today, with evolutionary geneticists often preferring the language of genetic ancestry groups [15, 16]. This phrase addresses genetic differences between people and variation of gene frequencies among human populations. There are protagonists and antagonists in this debate on population classification into racial groups or genetic ancestry groups.

5.3.1   Do Racial Groups Exist?

Genetic ancestry has been determined using two methods. Cavalli-Sforza synthesized knowledge from archeology, linguistics, history, and genetics to describe the migration of people out of Africa and into various regions of the world [17]. This was the first evidence-based refutation of morphology-based racial groups. Subsequent researchers combined anthropological information, obtained from various sources, with data on mitochondrial DNA (mtDNA) and showed migration out of Africa to other parts of the world, with the first stream of migrants moving along the seacoasts into Central and Western Asia, South Asia, and South-East Asia and subsequently into East Asia and Australasia. The second wave of migrants moved into Central Asia and Western Europe and then to East Asia and the American continents (Fig. 5.4) [18, 19].

Reich and colleagues used ancient DNA technology, developed in Germany and the USA, to trace the path of migration and evolution and identified multiple routes of migration out of Africa with significant admixture with Neanderthals in Europe, Denisovians in North and Eastern Asia, and Ancient clines in South and South-East Asia [14]. Asian races, which are the focus of this chapter, have seemingly evolved from Yamanaya cline estimated to be present in Caucasus Mountain region and Central Asia (Fig. 5.5) [14, 20]. More details of this ancient genetic ancestry group or cline are unknown because it has been extinct for at least 5000 years. There is

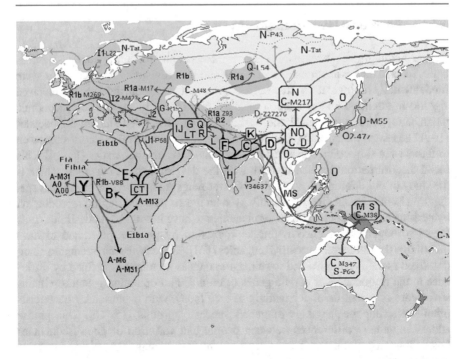

Fig. 5.4 Migration and the peopling of Asia. Evidence from mitochondrial DNA (mtDNA) studies [23]

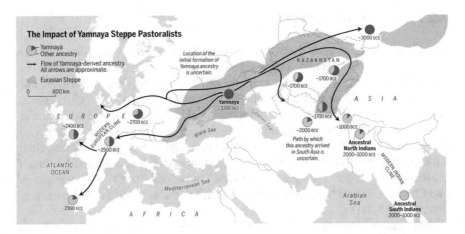

Fig. 5.5 Evidence of peopling of Asia by ancient Yamanaya Steppe Pastoralists from Central Asia and evolution of regional groups in Asia by admixture of genes from native populations [14, 20]

evidence of genetic admixture of this cohort with Neanderthals in West and Central Asia, Denisovians in Eastern Asia, and ancient Indian groups and others in South and South-East Asia. This could be responsible for the peopling of the Asian continent [14]. Importantly there is anthropological and genetic evidence of a bidirectional migration of populations and substantial admixture of genes from central regions to the peripheral regions and vice versa, especially in the Asian continent [14]. The regional differences in ancient DNA and differences in modern genetic architecture of the Asian populations is a work-in-progress and more studies are needed [21, 22].

Significant differences in health status of genetic ancestry and ethnic groups have been reported. Two chronic conditions have received attention. Clinical and epidemiological studies have consistently shown a higher prevalence of hypertension among African Americans in the US [12]. Multiple hypotheses have been raised to explain this difference, including social and economic factors and genetics. Similarly, in the past hundred years, diabetes has been identified with the Jews, American Indians and Mexican Americans in the US [9]. It is now understood that the health disparities that exist in various race-based or ethnic groups lead to racism, which is a much more fundamental cause of socioeconomic and health disparities in various genetic ancestry groups (races) [13, 15, 24]. Most scientists now believe that the differing rates of hypertension incidence, hypertension awareness, treatment and control, and hypertension-related adverse outcomes in diverse groups is manifestation of racism and not race [24, 25]. We shall follow this premise in highlighting differences in hypertension among various genetic ancestry groups in Asia.

5.4   African Americans and Primitive Populations in Asia

Hypertension is 42–45% more prevalent in African Americans than in Mexican Americans and White Americans [26]. Risk factors that predispose to higher hypertension prevalence in this group include advancing age, family history of hypertension, obesity, physical inactivity, high dietary sodium intake, low dietary potassium intake, low vitamin D intake, harmful use of alcohol, psychosocial stress, low socioeconomic status, low educational attainment, and harmful psychological traits [26]. Most scientists now believe that these factors are more important than race or genetics in contributing to the higher hypertension prevalence and complications in this group. Indeed, availability of better primary care and hypertension management in the Kaiser Permanent programme in the US has resulted in a narrowing of the uncontrolled hypertension gap between African Americans and Whites, from 8.1% in 2009 to 3.9% in 2014 [27]. Although, there are differences in the pharmacogenomics of hypertension in the two groups [28], data clearly show that tackling social determinants can lead to better awareness, treatment and control of hypertension in African Americans, and the so-called racial differences can equalize [26].

Hypertension prevalence has been studied in only a few tribal populations in Asia. In India, a meta-analysis and a multisite study reported higher hypertension prevalence in these populations [29, 30] than in the general population [31]. Tribal populations in Asia are considered a separate genetic ancestry group that belong to earlier migrants out of Africa (Fig. 5.4) [18]. Similar to African Americans, a higher prevalence of hypertension in more acculturated tribal populations could be due to greater prevalence of adverse socioeconomic, dietary, and other lifestyle factors along with poorer hypertension management [29, 32]. More studies are needed among these groups of Asians who are widely dispersed in India, China, and South-East Asian countries. Pharmacogenomic characteristics should also be evaluated in these populations.

5.5   East Asians, South Asians, South-East Asians, and West Asians

Asian populations have been arbitrarily divided into genetic ancestry groups or racial groups based on older classification of races into Caucasiods, Mongoloid, and Negroids [10, 17]. Recent anthropological data and genetic mapping of populations in Asia reveals less genetic diversity than believed earlier [14, 18]. The racial classification into Caucasiods (West, Central and South Asia), Mongoloids (East and South-East Asia) and Negroids (ancient populations, tribals) is arbitrary. To evaluate the association of geographical location and the racial classification with hypertension prevalence, data on age-adjusted hypertension in various Asian countries is presented.

Studies have reported the prevalence and disease burden from hypertension in almost all Asian countries [2, 3]. Based on age-adjusted hypertension prevalence data from the NCDRisC for the years 1990–2015 [4], the prevalence of hypertension varies from a low of 8–10% in high-income countries of Eastern and South-Eastern Asia to a high of 30–35% in countries of Central and Western Asia (Figs. 5.6 and 5.7). More detailed analysis reveals that variability in the prevalence of hypertension is related to the socioeconomic development of a particular country rather than geographic location. An important finding is that the prevalence of hypertension is higher in men versus women in almost all countries in Asia. This is in contrast to the female dominance of hypertension in most European countries and needs further studies. The prevalence of hypertension in various Asian countries shows a significant correlation with cardiovascular mortality. Countries with higher rates of hypertension also have higher rates of cardiovascular mortality. Hypertension is the most important risk factor for stroke, which is the predominant form of cardiovascular disease in East Asia, although the prevalence of hypertension in these countries is lower than in Southern and Central Asian countries. Again, this needs to be investigated further [33]. On the other hand, higher hypertension prevalence rates in China compared with other Asian countries have been reported in the PURE (Prospective Urban Rural Epidemiology) study [34]. This large study, using uniform methodology, reported that the hypertension prevalence in individuals aged 35–70 years was the highest in China (42%) and the South-East Asian region (47%) compared with South Asian (32%), and West Asian (30%) countries.

In conclusion, clinical epidemiology suggests that race is of questionable importance as a hypertension risk factor in Asia. Ancient DNA studies have identified that there are no major racial differences in Asian populations, although ethnic differences are present. Geographic epidemiology of hypertension in Asia and ethnicity-based distribution does not show significant correlation [32, 33]. China is an outlier in many such studies. More studies are needed to clarify this finding.

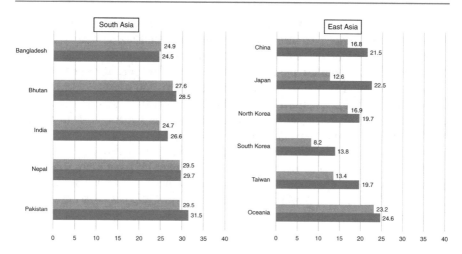

Fig. 5.6 Age-adjusted hypertension prevalence (%) in men and women in South Asian and East Asian countries. (Data source: NCDRisC 2015)

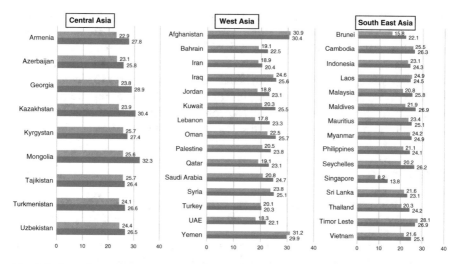

Fig. 5.7 Age-adjusted hypertension prevalence (%) in men and women in Central Asian, West Asian and South-East Asian counties. (Data source: NCDRisC 2015)

5.5.1 Race and Hypertension Management

The association between ethnic-racial differences and hypertension management is an important question [6, 7]. So far, the only relevant racial differences in hypertension management have been reported between White Caucasian Americans and African Americans in USA [24, 26]. It has been argued that African Americans are not a homogenous group and multiple factors could be responsible for hypertension. The low socioeconomic status of this group in USA is an important risk factor [9].

Multiple biological factors have also been identified, including: high sympathetic nervous system activity, alteration in renin-angiotensin-aldosterone system (RAAS), neurohormonal influences, alterations in circadian control of BP, exaggerated BP responses to various stimuli, increased sodium sensitivity, impaired renal handling of sodium, endothelial dysfunction, and chronic alternations in vascular structure and function [26]. Clinical trials have reported that dihydropyridine calcium channel blockers and diuretics are better drugs for treating hypertension in African Americans than beta-blockers and RAAS blockers. [26]

No similar race-based differences in hypertension pathophysiology have been identified in Asians. Salt sensitivity could be important [33], but more studies are needed. Studies on hypertension management among various racial groups – Caucasians, East Asians, South Asians, and the Afro-Caribbean—in the UK and Canada have failed to show variable responses to different hypertension management strategies, including drugs [35–37].

5.6   Conclusions

Hypertension is the most important cause of disease burden and deaths in Asia. However, there is heterogeneity within Asia, where the highest mortality burden from high BP is in Central and South-East Asian countries (Table 5.1). Epidemiological studies show that there are no differences in various genetic ancestry groups. The adult prevalence of hypertension varies from 10 to 35% in men and 8 to 30% in women in different countries (Figs. 5.6 and 5.7). Differences in hypertension prevalence are associated with socioeconomic development, with lower disease burden in more developed countries of East and South-East Asia.

References

1. Global Burden of Disease Collaborators. Global burden of 87 risk factors in 204 countries and territories, 1990–2019: a systematic analysis for the Global Burden of Disease Study 2019. Lancet. 2020;396(10258):1223–49.
2. Roth GA, Mensah GA, Johnson CO, Addolorato G, Ammirati E, Baddour LM, et al. Global burden of cardiovascular diseases and risk factors, 1990–2019: update from the GBD 2019 study. J Am Coll Cardiol. 2020;76(25):2982–3021.
3. NCD Risk Factor Collaboration (NCD-RisC). Worldwide trends in blood pressure from 1975 to 2015: a pooled analysis of 1479 population-based measurement studies with 19·1 million participants. Lancet. 2017;389(10064):37–55.
4. Non-Communicable Disease Risk Factor Collaboration. www.ncdrisc.org. Accessed 2 June 2021.
5. Global Burden of Disease. www.healthdata.org. Accessed 3 June 2021.
6. Chow CK, Gupta R. Blood pressure control: a challenge to global health systems. Lancet. 2019;394:613–5.
7. Schutte AE, Venkateshmurthy NS, Mohan S, Prabhakaran D. Hypertension in low- and middle-income countries. Circ Res. 2021;128(7):808–26.
8. Gupta R, Xavier D. Hypertension: the most important non-communicable disease risk factor in India. Indian Heart J. 2018;70:565–72.

9. Tuchman AM. Diabetes: a history of race and disease. New Haven: Yale University Press; 2020.
10. Nei M, Roychoudhury AK. Genetic variation within and between the three major races of man, Caucasoids, Negroids, and Mongoloids. Am J Hum Genet. 1974;26:421–43.
11. Tishkoff SA, Verrelli BC. Patterns of human genetic diversity: implications for human evolutionary history and disease. Annu Rev Genomics Hum Genet. 2003;4:293–340.
12. Burchard EG, Ziv E, Coyle N, Gomez SL, Tang H, Karter AJ, et al. The importance of race and ethnic background in biomedical research and clinical practice. N Engl J Med. 2003;348:1170–5.
13. Wade N. A troublesome inheritance: genes, race and human history. New York: Penguin; 2015.
14. Reich D. Who we are and how we got here: ancient DNA and the new science of the human past. Oxford: Oxford University Press; 2018.
15. Rotimi CN, Jorde LB. Ancestry and disease in the age of genomic medicine. N Engl J Med. 2010;363:1551–8.
16. Oni-Orisan A, Mavura Y, Banda Y, Thornton TA, Sebro R. Embracing genetic diversity to improve Black health. N Engl J Med. 2021;384(12):1163–7.
17. Cavalli-Sforza LL, Cavalli-Sforza L, Menozzi P, Piazza A. The history and geography of human genes. Princeton: Princeton University Press; 1994.
18. Wells S. The journey of man: a genetic odyssey. New Delhi: Penguin Books; 2002.
19. Meredith M. Born in Africa: the quest for the origins of human life. New York: Public Affairs; 2011.
20. Narasimhan VM, Patterson N, Moorjani P, Rohland N, Bernardos R, Mallick S, et al. The formation of human populations in South and Central Asia. Science. 2019;365:eaat7487.
21. Joseph T. Early Indians: the story of our ancestors and where we came from. New Delhi: Juggernaut Books; 2018.
22. Wanderers MP, Kings M. The story of India through its languages. Penguin Random House India Private Limited; 2021.
23. Maulucioni A. Successive dispersal of human lineages during the peopling of EURAFRASIA. https://commons.wikimedia.org/w/index.php?curid=12087218Migration. Accessed 5 June 2021.
24. US Department of Health and Human Services. The Surgeon General's call to action to control hypertension. Washington, DC: US Department of Health and Human Services, Office of the Surgeon General; 2020.
25. Cooper RS, Kaufman JS, Ward R. Race and genomics. N Engl J Med. 2003;348:1166–70.
26. Mensah GA. Hypertension in African Americans. In: Bakris GL, Sorrentino MJ, editors. Hypertension: a companion to Braunwald's heart disease. Philadelphia: Elsevier; 2018. p. 383–92.
27. Barolome RE, Chen A, Handler J, Platt ST, Gould B. Population care management and team-based approach to reduce racial disparities among African Americans/Blacks with hypertension. Perm J. 2016;20:53–9.
28. Iniesta R, Campbell D, Venturini C, Faconti L, Singh S, Irvin MR, et al. Gene variants at loci related to blood pressure account for variation in response to antihypertensive drugs between Black and White individuals. Hypertension. 2019;74:614–22.
29. Rizwan SA, Kumar R, Singh AK, Kusuma YS, Yadav K, Pandav CS. Prevalence of hypertension in Indian tribes: a systematic review and meta-analysis of observational studies. PLoS One. 2014;9:e95896.
30. Laxmaiah A, Meshram II, Arlappa N, et al. Socio-economic and demographic determinants of hypertension and knowledge, practices and risk behaviour of tribals in India. Indian J Med Res. 2015;141:697–708.
31. Gupta R, Gaur K, Ram CVS. Emerging trends in hypertension epidemiology in India. J Hum Hypertens. 2019;33:575–87.
32. Zeng Z, Chen J, Xiao C, Chen W. A global view on prevalence of hypertension and human development index. Ann Glob Health. 2021;86:67.
33. Zhao D. Epidemiological features of cardiovascular disease in Asia. JACC Asia. 2021;1:1–13.

34. Chow CK, Teo KK, Rangarajan S, Islam S, Gupta R, Avezum A, et al. Prevalence, awareness, treatment, and control of hypertension in rural and urban communities in high-, middle-, and low-income countries. JAMA. 2013;310:959–68.
35. Gupta AK, Poulter NR, Dobson J, Eldridge S, Cappuccio FP, Caulfield M, et al. Ethnic differences in blood pressure response to first and second line antihypertensive therapies in patients randomized in the ASCOT trial. Am J Hypertens. 2010;23:1023–30.
36. Wang JG, Huang QF. Hypertension in East Asians and Native Hawaiians. In: Bakris GL, Sorrentino MJ, editors. Hypertension: a companion to Braunwald's heart disease. Philadelphia: Elsevier; 2018. p. 21–6.
37. Joseph P, Gupta R, Yusuf S. Hypertension in South Asians. In: Bakris GL, Sorrentino MJ, editors. Hypertension: a companion to Braunwald's heart disease. Philadelphia: Elsevier; 2018. p. 27–31.

Relation of Genetics and Obesity with Hypertension: An Asian Perspective

Rajeev Gupta

6.1   Introduction

Ancient DNA studies have identified the course of human migration in Asia and reported basic genetic similarities among populations. However, an important concept is the presence of minor genetic variations that lead to major phenotypic changes. Two types of genetic variations are important in hypertension pathophysiology [1]. The first is uncommon and relates to rare monogenic forms, clinically identified as secondary hypertension. Multiple monogenic forms of hypertension exist and the allele frequency is below 1/1000 with a large effect size per variant. So far about 13 gene loci have been identified, although the estimated number of genes involved is likely to be 15–20. The effect size of these genetic variants is large, with systolic blood pressure (SBP) being 20 mmHg higher in those with dominant alleles [2–4]. Monogenic forms of hypertension identified to date are due to variation in genes involved in aldosterone, renin, and cortisol metabolism. Monogenic hypertensive syndromes include glucocorticoid remediable aldosteronism (GRA), Gordon syndrome, familial hyperaldosteronism III, pseudoaldosteronism, syndrome of apparent mineralocorticoid excess, Bilginturan syndrome, autosomal dominant hypertension with exacerbation in pregnancy, and congenital adrenal hyperplasia types (Table 6.1). These syndromes are rare in Asian populations are not discussed here.

R. Gupta (✉)
Department of Preventive Cardiology and Internal Medicine, Eternal Heart Care Centre and Research Institute, Mount Sinai New York Affiliate, Jaipur, India

Academic Research Development Unit, Rajasthan University of Health Sciences, Jaipur, India
e-mail: drrajeev.gupta@eternalheart.org

Table 6.1 Important monogenic hypertension syndromes

Pathogenesis	Syndrome	Genes	Inheritance
Elevated aldosterone	Glucocorticoid remediable aldosteronism	CYP11B2	AD
	Gordon syndrome	WNK1, WNK4, KLHL3, CUL3	AR and AD
	Familial hyperaldosteronism III	KCNJ5	AD
Low aldosterone	Liddle syndrome	SCNN1B, SCNN1G	AD
	Apparent mineralocorticoid excess	HSD11B2	AR
Low aldosterone and other features	Hypertension and brachydactyly	PDE3A	AD
	Autosomal dominant hypertension of pregnancy	NR3C2	AR
	Congenital adrenal hyperplasia	CYP11B1, CYP17a1	AR

AD autosomal dominant, *AR* autosomal recessive

The genomics of the more common primary hypertension are complex. Genes affect both SBP and diastolic blood pressure (DBP) and primary hypertension has a heritability of 30–50%. It has been estimated that more than 500 genes (loci) are involved in this form of hypertension and so far about 90 genes of major importance have been identified [1]. However, each of these genes has a small effect size, with a blood pressure (BP) increment of 0.5–1 mmHg per allele. The complex genetic architecture of hypertension and variations in various ethnic groups in Asia is discussed below.

6.1.1 Single Nucleotide Polymorphisms and Genome-Wide Association Studies

Single nucleotide polymorphisms (SNPs) are the basic unit of a gene. Initial studies on the genetics of hypertension focused on identification of individual SNPs that modulate a protein function and lead to hypertension using case-control study design [5, 6]. In previous decades, a number of gene-linkage and candidate-gene studies were performed to identify significant SNPs, but these yielded only a few reproducible genetic results, mainly due to small sample sizes [1]. With the development of modern microarray platforms where millions of genetic variants can be genotyped, research into genetics of common diseases (coronary artery disease, stroke, diabetes, hypertension, atrial fibrillation, etc.) has proliferated rapidly [7]. The technique that permits interrogation of almost the entire genome for association with a trait such as BP is known as genome-wide association studies (GWAS). Apart from multiple SNPs, this technique can also identify copy number polymorphisms and structural variants (copy number variations; CNVs), methylation marks (epigenetics) and some additional variants.

GWAS identifies both rare and common variants responsible for hypertension and while rare variants have a much greater influence on BP (Table 6.1), the common variants responsible for small effects are more important. Hypertension is a polygenic disorder in which genes act in combination with environmental exposures to make a modest contribution to BP. Furthermore, different subsets of genes may lead to different phenotypes associated with hypertension, such as obesity, dyslipidemia, and insulin resistance [7]. Candidate-gene studies and GWAS have identified a number of hypertension-related genes involved in pathways that regulate BP. Genes identified include those that encode components of renin-angiotensin-aldosterone system, atrial natriuretic peptide, beta-2 adrenoreceptor, alpha-adducin, and others (Table 6.2).

Earlier studies reported that genetic determinants account for about 1% of BP variance, while family studies have estimated the heritability of hypertension to be 30–40% [1]. One hypothesis accounting for missing heritability is the epigenetic modification of DNA due to environmental factors, and there also may be genetic determinants of target organ damage and vascular disease attributed to hypertension

Table 6.2 Examples of genes that influence blood pressure

Chromosome No.	Gene SNPs	Phenotype
1	CASZ1, MTHFR-NPPB, ST7L-CAPZA1-MOV10, MDM4, AGT	SBP, DBP, HTN
2	OSR1, KCNK3, FER1L5, FIGN-GRB14, STK39, PDE1A	PP, SBP, DBP, MAP
3	HRH1-ATG7, SLC4A7, ULK4, MAP4, CDC25A, MIR1263, MECOM	SBP, DBP, MAP
4	CHIC2, FGF5, ARHGAP24, SLC39A8, ENPEP, GUCY1A3-GUCY1B3	PP, multiphen, SBP, DBP
5	NPR3-C5orf23, GPR98/ARRDC3, PRDM6, ABLIM3-SH3TC2, EBF1	SBP, DBP
6	HFE, BAT2-BAT5, TTBK1-ZNF318, ZNF318-ABCC10, RSPO3, PLEKHG1	SBP, DBP
7	HDAC3, HOXA-EVX1, IGFBP1-IGFBP3, IGFBP3, CDK6, PIK3CG, NOS3, PRKAG2	PP, SBP, DBP, multiphen
8	BLK-GATA4, CDH17, NOV	SBP, DBP, multiphen
9	SMARCA2-VLDLR	SBP-age-spec
10	CACNB2, C10orf107, VCL, PLCE1, CYP17A1-NT5C2, ADRB1	SBP, DBP, MAP
11	LSP1-TNNT3, H19, ADM, PLEKHA7, NUCB2, LRRC10B-SYT7, RELA, FLJ32810-TMEM133, ADAMTS8	SBP, DBP, MAP, PP
12	PDE3A, HOXC4, ATP2B1, SH2B3, RPL6-PTPN11-ALDH2, TBX5-TBX3	SBP, DBP, HTN
15	FBN1, ITGA11, CYP1A1-ULK3, FURIN-FES	PP, SBP-age-spec
16	UMOD, NFATS	SBP-age-spec, DBP, PP
17	PLCD3, GOSR2, ZNF652, C17orf82-TBX2	SBP, DBP
19	AMH-SF3A2	PP
20	JAG1, GNAS-EDN3	SBP, DBP, MAP

Age-spec age-specific, *DBP* diastolic blood pressure, *MAP* mean arterial pressure, *multiphen* multiple phenotype, *PP* pulse pressure, *SBP* systolic blood pressure

[7]. On the other hand, recent reviews have reported that SNPs identified in GWAS explain approximately 27% of the 30–50% estimated heritability of BP [8]. A majority of BP SNPs show pleiotropic associations, and unraveling those signals and underpinning biological pathways offers potential opportunities for drug repurposing [9]. The main limitation of these data are that the studies have mostly been performed in White Caucasians and there is almost no representation of Asian populations in the large hypertension genetic collaborations [10].

In East Asian populations, earlier GWAS confirmed the presence of SNPs found in other populations—CSK, CYP17A1, MTHFR and FGF5 [11]. The Asian Genetic Epidemiology Network consortium is a large meta-analysis of GWAS on BP traits among East Asians that included 19,608 individuals. After de novo genotyping, seven loci previously identified in European populations were confirmed as important in East Asians, plus identification of an additional six novel loci: ST7L-CAPZA1, FIGN-GRB14, ENPEP, NPR3, TBX3, and ALDH2 [12, 13]. In another meta-analysis of a Han Chinese population ($n = 11,816$ discovery, $n = 69,146$ replication), twelve novel loci (CASZ1, MOV10, FGF5, CYP17A1, SOX6, ATP2B1, ALDH2, JAG1, CACNA1D, CYP21A2, MED13L, and SLC4A7) were identified [14]. These findings suggest presence of some allelic heterogeneity between Europeans and Asians. A study from Korea reported the importance of genetic variations in ATP2B1, CSK, ARSG, and CSMD1 loci in relation to BP and hypertension [15]. A Japanese study identified ATP2B1 as important gene responsible for hypertension, which is similar to Caucasians [16]. More studies are needed to evaluate important genetic determinants of hypertension in East Asian populations by creating international consortia.

Large GWAS have not been conducted in populations from Southern and Western Asian countries. A trans-ancestry meta-analysis including 320,251 individuals of East Asian, European, and South Asian ancestry identified 12 new loci that were associated with BP and highlighted genes involved in vascular smooth muscle (IGFBP3, KCNK3, PDE3A, and PRDM6) and renal function (ARHGAP24, OSR1, SLC22A7, and TBX2) [17]. In addition, an important role of DNA methylation was identified. A large meta-analysis of one million people of European ancestry identified 535 novel BP loci that highlighted shared genetic architecture between BP and lifestyle exposures [18]. Such studies are important to provide biological insights and guide drug development. These should be done in Asian populations.

6.1.2   Polygenic Risk Scores

Essential hypertension is influenced by multiple genetic variants with small individual effect sizes. Meaningful risk prediction necessitates examination of the aggregated impact of these multiple variants [10]. This is through calculation of a polygenic risk score (PRS), which is a mathematical aggregate of risk conferred by all of the SNPs significantly associated with BP [19]. The risk information provided by the PRS is different from that obtained from genetic markers of monogenic disorders. The latter is a dichotomous result (either high or low probability of disease), whereas the former provides a wider range of probabilistic risk [19]. In addition, the rare variant genotype points to specific biological impact of the variant, whereas the PRS is an amalgamation of numerous small-effect variants across the genome with no specific pathway implicated. A PRS constructed to use of all the significant GWAS BP SNPs showed a significant association with stroke, coronary artery disease, heart failure and left ventricular mass [20].

There is considerable interest in the use of PRS as a biomarker for early intervention, although currently there is no evidence for the clinical utility of PRS for intervention or disease prevention [19, 21]. It is likely that PRS that have been developed recently have limited utility because studies were conducted in adults aged >40 years where disease is already well established. However, PRS may have more value in identification of younger at-risk individuals [22]. This merits further study. PRS has limited utility for facilitating the personalization of treatment or new drug discovery, primarily because it is derived from an amalgamation of all genetic variants and does not represent unique pathways.

We calculated a hypertension gene risk score incorporating 23 SNPs in 181 patients with hypertension in India (Table 6.3) [23]. SNPs were identified using data from international studies (mentioned in Table 6.2) and from our previous studies in India in coronary artery disease and diabetes patients [24, 25]. Most of the well-known BP SNPs were present in significant proportions (GRB14-FIGN, ABLIM3-SH3TC2, PIK3CG, CYP17A1, CNNM2, NT5C2, LRRC108-SYT7,

Table 6.3 SNPs for calculation of hypertension gene risk score in India

Genes	SNP ID	Prevalence (%) $n = 181$
CASZ1	rs880315	119 (65.7)
KCNK3	rs1275988	109 (60.2)
GRB14, FIGN	rs16849225	168 (92.8)
SLC4A7	rs820430	128 (70.7)
CACNA1D	rs9810888	96 (53.0)
FGF5	rs1902859	46 (25.4)
ENPEP	rs6825911	67 (37.0)
ABLIM3, SH3TC2	rs9687065	168 (92.8)
ZNF318	rs1563788	71 (39.2)
PIK3CG	rs17477177	161 (88.9)
WBP1L	rs284844	58 (32.0)
CYP17A1	rs4409766	170 (93.9)
CNNM2	rs11191548	172 (95.0)
NT5C2	rs11191580	170 (93.9)
SOX6	rs4757391	42 (23.2)
LRRC10B, SYT7	rs751984	154 (85.1)
ATP2B1	rs17249754	143 (79.0)
ATXN2	rs653178	178 (98.3)
HECTD4	rs11066280	181 (100.0)
TBX3	rs35444	112 (61.9)
MED13L	rs11067763	127 (70.2)
C17orf82, TBX2	rs2240736	119 (65.7)
JAG1	rs1887320	109 (60.2)

ATXN2, and HECTD4) (Table 6.3). The PRS had a normal distribution, similar to studies performed in individuals of other ethnicities [19]. This is a small demonstration project and larger prospective studies are required to determine the diagnostic and prognostic significance of these findings. It has been speculated that PRS can provide important diagnostic value at younger age and can also risk stratify patients so that intensive control of BP and lifestyle messages are provided earlier in the course of the disease.

6.1.3 Pharmacogenomics

Pharmacogenomics has been defined as the study of the role of genomic variations in drug response. It is the science of combining pharmacology and genomics and analyses the influence of individual genes on the response to drugs in terms of pharmacokinetics and pharmacodynamics [26]. There are many examples where host genomics is important in drug response. In cardiovascular medicine, for example, there is the influence of the CYP2C19 gene on response to clopidogrel, the role of VKORC1, CYP2C9 and CYP4F2 genes in the response to vitamin K antagonists, and the influence of the SLCO1B1 and OATP1B1 genes in statin-induced myopathy [26].

Drugs used for hypertension are also influenced by pharmacogenomics. Metoprolol, carvedilol, and propranolol are metabolized by CYP2D6, which can influence drug clearance and responsiveness [26]. African Americans are known to harbor APOL1 gene variants, which are known to influence rate of progression of chronic kidney disease in patients with hypertension [27]. The GenHAT study (a sub-study of ALLHAT) was the largest randomized clinical trial to explore gene panel risk prediction with respect to antihypertensive therapies ($n = 39,000$) [28]. This trial failed to show any significant usefulness of genetic stratification in clinical outcomes for Caucasians and African American participants [29]. Whole-genome approaches and next-generation sequencing are likely to identify new pharmacogenes, including therapeutic target genes that determine drug efficacy and safety. Studies are needed in ethnically diverse Asian patients to identify the importance of carefully selected genes in BP responsiveness to various antihypertensive drugs. However, gene-modifying therapies remain a pipe dream for hypertension management and cardiovascular disease prevention [29].

6.2 Obesity

Obesity has been a defined as a state of excess adipose tissue mass [30].It has been estimated that about 60–65% of cases of essential hypertension are related to overweight and obesity [30]. The most widely used, though imperfect, measure of obesity is body mass index (BMI) which is equal to weight (kg)/height (m²). Other measures of obesity include anthropometry (skin-fold thickness, waist circumference, waist-hip ratio, etc.), densitometry, computed tomography or magnetic resonance imaging, and electrical impedance. Based on BMI, it has been determined that more than half of the global population is either overweight (BMI 25–29.9 kg/m²) or obese (BMI ≥30 kg/m²) [31]. The prevalence of obesity is significantly higher in high- and middle-income countries compared with low- and lower middle-income countries. In higher-income countries, obesity is more common among the socioeconomically deprived, while in lower-income countries it is more prevalent in the affluent [32].

Abdominal obesity has been defined by either waist size or waist-hip ratio, and different diagnostic criteria have been proposed for Asian populations (waist circumference: men >90 cm, women >80 cm; waist-hip ratio: men >0.9, women >0.8) compared with Caucasians in Europe and the Americas, West Asian and Eastern Mediterranean populations and other geographic groups (waist circumference: men >94 cm, women >80 cm; waist-hip ratio: men >1.0, women >0.9) [32]. The prevalence of abdominal obesity is significantly higher in Asian populations, using Asia-specific criteria, compared with other geographical groups, and in many countries more than half of the adult population is abdominally obese [33]. The etiology of obesity and abdominal obesity is complex and may vary from country to country. Risk factors contributing to obesity include social determinants and environmental factors, rising incomes, changing food supplies, and reduced physical activity [29].

Both generalized and abdominal obesity are associated with hypertension and there is a significant linear relationship between increasing SBP/DBP and BMI in populations across the continents. Data from an Indian study (India Heart Watch, $n \approx 6000$) showed a significant association between SBP and BMI ($R^2 = 0.028$), waist circumference ($R^2 = 0.050$), waist-hip ratio ($R^2 = 0.007$) and hip circumference ($R^2 = 0.035$) ($p < 0.001$) (Fig. 6.1) [34]. Mechanisms of hypertension differ

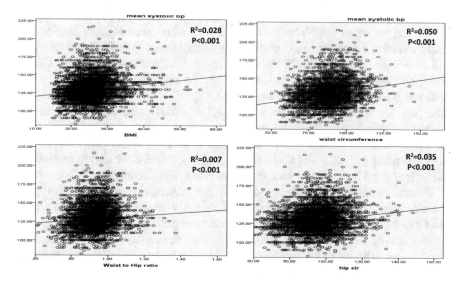

Fig. 6.1 Correlation between SBP and both BMI and waist circumference in the India Heart Watch Study [34]

between the two types of obesity: generalized obesity has predominant sympathetic nervous system overactivity while metabolic factors are as important as sympathetic nervous system activity in abdominal obesity [35].

6.2.1 Generalized Obesity, Abdominal Obesity and Hypertension

Reflex sympathetic overactivation to utilize (burn) excess fat is an important compensatory mechanism in obesity. This leads to sympathetic overactivity in tissues such as vascular smooth muscle and the kidneys and leads to hypertension [35]. Sympathetic activation is implicated in the development of insulin resistance and other features of the metabolic syndrome but the precise stimulus to sympathetic overflow is unknown. The adipocyte also has important endocrine functions that are beyond its primary function of fat storage. Multiple factors are released by adipose tissue, including hormones (leptin, adiponectin, resitin, etc.), cytokines (tumor necrosis factor-alpha, interleukin-6, etc.), substrates (free fatty acids, glycerol), enzymes (aromatase, 11-beta-hydroxysteroid dehydrogenase type 1), complement factors (adipsin) and others (plasminogen activator inhibitor-1, angiotensinogen, retinal binding protein-4) [32]. All these are important in hypertension pathophysiology.

Studies conducted in Asian populations have highlighted the importance of BMI and markers of abdominal obesity in hypertension. Utilizing data from the Chinese National Stroke Prevention project, Wang et al. reported a significant association between obesity markers (BMI, waist circumference, visceral adiposity index, body adiposity index and lipid accumulation product index) and the risk of hypertension ($p < 0.001$) [36]. BMI was found to be the strongest indicator of hypertension [36]. In addition, an Asian Collaborative study of 1.1 million persons from Eastern and Southern Asian countries found a variable association between BMI and all-cause and cardiovascular mortality [37]. Furthermore, in East Asian cohorts (from China, Japan, and Korea), there was an increased risk of cardiovascular mortality when BMI was below $22.6 \, kg/m^2$ and more than $27.5 \, kg/m^2$, while in South Asian cohorts there was a weak association between a BMI of $>20.0 \, kg/m^2$ and cardiovascular mortality. This greater risk is partially mediated via the presence of hypertension. In the PURE study, in which >50% participants were from Asian countries, higher BMI and waist-hip ratio was associated with greater SBP, DBP and hypertension, and an increased incidence of cardiovascular events and deaths [38].

In lower middle- and middle-income countries of Asia, changes in diet and physical activity are among the hypothesized leading contributors to obesity. Emerging risk factors include environmental contaminants, chronic psychosocial stress, neuroendocrine dysregulation, and genetic/epigenetic mechanisms [39]. The association between adipose tissue metabolism and the pathogenesis of hypertension has not been well studied among Asians, especially in East Asians (except Japan) and South Asians, and further studies are required.

6.2.2   Muscle Strength, Frailty, and Hypertension

An inverse correlate of obesity is frailty and muscular weakness. Frailty is defined as an aging-related syndrome of physiological decline, characterized by marked vulnerability to adverse health outcomes. Frailty can be assessed using a variety of tools such as a phenotype model, a cumulative deficit model, comprehensive geriatric assessment, and others [40]. Hypertension associated with frailty is widespread in Asia, especially in older individuals with low socioeconomic status. In the PURE study, with predominantly Asian participants, grip strength was used as a measure of frailty [38]. Individuals with grip strength in the lowest quintile were at significantly higher risk of both cardiovascular and non-cardiovascular death than those with grip strength in the highest quintile (hazard ratio [95% confidence interval] 1.78 [1.42–2.23] and 1.56 [1.36–1.79], respectively). Interestingly, associations between frailty (based on grip strength) and cardiovascular disease events/mortality were not seen in high-income countries from the region, significant associations were observed in middle-income countries (China, Southeast Asia, and West Asia participants) and low-income countries (South Asia and Africa participants) [38].

A meta-analysis of observational studies and also reported an association between frailty and outcomes in patients with hypertension [41]. Data from six longitudinal studies (two from China) and one cross-sectional study (Japan) were pooled and frailty was found to be a significant predictor of mortality (hazard ratio 2.45, 95% confidence interval 2.08–2.88). It has been suggested that more studies with larger sample sizes are needed using robust tools to identify frailty among individuals with hypertension, especially from low- and lower middle-income countries. These data support the observation that BMI at both extremes is associated with adverse cardiovascular outcomes in Asia [37], and both high and low BMI are important risk factors for hypertension and cardiovascular disease.

6.2.3   Interventions for Obesity, and Pharmaceutical Agents

Lifestyle interventions are crucial for addressing the obesity epidemic in Asia. These include focus on healthy diet and enhanced physical activity using macro-level interventions, and by addressing the social determinants [29]. Both are important in the management of generalized and abdominal obesity [42, 43]. However, the long-term effectiveness of an individualized lifestyle approach is marginal, as reported in the US-based LOOK-AHEAD trial and several trials in Asia [44, 45].

Pharmaceutical agents have also been studied for obesity management, but the outcomes are not satisfactory and most of the drugs used to date have been associated with major side effects (Table 6.4) [46]. There are no Asian-specific trials of any of these drugs, and larger and long-term trials should be conducted in Asian populations, especially focused on abdominal obesity, which is associated with a substantial risk of hypertension, diabetes, and cardiovascular disease in these populations.

Table 6.4 Pharmaceutical agents for obesity management

Strategy	Drugs	Usefulness
Metabolic drugs	GLP-1 receptor agonists (e.g., semaglutide)	+++
	Twincretin agents (e.g., tirzepatide)	+++
	SGLT2 inhibitors (e.g., dapagliflozin)	++
	Locaserin	+
Surgery	Bariatric surgery	++++
Older drugs	Anorexigenic drugs, antidepressants, cannabinoids	Side effects and toxicity

GLP-1 glucagon-like peptide-1, *SGLT2* sodium-glucose co-transporter 2

6.3   Social Determinant Approach for Hypertension Management

Prospective studies have reported a number of hypertension risk factors that encompass macrolevel social determinants of health to individual-level factors (Table 6.5). Many of these hypertension determinants and risk factors can be controlled by social, economic, and health policy and clinical interventions, as highlighted in the Lancet Commission on Hypertension [42]. The Lancet Non-Communicable Disease Countdown 2030 strategy document outlines multiple pathways for achieving the United Nations' (UN) Sustainable Development Goal (SDG) target 3.4—to reduce premature mortality from noncommunicable diseases by a third relative to 2015 levels by 2030 [47]. It confirms that globally, regionally, and nationally, the risk of dying from various noncommunicable diseases is marked by huge diversity in terms of magnitude. It also highlights the fact that no country can achieve this UN SDG target by addressing a single disease. Pathways to the SDG 3.4 target require accelerated reductions in several diseases (especially hypertension and cardiovascular diseases) to the rates of decline achieved in the best performing 10% of countries evaluated. Essential components of strategies to achieve SDG target 3.4 in most countries include effective health system interventions, the most important being hypertension management. Scaling up these interventions requires an accessible and equitable health system with the capacity for priority setting, and implementation of noncommunicable diseases care within the health system is especially important in Asia [48, 49].

Addressing social determinants is central to the prevention of hypertension and to improve its management and control [29]. Social and financial policies are important and can lead to significant change in population health. The policies relevant to Asia include focus on universal right to education, universal basic income, job guarantee schemes, social support, strict application of the World Health Organization framework convention on tobacco control guidelines, taxes on cigarettes, tobacco, alcohol and sugar-sweetened beverages, and a ban on trans fats [48]. Healthcare policies include promulgation of socialized medicine, state-funded health services, high-quality universal healthcare, health insurance for select groups (poor, marginalized, elderly, etc.), drug price control and production of high-quality generics, and free medicine supply in public health systems [49, 50]. Technological interventions could be important and need more Asia-specific studies before implementation [51].

Table 6.5 Macrolevel and microlevel social and clinical determinants of hypertension that need suitable interventions

Macrolevel and social determinants of health	Individual and clinical factors
Macrolevel social factors	Individual-level socioeconomic factors
Human and social development	Maternal nutrition and development origin
Societal evolution	Lifetime social class
Societal inequality	Education
Urbanization	Income
Migration	Occupation
Area-based measures	Employment status
Macroeconomic factors	Exclusion
Measures of income	Stress
Gini coefficient	Addiction
Education	Anthropometric factors
Food availability	Age
Transport	Sex
Political determinants	Body-mass index
Cultural determinants	Waist circumference, waist-hip ratio
Healthcare service delivery	Adherence to healthy lifestyles
Universal health coverage	Smoking/tobacco use
High quality primary care	Healthy/unhealthy diet
Availability of medicines	Alcohol abuse
Medicine cost	Primary prevention (risk factor management)
	Secondary prevention (disease management)

6.4 Conclusions

Genetic factors are important in hypertension pathophysiology, and a number of genes and polymorphisms that predispose to hypertension and its complications have been identified by genome-wide association studies. However, there are limited hypertension genome-wide and candidate-gene association studies in Asia. Hypertension polygenic risk scores should be evaluated for early identification of hypertension risk. Generalized and abdominal obesity are important hypertension risk factors and are widely prevalent in most Asian countries. Asia-specific hypertension pharmacogenomics and other management strategies, including technology-based interventions, need further evaluation. The burden of hypertension in Asia can be alleviated by implementation of a social determinant approach [52] to control proximate hypertension risk factors.

References

1. Ehret GB. Genetics of hypertension. In: Bakris GL, Sorrentino MJ, editors. Hypertension: a companion to Braunwald's heart disease. Philadelphia: Elsevier; 2018. p. 52–9.
2. Milford DV. Investigation of hypertension and the recognition of monogenic hypertension. Arch Dis Child. 1999;81(5):452–5.
3. Luft FC. Monogenic hypertension: lessons from the genome. Kidney Int. 2001;60:381–90.

4. Simonetti GD, Mohaupt MG, Bianchetti MG. Monogenic forms of hypertension. Eur J Pediatr. 2012;171(10):1433–9.
5. Kurland L, Liljedahl U, Lind L. Hypertension and SNP genotyping in antihypertensive treatment. Cardiovasc Toxicol. 2005;5(2):133–42.
6. Doris PA. Hypertension genetics, single nucleotide polymorphisms, and the common disease: common variant hypothesis. Hypertension. 2002;39(2 Pt 2):323–31.
7. Kotchen TA. Hypertensive vascular disease. In: Jameson JL, Fauci AS, Kasper DL, et al., editors. Harrison' principles of internal medicine. 20th ed. New York: McGraw Hill; 2018. p. 1890–906.
8. Padmanabhan S, Dominiczak AF. Genomics of hypertension: the road to precision medicine. Nat Rev Cardiol. 2021;18:235–50.
9. Lip S, Padmanabhan S. Genomics of blood pressure and hypertension: extending the mosaic theory toward stratification. Can J Cardiol. 2020;36:694–705.
10. Sharma S. Hypertension gene risk score in diagnosis and prediction of complications. RUHS J Health Sci. 2021;6:53–62.
11. Xi B, Shen Y, Reilly KH, Wang X, Mi J. Recapit ulation of four hypertension susceptibility genes (CSK, CYP17A1, MTHFR, FGF5) in east Asians. Metabolism. 2013;62:196–203.
12. Kato N, Takeuchi F, Tabara Y, Kelly TN, Go MJ, Sim X, et al. Meta-analysis of genome-wide association studies identifies common variants associated with blood pressure variation in East Asians. Nat Genet. 2011 Jun;43(6):531–8.
13. Kato N, Loh M, Takeuchi F, Verweij N, Wang X, Zhang W, et al. Trans-ancestry genome-wide association study identifies 12 genetic loci influencing blood pressure and implicates a role for DNA methylation. Nat Genet. 2015;47(11):1282–93.
14. Wang Y, Wang JG. Genome-wide association studies of hypertension and several other cardiovascular diseases. Pulse (Basel). 2019;6:169–86.
15. Hong KW, Go MJ, Jin HS, Lim JE, Lee JY, Han BG, et al. Genetic variations in ATP2B1, CSK, ARSG and CSMD1 loci are related to blood pressure and/or hypertension in two Korean cohorts. J Hum Hypertens. 2010;24:367–72.
16. Tabara Y, Kohara K, Miki T, Millennium Genome Project for Hypertension. Hunting for genes of hypertension: the Millennium Genome Project for Hypertension. Hypertens Res. 2012;35:567–73.
17. Kato N, Loh M, Takeuchi F, Verweij N, Wang X, Zhang W, et al. Trans-ancestry geneome-wide association study identifies 12 genetic loci influencing blood pressure and implicates a role for DNA methylation. Nat Genet. 2015;47:1282–93.
18. Evangelou E, Warren HR, Mosen-Ansorena D, Mifsud B, Pazoki R, Gao H, et al. Genetic analysis of over 1 million people identifies 535 new loci associated with blood pressure traits. Nat Genet. 2018;50:1412–25.
19. Torkmani A, Wineinger NE, Topol E. The personal and clinical utility of polygenic risk scores. Nat Rev Genet. 2018;19:581–90.
20. Ehret GB, Munroe PB, Rice KM, Bochud M, Johnson AD, Chasman DI, et al. Genetic variants in novel pathways influence blood pressure and cardiovascular disease risk. International Consortium for Blood Pressure Genome-Wide Association Studies. Nature. 2011;478:103–9.
21. Mars N, Koskela JT, Ripatti P, Kiiskinen TTJ, Havulinna AS, Lindbohm JV, et al. Polygenic and clinical risk scores and their impact on age at onset and prediction of cardiometabolic diseases and common cancers. Nat Med. 2020;26:549–57.
22. Kullo IJ, Dikilitas O. Polygenic risk scores for diverse ancestries. J Am Coll Cardiol. 2020;76:715–8.
23. G B, Sharma S. Hypertension gene risk score and influence of physical activity on cardiovascular risk in hypertension: Mendelian randomization study. PhD Thesis, Rajasthan University of Health Sciences, Jaipur; 2021.
24. Wang M, Menon R, Mishra S, Patel AP, Chaffin M, Tanneeru D, et al. Validation of a genome-wide polygenic risk score for coronary artery disease in South Asians. J Am Coll Cardiol. 2020;76:703–14.

25. Paul S, INDIGENIUS Consortium. Report submitted to Indian Council of Medical Research; 2021.
26. Magavern EF, Kaski JC, Turner RM, Drexel H, Janmohamed A, Scourfield A, et al. The role of pharmacogenomics in contemporary cardiovascular therapy: a position statement from the European Society of Cardiology Working Group on Cardiovascular Pharmacotherapy. Eur Heart J Cardiovasc Pharmacother. 2022;8:85.
27. Parsa A, Kao WHL, Xie D, Astor BC, Li M, Hsu CY, et al. APOL1 risk variants, race and progression of chronic kidney disease. N Engl J Med. 2013;369:2183–96.
28. Lynch AI, Eckfeldt JH, Davis BR, Ford CE, Boerwinkle E, Leiendecker-Foster C, et al. Gene panels to help identify subgroups at high and low risk of coronary heart disease among those randomized to antihypertensive treatment: the GenHAT study. Pharmacogenet Genomics. 2012;22:355–66.
29. Gupta R, Wood DA. Primary prevention of ischemic heart disease: populations, individuals and healthcare professionals. Lancet. 2019;394:685–96.
30. Landsberg L. Obesity. In: Bakris GL, Sorrentino MJ, editors. Hypertension: a companion to Braunwald's heart disease. Philadelphia: Elsevier; 2018. p. 328–34.
31. NCD Risk Factor Collaboration (NCD-RiSC). Rising rural body-mass index is the main driver of global obesity epidemic in adults. Nature. 2019;569:260–4.
32. Kushner RF. Evaluation and management of obesity. In: Jameson JL, Kasper DL, Longo DL, Fauci AS, Hauser SL, Loscalzo J, editors. Harrison's principles of internal medicine. 20th ed. New York: McGraw Hill; 2018. p. 2843–50.
33. Misra A, Jayawardene R, Anoop S. Obesity in South Asia: phenotype, morbidities and mitigation. Curr Obes Rep. 2019;8:43–52.
34. Guptha LS. A cross-sectional epidemiology study of the relationships between body mass index and the risk of diabetes, and diabetes and the QRISK2 10-Year cardiovascular risk score using India Heart Watch data. PhD thesis, Trident University International, San Diego; 2021. https://www.proquest.com/openview/8d3d1772bfee8ca075388311d98982e7/1?pq-origsite=gscholar&cbl=18750&diss=y. Accessed 26 Jun 2021.
35. Victor RG. Systemic hypertension: mechanisms and diagnosis. In: Zipes DP, Libby P, Bonow RO, Mann DL, Tomaselli GF, Braunwald E, editors. Braunwald's heart disease. 11th ed. New Delhi: Elsevier; 2019. p. 910–27.
36. Wang C, Fu W, Cao S, Xu H, Tian Q, Gan Y, et al. Association of adiposity indicators with hypertension among Chinese adults. Nutr Metab Cardiovasc Dis. 2021;31:1391–400.
37. Zheng W, McLerran DF, Rolland B, Zhang X, Inoue M, Matsuo K, et al. Association between body mass index and risk of death in more than 1 million Asians. N Engl J Med. 2011;364:719–29.
38. Yusuf S, Joseph P, Rangarajan S, Islam S, Mente A, Hystad P, et al. Modifiable risk factors, cardiovascular disease, and mortality in 155,722 individuals from 21 high-income, middle-income, and low-income countries (PURE): a prospective cohort study. Lancet. 2020;395:795–808.
39. Ford ND, Patel SA, Venkat Narayan KM. Obesity in low and middle-income countries: burden, drivers and emerging challenges. Annu Rev Public Health. 2017;38:145–64.
40. Apóstolo J, Cooke R, Bobrowicz-Campos E, Santana S, Marcucci M, Cano A, et al. Predicting risk and outcomes for frail older adults : an umbrella review of frailty screening tools. JBI Database Syst Rev Implement Rep. 2017;15:1154–208.
41. Hu K, Zhou Q, Jiang Y, Shang Z, Mei F, Gao Q, et al. Association between frailty and mortality, falls and hospitalization among patients with hypertension: a systematic review and meta-analysis. Biomed Res Int. 2021;2021:2690296.
42. Olsen MH, Angell SY, Asma S, Boutouyrie P, Burger D, Chirinos JA, et al. A call to action and a life-course strategy to address the global burden of raised blood pressure on current and future generations: the Lancet Commission on Hypertension. Lancet. 2016;388:2665–712.
43. Tronieri JS, Wadden TA, Chao AM, Tsai AG. Primary care interventions for obesity: review of the evidence. Curr Obes Rep. 2019;8:128–36.

44. Look AHEAD Research Group, Wing RR, Bolin P, Brancati FL, Bray GA, Clark JM, et al. Cardiovascular effects of intensive lifestyle intervention in type 2 diabetes. N Engl J Med. 2013;369:145–54.
45. LeBlanc EL, Patnode CD, Webber EM, Redmond N, Rushkin M, O'Connor EA. Behavioural and pharmacotherapy weight loss interventions to prevent obesity related morbidity and mortality in adults: an updated systematic review for the US Preventive Services Task Force. Report No. 18-05239-EF-1. Rockville: Agency for Healthcare Research and Quality (US); 2018.
46. Srivastava G, Apovian CL. Current pharmacotherapy for obesity. Nat Rev Endocrinol. 2018;4:12–24.
47. NCD Countdown 2030 Collaborators. NCD Countdown 2030: pathways to achieving Sustainable Development Goal target 3.4. Lancet. 2020;396:918–34.
48. Gupta R, Yusuf S. Challenges in management and prevention of ischemic heart disease in low socioeconomic status people in LLMICs. BMC Med. 2019;17(1):209.
49. Jeemon P, Gupta R, Onen C, Adler A, Gaziano TA, Prabhakaran D, et al. Management of hypertension and dyslipidemia for primary prevention of cardiovascular diseases. In: Prabhakaran D, Anand S, Gaziano TA, Mbanya JC, Wu Y, Nugent R, editors. Cardiovascular, respiratory, and related disorders, Chap. 22. 3rd ed. Washington, DC: The World Bank; 2017.
50. Gupta R, Yusuf S. Towards better hypertension management in India. Indian J Med Res. 2014;139(5):657–60.
51. Bhavnani SP, Parakh K, Atrej A, Druz R, Graham GN, Hayek SS, et al. Roadmap for innovation- ACC health policy statement of healthcare transformation in the era of digital big data and precision health: a report of the American College of Cardiology Task Force on health policy statements and systems of care. J Am Coll Cardiol. 2017;2017(70):2696–718.
52. Gupta R, Gupta S. Social determinants' approach for hypertension control. In: CVS R, editor. Cardiological Society of India: hypertension reviews 2020. Noida: Incessant Nature Science Publishers; 2020. p. 11–21.

Effect of Pollution and Environmental Factors on Hypertension and CVD

7

Suganthi Jaganathan and Dorairaj Prabhakaran

7.1   Introduction

Every individual is exposed to a multitude of environmental factors in their every-day lives. These can be classified into the following: [1] a general external environment, including factors such as the urban environment, climatic factors such as climate change, social capital, and stress; [2] an external environment that includes specific contaminants, diet, physical activity, tobacco, infections, etc.; and [3] an internal environment that includes internal biological factors such as metabolic factors, gut microflora, inflammation, and oxidative stress [1]. This chapter discusses general external environmental factors and their role in cardiovascular diseases (CVDs), with specific reference to hypertension. This includes exposures such as air pollution, green spaces and built environment, night light, noise, extremes of temperature, metals and metalloids and their imbalance, which are known to be major causes of chronic noncommunicable diseases globally. Climate change is evolving issue, and its effects on CVDs are not yet fully understood and this issue is therefore beyond the scope of this chapter.

S. Jaganathan
Public Health Foundation of India, Delhi-NCR, Delhi, India

Institute of Environmental Medicine, Karolinska Institutet, Stockholm, Sweden

D. Prabhakaran (✉)
Public Health Foundation of India, Delhi-NCR, Delhi, India

Centre for Chronic Disease Control, New Delhi, India

Department of Epidemiology, London School of Hygiene and Tropical Medicine, London, UK
e-mail: dprabhakaran@ccdcindia.org

© The Author(s), under exclusive license to Springer Nature Switzerland AG 2022
C. V. S. Ram et al. (eds.), *Hypertension and Cardiovascular Disease in Asia*,
Updates in Hypertension and Cardiovascular Protection,
https://doi.org/10.1007/978-3-030-95734-6_7

7.1.1 Air Pollution

Air pollution results from a complex blend of gaseous mixtures and particulate constituents that vary by time and location [2]. It is a function of chemistry, size, or mass, and specific constituent elements. Each air pollutant has a distinct source, which could be domestic, anthropogenic (a result of human activity), industrial, construction, secondary processes, and so on. The World Health Organization (WHO) has developed air quality guidelines with maximum threshold levels for these pollutants. The WHO air quality guidelines are designed to offer guidance in reducing the health impacts of air pollution based on expert evaluation of current scientific evidence, and these standards are not legally binding. Hence, nonadherence to these guidelines is not strictly enforced. In addition to the WHO air quality guidelines, individual countries set their own air quality guidelines unique to their local context. Table 7.1 lists air quality standards of various pollutants set by WHO and examples from other countries [3–6].

1. Particulate matter
   One of the major air pollutants, particulate matter (PM) is broadly categorized by its aerodynamic diameter: <10 μm (thoracic particles [PM_{10}]), <2.5 μm (fine particles [$PM_{2.5}$]), <0.1 μm (ultrafine particles), and between 2.5 and 10 μm (coarse [$PM_{2.5-10}$]). PM is quantified by the particles (mass) contained per cubic meter (μg/m^3) of air. PM is composed of both primary and secondary chemical components. Primary PM is derived from particle emissions from a specific source. Secondary PM originates from gas-phase chemical compounds present in the ambient atmosphere that have participated in new particle formation or condensed onto existing particles. Primary particles, and the gas-phase compounds that ultimately contribute to PM, are emitted by natural and anthropogenic (human) sources [7]. However, over 90% of the pollutant mass breathed in urban settings is from gases or vapor phase compounds, including volatile

Table 7.1 Overview of air quality guidelines

Major air pollutants	WHO standards	US standards	EU standards	Indian standards	Chinese standards
$PM_{2.5}$	15 μg/m^3 (D); 5 μg/m^3 (A)	35 μg/m^3 (D); 12 μg/m^3 (A)	25 μg/m^3 (A)	60 μg/m^3 (D); 40 μg/m^3 (A)	75 μg/m^3 (D); 35 μg/m^3 (A)
PM_{10}	45 μg/m^3 (D); 15 μg/m^3 (A)	150 μg/m^3 (D)	50 μg/m^3 (D); 40 μg/m^3 (A)	100 μg/m^3 (D); 60 μg/m^3 (A)	150 μg/m^3 (D); 70 μg/m^3 (A)
SO_2	20 μg/m^3 (D)	75 ppb (H)	350 μg/m^3 (H); 125 μg/m^3 (D)	80 μg/m^3 (D); 50 μg/m^3 (A)	150 μg/m^3 (D); 60 μg/m^3 (A)
NO_2	25 μg/m^3 (D); 10 μg/m^3 (A)	100 ppb (H); 53 ppb (A)	200 μg/m^3 (H); 40 μg/m^3 (A)	80 μg/m^3 (D); 40 μg/m^3 (A)	80 μg/m^3 (D); 40 μg/m^3 (A)
O_3	100 μg/m^3 (8h)	0.070 ppm (8 h)	120 μg/m^3 (8h)	100 μg/m^3 (8h)	160 μg/m^3 (8h)

A annual, *D* daily, *H* hourly, *8 h* 8-hourly, *$PM_{2.5}$* particles with a diameter of <2.5 μm, *PM_{10}* particles with a diameter of <10 μm, *SO_2* sulfur dioxide, *NO_2* nitric dioxide, *O_3* ozone, *ppm* parts per million organic carbons [8]. Anthropogenic sources can be divided into stationary and mobile sources. Stationary sources include fuel combustion for electricity production and other purposes, industrial processes, agricultural activities, road and building construction, demolition, and biomass combustion. Mobile sources include diesel- and gasoline-powered highway vehicles and other engine-driven sources such as locomotives, ships, aircraft, and construction and agricultural equipment [7]. According to revised 2021 WHO air quality guidelines for fine particulate matter ($PM_{2.5}$), the annual mean must be 5 μg/m^3 and the 24-h mean must be 15 μg/m^3 [9]. Global population-weighted $PM_{2.5}$ increased by 11.2% from 1990 to 2015. Countries like Bangladesh, China, India, Nepal, and Pakistan have higher population-weighted mean concentrations of $PM_{2.5}$ than other countries in the region as shown in Fig. 7.1 [10, 11].

2. Ozone (O_3)

Ozone is the most prevalent secondary pollutant, often referred to as "smog." Ozone develops in the atmosphere from gases that come out of tailpipes, smokestacks, and many other sources. When these gases come in contact with sunlight, they react and form ozone smog. This occurs via reaction of pollutants such as nitrogen oxides (NO_x) from the vehicle and industry emissions and volatile organic compounds (VOCs) emitted by vehicles, solvents, and industry with sunlight (photochemical reaction). As a result, the highest levels of ozone pollution occur during periods of sunny weather [8, 9]. The WHO-recommended limit for ozone is 100 μg/m^3 8-h mean. Global population-weighted O_3 increased by 7.2% from 1990 to 2015 [10].

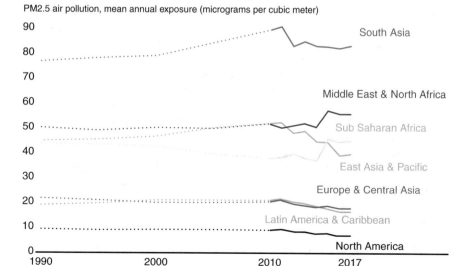

South Asia has the highest level of PM2.5 air pollution

PM2.5 air pollution, mean annual exposure (micrograms per cubic meter)

Fig. 7.1 PM2.5 levels in various regions of the world. Between 1990 and 2010 the data is available at 5 year intervals (1990, 1995, 2000, 2005, 2010). For the other years, the data points have been interpolated. (Source: World Development Indicators (EN.ATM.PM2.5.MC.M3))

3. Nitrogen dioxide (NO_2)

    NO_2 is the indicator for gaseous oxides of nitrogen (i.e., oxidized nitrogen compounds), including nitric oxide and gases produced from reactions involving NO_2 and nitric oxide [12]. The primary sources of anthropogenic emissions of NO_2 are combustion processes (heating, power generation, and engines in vehicles and ships). NO_2 is the main source of nitrate aerosols, which form an essential fraction of $PM_{2.5}$ and, in the presence of ultraviolet light, ozone. The current WHO guideline value of 10 µg/m^3 (annual mean) and 25 µg/m^3 (daily average) was set to protect the public from the health effects of NO_2 [9].

4. Sulfur dioxide (SO_2)

    The largest source of SO_2 in the atmosphere is the burning of fossil fuels by power plants and other industrial facilities. Smaller sources of SO_2 emissions include industrial processes such as extracting metal from ore, natural sources such as volcanoes, and locomotives, ships and other vehicles and heavy equipment that burn fuel with a high sulfur content [13]. As per WHO air quality guidelines, SO_2 concentrations should not exceed a 24-h mean of 20 µg/m^3 [9].

5. Household air pollution

    Sources of household air pollution (HAP) vary from country to country. They include cooking fuel, cooking method, heating, tobacco smoking, incense sticks, mosquito repellents, cleaning and personal care agents, pet animals, construction dust, and so on. The most significant contributor to HAP is unclean cooking fuel. Higher levels of HAP levels are typically encountered in most low- and middle-income countries such as India, and these are much higher than ambient outdoor levels in the exact geographic location, contributing to steep indoor-outdoor gradients [8]. Type of housing, including building materials and proper ventilation, is also a significant factor in HAP. Urban dwellings are often overcrowded in developing countries and do not have adequate ventilation in the kitchen/cooking area [14]. For instance, mean indoor 24-h PM_{10} levels of 200–2000 µg/m^3 are common. Peak exposures of >30,000 µg/m^3 during periods of cooking with exposure to low-efficiency combustion of biomass fuels have been reported [15]. From India state-level burden estimations, the 24-h mean concentration of $PM_{2.5}$ in households that use solid cooking fuel ranged from 163 µg/m^3 (95% confidence interval [CI] 143–183) in the living area to 609 µg/m^3 (95% CI 547–671) in the kitchen area [16]. The recommendations for regulating household air pollution focus on reducing emissions of pollutants as much as possible while also recognizing the importance of adequate ventilation and information and support for households to ensure the best use of technologies and fuels. Specific recommendations include household energy fuels and emphasis on accelerating access to clean household fuels [17].

7.1.1.1 Mechanisms Associated with Air Pollution and CVD

Environmental exposures have been associated with a range of cardiovascular health effects. Most environmental factors induce a similar response in the body when inhaled, ingested, and absorbed through skin contact. There are several pathways by

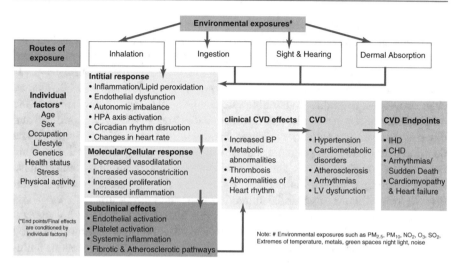

Fig. 7.2 Mechanisms by which environmental exposures are associated with cardiovascular disease

which environmental factors can influence CVDs and outcomes depending on their entry route into the body. There are a variety of mechanisms by which various inhaled air pollutants are associated with cardiovascular mortality and hypertension (Fig. 7.2).

After inhalation, air pollutants are deposited in the respiratory tract and lungs, and the smaller particles reach the alveolar region. These particles induce an inflammatory response and cross the alveolar membrane to enter the systemic circulation. Primary initiating pathways are pollution-mediated oxidative stress, local inflammation, and ion channel or receptor activation [18–20]. These trigger a chain of events that include signal transduction via the systemic release of numerous biological intermediates (such as oxidized lipids, cytokines, activated immune cells, microparticles, and endothelins), autonomic imbalance and activation of the hypothalamic-pituitary-adrenal (HPA) axis, and nanoparticles or pollutant constituents reaching the circulation or transmitted via neurological pathways. These pathways are associated with various effects on the cardiovascular system, including endothelial barrier disruption or dysfunction, tissue (vascular or adipocyte) inflammation, heightened coagulation and thrombosis, increased potential for cardiac arrhythmia and responses due to autonomic imbalance or HPA activation (e.g., vasoconstriction and increased blood pressure [BP]), secondary tissue damage (leading to plaque instability) and epigenomic changes. Finally, this paradigm of responsivity includes chronic end-organ changes resulting from long-term exposures that promote future susceptibility. These changes include the potentiation of cardiometabolic disorders, such as hypertension, diabetes mellitus, left ventricular hypertrophy or remodeling, vascular hypertrophy, proteinuria or renal disease, and the progression of atherosclerosis.

7.1.1.2 Exposure Temporality

The timing and scale of the CVD effects of pollution vary and might depend on pollutant characteristics, the duration of exposure and the degree of susceptibility, influenced by genetic predisposition and homeostatic pathways. Various CVD effects are triggered immediately after exposure to air pollution. CVD effects may even convergence, leading to additive effects of the pathways, contributing to the development of cardiovascular risk factors (such as high glucose levels and BP) and diseases (such as coronary artery disease, heart failure, arrhythmia, and stroke). The duration of exposure and pollutant characteristics are very important in determining the extent of health impacts and are less studied. Epidemiological studies have shown a stronger association between $PM_{2.5}$ and CVD mortality, ischemic heart disease, heart failure, myocardial infarction (MI), and stroke for both long- and short-term exposures. Short-term exposures are also associated with hospitalization, especially for heart failure, arrhythmias, and atrial fibrillation. It must be noted that there is also no safe threshold of air pollution below which long-term air pollution exposure has no health effects [19, 20].

7.1.1.3 Pollutant Interactions and Climate Change

Climate and weather conditions have a strong influence on the spatiotemporal distribution of air pollutants. Analyzing pollutant interactions means studying two or more pollutants for their association with the health outcome of interest. Effect modification of multiple pollutants may be interdependent, synergistic (positive interdependence) or antagonistic (negative interdependence). It is also essential to study their interactions to understand if they have certain risks that can help prevent adverse effects and for insights into impact mechanisms. Synergism may increase the disease burden beyond that anticipated from the threat of one pollutant alone and could place some people at particularly high risk (and vice versa for an antagonistic effect). Very few epidemiological studies undertake multipollutant analyses and study complex pollutant interactions. These complex interactions between air pollutants coupled with climate change suggest that future policies to mitigate these twin challenges will benefit from greater coordination. Assessing the health implications of alternative policy approaches toward climate and pollution mitigation will be a critical area of future work [21, 22].

7.1.1.4 Cardiovascular Diseases and Mortality

1. Particulate matter
   The link between $PM_{2.5}$ and cardiovascular mortality has been noted at both low and high levels of exposure. It was ranked as the fourth mortality risk factor in 2015 and causes 6.7 million deaths globally, with 59% of these deaths occurring in East and South Asia. Over 20% of these pollution-related deaths are due to CVD [23]. For example, short-term elevations in regions with low daily levels of exposure to $PM_{2.5}$ (<35 µg/m³) is associated with a 0.3–1.0% increase in the relative risk of cardiovascular mortality per 10 µg/m³ increase in $PM_{2.5}$ [24]. At higher levels of daily exposure (such as in China, where daily $PM_{2.5}$ levels are 39–177 µg/m³), a meta-analysis of seven studies reported that each 10 µg/m³

increase in $PM_{2.5}$ was associated with a 0.35% (95% CI 0.06–0.65) excess risk of cardiovascular death [25]. Although a growing body of studies supports the toxicity of ultrafine particles and possibly coarse PM [26], the growing burden of evidence highlights $PM_{2.5}$ as the principal air pollutant that poses the greatest threat to global public health, especially CVDs [27].

2. Ozone

Globally, exposure to ozone caused 254,000 (95% CI 97,000–422,000) deaths in 2015 [10]. However, evidence for the role of ozone in increasing CVD deaths is somewhat contradictory. Large registry-based studies from France and Sweden found associations between short-term increases in ozone levels and acute coronary events [28], out-of-hospital cardiac arrest [29] and stroke [30]. In contrast, a meta-analysis and systematic review reported no evidence of associations between long-term annual ozone concentrations and the risk of death from CVD [31].

3. Oxides of nitrogen

Evidence for the effects of short-term and long-term NO_2 exposure on cardiovascular outcomes (e.g., arrhythmia, cerebrovascular diseases, and hypertension), is inconsistent. There still is uncertainty about whether NO_2 exposure has products that are independent of other traffic-related pollutants. However, recent epidemiologic studies indicate that short-term NO_2 exposure may trigger MI and hospital admissions or emergency department (ED) visits for MI, angina, and their underlying cause, ischemic heart disease [32]. New findings from experimental studies point to the potential for NO_2 exposure to induce cardiovascular effects and diabetes but are not sufficient to address uncertainties in the epidemiologic evidence [12].

4. Sulfur dioxide

Sulfur dioxide (SO_2) is associated with changes in vasomotor tone [33], heart rate and cardiac function [34]. Such mechanisms may underlie the association between SO_2 and CVD [35, 36]. An occupational study found that a 10% decrease in SO_2 emission was associated with 0.28% (95% CI 0.39–0.95) and 1.69% (95% CI 0.99–2.38) reduction in CVD risk for males and females, respectively. Thus, enhancing regulations on SO_2 emission control represents a target for national and international intervention to prevent CVD [33].

5. Household air pollution

The Global Burden of Disease (GBD) study estimated that HAP was associated with 2.8 million deaths worldwide in 2015 [37]. In India, HAP is responsible for 0.61 million (95% CI 0.39–0.86 million) deaths annually [38]. A prospective study in China showed that the use of solid fuel was significantly linked with increased cardiovascular mortality (hazard ratio [HR] 1.20, 95% CI 1.02–1.41, compared to clean energy) [19]. Large-scale intervention trials are underway to investigate the effect of liquid petroleum gas (LPG) stove intervention (clean fuel alternatives) on HAP exposures and health across the lifespan [39, 40].

Besides clean fuel, air filters have also been widely studied. Small randomized, controlled trials of air filtration to improve indoor air quality have shown a consistent decrease in BP with a reduction in exposure to air pollution [41], confirming the causality of exposure in mediating the increase in BP. A trial enrolling elderly individual in the USA found that air filtration systems reduced $PM_{2.5}$ from 15 to 7.4–10.9 $\mu g/m^3$, which was associated with a systolic BP (SBP) reduction of 3.2 mmHg (95% CI −0.2, 6.1) [42]. The mechanisms underlying short-term BP increases in response to HAP might involve rapid alterations in autonomic tone that are exacerbated in the presence of endothelial dysfunction [43]. Emerging evidence shows that exposure to concentrated $PM_{2.5}$, but not ozone gas, led to a linear increase in diastolic BP (DBP), which correlated with changes in measures of autonomic function, such as heart rate variability [43]. Whether or not the cardiovascular effects of air pollution can be modulated by intensive BP control is yet to be confirmed [19].

7.1.1.5 Hypertension

1. Particulate matter
   Most studies, including controlled clinical trials and large-scale meta-analyses, support associations between $PM_{2.5}$ and higher SBP and DBP [44, 45]. In contrast, some studies have reported protective effects or no association [46–49]. Several studies have been published investigating associations between short-term $PM_{2.5}$ concentrations and BP, but, overall, recent data provide mixed evidence for associations [7]. Acute exposure to $PM_{2.5}$ at 10 $\mu g/m^3$ correlated with an increase in SBP of 1–3 mmHg [50]. The association between $PM_{2.5}$ and BP appears to be stronger among men, Asian individuals, those with higher body mass index (BMI), and individuals living in areas with high air pollution levels [51, 52]. Long-term exposure to $PM_{2.5}$ is associated with correspondingly more significant increases in BP and is also associated with incident hypertension [52]. Importantly, associations between $PM_{2.5}$ and hypertension have been observed in studies from countries with low levels of pollution (Canada and the USA) and those with extremely high levels of exposure to $PM_{2.5}$ (India, China), with no evidence for a flattening of the response at higher exposures [53]. A longitudinal study found that long- and short-term exposure had strong effects on SBP and DBP, and the risk of developing hypertension [52]. Higher average systolic BP differences were seen for monthly and annual exposures (1.8 and 3.3 mmHg per interquartile range; ~9–15 $\mu g/m^3$ of $PM_{2.5}$, respectively). Positive, but less pronounced associations, were observed for DBP. Average $PM_{2.5}$ exposure over durations of 1 year, 1.5 years and 2 years increased the risk of developing hypertension by 1.5 times (50% higher risk) 1.6 times (60% higher risk), and 1.2 times (20% higher risk), respectively. These associations were stronger in study participants who had higher waist-to-hip ratios (an indicator of central obesity) [52].

2. Oxides of nitrogen
   There is limited evidence from available epidemiologic studies and controlled human studies to suggest that short-term exposure to ambient NO_2 is associated with changes in BP or cardiac output in the population overall. However, it is important to note that there is no evidence of any association in longitudinal studies and but there is mixed evidence from cross-sectional studies [12].

3. Ozone
   The association between ozone and BP is not consistent and needs to be investigated further. Some previous studies found that ozone exposure was negatively (beneficially) or not associated with BP [54–56], but a recent study has reported a positive association between ozone and both SBP and DBP measurements at slightly higher exposure concentrations [43, 57].

7.1.2   Green Spaces and Built Environment

Green space is land partly or entirely covered with grass, trees, shrubs, or other vegetation. These are generally described as man-made or modified structures that provide people with living, working and recreational spaces. In general, parks, playgrounds, vegetation, community gardens and forests are considered green spaces [58]. They are believed to increase the quality of urban settings and promote sustainable lifestyles, improving both the health and well-being of the population [59]. The built environment is closely related to green spaces which includes all aspects of our lives, encompassing the buildings, water and electricity distribution systems, and roads, bridges, and transportation systems. However, the built environment is confounded by the socioeconomic status (SES) of an individual or family.

There is evidence that the built environment, an important determinant of health and well-being, may contribute to CVD risk [60]. Green space in urban areas lowers air temperature [61, 62] and can improve air quality [63, 64]. Given that long-term exposure to air pollution is related to higher CVD mortality [65, 66], green spaces play a pivotal role in combating air pollution and improving CV health. Socio-environmental characteristics such as social cohesion, neighborhood identity, and stigmatization vary between communities with different area characteristics, contributing to CVD risk factors and mortality. Social inequalities such as living in deprived neighborhoods are associated with increased CVD prevalence and mortality, independent of personal income, education, and occupation or established CVD risk factors [67].

The following are possible causal mechanisms through which green spaces and built environment contribute to CVDs: psychological and physiological effects of contact with the natural environment; reduction of pollutants; and opportunities for social contacts and outdoor physical activity [68]. Some evolutionary hypotheses state that human beings have a genetic need for nature. The impact of green spaces on health is also often explained by green space-obesity and green space-physical activity associations [69]. Green space has also been related to reduced stress and better self-rated health, leading to lower CVD mortality [70].

7.1.2.1 Cardiovascular Mortality and Hypertension
A review of studies found a significant reduction in the risk of CVD mortality among those living in areas with greater amounts of green space [71]. In a cross-sectional study, the most versus least income-deprived group had a higher risk of CVD mortality in areas with the lowest amounts of green space coverage (incidence

rate ratio [IRR] 2.19, 95% CI 2.04–2.34) and those with the highest green space coverage (IRR 1.54, 95% CI 1.38–1.73) [72]. A study that focused on parks and recreation spaces found that proximity to a major road was independently associated with increased CVD risk [73]. Data from a US study showed that smaller amounts of green space were associated with higher stroke mortality [74]. CVD mortality in the least green areas was twice that in the greenest areas [75] and the odds of CVD-related hospitalization were 37% lower [76]. In a longitudinal study conducted in Ontario, Canada, higher levels of greenness were associated with a lower risk of CVD and stroke mortality [77]. In the US, residential proximity to green spaces was found to be associated with higher survival rates after ischemic stroke, even after adjustment for socioeconomic factors [78]. In a Japanese study [79], 5-year survival of the elderly was higher in those living near accessible green space. Studies from Hong Kong also showed that greater amounts of natural green space are associated with lower CVD mortality [80, 81]. A study conducted in Australian adults reported that people who spent more time in green spaces had lower rates of high BP [82, 83]. Furthermore, a Chinese study found that BP and hypertension prevalence were lower in people living in greener areas [84].

In summary, these data show that exposure to vegetation decreases CVD risk, mortality, and severity. However, most existing studies have a cross-sectional design and therefore cannot provide information about causality. In addition, most were conducted in developed countries, highlighting a need for population-based longitudinal studies investigating the association between urban green space and CVD in developing countries.

7.1.3   Night Light

Artificial light and technology (such as televisions, computer screens, and smartphones) expose people to light in the evening, opposing and desynchronizing intrinsic clocks. Night light has been extensively researched in human circadian and sleep patterns, and has been shown to cause acute suppression of melatonin in response to light exposure and a shift in circadian phases. The circadian rhythm controls internal process of the sleep-wake cycle, which is linked to metabolism-regulating blood sugar and cholesterol [85] and regulating heart rate. Light cues, the primary clock input signal (synchronizer), are received through the retina and transmitted to the master clock, which synchronizes peripheral clocks throughout the body via various neurohumoral signals [86]. Circadian rhythms are essential regulators of cardiovascular physiology and disease [87, 88]. Chronic circadian misalignment is associated with elevated BP and increases overall cardiovascular risk [89]. There is sufficient evidence that common CVD events, including sudden cardiac arrest, stroke, and MI, predominantly occur in the early hours of the morning in response to rising heart rate and BP [90] and early morning release of hormones such as cortisol [91].

Night light and shift work, a well-studied domain among occupational research, are known to disturb the circadian rhythm and alter BP, increasing the risk for CVD and mortality [92]. It is also widely accepted that circadian rhythm disruption can increase the inflammatory responses leading to major CVDs [93].

7.1.3.1 Hypertension

Studies show that short-wavelength light (blue light and light-emitting diode [LED] light) is a particularly strong desynchronizer and is associated with metabolic syndrome, sleepiness, and reduced quality of life [94]. Another well-studied example of circadian disruption is in the intensive care unit. Critically ill patients in the intensive care unit are often exposed to many factors such as parenteral feeding and 24/7 artificial lights. Cardiovascular functions with a 24-h rhythm, such as BP and heart rate, and rhythms in core temperature, hormone secretion and activity, are disrupted in these patients [95].

7.1.4 Noise

The definition of noise is "unwanted or disturbing sound." Sound becomes unwanted when it interferes with normal activities such as sleeping, conversation, or disrupts or diminishes quality of life. Like air pollution, noise is another environmental factor that has an important bearing on cardiovascular health and disease. Noise generated from several sources, such as roadway traffic, railroads, construction activities and aircrafts, interferes with communication, causes annoyance, and disturbs sleep. The WHO reports an onset of adverse health effects in humans exposed to noise levels above 40 decibels (dB) at night [89]. Direct exposure studies in humans have shown that simulated traffic noise increases BP, heart rate and cardiac output, which are likely to be mediated by the release of catecholamines, cortisol, and other stress hormones. Similarly, exposure to aircraft noise, particularly at night, induces endothelial dysfunction measured by flow-dependent dilation, increases BP [67]. Acute effects occur not only at high sound levels in occupational settings but also at relatively low environmental noise levels when concentration, relaxation, or sleep is disturbed [18].

7.1.4.1 Cardiovascular Morbidity and Mortality

Insights from epidemiological studies show that exposure to traffic noise (aircraft, road vehicles and trains) is associated with increased cardiovascular morbidity and mortality [96]. In Europe, 40% of the population are exposed to road traffic noise exceeding a daytime average sound level (L_{DN}) of 55 dBA L_{DN} and >30% population are exposed to >55 dB at night which can lead to sleep disturbances [97]. Traffic noise in Europe is responsible for 18,000 premature deaths, 1.7 million cases of hypertension, and 80,000 hospitalizations each year [98]. In the US, nearly 46% of the population (145.5 million individuals) are exposed to noise at levels exceeding 55 dBA L_{DN}, and 43.8 million individuals are exposed to noise levels exceeding 65 dBA L_{DN} [99]. Constant noise exposure induces stress, and affects cognitive function, autonomic homeostasis, and sleep quality, all of which could increase CVD risk.

7.1.5 Extremes of Temperature

The recent Intergovernmental Panel on Climate Change Sixth Assessment Report [100] highlighted that the global mean surface temperature on earth is projected to rise in the future. Global surface temperature between 2001 and 2020 was 0.99 [0.84–1.10] °C higher than in the period 1850–1900. Last decade (2011–2020), the global surface temperature was 1.09 [0.95–1.20] °C higher than in 1850–1900, with larger increases over land (1.59 [1.34–1.83] °C) than over the ocean (0.88 [0.68–1.01] °C) (IPCC 2021). Outdoor temperatures are associated with worsening of CVDs [101], but the effect of extreme temperatures on CVDs remains unclear [102]. There are limited studies assessing the impact of hot temperatures and heat-waves on cardiovascular morbidity. Similarly, extreme cold increases the risk of cardiovascular hospitalization at that time and for some days after [103, 104]. The impact of cold spells on health has studied in cold-weather regions of Russia [105], Canada [106], China [107], the USA [108] and the United Kingdom [109]. It has been reported that both people and housing conditions are not well adapted to extreme cold-weather events, and hence the increased adverse health events [110, 111].

Cold weather is thought to be responsible for the greatest proportion of temperature-related CVD mortality. Cold seasons cause physiological changes, including increases in blood sugar, cholesterol levels, fibrinogen concentration and platelet aggregation, predisposing to formation of clots in the coronary artery, the start of acute MI and life-threatening arrhythmias [108]. Cold temperature also causes peripheral vasoconstriction and an increase of cardiac afterload in patients with preexisting CVDs. Furthermore, cold weather may contribute to the spread of respiratory infections, with an indirect effect on cardiovascular performance [112].

On the other hand, physiological reactions to heat have been studied and ana-lyzed in experimental settings. The studies that have been referenced below are conducted in young and healthy volunteers, especially sports people and military personnel. It has been shown that cutaneous and systemic vascular conductance is significantly increased during heat exposure, thus ameliorating the increase in core temperature [113]. On the other hand, this imposes an acute demand on the cardio-vascular system by recruitment of cardiac reserve and increases heart rate and car-diac output by up to 7–10 L/min [113] and decreases central blood volume, right atrial pressure, mean pulmonary artery pressure, left atrial volumes, aortic mean pressure, stroke volume, and total peripheral resistance (−52%) [114, 115].

7.1.5.1 Cardiovascular Mortality

Globally, increased mortality attributed to extreme outdoor temperatures has been reported. This is exemplified by an increase in all-cause mortality of 2.1% (95% CI 1.6–2.6) when temperature increases by 1 °C above the regional heat threshold. Smaller increases have been reported in mortality from cardiovascular causes (1.8% [1.2–2.5]) and MI (1.1% [0.7–1.5]), while there was a greater increase in deaths attributable to arrhythmias by (5.0% [3.2–6.9]). Overall, 33.9% of heat-related deaths were attributable to cardiovascular causes [116, 117].

7.1.5.2 Hypertension

Indoor and outdoor temperatures have independent effects on SBP, and both temperatures must be controlled for in studies that measure BP. A study that used data from the WHO MONICA Project risk factor surveys from 25 populations in 16 countries found that populations closer to the equator showed larger seasonal changes in BP [114]. A 1 °C increase in indoor temperature reduced SBP by an average of 0.31 mmHg (95% CI −0.44, −0.19), and a 1 °C increase in outdoor temperature reduced BP by an average of 0.19 mmHg (95% CI −0.26, −0.11); the effect of outdoor temperature remained after controlling for indoor temperature [118].

7.1.6 Metals and Metalloids (Arsenic, Cadmium, and Lead)

Metals and metalloids are widely distributed in the earth's crust and are often detected at very low concentrations in the human body. On the other hand, their presence in the air, soil, and water, even in traces, can cause serious problems. Multiple metals can induce oxidative stress, potential atherogenic effects, endocrine disruption, and abnormal endothelial vascular functions. Specifically, metals can produce reactive radicals, and deplete glutathione and other proteins involved in redox balance. Arsenic, cadmium, and lead have consistently been associated with increased risk of CVD and are included in the top ten environmental chemicals of concern compiled by the WHO and the Agency for Toxic Substances and Disease Registry (ATSDR) [119].

7.1.6.1 Arsenic

Prospective cohort studies evaluating arsenic exposure (using biomarkers or drinking water measures) have found associations between arsenic levels in the low-to-moderate range and the occurrence of CVD events, particularly coronary heart disease. In addition to clinical CVD outcomes, arsenic exposure is associated, although inconsistently, with established CVD risk factors such as hypertension [120] and electrocardiographic abnormalities (including prolonged QT interval) [121] in experimental and epidemiological studies. Clinical studies have also reported dose-dependent arsenic-induced cardiovascular effects, including hypertension, atherosclerosis, coronary heart disease, and stroke [122, 123].

7.1.6.2 Cadmium

Cadmium damages vascular tissues, induces endothelial dysfunction, and promotes atherosclerosis via oxidative mechanisms [124]. It can also disrupt physiological responses to oxidative stress by competing with zinc for binding to metallothionein, a group of cysteine-rich proteins involved in antioxidant reactions, and by binding glutathione [125]. It can also contribute to atherosclerosis indirectly through hypertension and renal effects.

Chronic cadmium exposure is associated with hypertension [126]. However, the exact influence of cadmium on the cardiovascular system remains controversial.

More importantly, the data show that cadmium may exert effects on the cardiovascular system at extremely low exposure levels. Data from in vitro showed that low-dose cadmium levels may contribute to the initiation of pathophysiological changes in the vessel wall [127].

7.1.6.3 Lead

Humans experience environmental exposure to lead mostly from air and dust, and sometimes from food, soil, and drinking water [128]. Broadly, lead is associated with cardiovascular outcomes [20]. High blood lead levels have been associated with mortality due to coronary heart disease, peripheral artery diseases, high BP, ventricular hypertrophy, and chronic kidney diseases. Lead promotes oxidative stress, inflammation, endothelial dysfunction, and proliferation of vascular smooth muscle cells and fibroblasts, and inhibits heart rate variability [129].

Chronic low-level lead exposures are associated with the pathogenesis of CVD [130]. Population-based studies have found that lead is associated with hypertension [131] and high BP [132], and atherosclerosis and increased cardiovascular mortality [133]. The precise mechanism explaining the hypertensive effect of lead exposure is unknown. An inverse association between estimated glomerular filtration rate and blood lead levels below 5 g/dL has been observed in general population studies [134], indicating that lead-induced reductions in renal function could play a major role in the development of hypertension in this setting.

7.2   What Needs to Be Done

Complex issues such as air pollution, extremes of temperature, green spaces, and other exposures as discussed earlier in this chapter require comprehensive evaluation to facilitate understanding of the health impacts and take necessary action. Most of the epidemiological studies discussed in this chapter are based on high-income countries, with a few exceptions. There are several understudied areas of research, and several successful policies and interventions that could help to improve the population health.

7.2.1   Research and Capacity Building

There are still many areas where environmental exposures and their impacts on CVD are yet to be explored. Some areas where further research is required are described below, although the list is not exhaustive [8, 135].

1. **Pollutant mixtures and interaction**: Traditionally, studies have only examined one pollutant at a time. This means that there are very few multipollutant studies, especially in the Asian context. This is important because pollutant interactions play a key role in health and disease.

2. **Longitudinal studies for cardiovascular risk factors**: There is a need to investigate CVD risk factors in longitudinal studies to allow better appreciation of the exposure-response function in diverse exposure settings.
3. **Time-series studies for cardiovascular mortality**: CVD mortality in Asian cities and environmental exposures is not well documented. These data would be important to help establish stricter guidelines around dangerous environmental exposures.
4. **Context specific studies in high/low exposure and rural/urban areas**: Given that environmental pollutants are so diverse in spatiotemporal scale, they need to be studied for specific context, including in both rural and urban areas, and in high and low exposure settings. Exposure levels can vary even within a given city/state/province.
5. **Studies examining both outdoor and indoor air pollution**: There is a school of thought that studying outdoor or indoor air pollution in isolation could lead to either under- or over-estimation of exposures. Therefore, detailed exposure assessment including both indoor and outdoor spaces should be considered. In addition, standard procedures for ascertaining exposures need to be used to facilitate comparison between studies.
6. **Speciation analysis**: Some of the studies that have assessed the composition of PM indicated that composition is important factor for specific health impacts.
7. **Intervention trials for specific population and health outcomes**: Research on reducing various environmental exposures and documenting the effectiveness of interventions will be informative. Given that most listed environmental exposures have a cultural angle, it is imperative to conduct intervention trials so that the data can be used to help inform local policies.
8. **Health impact assessments**: These assessments form an important part of the evidence base about various environmental exposures. Many Asian countries lack continuous monitoring data relating to environmental exposures and robust health impact assessments.
9. **Capacity building**: There are two aspects to this. Firstly, environmental health research is an emerging field in several Asian countries, and therefore there is a need to build capacity at various levels, such as monitoring, research, policies, setting regulations, health association studies, etc. Secondly, healthcare professionals need to have knowledge and awareness about the potential health risks associated with environmental exposures. This allows them to better inform patients with specific health conditions to prevent acute exposures.

7.2.2 Policies and Intervention

Some historic policies have contributed to reductions in air pollution and prevented CVD mortality in different parts of the world [131, 132]. These are some examples:

1. 1952 London fog: This occurred when coal consumption in the UK skyrocketed and smoke enveloped cities like Glasgow, Leeds, London, and Manchester,

blocking out the sun, blackening buildings, increasing the severity of fog, and contributing to 12,000 excess deaths. Subsequently, the UK government passed the Clean Air Act of 1956 which regulated both domestic and industrial smoke emissions for the first time [136].

2. Air pollution reached its worst levels in Los Angeles during the 1940s and 1950s. Millions of people driving millions of cars plus temperature inversion provided Los Angeles with a near perfect environment for the production and containment of photochemical smog. In 1947, the Los Angeles County Board of Supervisors established the nation's first air pollution control program by creating the Los Angeles County Air Pollution Control District. A children's health study captured lung function data from three different cohorts of children born in the 1980s, 1990s, and 2000s, and showed significant improvements in lung function development, irrespective of asthma status [137].

3. Beijing Olympics: This is a successful example of reducing air pollution through traffic control, closing highly polluting factories and use of natural gas in place of coal-fired boilers, which was associated with decreased cardiovascular mortality [138].

4. A ban on coal sales in Dublin was associated with decreased cardiovascular deaths [139].

1. **Monitoring of environmental exposures and data availability**: Strengthening of the existing monitoring network and expansion to cover areas that are sparsely monitored. In addition to monitoring, improving data availability for researchers and policy makers is important.

2. Given the diversity of exposures and their sources, policy action and interventions should take a **multidisciplinary approach** including experts such as engineers, urban planners, transport and health researchers, social scientists, intervention specialists, economists, and others. Some of the best policy approaches are to reduce industrial emissions, improve energy options, reduce traffic congestion, and remove sources of pollutant.

3. **Vehicular and fuel standards**: Based on knowledge that the major contributing sources of PM and ozone are poor vehicular and fuel standards, along with incomplete combustion, there is need to work in this area and ensure strict regulations for adapting to newer vehicular and fuel standards. Best practices with vehicular emission and standards should be improved, and increased use of public transport and strict policies related to replacement of older diesel vehicles are needed.

4. **Construction activities with improved materials**: Research is also required in advanced construction materials and technologies which should be designed to help reduce the air pollution and improve cost-effectiveness.

5. **Household clean fuel with adequate ventilation**: There are a few intervention studies that have promoted the uptake of clean cooking fuel in developing countries. More research is required to facilitate understanding of the health impact and cost-effectiveness of these interventions in various settings. Houses must be

designed with special attention to exposure-ventilation and have dust suppression techniques.

6. **Personal protection:** This includes anti-pollution masks, air purifiers, and noise-canceling technologies. It is an emerging field where these are tested in various settings to determine their effectiveness in improving the health. Personal protection is design to personal exposure to air pollution and noise. Very few studies have been conducted in Asian cities, which have higher loads of both indoor and outdoor air pollution. It is also important to evaluate the health impacts of using personal protection interventions in occupational settings and high-risk groups.

7. **Planned cities with more access to green spaces**: Emerging economies need to plan their cities for future generations, with more focus on green spaces. This requires improved urban planning, and more localized studies will help inform policy makers about the benefits of living closer to green spaces.

8. **Regulating noise levels and metals**: Regulating noise levels and exposures to metals and metalloids need to be contextualized for specific settings. This is only possible when there is robust evidence about their health impacts.

9. **Climate resilient households**: With looming challenges of extreme temperatures, development of climate-friendly policies is needed to minimize the health impacts of higher temperatures. Again, this also specific for different cities and countries, which all have different challenges. Some cities are being affected by heat waves, while others have extreme cold, some have drought conditions, and others are experiencing more floods and hurricanes.

References

1. Vrijheid M. The exposome: a new paradigm to study the impact of environment on health. Thorax. 2014;69(9):876–8.
2. IARC Working Group on the Evaluation of Carcinogenic Risks to Humans. Outdoor air pollution. IARC Monogr Eval Carcinog Risks Hum. 2016;109:9–444.
3. US Standards. https://www.epa.gov/criteria-air-pollutants/naaqs-table#2. Accessed 18 Aug 2021.
4. WHO & EU Standards. https://www.eea.europa.eu/themes/air/air-quality-concentrations/air-quality-standards. Accessed 18 Aug 2021.
5. Indian Standards. http://cpcbenvis.nic.in/air_pollution_main.html#. Accessed 18 Aug 2021.
6. Chinese standards. http://english.mee.gov.cn/Resources/standards/Air_Environment/quality_standard1/201605/W020160511506615956495.pdf. Accessed 18 Aug 2021.
7. U.S. EPA. Integrated science assessment (ISA) for particulate matter (Final Report, Dec 2019). https://cfpub.epa.gov/ncea/isa/recordisplay.cfm?deid=347534.
8. Rajagopalan S, Al-Kindi SG, Brook RD. Air pollution and cardiovascular disease: JACC state-of-the-art review. J Am Coll Cardiol. 2018;72(17):2054–70.
9. WHO. https://www.who.int/news-room/fact-sheets/detail/ambient-(outdoor)-air-quality-and-health.
10. Cohen AJ, Brauer M, Burnett R, Anderson HR, Frostad J, Estep K, et al. Estimates and 25-year trends of the global burden of disease attributable to ambient air pollution: an analysis of data from the Global Burden of Diseases Study 2015. Lancet. 2017;389(10082):1907–18.

11. World Development Indicators. https://datatopics.worldbank.org/world-development-indicators/stories/the-global-distribution-of-air-pollution.html. Accessed 18 Aug 2021.
12. EPA US. Integrated science assessment for oxides of nitrogen–health criteria. Washington, DC: US Environmental Protection Agency [Google Scholar]; 2016.
13. https://www.epa.gov/so2-pollution/sulfur-dioxide-basics.
14. Apte K, Salvi S. Household air pollution and its effects on health. F1000Res. 2016;5:F1000 Faculty Rev-2593.
15. Fullerton DG, Bruce N, Gordon SB. Indoor air pollution from biomass fuel smoke is a major health concern in the developing world. Trans R Soc Trop Med Hyg. 2008;102(9):843–51.
16. Balakrishnan K, Ghosh S, Ganguli B, Sambandam S, Bruce N, Barnes DF, et al. State and national household concentrations of PM2.5 from solid cookfuel use: results from measurements and modeling in India for estimation of the global burden of disease. Environ Health. 2013;12(1):77.
17. WHO Indoor Air Quality Guidelines. Household fuel combustion. Geneva: World Health Organization; 2014.
18. Münzel T, Gori T, Al-Kindi S, Deanfield J, Lelieveld J, Daiber A, et al. Effects of gaseous and solid constituents of air pollution on endothelial function. Eur Heart J. 2018;39(38): 3543–50.
19. Al-Kindi SG, Brook RD, Biswal S, Rajagopalan S. Environmental determinants of cardiovascular disease: lessons learned from air pollution. Nat Rev Cardiol. 2020;17(10):656–72.
20. Cosselman KE, Navas-Acien A, Kaufman JD. Environmental factors in cardiovascular disease. Nat Rev Cardiol. 2015;12(11):627–42.
21. Kinney PL. Interactions of climate change, air pollution, and human health. Curr Environ Health Rep. 2018;5(1):179–86.
22. WHO. 2003. https://www.euro.who.int/__data/assets/pdf_file/0005/112199/E79097.pdf.
23. Roth GA, Mensah GA, Johnson CO, Addolorato G, Ammirati E, Baddour LM, et al. Global burden of cardiovascular diseases and risk factors, 1990-2019: update from the GBD 2019 study. J Am Coll Cardiol. 2020;76(25):2982–3021.
24. Hayes RB, Lim C, Zhang Y, Cromar K, Shao Y, Reynolds HR, Silverman DT, Jones RR, Park Y, Jerrett M, Ahn J, Thurston GD. PM2.5 air pollution and cause-specific cardiovascular disease mortality. Int J Epidemiol. 2020;49(1):25–35.
25. Zhao L, Liang HR, Chen FY, Chen Z, Guan WJ, Li JH. Association between air pollution and cardiovascular mortality in China: a systematic review and meta-analysis. Oncotarget. 2017;8(39):66438–48.
26. Schraufnagel DE. The health effects of ultrafine particles. Exp Mol Med. 2020;52(3): 311–7.
27. Brook RD, Newby DE, Rajagopalan S. Air pollution and cardiometabolic disease: an update and call for clinical trials. Am J Hypertens. 2017;31(1):1–10.
28. Ruidavets JB, Cournot M, Cassadou S, Giroux M, Meybeck M, Ferrières J. Ozone air pollution is associated with acute myocardial infarction. Circulation. 2005;111(5):563–9.
29. Raza A, Bellander T, Bero-Bedada G, Dahlquist M, Hollenberg J, Jonsson M, et al. Short-term effects of air pollution on out-of-hospital cardiac arrest in Stockholm. Eur Heart J. 2014;35(13):861–8.
30. Henrotin JB, Zeller M, Lorgis L, Cottin Y, Giroud M, Béjot Y. Evidence of the role of short-term exposure to ozone on ischaemic cerebral and cardiac events: the Dijon Vascular Project (DIVA). Heart. 2010;96(24):1990–6.
31. Atkinson RW, Butland BK, Dimitroulopoulou C, Heal MR, Stedman JR, Carslaw N. Long-term exposure to ambient ozone and mortality: a quantitative systematic review and meta-analysis of evidence from cohort studies. BMJ Open. 2016;6(2):e009493.
32. Mustafic H, Jabre P, Caussin C, Murad MH, Escolano S, Tafflet M, et al. Main air pollutants and myocardial infarction: a systematic review and meta-analysis. JAMA. 2012;307(7):713–21.
33. Lin CK, Lin RT, Chen PC, Wang P, De Marcellis-Warin N, Zigler C, et al. A global perspective on sulfur oxide controls in coal-fired power plants and cardiovascular disease. Sci Rep. 2018;8(1):2611.

34. Routledge HC, Manney S, Harrison RM, Ayres JG, Townend JN. Effect of inhaled sulphur dioxide and carbon particles on heart rate variability and markers of inflammation and coagulation in human subjects. Heart. 2006;92(2):220–7.
35. Shah AS, Langrish JP, Nair H, McAllister DA, Hunter AL, Donaldson K, et al. Global association of air pollution and heart failure: a systematic review and meta-analysis. Lancet. 2013;382(9897):1039–48.
36. Hoek G, Brunekreef B, Fischer P, van Wijnen J. The association between air pollution and heart failure, arrhythmia, embolism, thrombosis, and other cardiovascular causes of death in a time series study. Epidemiology. 2001;12(3):355–7.
37. Cohen AJ, Brauer M, Burnett R, Anderson HR, Frostad J, Estep K, et al. Estimates and 25-year trends of the global burden of disease attributable to ambient air pollution: an analysis of data from the Global Burden of Diseases Study 2015. Lancet. 2017;389(10082):1907–18.
38. India State-Level Disease Burden Initiative Air Pollution Collaborators. The impact of air pollution on deaths, disease burden, and life expectancy across the states of India: the Global Burden of Disease Study 2017. Lancet Planet Health. 2019;3(1):e26–39.
39. Simkovich SM, Underhill LJ, Kirby MA, Goodman D, Crocker ME, Hossen S, et al. Design and conduct of facility-based surveillance for severe childhood pneumonia in the Household Air Pollution Intervention Network (HAPIN) trial. ERJ Open Res. 2020;6(1):00308–2019.
40. Checkley W, Williams KN, Kephart JL, Fandiño-Del-Rio M, Steenland NK, Gonzales GF, et al. Effects of a household air pollution intervention with liquefied petroleum gas on cardiopulmonary outcomes in Peru. A randomized controlled trial. Am J Respir Crit Care Med. 2021;203(11):1386–97.
41. Walzer D, Gordon T, Thorpe L, Thurston G, Xia Y, Zhong H, et al. Effects of home particulate air filtration on blood pressure: a systematic review. Hypertension. 2020;76(1):44–50.
42. Morishita M, Adar SD, D'Souza J, Ziemba RA, Bard RL, Spino C, et al. Effect of portable air filtration systems on personal exposure to fine particulate matter and blood pressure among residents in a low-income senior facility: a randomized clinical trial. JAMA Intern Med. 2018;178(10):1350–7.
43. Brook RD, Urch B, Dvonch JT, Bard RL, Speck M, Keeler G, et al. Insights into the mechanisms and mediators of the effects of air pollution exposure on blood pressure and vascular function in healthy humans. Hypertension. 2009;54(3):659–67.
44. Liang R, Zhang B, Zhao X, Ruan Y, Lian H, Fan Z. Effect of exposure to PM2.5 on blood pressure: a systematic review and meta-analysis. J Hypertens. 2014;32(11):2130–40; discussion 2141.
45. Lin H, Guo Y, Zheng Y, Di Q, Liu T, Xiao J, et al. Long-term effects of ambient PM2.5 on hypertension and blood pressure and attributable risk among older Chinese adults. Hypertension. 2017;69(5):806–12.
46. Brook RD, Bard RL, Burnett RT, Shin HH, Vette A, Croghan C, et al. Differences in blood pressure and vascular responses associated with ambient fine particulate matter exposures measured at the personal versus community level. Occup Environ Med. 2011;68(3):224–30.
47. Chung M, Wang DD, Rizzo AM, Gachette D, Delnord M, Parambi R, et al. Association of PNC, BC, and PM2.5 measured at a central monitoring site with blood pressure in a predominantly near highway population. Int J Environ Res Public Health. 2015;12(3):2765–80.
48. Weichenthal S, Hatzopoulou M, Goldberg MS. Exposure to traffic-related air pollution during physical activity and acute changes in blood pressure, autonomic and micro-vascular function in women: a cross-over study. Part Fibre Toxicol. 2014;11:70.
49. Morishita M, Bard RL, Kaciroti N, Fitzner CA, Dvonch T, Harkema JR, et al. Exploration of the composition and sources of urban fine particulate matter associated with same-day cardiovascular health effects in Dearborn, Michigan. J Expo Sci Environ Epidemiol. 2015;25(2):145–52.
50. Lee BJ, Kim B, Lee K. Air pollution exposure and cardiovascular disease. Toxicol Res. 2014;30(2):71–5.
51. Du Y, Xu X, Chu M, Guo Y, Wang J. Air particulate matter and cardiovascular disease: the epidemiological, biomedical and clinical evidence. J Thorac Dis. 2016;8(1):E8–E19.

52. Prabhakaran D, Mandal S, Krishna B, Magsumbol M, Singh K, Tandon N, et al. GeoHealth Hub Study Investigators, COE-CARRS Study Investigators. Exposure to particulate matter is associated with elevated blood pressure and incident hypertension in urban India. Hypertension. 2020;76(4):1289–98.
53. Giorgini P, Di Giosia P, Grassi D, Rubenfire M, Brook RD, Ferri C. Air pollution exposure and blood pressure: an updated review of the literature. Curr Pharm Des. 2016;22(1): 28–51.
54. Hoffmann B, Luttmann-Gibson H, Cohen A, Zanobetti A, de Souza C, Foley C, et al. Opposing effects of particle pollution, ozone, and ambient temperature on arterial blood pressure. Environ Health Perspect. 2012;120:241–6.
55. Brook RD, Urch B, Dvonch JT, Bard RL, Speck M, Keeler G, et al. Insights into the mechanisms and mediators of the effects of air pollution exposure on blood pressure and vascular function in healthy humans. Hypertension. 2009;54(3):659–67.
56. Delfino RJ, Tjoa T, Gillen DL, Staimer N, Polidori A, Arhami M, et al. Traffic-related air pollution and blood pressure in elderly subjects with coronary artery disease. Epidemiology. 2010;21(3):396–404.
57. Day DB, Xiang J, Mo J, Li F, Chung M, Gong J, et al. Association of ozone exposure with cardiorespiratory pathophysiologic mechanisms in healthy adults. JAMA Intern Med. 2017;177:1344–53.
58. https://www3.epa.gov/region1/eco/uep/openspace.html.
59. WHO Regional Office for Europe. Urban green spaces and health: a review of evidence. Copenhagen: WHO Regional Office for Europe (WHO); 2016. http://www.euro.who.int/en/health-topics/environment-and-health/urban-health/publications/2016/urban-green-spaces-and-health-a-review-of-evidence-2016. Accessed 4 Aug 2021.
60. Chow CK, Lock K, Teo K, Subramanian SV, McKee M, Yusuf S. Environmental and societal influences acting on cardiovascular risk factors and disease at a population level: a review. Int J Epidemiol. 2009;38(6):1580–94.
61. Bowler DE, Buyung-Ali LM, Knight TM, Pullin AS. A systematic review of evidence for the added benefits to health of exposure to natural environments. BMC Public Health. 2010;10:456.
62. Takebayashi H. Influence of urban green area on air temperature of surrounding built-up area. Climate. 2017;5:60.
63. Liu H-L, Shen Y-S. The impact of green space changes on air pollution and microclimates: a case study of the Taipei metropolitan area. Sustainability. 2014;6:8827.
64. Selmi W, Weber C, Rivière E, Blond N, Mehdi L, Nowak D. Air pollution removal by trees in public green spaces in Strasbourg city, France. Urban For Urban Green. 2016;17:192–201.
65. Pope CA 3rd, Burnett RT, Thurston GD, Thun MJ, Calle EE, Krewski D, et al. Cardiovascular mortality and long-term exposure to particulate air pollution - epidemiological evidence of general pathophysiological pathways of disease. Circulation. 2004;109(1):71–7.
66. Shen YS, Lung SC. Can green structure reduce the mortality of cardiovascular diseases? Sci Total Environ. 2016;566-567:1159–67.
67. Bhatnagar A. Environmental determinants of cardiovascular disease. Circ Res. 2017;121(2):162–80.
68. Frumkin H, Bratman GN, Breslow SJ, Cochran B, Kahn PH Jr, Lawler JJ, et al. Nature contact and human health: a research agenda. Environ Health Perspect. 2017;125(7):075001.
69. Klompmaker JO, Hoek G, Bloemsma LD, Gehring U, Strak M, Wijga AH, et al. Green space definition affects associations of green space with overweight and physical activity. Environ Res. 2018;160:531–40.
70. Tamosiunas A, Grazuleviciene R, Luksiene D, Dedele A, Reklaitiene R, Baceviciene M, et al. Accessibility and use of urban green spaces, and cardiovascular health: findings from a Kaunas cohort study. Environ Health. 2014;13(1):20.
71. Seo S, Choi S, Kim K, Kim SM, Park SM. Association between urban green space and the risk of cardiovascular disease: a longitudinal study in seven Korean metropolitan areas. Environ Int. 2019;125:51–7.

72. Mitchell R, Popham F. Effect of exposure to natural environment on health inequalities: an observational population study. Lancet. 2008;372(9650):1655–60.
73. Chum A, O'Campo P. Cross-sectional associations between residential environmental exposures and cardiovascular diseases. BMC Public Health. 2015;15:438.
74. Hu Z, Liebens J, Rao KR. Linking stroke mortality with air pollution, income, and greenness in Northwest Florida: an ecological geographical study. Int J Health Geogr. 2008;7:20.
75. Mytton OT, Townsend N, Rutter H, Foster C. Green space and physical activity: an observational study using health survey for England data. Health Place. 2012;18(5):1034–41.
76. Pereira G, Foster S, Martin K, Christian H, Boruff BJ, Knuiman M, et al. The association between neighborhood greenness and cardiovascular disease: an observational study. BMC Public Health. 2012;12:466.
77. Villeneuve PJ, Jerrett M, Su JG, Burnett RT, Chen H, Wheeler AJ, et al. A cohort study relating urban green space with mortality in Ontario, Canada. Environ Res. 2012;115:51–8.
78. Wilker EH, Wu CD, McNeely E, Mostofsky E, Spengler J, Wellenius GA, et al. Green space and mortality following ischemic stroke. Environ Res. 2014;133:42–8.
79. Takano T, Nakamura K, Watanabe M. Urban residential environments and senior citizens' longevity in megacity areas: the importance of walkable green spaces. J Epidemiol Community Health. 2002;56:913–8.
80. Wang D, Lau KK, Yu R, Wong SYS, Kwok TTY, Woo J. Neighbouring green space and mortality in community-dwelling elderly Hong Kong Chinese: a cohort study. BMJ Open. 2017;7(7):e015794.
81. Xu L, Ren C, Yuan C, Nichol JE, Goggins WB. An Ecological Study of the Association between Area-Level Green Space and Adult Mortality in Hong Kong. Climate. 2017;5(3):55.
82. Dzhambov AM, Markevych I, Lercher P. Greenspace seems protective of both high and low blood pressure among residents of an Alpine valley. Environ Int. 2018;121(Pt 1):443–52.
83. Shanahan DF, Bush R, Gaston KJ, Lin BB, Dean J, Barber E, et al. Health benefits from nature experiences depend on dose. Sci Rep. 2016;6:28551.
84. Yang BY, Markevych I, Bloom MS, Heinrich J, Guo Y, Morawska L, et al. Community greenness, blood pressure, and hypertension in urban dwellers: the 33 communities Chinese Health Study. Environ Int. 2019;126:727–34.
85. Jagannath A, Taylor L, Wakaf Z, Vasudevan SR, Foster RG. The genetics of circadian rhythms, sleep and health. Hum Mol Genet. 2017;26(R2):R128–38.
86. Husse J, Eichele G, Oster H. Synchronization of the mammalian circadian timing system: light can control peripheral clocks independently of the SCN clock: alternate routes of entrainment optimize the alignment of the body's circadian clock network with external time. BioEssays. 2015;37(10):1119–28.
87. Crnko S, Du Pré BC, Sluijter JPG, Van Laake LW. Circadian rhythms and the molecular clock in cardiovascular biology and disease. Nat Rev Cardiol. 2019;16(7):437–47.
88. Blume C, Garbazza C, Spitschan M. Effects of light on human circadian rhythms, sleep and mood. Somnologie (Berl). 2019;23(3):147–56.
89. Morris CJ, Purvis TE, Mistretta J, Hu K, Scheer FAJL. Circadian misalignment increases C-reactive protein and blood pressure in chronic shift workers. J Biol Rhythm. 2017;32(2):154–64.
90. William J. Elliott, cyclic and circadian variations in cardiovascular events. Am J Hypertens. 2001;14(9 Pt 2):291S–5S.
91. Iob E, Steptoe A. Cardiovascular disease and hair cortisol: a novel biomarker of chronic stress. Curr Cardiol Rep. 2019;21(10):116.
92. James SM, Honn KA, Gaddameedhi S, Van Dongen HPA. Shift work: disrupted circadian rhythms and sleep-implications for health and Well-being. Curr Sleep Med Rep. 2017;3(2):104–12.
93. Thosar SS, Butler MP, Shea SA. Role of the circadian system in cardiovascular disease. J Clin Invest. 2018;128(6):2157–67.
94. Tosini G, Ferguson I, Tsubota K. Effects of blue light on the circadian system and eye physiology. Mol Vis. 2016;22:61–72.

95. Potter GDM, Skene DJ, Arendt J, Cade JE, Grant PJ, Hardie LJ. Circadian rhythm and sleep disruption: causes, metabolic consequences, and countermeasures. Endocr Rev. 2016;37:584–608.
96. Münzel T, Gori T, Babisch W, Basner M. Cardiovascular effects of environmental noise exposure. Eur Heart J. 2014;35(13):829–36.
97. Berglund B, Lindvall T, Schwela DH, World Health Organization, Occupational and Environmental Health Team. Guidelines for community noise. World Health Organization; 1999. https://apps.who.int/iris/handle/10665/66217.
98. Hahad O, Kröller-Schön S, Daiber A, Münzel T. The cardiovascular effects of noise. Dtsch Arztebl Int. 2019;116(14):245–50.
99. Hammer MS, Swinburn TK, Neitzel RL. Environmental noise pollution in the United States: developing an effective public health response. Environ Health Perspect. 2014;122(2):115–9.
100. IPCC. Summary for policymakers. In: Climate change 2021: the physical science basis. Contribution of Working Group I to the sixth assessment report of the Intergovernmental Panel on Climate Change. Cambridge University Press; 2021 (In Press).
101. Ponjoan A, Blanch J, Alves-Cabratosa L, Martí-Lluch R, Comas-Cufí M, Parramon D, et al. Effects of extreme temperatures on cardiovascular emergency hospitalizations in a Mediterranean region: a self-controlled case series study. Environ Health. 2017;16(1):32.
102. Turner LR, Barnett AG, Connell D, Tong S. Ambient temperature and cardiorespiratory morbidity: a systematic review and meta-analysis. Epidemiology (Camb Mass). 2012;23:594–606.
103. Ryti NRI, Guo Y, Jaakkola JJK. Global Association of Cold Spells and Adverse Health Effects: a systematic review and meta-analysis. Environ Health Perspect. 2016;124:12–22.
104. Phung D, Thai PK, Guo Y, Morawska L, Rutherford S, Chu C. Ambient temperature and risk of cardiovascular hospitalization: an updated systematic review and meta-analysis. Sci Total Environ. 2016;550:1084–102.
105. Shaposhnikov D, Revich B, Gurfinkel Y, Naumova E. The influence of meteorological and geomagnetic factors on acute myocardial infarction and brain stroke in Moscow, Russia. Int J Biometeorol. 2014;58:799–808.
106. Lavigne E, Gasparrini A, Wang X, Chen H, Yagouti A, Fleury MD, et al. Extreme ambient temperatures and cardiorespiratory emergency room visits: assessing risk by comorbid health conditions in a time series study. Environ Health Glob Access Sci Source. 2014;13:5.
107. Ma W, Xu X, Peng L, Kan H. Impact of extreme temperature on hospital admission in Shanghai, China. Sci Total Environ. 2011;409:3634–7.
108. Madrigano J, Mittleman MA, Baccarelli A, Goldberg R, Melly S, von Klot S, et al. Temperature, myocardial infarction, and mortality: effect modification by individual- and area-level characteristics. Epidemiology (Camb Mass). 2013;24:439–46.
109. Sartini C, Barry SJE, Wannamethee SG, Whincup PH, Lennon L, Ford I, et al. Effect of cold spells and their modifiers on cardiovascular disease events: evidence from two prospective studies. Int J Cardiol. 2016;218:275–83.
110. Ezekowitz JA, Bakal JA, Westerhout CM, Giugliano RP, White H, Keltai M, et al. The relationship between meteorological conditions and index acute coronary events in a global clinical trial. Int J Cardiol. 2013;168:2315–21.
111. Aubinière-Robb L, Jeemon P, Hastie CE, Patel RK, McCallum L, Morrison D, et al. Blood pressure response to patterns of weather fluctuations and effect on mortality. Hypertension. 2013;62(1):190–6.
112. Moghadamnia MT, Ardalan A, Mesdaghinia A, Keshtkar A, Naddafi K, Yekaninejad MS. Ambient temperature and cardiovascular mortality: a systematic review and meta-analysis. PeerJ. 2017;5:e3574.
113. Crandall CG, Wilson TE. Human cardiovascular responses to passive heat stress. Compr Physiol. 2015;5(1):17–43.
114. De Blois J, Kjellstrom T, Agewall S, Ezekowitz JA, Armstrong PW, Atar D. The effects of climate change on cardiac health. Cardiology. 2015;131(4):209–17.

115. Wilson TE, Tollund C, Yoshiga CC, et al. Effects of heat and cold stress on central vascular pressure relationships during orthostasis in humans. J Physiol. 2007;585:279–85.

116. Rowell LB, Brengelmann GL, Murray JA. Cardiovascular responses to sustained high skin temperature in resting man. J Appl Physiol. 1969;27:673–80.

117. Gasparrini A, Armstrong B, Kovats S, Wilkinson P. The effect of high temperatures on cause-specific mortality in England and Wales. Occup Environ Med. 2012;69(1):56–61.

118. Barnett AG. Temperature and cardiovascular deaths in the US elderly: changes over time. Epidemiology. 2007;18(3):369–72.

119. ATSDR (Agency for Toxic Substances and Disease Registry). CERCLA priority list of hazardous substances. 2005. http://www.atsdr.cdc.gov/cercla/05list.html/, https://www.atsdr.cdc.gov/. Accessed 17 Aug 2021.

120. Abhyankar LN, Jones MR, Guallar E, Navas-Acien A. Arsenic exposure and hypertension: a systematic review. Environ Health Perspect. 2012;120(4):494–500.

121. Mumford JL, Wu K, Xia Y, Kwok R, Yang Z, Foster J, et al. Chronic arsenic exposure and cardiac repolarization abnormalities with QT interval prolongation in a population-based study. Environ Health Perspect. 2007;115(5):690–4.

122. Tseng CH, Chong CK, Tseng CP, Hsueh YM, Chiou HY, Tseng CC, et al. Long-term arsenic exposure and ischemic heart disease in arseniasis-hyperendemic villages in Taiwan. Toxicol Lett. 2003;137(1–2):15–21.

123. Chiou HY, Huang WI, Su CL, Chang SF, Hsu YH, Chen CJ. Dose-response relationship between prevalence of cerebrovascular disease and ingested inorganic arsenic. Stroke. 1997;28(9):1717–23.

124. Nawrot TS, Staessen JA, Roels HA, Munters E, Cuypers A, Richart T, et al. Cadmium exposure in the population: from health risks to strategies of prevention. Biometals. 2010;23(5):769–82.

125. Valko M, Morris H, Cronin MT. Metals, toxicity and oxidative stress. Curr Med Chem. 2005;12(10):1161–208.

126. Gallagher CM, Meliker JR. Blood and urine cadmium, blood pressure, and hypertension: a systematic review and meta-analysis. Environ Health Perspect. 2010;118(12):1676–84.

127. Bernhard D, Rossmann A, Henderson B, Kind M, Seubert A, Wick G. Increased serum cadmium and strontium levels in young smokers: effects on arterial endothelial cell gene transcription. Arterioscler Thromb Vasc Biol. 2006;26(4):833–8.

128. Tchounwou PB, Yedjou CG, Patlolla AK, Sutton DJ. Heavy metal toxicity and the environment. Exp Suppl. 2012;101:133–64.

129. Alissa EM, Ferns GA. Heavy metal poisoning and cardiovascular disease. J Toxicol. 2011;2011:870125.

130. Staessen JA, Bulpitt CJ, Fagard R, Lauwerys RR, Roels H, Thijs L, et al. Hypertension caused by low-level lead exposure: myth or fact? J Cardiovasc Risk. 1994;1(1):87–97.

131. Schwartz J. Lead, blood pressure, and cardiovascular disease in men. Arch Environ Health. 1995;50(1):31–7.

132. Nawrot TS, Thijs L, Den Hond EM, Roels HA, Staessen JA. An epidemiological re-appraisal of the association between blood pressure and blood lead: a meta-analysis. J Hum Hypertens. 2002;16(2):123–31.

133. Schober SE, Mirel LB, Graubard BI, Brody DJ, Flegal KM. Blood lead levels and death from all causes, cardiovascular disease, and cancer: results from the NHANES III mortality study. Environ Health Perspect. 2006;114(10):1538–41.

134. Muntner P, Menke A, DeSalvo KB, Rabito FA, Batuman V. Continued decline in blood lead levels among adults in the United States: the National Health and Nutrition Examination Surveys. Arch Intern Med. 2005;165(18):2155–61.

135. Brauer M, Casadei B, Harrington RA, Kovacs R, Sliwa K, WHF Air Pollution Expert Group. Taking a stand against air pollution-the impact on cardiovascular disease: a joint opinion from the World Heart Federation, American College of Cardiology, American Heart Association, and the European Society of Cardiology. Circulation. 2021;143(14):e800–4.

136. LOGAN WP. Mortality in the London fog incident, 1952. Lancet. 1953;1(6755):336–8.
137. Gauderman WJ, Urman R, Avol E, Berhane K, McConnell R, Rappaport E, et al. Association of improved air quality with lung development in children. N Engl J Med. 2015;372(10):905–13.
138. Su C, Hampel R, Franck U, Wiedensohler A, Cyrys J, Pan X, et al. Assessing responses of cardiovascular mortality to particulate matter air pollution for pre-, during- and post-2008 Olympics periods. Environ Res. 2015;142:112–22.
139. Clancy L, Goodman P, Sinclair H, Dockery DW. Effect of air-pollution control on death rates in Dublin, Ireland: an intervention study. Lancet. 2002;360(9341):1210–4.

What Is New in the Non-pharmacological Approaches to Hypertension Control

8

Kwee Keng Kng and Ashish Anil Sule

Non-pharmacological therapy such as lifestyle modification is recommended to all patients with hypertension or high normal blood pressure as it may be sufficient to delay or prevent the need for drug therapy in patients with Grade I hypertension (SBP of 140–159 mmHg or DBP of 90–99 mmHg). Lifestyle modification can also augment the effects of antihypertensive agents. However, lifestyle changes should never delay the initiation of drug therapy in patients with high cardiovascular risk. A major limitation of lifestyle modification in hypertension control is that there is poor persistence over time [1].

All healthcare professionals have an important role in promoting lifestyle changes. Lifestyle modification that we are discussing include diet, dietary sodium restriction, smoking, alcohol intake, weight loss and physical activity (Table 8.1).

8.1 Diet

The concept that multiple dietary factors affect blood pressure is supported by a significant body of evidence. In non-hypertensive individuals, dietary modifications can lower blood pressure and prevent hypertension. In uncomplicated Grade I hypertension, dietary modifications serve as initial treatment before drug therapy.

The DASH diet is a combination diet rich in fruits, vegetables, legumes, and low-fat dairy products and low in snacks, sweets, meats, and saturated and total fat. The DASH diet comprised of four to five servings of fruit, four to five servings of

K. K. Kng
Ambulatory Care, Pharmacy, Singapore, Singapore
e-mail: kwee_keng_kng@ttsh.com.sg

A. A. Sule (✉)
Internal Medicine, Vascular Medicine and Hypertension, Singapore, Singapore
e-mail: ashish_anil@ttsh.com.sg

© The Author(s), under exclusive license to Springer Nature Switzerland AG 2022
C. V. S. Ram et al. (eds.), *Hypertension and Cardiovascular Disease in Asia*,
Updates in Hypertension and Cardiovascular Protection,
https://doi.org/10.1007/978-3-030-95734-6_8

Table 8.1 Guidelines for non-pharmacological control of hypertension

	ESC/ESH [1]	ACC/AHA [21]	Singapore [22]	China [23]	India [24]	Japan [25][a]	Korea [26]	Malaysia [27]	Thailand [28]
Diet	Increased consumption of vegetables, fresh fruits, fish, nuts, and unsaturated fatty acids (olive oil); low consumption of red meat; and consumption of low-fat dairy products are recommended	A heart-healthy diet, such as the DASH (Dietary Approaches to Stop Hypertension) diet, that facilitates achieving a desirable weight is recommended for adults with elevated BP or hypertension	Increase the consumption of vegetables, fruits, low-fat dairy products, and decrease the intake of saturated and total fats	Reasonable meal, balanced diet	The diet should be rich in whole grains, fruits, vegetables, and low-fat dairy products and avoid saturated fat and cholesterol. This eating plan is known as the Dietary Approaches to Stop Hypertension diet	–	Increased consumption of vegetables, fresh fruits, fish, nuts, and unsaturated fatty acids; low consumption of red meat; and consumption of low-fat dairy products	Encourage diet rich in fruits, vegetables, and dairy products with reduced saturated and total fat	Regular modification for consumption of healthy foods

	ESC/ESH [1]	ACC/AHA [21]	Singapore [22]	China [23]	India [24]	Japan [25][a]	Korea [26]	Malaysia [27]	Thailand [28]
Dietary sodium restriction	Salt restriction to <5 g per day is recommended	Sodium reduction is recommended for adults with elevated BP or hypertension	Advise patient to restrict salt intake to 5–6 g per day	To reduce sodium intake, gradually reduce the daily salt intake to <6 g, and increasing potassium intake	Salt restriction to <5 g per day is recommended	The target of salt intake restriction should be 6 g per day, but caution is needed because excessive salt intake restriction may cause dehydration on massive sweating	Salt restriction to <6 g per day is recommended	Reduce salt intake to <2 g of sodium or <5 g of salt a day (equivalent to 1 teaspoonful of salt)	Limiting the amount salt and sodium in food

(continued)

Table 8.1 (continued)

	ESC/ESH [1]	ACC/AHA [21]	Singapore [22]	China [23]	India [24]	Japan [25][a]	Korea [26]	Malaysia [27]	Thailand [28]
Smoking	Smoking cessation, supportive care, and referral to smoking cessation programs are recommended	–	Advise and offer assistance to all smokers to quit smoking	Do not smoke, completely quit smoking, and avoid passive smoking	Consumption of tobacco in any form is the single most powerful modifiable lifestyle factor for prevention of cardiovascular disease in hypertensives. E-cigarettes, are also harmful and their use needs to be strongly discouraged	Physicians should instruct smokers to quit smoking	Smoking cessation, supportive care, and referral to smoking cessation programs are recommended	Stop smoking to reduce overall cardiovascular risk	Stop smoking

| Alcohol intake | It is recommended to restrict alcohol consumption to: • Less than 14 units per week for men • Less than 8 units per week for women It is recommended to avoid binge drinking | Adult men and women with elevated BP or hypertension who currently consume alcohol should be advised to drink no more than 2 and 1 standard drinks[b] per day, respectively | Moderate alcohol consumption to no more than 2 standard drinks per day for men, and to no more than 1 standard drink per day for women | Do not drink or restrict alcohol | Alcohol consumption should be limited to no more than 2 drinks per day (24 oz beer, 10 oz wine, 3 oz of 80-proof whiskey) for most men and no more than one drink per day for women and lighter weight people | Physicians should instruct patients routinely taking a moderate or larger amount of alcohol to restrict alcohol intake | It is recommended to moderate alcohol consumption to less than 2 drinks per day | Refrain from alcohol intake. Advise patient who insists to continue drinking to consume ≤2 drinks per day | Limiting alcoholic beverages |

(continued)

Table 8.1 (continued)

	ESC/ESH [1]	ACC/AHA [21]	Singapore [22]	China [23]	India [24]	Japan [25][a]	Korea [26]	Malaysia [27]	Thailand [28]
Weight loss	Body weight control is indicated to avoid obesity (BMI >30 kg/m^2 or waist circumference >102 cm in men and >88 cm in women), as is aiming at healthy BMI (about 20–25 kg/m^2) and waist circumference values (<94 cm in men and <80 cm in women) to reduce blood pressure and cardiovascular risk	Weight loss is recommended to reduce BP in adults with elevated BP or hypertension who are overweight or obese	Unless contraindicated, advise patients to reduce weight to a body mass index (BMI) below 23 kg/m^2 and to a waist circumference below 90 cm in men, and below 80 cm in women (for Asians)	To control body weight to make BMI <24, and to make waist circumference <90 cm for male and <85 cm for female	Weight reduction of as little as 4.5 kg has been found to reduce blood pressure in a large proportion of overweight persons with hypertension	In obese patients, the desirable body weight should be targeted, but rapid weight loss may be harmful; therefore, long-term reasonable weight control should be individually promoted	Body weight control to reduce BMI <25 kg/m^2 is recommended for BP reduction	Achieve a weight loss of as little as 1 kg from baseline to reduce blood pressure by 1 mmHg SBP	Weight reduction in overweight and obese individuals

Physical activity	Regular aerobic exercise (e.g., at least 30 min of moderate dynamic exercise on 5–7 days per week) is recommended	Increased physical activity with a structured exercise program is recommended for adults with elevated BP or hypertension	Advise patients to do at least 30 min of moderate dynamic exercise 5–7 days per week. Any physical exercise above the basal level, up to 150 min/week, confers incremental cardiovascular and metabolic benefits, including blood pressure reduction	To increase exercise, medium intensity; 4–7 times per week; 30–60 min each time	A program of 30–45 min of brisk walking or swimming 3–4 times a week can lower SBP by 7–8 mmHg	Aerobic exercise is recommended, but we recommend walking at a standard speed, but not fast walking, considering the risk of falling, an increase in the risk of arthropathy, and cardiac load	Regular aerobic exercise (e.g., at least 30 min of moderate dynamic exercise 5–7 days per week) is recommended	Advise patients to perform physical activity (e.g., moderate intensity aerobic exercise of at least 150 min per week)	Increasing regular physical activity and/or aerobic exercise

[a] For Hypertension in older persons

[b] In the United States, 1 "standard" drink contains roughly 14 g of pure alcohol, which is typically found in 12 oz of regular beer (usually about 5% alcohol), 5 oz of wine (usually about 12% alcohol), and 1.5 oz of distilled spirits (usually about 40% alcohol)

~ 1 teaspoon of salt = 5.69 g salt = about 100 mmol sodium

vegetables, two to three servings of low-fat dairy per day, and <25% fat. It was observed that the DASH diet reduced the blood pressure by 5.5/3.0 mmHg more. In hypertensive patients, the DASH diet reduced the blood pressure by 11.4/5.5 mmHg more. In conclusion, the DASH diet can substantially lower blood pressure and offers a nutritional approach to preventing and treating hypertension [2].

ESC/ESH recommends a healthy balanced diet containing vegetables, legumes, fresh fruits, low-fat dairy products, wholegrains, fish, and unsaturated fatty acids (especially olive oil), and to have a low consumption of red meat and saturated fatty acids [1].

8.2   Dietary Sodium Restriction

There is a direct relationship between sodium intake and elevated blood pressure [2, 3]. Dietary modifications are established to lower blood pressure and one of the ways is to reduce salt intake [2]. Sodium is commonly consumed as sodium chloride (table salt). It is uncommon not to find salt in food. Though population-wide sodium reduction is a means to prevent cardiovascular disease and stroke [4], it is challenging to reduce and limit salt intake.

As sodium intake is reduced, blood pressure is also reduced. A few studies also showed that these effects in Asian populations were greater. A Cochrane review has shown that a reduction of sodium intake from a high average of 201 mmol/day (11.6 g of salt) to an average level of 66 mmol/day (3.8 g of salt), resulted in a decrease in blood pressure of 7.8/2.7 mmHg in Asian people with hypertension [3]. In those hypertensive patients already on drug therapy, dietary sodium restriction can further lower blood pressure [2]. Hence, dietary sodium restriction is recommended in major international and national guidelines as a component of nonpharmacologic therapy of hypertension.

ESC/ESH recommends sodium intake to be limited to approximately 2.0 g per day (equivalent to approximately 5.0 g salt per day) in the general population and to try to achieve this goal in all hypertensive patients [1].

8.3   Smoking

Cigarette smoking, including second-hand smoking, [5] increases blood pressure [6]. In a study on normotensive smokers, the average systolic blood pressure increased by approximately 20 mmHg after the first cigarette. Blood pressure began to fall 10–15 min after smoking was stopped. However, if smoking was continued, blood pressure remained elevated [7].

Smoking cessation is the single most effective lifestyle modification for the prevention of a large number of cardiovascular diseases [6]. Smoking cessation can be achieved by non-pharmacological methods such as cold turkey and gradual reduction method. Pharmacological agents approved for smoking cessation include nicotine replacement therapy, bupropion and varenicline.

Bupropion was effective for smoking cessation in 248 Chinese patients treated in the outpatient setting. At the end of 12 weeks, the abstinence rate for Bupropion and that of placebo was 39.8% and 8.0%, respectively (both $p < 0.001$) [8].

In a narrative review, it was concluded that varenicline was an effective medication that could assist smoking cessation in the Asians. The results in Asians (Japan, Korea, and Taiwan) were about the same as those for western populations [9]. A study involving 618 Japanese smokers reported a complete abstinence rate of 65.4% for varenicline at weeks 9–12, which was comparable to the 44% abstinence rate reported in the US studies [10].

8.4   Alcohol Intake

Study has shown an association between excess alcohol intake and the development of hypertension [11].

In a Cochrane review of alcohol consumption in healthy persons, alcohol was reported to have a biphasic effect on blood pressure after consumption. Both medium- (14–28 g) and high-dose alcohol (greater than 30 g) decreased blood pressure within 6 h of consumption while high-dose alcohol increased in blood pressure after 12 h [12].

In a systematic review and meta-analysis, it was found that a reduction in alcohol intake was associated with increased blood pressure reduction in people who drank more than two drinks per day. On the other hand, a reduction in blood pressure was not seen with a reduction in alcohol in people who drank 2 or fewer drinks per day. In people who drank 6 or more drinks per day, reduction in systolic blood pressure of mean difference −5.50 mmHg (95% CI −6.70 to −4.30) and reduction in diastolic blood pressure of mean difference −3.97 mmHg, 95% CI −4.70 to −3.25) was significant if they reduced their intake by about 50% [13].

Conversely, in another meta-analysis, moderate alcohol intake seems to have cardioprotective effect where people who consumed 8–10 g of alcohol per day had a decreased risk for all-cause mortality (RR 0.82, 95% CI 0.76–0.88) [14].

ESC/ESH recommends hypertensive men who drink alcohol should be advised to limit their consumption to 14 units per week and women to 8 units per week (1 unit is equal to 125 mL of wine or 250 mL of beer) [1].

8.5   Weight Loss

Excess weight usually raises blood pressure while weight loss usually lowers blood pressure [2]. Every 1 kg weight loss can lead to blood pressure lowering by 0.5–2 mmHg [15]. Weight loss diets reduced body weight and blood pressure [16].

ESC/ESH recommends maintenance of a healthy body weight (BMI of approximately 20–25 kg/m^2 in people <60 years of age; higher in older patients) and waist circumference (<94 cm for men and <80 cm for women) is recommended for non-hypertensive individuals to prevent hypertension, and for hypertensive patients to reduce blood pressure [1].

8.6 Physical Activity

Regular exercise is recommended for lowering blood pressure [17]. There is an inverse relationship between higher exercise dose with a lower incidence of hypertension incidence [18] and a lower rate of mortality [19]. In a meta-analysis, aerobic exercise was found to significantly lower both resting systolic/diastolic blood pressure and daytime ambulatory blood pressure by 3.0/2.4 mmHg and 3.3/3.5 mmHg, respectively [20]. There is no one exercise prescription that is suitable for all adults and the prescription should be individualized.

ESC/ESH recommends hypertensive patients should be advised to participate in at least 30 min of moderate intensity dynamic aerobic exercise (walking, jogging, cycling, or swimming) on 5–7 days per week [1].

References

1. Williams B, Mancia G, Spiering W, Agabiti Rosei E, Azizi M, Burnier M, et al. 2018 ESC/ESH guidelines for the management of arterial hypertension. Eur Heart J. 2018;39(33):3021–104.
2. Appel LJ, Brands MW, Daniels SR, Karanja N, Elmer PJ, Sacks FM. Dietary approaches to prevent and treat hypertension. Hypertension. 2006;47(2):296–308.
3. Graudal NA, Hubeck-Graudal T, Jurgens G. Effects of low sodium diet versus high sodium diet on blood pressure, renin, aldosterone, catecholamines, cholesterol, and triglyceride. Cochrane Database Syst Rev. 2017;4:CD004022.
4. Whelton PK, Appel LJ, Sacco RL, Anderson CAM, Antman EM, Campbell N, et al. Sodium, blood pressure, and cardiovascular disease: further evidence supporting the American Heart Association sodium reduction recommendations. Circulation. 2012;126(24):2880–9.
5. Yarlioglues M, Kaya MG, Ardic I, Calapkorur B, Dogdu O, Akpek M, et al. Acute effects of passive smoking on blood pressure and heart rate in healthy females. Blood Press Monit. 2010;15(5):251–6.
6. Virdis A, Giannarelli C, Neves MF, Taddei S, Ghiadoni L. Cigarette smoking and hypertension. Curr Pharm Des. 2010;16(23):2518–25.
7. Groppelli A, Giorgi D, Omboni S, Parati G, Mancia G. Persistent blood pressure increase induced by heavy smoking. J Hypertens. 1992;1(10):495–9.
8. Sheng L, Tang Y, Jiang Z, Yao C, Gao J, Xu G-Z, et al. Sustained-release bupropion for smoking cessation in a Chinese sample: a double-blind, placebo-controlled randomized trial. Nicotine Tob Res. 2013;15(2):320–5.
9. Xiao D, Chu S, Wang C. Smoking cessation in Asians: focus on varenicline. Patient Prefer Adherence. 2015;9:579–84.
10. Gonzales D, Rennard SI, Nides M, Oncken C, Azoulay S, Billing CB, et al. Varenicline, an $\alpha 4 \beta 2$ nicotinic acetylcholine receptor partial agonist, vs sustained-release bupropion and placebo for smoking cessation: a randomized controlled trial. JAMA. 2006;296(1):47–55.
11. Stranges S, Wu T, Dorn JM, Freudenheim JL, Muti P, Farinaro E, et al. Relationship of alcohol drinking pattern to risk of hypertension. Hypertension. 2004;44(6):813–9.
12. Tasnim S, Tang C, Musini VM, Wright JM. Effect of alcohol on blood pressure. Cochrane Database Syst Rev. 2020;7:CD012787.

13. Roerecke M, Kaczorowski J, Tobe SW, Gmel G, Hasan OSM, Rehm J. The effect of a reduction in alcohol consumption on blood pressure: a systematic review and meta-analysis. Lancet Public Health. 2017;2(2):e108–20.
14. Huang C, Zhan J, Liu Y-J, Li D-J, Wang S-Q, He Q-Q. Association between alcohol consumption and risk of cardiovascular disease and all-cause mortality in patients with hypertension: a meta-analysis of prospective cohort studies. Mayo Clin Proc. 2014;89(9):1201–10.
15. Stevens VJ, Corrigan SA, Obarzanek E, Bernauer E, Cook NR, Hebert P, et al. Weight loss intervention in phase 1 of the trials of hypertension prevention. Arch Intern Med. 1993;153(7):849–58.
16. Semlitsch T, Jeitler K, Berghold A, Horvath K, Posch N, Poggenburg S, et al. Long-term effects of weight-reducing diets in people with hypertension. Cochrane Database Syst Rev. 2016;3:CD008274.
17. Brook RD, Appel LJ, Rubenfire M, Ogedegbe G, Bisognano JD, Elliott WJ, et al. Beyond medications and diet: alternative approaches to lowering blood pressure: a scientific statement from the American Heart Association. Hypertension. 2013;61(6):1360–83.
18. Warburton DE, Charlesworth S, Ivey A, Nettlefold L, Bredin SS. A systematic review of the evidence for Canada's Physical Activity Guidelines for Adults. Int J Behav Nutr Phys Act. 2010;7:39.
19. Wen CP, Wai JPM, Tsai MK, Yang YC, Cheng TYD, Lee M-C, et al. Minimum amount of physical activity for reduced mortality and extended life expectancy: a prospective cohort study. Lancet. 2011;378(9798):1244–53.
20. Fagard RH, Cornelissen VA. Effect of exercise on blood pressure control in hypertensive patients. Eur J Cardiovasc Prev Rehabil. 2007;14(1):12–7.
21. Whelton PK, Carey RM, Aronow WS, Casey DE, Collins KJ, Dennison Himmelfarb C, et al. 2017 ACC/AHA/AAPA/ABC/ACPM/AGS/APhA/ASH/ASPC/NMA/PCNA guideline for the prevention, detection, evaluation, and management of high blood pressure in adults. J Am Coll Cardiol. 2018;71(19):e127–248.
22. Clinical practice guidelines. Singapore: Ministry of Health; 2017.
23. Liu L. 2018 Chinese Guidelines for Prevention and Treatment of Hypertension—a report of the Revision Committee of Chinese Guidelines for Prevention and Treatment of Hypertension. J Geriatr Cardiol JGC. 2019;16(3):182–241.
24. Shah SN, Munjal YP, Kamath SA, Wander GS, Mehta N, Mukherjee S, et al. Indian guidelines on hypertension-IV (2019). J Hum Hypertens. 2020;34(11):745–58.
25. Umemura S, Arima H, Arima S, Asayama K, Dohi Y, Hirooka Y, et al. The Japanese Society of hypertension guidelines for the management of hypertension (JSH 2019). Hypertens Res. 2019;42(9):1235–481.
26. Lee H-Y, Shin J, Kim G-H, Park S, Ihm S-H, Kim HC, et al. 2018 Korean Society of Hypertension guidelines for the management of hypertension: part II-diagnosis and treatment of hypertension. Clin Hypertens. 2019;25:20.
27. Ak P. Clinical practice guidelines. 2018;160.
28. ResearchGate [Internet]. [cited 2021 Sep 10]. https://www.researchgate.net/publication/341755994_2019_Thai_Guidelines_on_The_Treatment_of_Hypertension/link/6002e678a6fdccdcb8590cb0/download.

Special Diets in Various Asian Populations and Their Effect on Hypertension and Cardiovascular Disease

9

Neeta Deshpande

9.1 Introduction

Globally, cardiometabolic deaths can predominantly be attributed to dietary habits. The risk of cardiovascular disease (CVD) increases in direct proportion to hypertension across all ranges of blood pressure (BP) [1]. Although there are several risk factors for hypertension, including genetic and behavioural aspects, there is also evidence that diet plays an important role in the increasing prevalence of this disease. It is well known that chronic overconsumption of sodium-rich food, excessive calories, fats, and carbohydrates combined with reduced intake of plant-based food increases the risk of developing hypertension and CVD. From a public health and economic perspective, dietary strategies are viable options to mitigate the risks. To reiterate the importance that diet plays in the entire cardiometabolic conundrum, optimal lifestyle modification can reduce the risk of myocardial infarction by 81–94% [2, 3] compared to just 20–30% with the use of pharmacotherapy [4]. Of course, cultural norms and a population's traditional dietary habits may prove to be a challenge in the implementation of such strategies. Several studies explore the relationship between diet and these conditions. In some studies, the Dietary Approaches to Stop Hypertension (DASH) diet in a variety of versions [5–9], plus reduced salt intake [10–12] has been shown to be effective in reducing BP. Both individual foods and specific dietary patterns of certain geographical areas assume importance when we assess the relationship between hypertension and diet. However, a lot of existing data come from cross-sectional studies. Given that hypertension and CVD are major contributors to morbidity and mortality globally, it is imperative that we find preventive strategies, especially in the form of nutritional interventions, to try and reduce the prevalence of these diseases.

N. Deshpande (✉)
Belgaum Diabetes Centre, Weight Watch Centre, MM Dental College,
Belgaum & USM-KLE International Medical Program, Belgaum, Karnataka, India

9.2   Diet and Hypertension

9.2.1   The Relationship

Several nutrients, including sodium, potassium, fibre, whole grain, magnesium, nuts, protein, calcium, dairy products, and vitamin D, have all been shown to favourably or unfavourably affect BP, some with sufficient evidence and some without. The effects of sodium, potassium, fruits, vegetables, and fibre on BP are well known. However, there are many nutrients for which the relationship is not yet clear. More studies are needed to establish the relationship between nutrients such as magnesium, calcium and nuts, and hypertension.

Vegetarianism: There are convincing data supporting the beneficial effects of plant-based diets in the therapy of hypertension, as recently reviewed [13]. Historically, population studies and cross-sectional studies conducted as far back as 1929–1959 have shown a lower prevalence of hypertension in the indigenous populations of Africa, China, Germany, and Australia [14–16]. These were populations that consumed predominantly plant-based food. Vegetarianism is also practiced as a part of religious choices, and the effects are similar. For example, Seventh Day Adventists (vegetarians) had lower BP than Mormons (omnivores), although their lifestyles in other respects were comparable [17]. Similar results were obtained in another study, and the prevalence of hypertension was 8.5–10% in the omnivores compared with just 1–2% in the vegetarians [18]. A more recent meta-analysis of data from 32 studies showed that vegetarians had significantly lower BP than omnivores [19]. Data from prospective cohort studies have corroborated these findings, and all reported the significant superiority of plant-based versus meat-based diets with respect to BP levels and rates of hypertension [20, 21].

Several hypotheses could explain why plant-based diets are more successful in reducing BP. Predominantly plant-based diets are obviously higher in fibre and lower in calories, which facilitates weight loss [22], something that is known to lower BP. Secondly, many studies have noted an inverse relationship between high potassium intake and BP [23]. It is possible that vegetarians have a higher potassium intake than meat eaters [24], most likely due to the consumption of more fruits and vegetables by vegetarians. Another hypothesis explores the relationship between gut microbiota and hypertension but although there are many theories to explaining this connection, human studies on the subject are scarce.

DASH Diet: All the above studies paved the way for the formulation of the very effective DASH diet. It was used in the first ever randomized controlled trial to assess the effectiveness of a plant-based diet in hypertension, although limited amounts of lean meat were incorporated into this diet. To minimize confounding factors, dietary sodium was kept uniform across all the study groups. The results were very encouraging because the DASH diet significantly reduced systolic BP (SBP) by 5.5 mmHg and mean diastolic BP (DBP) by 3.3 mmHg compared with the control group [5]. Reductions in both SBP and DBP during consumption of the DASH diet were greater in patients with known hypertension than in those with normal BP at baseline. The combination of the DASH diet and sodium restriction had additive effects on BP reduction in a later trial. Perhaps both these trials could have shown even better effects on BP had the nonvegetarian component been eliminated entirely [6].

9.2.2   Role of Diet in Treatment

Even small reductions in BP can significantly reduce CVD risk and events [25]. Regardless of the need for pharmacotherapy, lifestyle modification is an important part of disease management across all stages of hypertension. As noted above, the DASH diet and sodium restriction are two dietary approaches that have gained a lot of attention in the treatment of hypertension [5].

Salt intake: The recent SOTRUE randomized, double-blind feasibility study examined the effects of a low sodium diet on BP in older adults (age >60 years) [10]. Meals in both arms of the study were matched for all nutrients, including potassium and macronutrients, and differed only with respect to sodium. Compared with the typical sodium meal plan, the low sodium meal plan decreased SBP by 4.8 mmHg, but this was not statistically significant.

A recent meta-analysis of data from 185 trials investigated the effect of a low sodium diet on BP, hormones, and lipids in both normotensive and hypertensive individuals [26]. Most of the included studies were conducted in White participants and therefore the results were stratified by race. Although some studies provided only weak evidence, in general, the BP-lowering effects of a low sodium diet were more pronounced in Black populations and Asians than in the White population. However, this effect was more pronounced in patients with hypertension compared with normal BP, suggesting that low sodium diets may not prevent hypertension but could be an effective treatment. Importantly, the meta-analysis showed consistent potential harmful effects of a low sodium diet on hormones such as renin, aldosterone, noradrenaline, adrenaline and on lipids, especially in normotensive individuals.

Overall, data show that salt restriction is an important component of the non-pharmacological treatment of hypertension. Physicians need to be aware of the salt content of common diets. The sodium content of foods (per 100 g) in common Indian diets is shown in Table 9.1.

Combined DASH diet and salt restriction: In 2001, the DASH Collaborative Research Group published the DASH-Sodium Study, which investigated the combination of a DASH diet with different dietary sodium levels in patients with or without hypertension [6]. Thus, the two important modalities of lifestyle modification and diet to control BP, namely sodium intake and the DASH diet, were combined in this study. The results very clearly showed that a combination of the DASH diet and low sodium intake had a more profound BP-lowering effect in subjects with and without hypertension. Another important finding was that reducing dietary sodium could have a beneficial effect on BP independently of the DASH diet.

A more recent study examined the effects of the low sodium and DASH diets alone and in combination based on baseline BP [7]. Although reducing salt alone was beneficial across BP categories, there was a dramatic reduction in BP when the

Table 9.1 Sodium content of foods per 100 g—common Indian diets

<25 mg Low	25–50 mg Moderate	50–100 mg Moderately high	>100 mg High
• Amla	• Raisins	• Cauliflower	• Amaranth
• Bitter gourd (Karela)	• Carrots	• Fenugreek (Methi)	(Rajgira)
• Bottle gourd (Laukee)	• Black gram (Urad) dal	• Lettuce (Salad Patta)	• Bacon
• Brinjal (Baingan)	• Green gram (Moth) dal	• Field beans Beetroot	• Egg
• Cabbage	• Bengal (Chana) gram	• Watermelon	• Lobster
• Lady finger (Bhindi)	• Banana	• Bengal gram dal	
• Cucumber	• Pineapple	• Tender liver	
• Peas, onion, potato	• Apple	• Prawns	
• Tomato ripe	• Mutton	• Beef	
• Milk		• Chicken	
• Wheat			

two dietary approaches were combined, especially in subjects with a baseline SBP ≥150 mmHg (−20.8 mmHg).

High fat DASH diet: More recent thinking has been that high fat diets do not have as much of a negative effect on metabolism as was thought at the time the DASH diet was conceived. Therefore, a recent study used a modification of the DASH diet that replaced low-fat components with high fat foods while all other components of the DASH diet were retained [27]. The results showed that the high fat DASH diet reduced BP to the same extent as the traditional DASH diet, and reduced plasma triglyceride and very low-density lipoprotein (VLDL) levels and not increasing low-density lipoprotein (LDL) cholesterol. The results of this study are particularly significant because they allow variety in the diet because there is variation in the macronutrient composition, allowing for some substitution of carbohydrates for fat. This could facilitate better adherence to the diet because it gives the individual more options.

Mediterranean diet: Another healthy diet that has possible beneficial effects on BP is the Mediterranean diet. This describes the staple foods of countries that border the Mediterranean Sea, such as Greece and Italy. The diet itself is mostly plant-based, much like the DASH diet. The main difference is a much higher fat content, mostly sourced from olive oil. In addition, seafood, dairy, and poultry are consumed in moderation. There are several randomized clinical trials on the subject [28–30], but with varied methodology.

9.2.3 Asian Perspective

It is well known that Asian communities living in the UK and other Western nations have a higher prevalence of hypertension and CVD than their Caucasian counterparts [31–33]. Unlike a few decades ago, the prevalence of hypertension has now increased in the countries of their origin, perhaps attributable to the increasing adoption of Western lifestyles and dietary patterns, which increases the prevalence of obesity and its consequences, including hypertension and CVD [34]. All published guidelines for treatment of hypertension emphasize the role of dietary therapy [34–36]. However, the implications and implementation of this strategy need to be examined for Asian populations. Added to the environmental insult, undesirable dietary patterns are harmful for a thrifty genotype. Another racial difference seen is renin suppression that has been seen in Japanese and Chinese individuals with hypertension [37]. All of this is conducive to an explosion of metabolic disorders in the Asian population. The various aspects of diet that influence hypertension from an Asian perspective are described below.

DASH diet: As one of the two well-known diets that have been well studied and known to impact hypertension favourably, the DASH diet appears to be closer to Asian cultural norms in that it advocates the consumption of fruits, vegetables, low-fat dairy, and reduced amounts of fat and cholesterol. On the other hand, not all Asian communities would be able to adopt the Mediterranean diet in entirety because it includes the use of olive oil with moderate intake of fish and wine in addition to the plant-based food that the DASH diet is based on. A study conducted in Japan showed that a modified version of the DASH diet improved cardiovascular risk factors when followed for 2 months [8].

Salt: In general, the sodium content of Asian cuisines is higher than recommended by guidelines. For example, Chinese cuisine often uses monosodium glutamate, which makes a significant contribution to sodium intake. In Asian population reducing salt intake to acceptable limits has been shown to have a beneficial effect on BP [11]. Although there is some evidence to show that potassium supplementation can help reduce BP [38, 39], none of the Asian hypertension guidelines specifically address the issue. Therefore, consumption of potassium-rich foods needs to be encouraged. Of course, the DASH and DASH-like diets that have been advocated by Asian guidelines do provide the requisite amount of dietary potassium.

A cross-sectional study conducted in China found that dietary intakes in middle-aged adults fell into four distinct patterns: traditional Chinese; Western style; animal food; and high-salt diets [40]. The results showed that those who consumed the animal food and high-salt dietary patterns were at higher risk of developing hypertension compared with those with the other patterns of dietary intake. Data from a South Indian population ($n = 8080$) showed that sodium intake was higher than that advocated by guidelines and was an independent predictor of higher SBP [41]. Salt intake was higher in men than in women, with a correspondingly higher prevalence of hypertension in men versus women.

Modified versions of these two dietary patterns (vegetable-based diets and salt-restricted diets) have been advocated by guidelines from China, Japan, Korea, Taiwan, and India [34, 42–45]. These guidelines recommend that salt consumption should not exceed 5–6 g/day. However, they do not specify the potassium intake. The Indian guidelines (2019) specify alcohol consumption: no more than two drinks per day for most men and no more than one drink a day for women [34].

Whole grains: The nutritional benefits of whole grains are well known, especially in terms of preventing metabolic diseases including hypertension. However, the consumption of whole grains is limited mostly to the affluent classes in the Asian region. A 3-year survey of Japanese adults showed that the chances of developing hypertension were considerably lower in those who consumed whole grain foods versus those who did not [46]. A population-based study performed in India showed that, among other variables, high dietary fat and low fibre intake were significant determinants of hypertension, as were urban location, obesity, and truncal obesity [47].

Monounsaturated fatty acids (MUFA): In another Asian study, 1529 Korean subjects without hypertension were followed for 4 years to investigate the effects of consuming MUFA and their metabolites on the incidence of new-onset hypertension. The results showed that MUFA and its metabolites had a protective effect against onset of hypertension in this population [48].

Obesity pattern: Higher BMI has been linked to higher BP [49]. The correlation is even stronger for visceral fat, which is reflected by waist circumference [50]. The Asian Indian phenotype with increased body fat is now a well-accepted entity. Managing and treating obesity optimally is an important component of dietary strategies to treat hypertension, especially in Asian populations.

9.3   Diet and Cardiovascular Disease (CVD)

9.3.1   The Relationship

There is a good body of evidence showing a correlation between healthy eating and a lower risk of CVD in Western populations. However, such evidence is quite scarce in Asian populations. It appears that Westernized diets high in sodium and low in potassium could cause oxidative stress, resulting in damage to the vascular endothelium. The underlying mechanism could be reduced availability of nitric oxide leading to defective relaxation of the vascular smooth muscle [51]. Westernized diets that are predominantly meat based and low in fruits and vegetables, as well as potassium, also lack phytochemicals, carotenoids, and other minerals present in plant-based foods. These substances are natural antioxidants that help to reduce oxidative stress. Indeed, the phytochemical content of the DASH diet was found it to be significantly higher than in the control group [9].

Another diet that has been shown to reduce cardiovascular outcomes is the Mediterranean diet. The ATTICA study conducted in Greece is worth mentioning in this regard [52]. It included 3042 individuals without CVD who were followed up for 10 years, at which time 2583 individuals could be assessed. Independent of any other factors, intake of the Mediterranean diet conferred considerable protection from CVD that was evident even in the presence of other risk factors, such as obesity and smoking. Reductions were also seen in levels of C-reactive protein and interleukin-6. Another analysis of the same study showed that a Mediterranean diet reduced cardiovascular risk by 29.3%, irrespective of statin use [53]. However, cardio protection was not seen in a subgroup of this population who followed a

DASH-like diet [54]. In a systematic review, the Mediterranean diet also protected against stroke [55].

Preliminary studies suggest that the gut microbiome could be altered by the consumption of soluble fibre to have a favourable effect on cardiovascular risk factors [56].

9.3.2 Role of Diet in Treatment

The DASH diet has been one of the mainstays of dietary treatment of CVD. The Nurse's Health Study showed that a DASH-type diet could reduce the risk of atherosclerotic heart disease and stroke [57]. One variation of the HF-DASH diet mentioned above replaces 10% of carbohydrates in the traditional low-fat DASH diet with unsaturated fats [58]. This was shown to lower triglycerides and improve HDL cholesterol, without any effect on LDL cholesterol, thereby improving the Framingham Risk Score. Given that adherence to very low-fat diets may be poor [59], this modification of the DASH diet may be beneficial in improving compliance while retaining the beneficial effects on cardiovascular risk.

Endothelial health is vital to the prevention of CVD and attenuation of CVD progression, and this is an area where diet can play a part. A meta-analysis of data from 1930 patients showed that the Mediterranean diet had a beneficial effect on endothelial health [28]. One measure of endothelial health, flow-mediated dilatation (FMD), was improved by 1.66%. This suggests that a Mediterranean diet could contribute to the early prevention of atherosclerotic vascular disease.

The CORDIOPREV study studied the comparative effects of a Mediterranean diet rich in olive oil and a low-fat diet on endothelial dysfunction and included some patients with diabetes [60]. Compared with baseline, FMD was significantly improved after 1.5 years in patients with type 2 diabetes or prediabetes who consumed a Mediterranean diet. Additionally, the FMD was significantly better in the Mediterranean versus low-fat diet group. This has implications for slowing the progression of CVD in patients with type 2 diabetes.

9.3.3 Asian Perspective

There is a comparative lack of studies that have evaluated the correlation between eating patterns and CVD in Asian populations. A recent cross-sectional study used food frequency questionnaires to explore the correlation between healthy eating patterns and cardiovascular risk factors in Chinese, Malay, and Indian individuals living in Singapore [61]. In all groups, healthy dietary patterns seemed most similar to the DASH diet and were associated with the healthiest individuals in terms of cardiovascular risk factors (i.e., BMI and serum lipid levels).

Many South Asians reside in the USA. A study of these individuals found that those who consumed meat, fried snacks and high fat food had a worse cardiovascular risk profile than those who consumed a DASH-like diet [62]. The latter group also had a lower prevalence of hypertension and other metabolic risk factors.

Vegetarianism: It has always been thought that vegetarianism protects against the development of cardiometabolic disorders. A vegetarian diet appears to be quite similar to the DASH diet in terms of content. However, South Asia, which has a large population that practices vegetarianism, has recently seen a significant rise in the prevalence of cardiometabolic disorders. It is possible that the vegetarianism practised in South Asia is not optimal in terms of the food groups consumed. Indeed, differences have been found between food groups consumed by vegetarian adults in South Asia and the US [63]. The number of practicing vegetarians in South Asia was much higher than in the US (33% vs 2.4%). Common food groups in the two vegetarian populations were legumes, fruit, and vegetables. However, consumption of desserts and fried foods was much higher in vegetarians from South Asian versus the US, while the consumption of refined cereals, juices, and sodas was much lower in vegetarians from the US versus South India. These differences resulted in American vegetarians being considerably less overweight/obese and less likely to have central obesity compared with their nonvegetarian counterparts. In contrast, South Asian vegetarians were only slightly less overweight/obese than nonvegetarian individuals from the same region. These data suggest that being vegetarian alone may not ensure a better cardiometabolic risk profile, and that it is actually the food groups consumed by vegetarians that are most important.

Indian diets: Indian diets are heterogeneous in terms of cuisines across different states and have the common issue of being high in carbohydrates. There is recent data which shows a strong association between high carbohydrate intake and the incidence of cardiometabolic diseases in India [64]. India's consumption of sugar is huge and is the largest in the world [65]. In most states, carbohydrates can contribute as much as 70% of the total daily calorie intake. This increases the risk of type 2 diabetes and other cardiovascular risk factors. Also, Asian Indians are ethnically prone to have insulin resistance [66]. This results in post-prandial spikes in blood sugar, even in non-diabetic individuals, which are responsible for an increased burden of oxidative stress. This is known to adversely affect the cardiovascular risk profile [67].

9.4 Conclusions

Dietary strategies to alleviate cardiometabolic diseases and hypertension, and their risk factors, are an important adjunct to pharmacotherapy. Dietary strategies are cost-effective from a public health perspective, especially in developing countries, such as many of those in Asia. In general, a diet that is mainly plant-based, such as the DASH diet and its modifications, along with a salt intake that does not exceed 5 g/day must be implemented to reduce the burden of hypertension in Asian countries. Diets rich in potassium should also be advocated. Many traditional Asian diets fulfil these criteria and must be actively propagated as lifestyle interventions for the prevention and management of hypertension. Many mechanisms explain the connection, but more studies are required to elucidate the association. Plant-based diets not only reduce BP, but also oxidative stress that impairs endothelial function, thereby reducing the intensity of major cardiovascular risk factors. More population-based studies are necessary in Asia to determine the dietary patterns and food groups consumed. Additionally, the focus should also be on attaining optimal BMI to achieve BP targets.

References

1. Lewington S, Clarke R, Qizilbash N, Peto R, Collins R, Prospective Studies Collaboration. Age-specific relevance of usual blood pressure to vascular mortality: a meta-analysis of individual data for one million adults in 61 prospective studies. Lancet. 2002;360(9349):1903–13.
2. Yusuf S, Hawken S, Ounpuu S, Dans T, Avezum A, Lanas F, et al. INTERHEART Study Investigators. Effect of potentially modifiable risk factors associated with myocardial infarction in 52 countries (the INTERHEART study): case-control study. Lancet. 2004;364(9438):937–52.
3. Ford ES, Bergmann MM, Kröger J, Schienkiewitz A, Weikert C, Boeing H. Healthy living is the best revenge: findings from the European Prospective Investigation Into Cancer and Nutrition-Potsdam study. Arch Intern Med. 2009;169(15):1355–62.
4. Chiuve SE, McCullough ML, Sacks FM, Rimm EB. Healthy lifestyle factors in the primary prevention of coronary heart disease among men: benefits among users and nonusers of lipid-lowering and antihypertensive medications. Circulation. 2006;114(2):160–7.
5. Appel LJ, Moore TJ, Obarzanek E, Vollmer WM, Svetkey LP, Sacks FM, et al. A clinical trial of the effects of dietary patterns on blood pressure. DASH Collaborative Research Group. N Engl J Med. 1997;336(16):1117–24.
6. Sacks FM, Svetkey LP, Vollmer WM, Appel LJ, Bray GA, Harsha D, et al. Effects on blood pressure of reduced dietary sodium and the Dietary Approaches to Stop Hypertension (DASH) diet. DASH-Sodium Collaborative Research Group. N Engl J Med. 2001;344(1):3–10.
7. Juraschek SP, Miller ER 3rd, Weaver CM, Appel LJ. Effects of sodium reduction and the DASH diet in relation to baseline blood pressure. J Am Coll Cardiol. 2017;70(23):2841–8.
8. Kawamura A, Kajiya K, Kishi H, Inagaki J, Mitarai M, Oda H, et al. Effects of the DASH-JUMP dietary intervention in Japanese participants with high-normal blood pressure and stage 1 hypertension: an open-label single-arm trial. Hypertens Res. 2016;39:777–85.
9. Most MM. Estimated phytochemical content of the dietary approaches to stop hypertension (DASH) diet is higher than in the Control Study Diet. J Am Diet Assoc. 2004;104(11):1725–7.
10. Juraschek SP, Millar CL, Foley A, Shtivelman M, Cohen A, McNally V, et al. The effects of a low sodium meal plan on blood pressure in older adults: the SOTRUE randomized feasibility trial. Nutrients. 2021;13(3):964.
11. de Brito-Ashurst I, Perry L, Sanders TA, Thomas JE, Dobbie H, Varagunam M, et al. The role of salt intake and salt sensitivity in the management of hypertension in South Asian people with chronic kidney disease: a randomised controlled trial. Heart. 2013;99:1256–60.
12. Graudal NA, Hubeck-Graudal T, Jürgens G. Effects of low-sodium diet vs. high-sodium diet on blood pressure, renin, aldosterone, catecholamines, cholesterol, and triglyceride (Cochrane Review). Am J Hypertens. 2012;25(1):1–15.
13. Joshi S, Ettinger L, Liebman SE. Plant-based diets and hypertension. Am J Lifestyle Med. 2019;14(4):397–405.
14. Donnison CP. Blood pressure in the African native. Its bearing upon the aetiology of hyperpiesia and arterio-sclerosis. Lancet. 1929;1:6–7.
15. Morse WR, Beh YT. Blood pressure amongst aboriginal ethnic groups of Szechwan Province, West China. Lancet. 1937;229(5929):966–8.

16. Casley-Smith JR. Blood pressures in Australian aborigines. Med J Aust. 1959;49:627–33.
17. Anholm AC. Relationship of a vegetarian diet to blood-pressure. Prev Med. 1978;7:35–55.
18. Rouse IL, Armstrong BK, Beilin LJ. The relationship of blood pressure to diet and lifestyle in two religious populations. J Hypertens. 1983;1(1):65–71.
19. Yokoyama Y, Nishimura K, Barnard ND, Takegami M, Watanabe M, Sekikawa A, et al. Vegetarian diets and blood pressure: a meta-analysis. JAMA Intern Med. 2014;174(4):577–87.
20. Steffen LM, Kroenke CH, Yu X, Pereira MA, Slattery ML, Van Horn L, et al. Associations of plant food, dairy product, and meat intakes with 15-y incidence of elevated blood pressure in young black and white adults: the Coronary Artery Risk Development in Young Adults (CARDIA) Study. Am J Clin Nutr. 2005;82(6):1169–77.
21. Borgi L, Curhan GC, Willett WC, Hu FB, Satija A, Forman JP. Long-term intake of animal flesh and risk of developing hypertension in three prospective cohort studies. J Hypertens. 2015;33(11):2231–8.
22. Barnard ND, Kahleova H, Levin SM. The use of plant-based diets for obesity treatment. Int J Dis Reversal Prev. 2019;1(1):12.
23. Aburto NJ, Hanson S, Gutierrez H, Hooper L, Elliott P, Cappuccio FP. Effect of increased potassium intake on cardiovascular risk factors and disease: systematic review and meta-analyses. BMJ. 2013;346:f1378.
24. Li D, Sinclair AJ, Mann NJ, Turner A, Ball MJ. Selected micronutrient intake and status in men with differing meat intakes, vegetarians and vegans. Asia Pac J Clin Nutr. 2000;9(1):18–23.
25. Cook NR, Cohen J, Hebert PR, Taylor JO, Hennekens CH. Implications of small reductions in diastolic blood pressure for primary prevention. Arch Intern Med. 1995;155(7):701–9.
26. Graudal NA, Hubeck-Graudal T, Jurgens G. Effects of low sodium diet versus high sodium diet on blood pressure, renin, aldosterone, catecholamines, cholesterol, and triglyceride. Cochrane Database Syst Rev. 2020;12(12):CD004022.
27. Chiu S, Bergeron N, Williams PT, Bray GA, Sutherland B, Krauss RM. Comparison of the DASH (Dietary Approaches to Stop Hypertension) diet and a higher-fat DASH diet on blood pressure and lipids and lipoproteins: a randomized controlled trial. Am J Clin Nutr. 2016;103(2):341–7.
28. Shannon OM, Mendes I, Köchl C, Mazidi M, Ashor AW, Rubele S, et al. Mediterranean diet increases endothelial function in adults: a systematic review and meta-analysis of randomized controlled trials. J Nutr. 2020;150(5):1151–9.
29. Estruch R, Ros E, Salas-Salvadó J, Covas M-I, Corella D, Arós F, et al. Primary prevention of cardiovascular disease with a Mediterranean diet supplemented with extra-virgin olive oil or nuts. N Engl J Med. 2018;378:e34.
30. Davis CR, Hodgson JM, Woodman R, Bryan J, Wilson C, Murphy KJ. A Mediterranean diet lowers blood pressure and improves endothelial function: results from the MedLey randomized intervention trial. Am J Clin Nutr. 2017;105:1305–13.
31. Enas EA, Garg A, Davidson MA, Nair VM, Huet BA, Yusuf S. Coronary heart disease and its risk factors in first-generation immigrant Asian Indians to the United States of America. Indian Heart J. 1996;48(4):343–53.
32. Hughes LO, Raval U, Raftery EB. First myocardial infarctions in Asian and white men. BMJ. 1989;298(6684):1345–50.
33. Palaniappan L, Wang Y, Fortmann SP. Coronary heart disease mortality for six ethnic groups in California, 1990-2000. Ann Epidemiol. 2004;14(7):499–506.
34. Shah SN, Core Committee Members: Billimoria AB, Mukherjee S, Kamath S, Munjal YP, Maiya M, Wander GS, Mehta N. Indian guidelines on management of hypertension (I.G.H) – IV 2019. Suppl J Assoc Phys India (JAPI). 2019;67(9):1–48.
35. Whelton PK, Carey RM, Aronow WS, et al. 2017ACC/AHA/AAPA/ABC/ACPM/AGS/APhA/vASH/ASPC/NMA/PCNA guideline for the prevention, detection, evaluation, and management of high blood pressure in adults. Hypertension. 2018;71:e13–e115.
36. Williams B, Mancia G, Spiering W, et al. 2018 ESC/ESH guidelines for the management of arterial hypertension. The Task Force for the management of arterial hypertension of the

European Society of Cardiology (ESC) and the European Society of Hypertension (ESH). Eur Heart J. 2018;39:3021–104.

37. Te Riet L, van Esch JH, Roks AJ, van den Meiracker AH, Danser AH. Hypertension: renin-angiotensin-aldosterone system alterations. Circ Res. 2015;116(6):960–75.

38. Kawano Y, Minami J, Takishita S, Omae T. Effects of potassium supplementation on office, home, and 24-h blood pressure in patients with essential hypertension. Am J Hypertens. 1998;11:1141–6.

39. Gu D, He J, Wu X, Duan X, Whelton PK. Effect of potassium supplementation on blood pressure in Chinese: a randomized, placebo-controlled trial. J Hypertens. 2001;19:1325–31.

40. Zheng PF, Shu L, Zhang XY, Si CJ, Yu XL, Gao W, et al. Association between dietary patterns and the risk of hypertension among Chinese: a cross-sectional study. Nutrients. 2016;8(4):239.

41. Ravi S, Bermudez OI, Harivanzan V, Kenneth Chui KH, Vasudevan P, Must A, et al. Sodium intake, blood pressure, and dietary sources of sodium in an adult south Indian population. Ann Glob Health. 2016;82(2):234–42.

42. Liu J. Highlights of the 2018 Chinese hypertension guidelines. Clin Hypertens. 2020;26:8. https://doi.org/10.1186/s40885-020-00141-3.

43. Umemura S, Arima H, Arima S, Asayama K, Dohi Y, Hirooka Y, et al. The Japanese Society of Hypertension guidelines for the management of hypertension (JSH 2019). Hypertens Res. 2019;42(9):1235–481.

44. Lee HY, Shin J, Kim GH, Park S, Ihm SH, Kim HC, et al. 2018 Korean Society of Hypertension guidelines for the management of hypertension: part II-diagnosis and treatment of hypertension. Clin Hypertens. 2019;25:20.

45. Lin HJ, Wang TD, Yu-Chih Chen M, Hsu CY, Wang KL, Huang CC, et al. 2020 Consensus Statement of the Taiwan Hypertension Society and the Taiwan Society of Cardiology on home blood pressure monitoring for the management of arterial hypertension. Acta Cardiol Sin. 2020;36(6):537–61.

46. Kashino I, Eguchi M, Miki T, Kochi T, Nanri A, Kabe I, et al. Prospective association between whole grain consumption and hypertension: the Furukawa Nutrition and Health Study. Nutrients. 2020;12(4):902.

47. Gupta R, Pandey RM, Misra A, Agrawal A, Misra P, Dey S, et al. High prevalence and low awareness, treatment and control of hypertension in Asian Indian women. J Hum Hypertens. 2012;26(10):585–93.

48. Lee H, Jang HB, Yoo MG, Chung KS, Lee HJ. Protective effects of dietary MUFAs mediating metabolites against hypertension risk in the Korean Genome and Epidemiology Study. Nutrients. 2019;11(8):1928.

49. Hubert HB, Feinleib M, McNamara PM, Castelli WP. Obesity as an independent risk factor for cardiovascular disease: a 26-year follow-up of participants in the Framingham Heart Study. Circulation. 1983;67:968–77.

50. Chuang SY, Chou P, Hsu PF, Cheng HM, Tsai ST, Lin IF, et al. Presence and progression of abdominal obesity are predictors of future high blood pressure and hypertension. Am J Hypertens. 2006;19(8):788–95.

51. Ávila-Escalante ML, Coop-Gamas F, Cervantes-Rodríguez M, Méndez-Iturbide D, Aranda-González II. The effect of diet on oxidative stress and metabolic diseases-clinically controlled trials. J Food Biochem. 2020;44(5):e13191.

52. Panagiotakos DB, Georgousopoulou EN, Pitsavos C, Chrysohoou C, Skoumas I, Pitaraki E, et al. Exploring the path of Mediterranean diet on 10-year incidence of cardiovascular disease: the ATTICA study (2002-2012). Nutr Metab Cardiovasc Dis. 2015;25(3):327–35.

53. Panagiotakos DB, Georgousopoulou EN, Georgiopoulos GA, Pitsavos C, Chrysohoou C, Skoumas I, et al. Adherence to Mediterranean diet offers an additive protection over the use of statin therapy: results from the ATTICA study (2002-2012). Curr Vasc Pharmacol. 2015;13(6):778–87.

54. Bathrellou E, Kontogianni MD, Chrysanthopoulou E, Georgousopoulou E, Chrysohoou C, Pitsavos C, et al. Adherence to a DASH-style diet and cardiovascular disease risk: the 10-year follow-up of the ATTICA study. Nutr Health. 2019;25(3):225–30.

55. Saulle R, Lia L, De Giusti M, La Torre G. A systematic overview of the scientific literature on the association between Mediterranean diet and the stroke prevention. Clin Ter. 2019;170(5):e396–408.
56. Wang Y, Ames NP, Tun HM, Tosh SM, Jones PJ, Khafipour E. High molecular weight barley β-glucan alters gut microbiota toward reduced cardiovascular disease risk. Front Microbiol. 2016;7:129.
57. Fung TT, Chiuve SE, McCullough ML, Rexrode KM, Logroscino G, Hu FB. Adherence to a DASH-style diet and risk of coronary heart disease and stroke in women. Arch Intern Med. 2008;168(7):713–20.
58. Appel LJ, Sacks FM, Carey VJ, Obarzanek E, Swain JF, Miller ER 3rd, et al. Effects of protein, monounsaturated fat, and carbohydrate intake on blood pressure and serum lipids: results of the OmniHeart randomized trial. JAMA. 2005;294(19):2455–64.
59. Henkin Y, Garber DW, Osterlund LC, Darnell BE. Saturated fats, cholesterol, and dietary compliance. Arch Intern Med. 1992;152(6):1167–74.
60. Torres-Peña JD, Garcia-Rios A, Delgado-Casado N, Gomez-Luna P, Alcala-Diaz JF, Yubero-Serrano EM, et al. Mediterranean diet improves endothelial function in patients with diabetes and prediabetes: a report from the CORDIOPREV study. Atherosclerosis. 2018;269:50–6.
61. Whitton C, Rebello SA, Lee J, Tai ES, van Dam RM. A healthy Asian a posteriori dietary pattern correlates with a priori dietary patterns and is associated with cardiovascular disease risk factors in a multiethnic Asian population. J Nutr. 2018;148(4):616–23.
62. Gadgil MD, Anderson CA, Kandula NR, Kanaya AM. Dietary patterns are associated with metabolic risk factors in South Asians living in the United States. J Nutr. 2015;145(6):1211–7.
63. Jaacks LM, Kapoor D, Singh K, Narayan KM, Ali MK, Kadir MM, et al. Vegetarianism and cardiometabolic disease risk factors: differences between South Asian and US adults. Nutrition. 2016;32(9):975–84.
64. Dehghan M, Mente A, Zhang X, Swaminathan S, Li W, Mohan V, et al.; Prospective Urban Rural Epidemiology (PURE) study investigators. Associations of fats and carbohydrate intake with cardiovascular disease and mortality in 18 countries from five continents (PURE): a prospective cohort study. Lancet. 2017;390(10107):2050–62.
65. Gulati S, Misra A. Sugar intake, obesity, and diabetes in India. Nutrients. 2014;6(12):5955–74.
66. Bhatnagar D, Anand IS, Durrington PN, Patel DJ, Wander GS, Mackness MI, et al. Coronary risk factors in people from the Indian subcontinent living in West London and their siblings in India. Lancet. 1995;345(8947):405–9.
67. O'Keefe JH, Gheewala NM, O'Keefe JO. Dietary strategies for improving post-prandial glucose, lipids, inflammation, and cardiovascular health. J Am Coll Cardiol. 2008;51(3):249–55.

Heart Rate Variability, Blood Pressure Variability: What Is Their Significance in Hypertension

10

Uday M. Jadhav and Shilpa A. Kadam

10.1 Introduction

Heart rate variability (HRV) represents changes in the RR interval and instantaneous heart rate (HR). The interval between successive beats is analysed. HRV is recognised as an important marker of autonomic activity imbalance, and indicates reduced vagal activity and increased sympathetic activity [1, 2]. Reduced HRV has been shown to be a marker of increased risk after acute myocardial infarction (AMI), and can provide an early indication of diabetic neuropathy [3].

Blood pressure variability (BPV) is also being increasingly recognised as an independent risk factor for target organ damage and cardiovascular events. Variations in BP can occur over the very short term (beat-to-beat), short term (over 24 h) or long term (between visits or seasons). There is increasing evidence showing that both short-term and long-term BPV correlate with target organ damage (TOD) and CV events in patients with hypertension [4]. Increase BPV is also associated with increased microvascular complications in diabetes mellitus (DM) and progression of renal failure in patients with chronic kidney disease (CKD) [5].

10.2 Heart Rate Variability

HRV is a non-invasive clinical tool that can help to detect cardiac autonomic dysregulation in hypertension. HRV is generally assessed by 2-min and 6-min beat-to-beat heart rate recordings [1, 6, 7]. Common HRV measures are summarised in Table 10.1.

U. M. Jadhav (✉) · S. A. Kadam
Cardiology Department, MGM New Bombay Hospital, New Mumbai, India

© The Author(s), under exclusive license to Springer Nature Switzerland AG 2022
C. V. S. Ram et al. (eds.), *Hypertension and Cardiovascular Disease in Asia*,
Updates in Hypertension and Cardiovascular Protection,
https://doi.org/10.1007/978-3-030-95734-6_10

Table 10.1 Commonly used measures of heart rate variability

Variable, units	Description
Statistical measures	
SDNN, ms	SD of all NN intervals
SDANN, ms	SD of the averages of NN intervals in all 5-min segments of the entire recording
RMSSD, ms	The square root of the mean of the sum of the squares of differences between adjacent NN intervals
SDNN index, ms	Mean of the SD of all NN intervals for all 5-min segments of the entire recording
SDSD, ms	SD of differences between adjacent NN intervals
Geometric measures	
HRV triangularindex, ms	Total number of all NN intervals divided by the height of the histogram of all NN intervals measured on a discrete scale
Differential index, ms	Difference between the widths of the histogram of differences between adjacent NN intervals measured at selected heights

NN normal to normal, *SD* standard deviation

Some of these measures are described below:

1. Mean normal-to-normal (NN) RR interval length: the RR interval reflects the sum of parasympathetic and sympathetic influences.
2. Standard deviation of NN RR intervals (SDNN): SDNN reflects total variability.
3. Root mean square of successive differences in NN RR intervals (RMSSD): RMSSD reflects high-frequency variations in heart rate and is a marker for the actions of the parasympathetic nervous system.

10.2.1 Reduced HRV: A Risk Factor

The autonomic background of HRV is well recognised. It reflects decreased vagal and increased sympathetic influences on the heart which cause electrical instability. Reduced HRV is a strong predictor of mortality after AMI [8]. Reduced HRV can also predict the development of diabetic neuropathy, especially the autonomic involvement [9]. Reduced HRV has been observed in heart failure, hypertension, mitral valve prolapse and hypertrophic cardiomyopathy [2]. Reduced parasympathetic and increased sympathetic activity has been shown in hypertension [10].

The autonomic nervous system and alteration of its regulation can potentially contribute to the development of hypertension [10]. The temporal sequence of HRV and blood pressure (BP) rise of interest to determine whether decreased HRV can predict incident hypertension or whether hypertension can alter HRV.

In a hypothesis proposed by Julius et al., mild hypertension is characterised by high heart rate and cardiac output, and normal vascular resistance [11, 12]. Over time, HRV of normotensive individuals and patients with persistent hypertension tends to converge. The heart rate will then decrease, cardiac output normalises, and vascular resistance increases. This is due to the combined effect of increased sympathetic activity and decreased parasympathetic activity, and can be labeled as the

"BP-seeking property" of the central nervous system [11–13]. Resistance to antihypertensive therapy may be associated with activation of the neurohormonal system as indicated by reduced SDANN and RMSSD [14, 15].

HRV may precede clinical hypertension, and low values of HRV indices such as SDNN and RMSSD at very low frequency (VLF), low frequency (LF) and high frequency (HF), have been shown to be related to the incidence of hypertension in normotensive subjects (BP <120/80 mmHg) and those with pre-hypertension (BP 120/80–139/89 mmHg) over 4 years of follow-up [16].

Sympathetic overactivity and parasympathetic withdrawal have been proposed in the development of clinical hypertension. Reduced HRV on 24-h Holter monitoring was significantly associated with hypertension, and was more common in patients with uncontrolled versus controlled BP [17].

10.2.2 Some Evidence

10.2.2.1 International Studies

The Framingham study reported an association between logarithmically transformed low-frequency power among men and no association for SDNN and high-frequency power in either sex [14]. The Atherosclerosis Risk in Communities (ARIC) substudy found that the incidence of hypertension was significantly increased in patients with NN RR intervals in the lowest versus highest quartile (hazard ratio 1.24, 95% confidence interval 1.10–1.40) [18]. For these individuals with normal BP at baseline ($n = 7099$), low heart rate variability predicted greater risk of incident hypertension over 9 years of follow-up. However, the rate of change in HRV over time did not differ between those with versus without hypertension [18]. A study conducted in India that enrolled 30 patients with hypertension and 30 individuals with normal BP showed that SDNN, RMSSD and pNN50 were significantly lower, and the LF-HF ratio was significantly higher those with versus without hypertension [19].

HRV is related to vagal tone in atrial fibrillation (AF), irrespective of hypertension. Individuals with permanent AF have higher HRV than those with paroxysmal AF, probably due to autonomic dysregulation [20]. Interestingly, medications designed to control AF or BP did not improve HRV [20].

Twenty-four-hour recording of SDNN is proposed as the "gold standard" for medical stratification of cardiac risk. SDNN values of <50 ms, 50–100 ms and > 100 ms are classified as unhealthy, compromised health and healthy, respectively [2].

10.2.2.2 Asian Studies

Studies in Asian populations have shown that impaired autonomic nervous function in patients with hypertension is strongly associated with uncontrolled BP [2, 19, 21–24]. The Toon Health Study enrolled 1888 men and women aged 30–79 years, and participants self-monitored BP at home twice in the morning and evening for 1 week [23]. The results showed that the parasympathetic nervous system activity parameters, low HF and RMSSD, were associated with increased home mean arterial pressure (MAP) in the morning rather than in the evening. These associations were independent of sex, age, body mass index, smoking, alcohol consumption, use of antihypertensive agents, diabetes and physical activity. The study also emphasised that physical inactivity, insomnia and socioeconomic stress factors induce sympathovagal imbalance and higher home BP in the morning.

With respect to antihypertensive drug classes, users of beta-blockers have been shown to have equivalent or higher HRV than non-users, while those using diuretics or angiotensin-converting enzyme inhibitors (ACEIs) had a lower HRV compared with non-users [18], and users of captopril had increased HRV [25].

10.3   Blood Pressure Variability

In patients with hypertension, increased BPV contributes to future cardiovascular events [4, 5]. The white coat effect is one indicator of BPV in clinical practice. Day-to-day BPV, visit-to-visit SBP variability and long-term BPV have been shown to be associated with an increased risk of stroke [4], cardiovascular events and mortality [1]. Short-term BPV can be measured by 24-h ambulatory BP monitoring [5] and long-term BPV can be determined based on visit-to-visit assessments [26]. Unlike morning BPV, evening BPV significantly predicted cardiovascular events independent of the corresponding home BP readings [27]. Lower nocturnal SBP, and non-dipper and reverse dipper patterns of nocturnal hypertension were associated with a higher risk of cerebral small vessel disease [28]. A summary of different BPV measures, how they are assessed and relevant influencing factors are summarised in Table 10.2.

Data from the Valsartan Antihypertensive Long-term Use Evaluation trial (VALUE) involving 14,000 hypertensive middle-aged and older subjects reported that there was a 10% increase in the risk of death and a 15% increase in risk of CV events for each 5 mmHg increase in the standard deviation (SD) of visit-to-visit and within-visit systolic BPV, respectively [29].

Rates of fatal and non-fatal cardiovascular events over a >15-year follow-up were significantly higher in the presence of high short-term (24-h) systolic BPV in

Table 10.2 Types of blood pressure variability

Type of BPV	Assessment	Measurement	Influencing factors
Very short term	Beat to beat	Intra-arterial recording with spectral analysis	Baroreceptor and chemoreceptor activity
Short term	Within 24 h	ABPM	↑ sympathetic activity, sleep, activity behavioral factors
Medium term	Day to day	HBPM	Age, increased arterial stiffness, emotional factor
Long term	Visit to visit	Repeat visit to the office	Behavioral change, drug choice and compliance, environmental factor

ABPM ambulatory blood pressure monitoring, *BPV* blood pressure variability, *HBPM* home blood pressure monitoring

1206 young patients with stage 1 hypertension (mean age 33 ± 8 years) [30]. In addition, BPV has been associated with arterial stiffness, left ventricular hypertrophy, decline in renal function, subclinical brain small vessel disease and the risk of developing foot ulcers in diabetes [31–36].

10.3.1 BPV and Antihypertensive Treatment

Short-term BPV is calculated as the standard deviation of 24-h, daytime or night-time systolic BP and diastolic BP. Different antihypertensive drugs classes may have a differential impact on BPV and this is probably more apparent on visit-to-visit BPV [37]. Short-term BPV is decreased by calcium channel blockers (CCBs), diuretics, and their combination [38]. Angiotensin receptor blockers (ARBs), beta-blockers and ACEIs may increase short-term BPV, as shown in a recent meta-analysis [38]. White coat effect was smaller in patients with hypertension treated mainly with the CCB amlodipine compared to those treated mainly with the beta-blocker atenolol. The addition of a CCB or diuretic to any antihypertensive treatment regimen may reduce BPV to the same extent as seen with monotherapy of these drugs. The addition of an ACEI or ARB agent did not have a similar impact [39, 40]. Amlodipine, which has a long elimination half-life of 34 h, has a positive effect on morning BPV [41].

The mechanism(s) underlying changes in BPV with different antihypertensive drug classes is not completely understood. Changes in peripheral vascular distensibility and differential effects on pulse wave velocity may explain some of the differences, such as the reduction in arterial compliance during treatment with beta-blockers and increased vascular compliance during CCB therapy [42, 43].

10.3.1.1 Circadian BPV

Recently, there has been renewed interest in bedtime chronotherapy after the results of Monitorización Ambulatoria para Predicción de Eventos Cardiovasculares (MAPEC) [44]. In this study, ≥1 antihypertensive drug was given at bedtime to provide better control of nighttime BP. After a median follow-up of 5.6 years, participants who had bedtime antihypertensive dosing had a significantly lower relative risk of cardiovascular event than those who took all their antihypertensive therapy in the morning [44]. However, current guidelines do not recommend evening dosing of antihypertensive and additional studies are needed to determine the validity of this approach [45].

10.3.1.2 Non-circadian BPV

Three months of treatment with amlodipine and indapamide sustained release was associated with greater reductions in daytime, nighttime and 24-h systolic BPV in the X-CELLENT (Natrilix SR Versus Candesartan and Amlodipine in the Reduction of Systolic Blood Pressure in Hypertensive Patients) study of 577 middle-aged patients with hypertension [46]. Two large analyses involving more than 4000 patients reported lower daytime BPV in those treated with telmisartan/amlodipine versus telmisartan/hydrochlorothiazide [46–48]. The triple therapy combination of olmesartan plus a dihydropyridine CCB and a thiazide diuretic, and dual combinations of olmesartan plus a dihydropyridine CCB or a dihydropyridine CCB plus a thiazide diuretic were associated with greater decreases in BPV compared with placebo and monotherapies [49].

10.3.1.3   Mid-Term BPV
Greater decreases in day-to-day BPV were seen during treatment with a CCB/ARB combination compared with a diuretic/ARB combination in the Japan Combined Treatment with Olmesartan and a Calcium Channel Blocker Versus Olmesartan and Diuretics Randomized Efficacy Study, despite similar reductions in systolic BP [49].

10.3.1.4   Long-Term BPV
The superiority of CCBs for reducing BPV compared with ARBs, beta-blockers or diuretics was reported in the COLM (Combination of OLMesartan) [50], COPE (Combination Therapy of Hypertension to Prevent Cardiovascular Events) [51] and ASCOT-BPLA studies [52], but not in the ELSA (European Lacidipine Study on Atherosclerosis) trial [53]. Thiazide-like diuretics were shown to be more effective than beta-blockers in reducing long-term BPV in the MRC trial [54].

10.4   Conclusion

Autonomic nervous system dysregulation with reduced vagal effects and increased sympathetic activity plays a potential role in the development of hypertension. Reduced HRV is a marker of this autonomic imbalance and has been shown to be significantly associated with hypertension. Reduced HRV is now recognised as marker of increased mortality in AMI and is a useful marker of development of diabetic neuropathy. As a marker of autonomic imbalance suggesting increased sympathetic activity, it also is increasingly being recognised as an important marker in hypertension, heart failure and mitral valve prolapse.

   Increased short- and long-term BPV has been shown to be a marker of increased TOD and cardiovascular risk in patients with hypertension, in addition to mean BP. The white coat effect and early morning rise in BP are markers of this BPV. Increased visit-to-visit BPV is also a marker of complications and TOD in hypertension. Some antihypertensive drugs, especially long-acting CCBs, can reduce BPV.

References

1. Mejía-Mejía E, Budidha K, Abay TY, May JM, Kyriacou PA. Heart rate variability (HRV) and pulse rate variability (PRV) for the assessment of autonomic responses. Front Physiol. 2020;11:779.

2. Heart rate variability: standards of measurement, physiological interpretation and clinical use. Task Force of the European Society of Cardiology and the North American Society of Pacing and Electrophysiology. Circulation. 1996;93(5):1043–65.
3. Blood Pressure Lowering Treatment Trialists' Collaboration. Pharmacological blood pressure lowering for primary and secondary prevention of cardiovascular disease across different levels of blood pressure: an individual participant-level data meta-analysis. Lancet. 2021;397(10285):1625–36.
4. Höcht C. Blood pressure variability: prognostic value and therapeutic implications. Int Sch Res Notices. 2013;2013:398485. https://doi.org/10.5402/2013/398485.
5. Rosei EA, Chiarini G, Rizzoni D. How important is blood pressure variability? Eur Heart J Suppl. 2020;22(Suppl E):E1–6.
6. Shaffer F, Ginsberg JP. An overview of heart rate variability metrics and norms. Front Public Health. 2017;5:258.
7. Grassi G, Mark A, Esler M. The sympathetic nervous system alterations in human hypertension. Circ Res. 2015;116:976–90.
8. Song T, Qu XF, Zhang YT, Cao W, Han BH, Li Y, Piao JY, Yin LL, Da Cheng H. Usefulness of the heart-rate variability complex for predicting cardiac mortality after acute myocardial infarction. BMC Cardiovasc Disord. 2014;14:59.
9. Chessa M, Butera G, Lanza GA, Bossone E, Delogu A, De Rosa G, et al. Role of heart rate variability in the early diagnosis of diabetic autonomic neuropathy in children. Herz. 2002;27(8):785–90.
10. Mancia G, Grassi G. The autonomic nervous system and hypertension. Circ Res. 2014;114(11):1804–14.
11. Julius S, Nesbitt S. Sympathetic overactivity in hypertension: a moving target. Am J Hypertens. 1996;9:113S–20S.
12. Julius S, Majahalme S. The changing face of sympathetic over activity in hypertension. Ann Med. 2000;32:365–70.
13. Palatini P, Julius S. Heart rate and the cardiovascular risk. J Hypertens. 1997;15:3–17.
14. Singh JP, Larson MG, Tsuji H, Evans JC, O'Donnell CJ, Levy D. Reduced heart rate variability and new-onset hypertension: insights into pathogenesis of hypertension: the Framingham Heart Study. Hypertension. 1998;32(2):293–7.
15. Logvinenko A, Mishchenko L, Kupchynskaja E, Gulkevych O, Ovdiienko T, Bezrodnyi V, et al. Heart rate variability in patients with resistant arterial hypertension. J Hypertens. 2017;35:e223.
16. Hoshi RA, Santos IS, Dantas EM, Andreão RV, Mill JG, Lotufo PA, Bensenor I. Reduced heart-rate variability and increased risk of hypertension—a prospective study of the ELSA-Brasil. J Hum Hypertens. 2021;35(12):1088–97. https://doi.org/10.1038/s41371-020-00460-w.
17. Julario R, Mulia E, Rachmi DA, A'yun MQ, Septianda I, Dewi IP, Juwita RR, Dharmadjati BB. Evaluation of heart rate variability using 24-hour Holter electrocardiography in hypertensive patients. J Arrhythmia. 2020;37(1):157–64.
18. Schroeder EB, Liao D, Chambless LE, Prineas RJ, Evans GW, Heiss G. Hypertension, blood pressure, and heart rate variability: the Atherosclerosis Risk in Communities (ARIC) study. Hypertension. 2003;42(6):1106–11.
19. Natarajan N, Balakrishnan AK, Ukkirapandian K. A study on analysis of heart rate variability in hypertensive individuals. Int J Biomed Adv Res. 2014;5:109–11.
20. Khan AA, Junejo RT, Thomas GN, Fisher JP, Lip GYH. Heart rate variability in patients with atrial fibrillation and hypertension. Eur J Clin Investig. 2021;51(1):e13361.
21. Yu Y, Xu Y, Zhang M, Wang Y, Zou W, Gu Y. Value of assessing autonomic nervous function by heart rate variability and heart rate turbulence in hypertensive patients. Int J Hypertens. 2018;2018:4067601.
22. Mori H, Saito I, Eguchi E, Maruyama K, Kato T, Tanigawa T. Heart rate variability and blood pressure among Japanese men and women: a community-based cross-sectional study. Hypertens Res. 2014;37(8):779–84.

23. Saito I, Takata Y, Maruyama K, Eguchi E, Kato T, Shirahama R, et al. Association between heart rate variability and home blood pressure: the toon health study. Am J Hypertens. 2018;31(10):1120–6.
24. Koichubekov BK, Sorokina MA, Laryushina YM, Turgunova LG, Korshukov IV. Nonlinear analyses of heart rate variability in hypertension. Ann Cardiol Angeiol (Paris). 2018;67(3):174–9.
25. Jansson K, Östlund R, Nylander E, Dahlström U, Hagerman I, Karlberg K-E, et al. The effects of metoprolol and captopril on heart rate variability in patients with idiopathic dilated cardiomyopathy. Clin Cardiol. 1999;22(6):397–402.
26. Ma W, Yang Y, Qi L, Zhang B, Meng L, Zhang Y, Li M, Huo Y. Relation between blood pressure variability within a single visit and stroke. Int J Hypertens. 2021;2021:2920140.
27. Asayama K, Ohkubo T, Hanazawa T, Watabe D, et al.; Hypertensive Objective Treatment Based on Measurement by Electrical Devices of Blood Pressure (HOMED-BP) Study Investigator. Association between amplitude of seasonal variation in self-measured home blood pressure and cardiovascular outcomes: HOMED-BP (Hypertension Objective Treatment Based on Measurement By Electrical Devices of Blood Pressure) Study. J Am Heart Assoc. 2016;5:e002995.
28. Chen YK, Ni ZX, Li W, Xiao WM, Liu YL, Liang WC, Qu JF. Diurnal blood pressure and heart rate variability in hypertensive patients with cerebral small vessel disease: a case-control study. J Stroke Cerebrovasc Dis. 2021;30(5):105673.
29. Mehlum MH, Liestøl K, Kjeldsen SE, Julius S, Hua TA, Rothwell PM, et al. Blood pressure variability and risk of cardiovascular events and death in patients with hypertension and different baseline risks. Eur Heart J. 2018;39(C):2243–51.
30. Palatini P, Saladini F, Mos L, Fania C, Mazzer A, Cozzio S, et al. Short-term blood pressure variability outweighs average 24-h blood pressure in the prediction of cardiovascular events in hypertension of the young. J Hypertens. 2019;37:1419–26.
31. Zhou TL, Henry RMA, Stehouwer CDA, Van Sloten TT, Reesink KD, Kroon AA. Blood pressure variability, arterial stiffness, and arterial remodeling: the Maastricht study. Hypertension. 2018;72:1002–10.
32. Kim JS, Park S, Yan P, Jeffers BW. Effect of inter-individual blood pressure variability on the progression of atherosclerosis in carotid and coronary arteries: a post hoc analysis of the NORMALISE and PREVENT studies. Eur Hear J Cardiovasc Pharmacother. 2017;3:82–9.
33. Mustafa ER, Istrătoaie O, Muşetescu R. Blood pressure variability and left ventricular mass in hypertensive patients. Curr Health Sci J. 2016;42(1):47–50.
34. Wang X, Wang F, Chen M, Wang X, Zheng J, Qin A. Twenty-four-hour systolic blood pressure variability and renal function decline in elderly male hypertensive patients with well-controlled blood pressure. Clin Interv Aging. 2018;13:533–40.
35. Filomena J, Riba-Llena I, Vinyoles E, Tovar JL, Mundet X, Castañé X, et al. Short-term blood pressure variability relates to the presence of subclinical brain small vessel disease in primary hypertension. Hypertension. 2015;66(3):634–40.
36. Palatini P. Risk of developing foot ulcers in diabetes: contribution of high visit-to-visit blood pressure variability. J Hypertens. 2018;36(11):2132–4.
37. Levi-Marpillat N, Macquin-Mavier I, Tropeano AI, et al. Antihypertensive drug classes have different effects on short-term blood pressure variability in essential hypertension. Hypertens Res. 2014;37:585–90.
38. Robinson TG, Davison WJ, Rothwell PM, Potter JF. Randomised controlled trial of a Calcium Channel or Angiotensin Converting Enzyme Inhibitor/Angiotensin Receptor Blocker Regime to Reduce Blood Pressure Variability following Ischaemic Stroke (CAARBS): a protocol for a feasibility study. BMJ Open. 2019;9(2):e025301.
39. Webb AJ, Fischer U, Mehta Z, Rothwell PM. Effects of antihypertensive-drug class on interindividual variation in blood pressure and risk of stroke: a systematic review and meta-analysis. Lancet. 2010;375:906–15.
40. Webb AJ, Rothwell PM. Effect of dose and combination of antihypertensives on interindividual blood pressure variability: a systematic review. Stroke. 2011;42:2860–5.

41. Ichihara A, Kaneshiro Y, Takemitsu T, Sakoda M. Effects of amlodipine and valsartan on vascular damage and ambulatory blood pressure in untreated hypertensive patients. J Hum Hypertens. 2006;20:787–94.
42. Rothwell PM. Limitations of the usual blood-pressure hypothesis and importance of variability, instability, and episodic hypertension. Lancet. 2010;375:938–48.
43. Lacolley P, Bezie Y, Girerd X, Challande P, Benetos A, Boutouyrie P, Ghodsi N, Lucet B, Azoui R, Laurent S. Aortic distensibility and structural changes in sinoaortic-denervated rats. Hypertension. 1995;26:337–40.
44. Hermida RC, Ayala DE, Mojón A, Fernández JR. Influence of circadian time of hypertension treatment on cardiovascular risk: results of the MAPEC study. Chronobiol Int. 2010;27(8):1629–51.
45. Hermida RC, Ayala DE, Fernández JR, Mojón A, Smolensky MH. Hypertension: new perspective on its definition and clinical management by bedtime therapy substantially reduces cardiovascular disease risk. Eur J Clin Investig. 2018;48:e12909.
46. Zhang Y, Agnoletti D, Safar ME, Blacher J. Effect of antihypertensive agents on blood pressure variability: the Natrilix SR versus candesartan and amlodipine in the reduction of systolic blood pressure in hypertensive patients (X-CELLENT) study. Hypertension. 2011;58(2):155–60.
47. Parati G, Dolan E, Ley L, Schumacher H. Impact of antihypertensive combination and monotreatments on blood pressure variability: assessment by old and new indices. Data from a large ambulatory blood pressure monitoring database. J Hypertens. 2014;32(6):1326–33.
48. Parati G, Schumacher H, Bilo G, Mancia G. Evaluating 24-h antihypertensive efficacy by the smoothness index: a meta-analysis of an ambulatory blood pressure monitoring database. J Hypertens. 2010;28(11):2177–83.
49. Omboni S, Kario K, Bakris G, Parati G. Effect of antihypertensive treatment on 24-h blood pressure variability: pooled individual data analysis of ambulatory blood pressure monitoring studies based on olmesartan mono or combination treatment. J Hypertens. 2018;36(4):720–33.
50. Ogihara T, Saruta T, Rakugi H, Saito I, Shimamoto K, Matsuoka H, et al.; COLM Investigators. Combination therapy of hypertension in the elderly: a subgroup analysis of the Combination of OLMesartan and a calcium channel blocker or diuretic in Japanese elderly hypertensive patients trial. Hypertens Res. 2015;38(1):89–96.
51. Ogihara T, Matsuzaki M, Matsuoka H, Shimamoto K, Shimada K, Rakugi H, et al.; COPE Trial Group. The combination therapy of hypertension to prevent cardiovascular events (COPE) trial: rationale and design. Hypertens Res. 2005;28(4):331–8. https://doi.org/10.1291/hypres.28.331. PMID: 16138563
52. Dahlöf B, Sever PS, Poulter NR, Wedel H, Beevers DG, Caulfield M, et al.; ASCOT Investigators. Prevention of cardiovascular events with an antihypertensive regimen of amlodipine adding perindopril as required versus atenolol adding bendroflumethiazide as required, in the Anglo-Scandinavian Cardiac Outcomes Trial-Blood Pressure Lowering Arm (ASCOT-BPLA): a multicentre randomised controlled trial. Lancet. 2005;366(9489):895–906. https://doi.org/10.1016/S0140-6736(05)67185-1. PMID: 16154016.
53. Zanchetti A, Bond MG, Hennig M, Neiss A, Mancia G, Dal Palù C, et al. Calcium antagonist lacidipine slows down progression of asymptomatic carotid atherosclerosis: principal results of the European Lacidipine Study on Atherosclerosis (ELSA), a randomized, double-blind, long-term trial. Circulation. 2002;106:2422–7.
54. Rothwell PM, Howard SC, Dolan E, O'Brien E, Dobson JE, Dahlöf B, et al. Effects of β blockers and calcium-channel blockers on within-individual variability in blood pressure and risk of stroke. Lancet Neurol. 2010;9(5):469–80.

Central Aortic Blood Pressure: Measurement and Clinical Significance

11

Upendra Kaul

11.1 Introduction

Detection and management of hypertension has traditionally been based on measuring the brachial artery pressure (BAP) using various types of instruments. These instruments need periodic caliberation. Blood pressure (BP) in the aorta and the brachial artery can differ from BAP due to reflected waves that modify the values. Systolic BP (SBP) and pulse pressure (PP) increase from the aorta towards the periphery, whereas diastolic BP (DBP) and mean artery pressure (MAP) decrease slightly towards the periphery.

Central aortic pressure (CAP) is a better indicator of the dynamics of blood flow supplying target organs, especially the brain and kidney [1]. In contrast to BAP, target organs are directly exposed to CAP. CAP has been shown to have a closer relationship with the occurrence of cardiovascular events than BAP [2, 3]. In addition, the effects of pharmacological therapy on CAP differ from the effect of treatment on BAP. The Conduit Artery Function Evaluation (CAFE) study is a good example of differential effects of drug interventions on central and peripheral pressure [4]. It showed that CAP was a better measure of hemodynamic burden on the heart and central organs, and a better predictor of stroke, myocardial infarction and heart failure, compared with BAP [4].

The best and most reproducible way of measuring CAP is a direct and invasive method, based on insertion of a catheter in the aorta. However, this is not suitable for application in routine clinical practice. Recently there has been considerable interest in the development of tools to non-invasively measure CAP that would be better suited to routine clinical use.

U. Kaul (✉)
Batra Heart Centre, BHMRC, New Delhi, India

© The Author(s), under exclusive license to Springer Nature Switzerland AG 2022
C. V. S. Ram et al. (eds.), *Hypertension and Cardiovascular Disease in Asia*,
Updates in Hypertension and Cardiovascular Protection,
https://doi.org/10.1007/978-3-030-95734-6_11

11.2   The Basis: Reflected Pulse Waves

The cardiac pump transfers blood from the centre towards the aorta and the peripheral vessels. Along these vessels, the pressure waves generated by the heart transmit to the periphery (forward wave) and are reflected back (reflected/backward wave). These reflected waves meet at different times and at different sites of the arterial tree during the cardiac cycle, forming different pressure waveforms and contours. This explains the differences in pulse wave contour along the arterial tree [5, 6]. The measure of the time when forward and reflected waves meet is the pulse wave velocity (the speed at which the pressure wave transmits in the artery). Pulse wave velocity is determined by arterial stiffness, which does not change significantly in the brachial artery with ageing. This is the basis of using a generalised transfer function to estimate aortic pulse wave from radial/brachial pulse wave [7]. However, there is change in the central arteries with ageing, hypertension, different disease states and exercise, which led many researchers to seek accurate and adaptive or individualised methods to estimate CAP [8, 9].

11.3   Devices and Methods

Many devices using different methods are available for measurement of CAP. Of these, the pulse waveform of carotid arteries should be the best surrogate for CAP [10]. However, waveforms obtained from peripheral arteries are most commonly used for non-invasive central BP estimation because they are easier to apply. These techniques can be either 'tonometry-based' [11, 12] or 'cuff-based' [13, 14]. The working methods of these central BP measurement devices include pulse waveform analysis, transfer function and N-point moving average (NPMA) [12]. Transfer function is a relationship between two physical properties [15, 16] and has been an accepted central BP measurement method. Pulse waveform analysis can be used to identify waveform characteristics. The peak of second peak of radial or brachial pulse wave (SBP2) in late systole is a pressure waveform resulting from distal pulse wave reflections. This correlates well with the peak of central aortic pressure waveforms (central SBP) [16, 17]. Central SBP and PP can be estimated with reasonable accuracy if the comprehensive waveforms, including SBP2, are used in regression equations.

The NPMA method (a mathematical low pass filter to remove random noise) to obtain a good high frequency waveform, which eliminates distortions from peripheral arterial waves and provides a good estimate of CAP, has also been reported [18, 19].

It is very important to test the accuracy of non-invasive measurements of CAP by simultaneously comparing these with invasive measurements (the current gold standard method). The accuracy determined from these comparisons is device specific, and a major limitation is that it is dependent on the accuracy of the cuff BP measured waveform used for calibration. There are a number of devices available for measuring CAP (Table 11.1).

Table 11.1   Invasively validated central blood pressure measuring devices

Central aortic pressure	Recording site	Method of recording (sensor)	Method	Calibration
Office measurement				
ArteriographTensioMed Ltd., Hungary	Brachial	Supra-systolic brachial cuff plethysmography	SBP2 + regression	Brachial cuff MAP/DBP
ARCsolver + VaSeraVS-1500Austrian Institute of Technology, Austria	Brachial	Brachial cuff pulse volume plethysmography	GTF	Brachial cuff SBP/DBP
BPLab Petr Telegin, Russia	Brachial	Brachial cuff pulse volume plethysmography	GTF	Brachial cuff SBP/DBP
BP + Uscom Ltd., Australia (acquire Pulsecor Ltd., Cardioscope II)	Brachial	Supra-systolic brachial cuff plethysmography	Physical model Brachial supra-systolic waveform	Brachial cuff SBP/DBP
Complior Alam Medical, France	Carotid	Applanation tonometry, Single, fixed	Simple substitution	Brachial cuff MAP/DBP
cBP301Centron Diagnostics, UK (acquired by SunTech Medical)	Brachial	Brachial cuff plethysmography	GTF	Brachial cuff SBP/DBP
DynaPulse Pulse Metric Inc., USA	Brachial	Supra-systolic brachial cuff plethysmography	Physical model	Brachial cuff SBP/DBP
Gaon Hanbyul Meditech, Korea	Radial	Applanation tonometry, Single, fixed	GTF	Brachial cuff SBP/DBP
HEM-9000AI Omron Healthcare, Japan	Radial	Applanation tonometry, Arrayed, fixed	SBP2 + regression	Brachial cuff SBP/DBP
Mobil-O-GraphI.EM GmbH, Germany Brachial Artery	Brachial	Brachial cuff pulse volume plethysmography	GTF	Brachial cuff SBP/DBP
NIHem Cardiovascular Engineering Inc., USA	Carotid	Applanation tonometry, Single, manual	Simple substitution	Brachial cuff MAP/DBP
Oscar 2 with SphygmoCor SunTech Medical, USA Brachial Artery	Brachial	Subdiastolic brachial cuff plethysmography	GTF	Brachial cuff SBP/DBP
PulsePen DiaTecne srl., Italy	Carotid	Applanation tonometry, Single, manual	Simple substitution	Brachial cuff MAP/DBP
SphygmoCor CVMS, AtCor Medical, Australia	Radial	Applanation tonometry, Single, manual	GTF	Brachial cuff SBP/DBP
SphygmoCor XCELAtCor Medical, Australia	Brachial	Subdiastolic brachial cuff plethysmography	GTF	Brachial cuff SBP/DBP

(continued)

Table 11.1 (continued)

Central aortic pressure	Recording site	Method of recording (sensor)	Method	Calibration
Vicorder Skidmore Medical Ltd., UK	Brachial	Brachial cuff pulse volume plethysmography	GTF	Brachial cuff MAP/DBP
WatchBP Microlife Corp, Taiwan	Brachial	Brachial cuff pulse volume plethysmography	(SBP2, DBP, As, Ad) + regression	Brachial cuff SBP/DBP
Ambulatory BP measurement				
Arteriograph 24 h, TensioMED Ltd., Hungary	Brachial	Supra-systolic brachial cuff plethysmography	SBP2 + regression	Brachial cuff MAP/DBP
ABPM 7100Welch Allyn, Inc. (acquired by Hillrom)	Brachial	Brachial cuff pulse volume plethysmography	GTF	Brachial cuff SBP/DBP
BPro + A-Pulse, HealthSTATS, Singapore (acquired by Hillrom)	Radial	Applanation tonometry, Single, fixed (watch type)	N-point moving average	Brachial cuff SBP/DBP
Mobil-O-Graph NGI.EM GmbH, Germany	Brachial	Brachial cuff pulse volume plethysmography	GTF	Brachial cuff SBP/DBP
Oscar 2 with SphygmoCor, SunTech Medical	Brachial	Subdiastolic brachial cuff plethysmography	GTF	Brachial cuff SBP/DBP
WatchBP O3 (2G), Microlife AG, Widnau, Switzerland	Brachial	Brachial cuff pulse volume plethysmography	(SBP2, DBP, As, Ad) + regression	Brachial cuff SBP/DBP

Ad area under pressure wave curve during diastole, *As* area under pulse wave during systole, *DBP* diastolic blood pressure, *GTF* generalised transfer function, *MAP* mean arterial blood pressure, *SBP* systolic blood pressure, *SBP2* second peak of radial or brachial pulse wave

11.4 Clinical Utility

The clinical utility of measuring CAP is increasingly being recognised. It has been shown that patients with normal BAP can have Stage I hypertension based on measurement of CAP [20]. If central BP is a better target for therapy, misclassification or diagnosis based on brachial BP may result in over- or undertreatment of hypertension, which is of clinical importance [21]. The diagnosis of hypertension based on office, home or ambulatory BP readings is currently based on BAP measurements. Because of the phenomenon of PP amplification, brachial SBP and PP are usually higher than CAP measurements made at the same time [22].

PP amplification reflects the disagreement between central and peripheral BP, and varies within and between individuals [23]. More importantly, such variability depends on a number of factors, including age, sex, height, heart rate, medications and systemic vascular diseases [24–27]. Also, non-invasive measurement of brachial SBP as a surrogate for central SBP has been shown to have a significant random error [28].

11.5   Cardiovascular Outcomes

CAP has been shown to correlate better with left ventricular mass index, carotid intima-media thickness and pulse wave velocity than brachial pressure. In a number of international studies conducted on Anglo Cardiff trial participants, CAP was more closely associated with preclinical indices of target organ damage and cardiovascular events, including mortality, compared with brachial pressure [28–30]. Better correlation between central SBP and cardiovascular events and mortality was also reported in a large study of African American patients with stage 2 hypertension (the ATLAAST study) [31]. In addition, the clinical benefits of different antihypertensive agents observed in the CAFÉ study (a substudy of ASCOT) was better associated with the reduction of central rather than brachial BP [4]. This observation led to a keen interest in the application of CAP for clinical practice. One important issue has been between-device variability in central BP measurements. This needs to be improved so that measurement of CAP becomes more acceptable.

11.6   Outcomes of Treatment

The differential responses of central and brachial BP to various antihypertensive drugs have been emphasised in the literature [30–32]. Beta-blockers have been shown to lower CAP to a lesser extent than renin angiotensin system blockers and dihydropyridine calcium channel blockers. For that reason, in the presence of similar brachial BP reduction with all these agents, the reduction in vascular events (especially stroke) is less with beta-blockers such as atenolol. Another observation is that beta-blockers have an unfavourable effect on PP amplification, as documented in the CAFE substudy of the ASCOT trial [4], and the BP GUIDE study [23].

Previous studies have shown that central BP-guided hypertension management may allow fewer medications to be used to achieve BP control and reduce adverse effects [33]. A recent randomised controlled trial demonstrated that optimisation of goal-directed medical therapy in patients with heart failure could be better achieved using central BP compared with brachial BP. The additional titration of drug dosages enhanced afterload reduction and contributed to better reverse remodelling, without increased risk of hypotension or worsening renal function [34].

Treatment targets in patients with elevated CAP have yet to be defined. In low- to medium-risk subjects with uncomplicated hypertension, it is reasonable to lower central BP to <130/90 mmHg. However, there is a need for outcome-based studies to define the optimal CAP reduction in various vascular diseases.

11.7   Isolated Central Hypertension

The reported discrepancy between central and brachial BP has been reported to have different clinical connotations. The European Society of Hypertension/European Society of Cardiology hypertension guidelines define cohorts of 'isolated central' and 'isolated brachial' hypertension in adults as follows: brachial hypertension (brachial SBP $\geq$140 mmHg or brachial DBP $\geq$90 mmHg) and central hypertension (central SBP $\geq$130 mmHg or central DBP $\geq$90 mmHg) [35]. Therefore, phenotypes of hypertension can be defined based on differences between the brachial and central BP. These different phenotypes were also described in a national representative cohort from Shanghai, China [36]. Patients with isolated central hypertension had higher left ventricular mass index, carotid-femoral pulse wave velocity and urinary albumin-creatinine ratio than those without central or brachial hypertension [37].

11.8   Future Directions

There is clearly a theoretical advantage to using CAP measurements to inform decisions about the diagnosis and follow-up of hypertension. However, inconsistencies in the non-invasive measurement of central BP due to poor calibration of peripheral waveforms and the poor corelation of these readings with invasive CAP measurements are issues that need serious attention. The International Society for Hypertension and World hypertension League, along with several other national hypertension councils, have called for regulation of the manufacture and marketing of BP instruments and cuffs [37]. With these joint efforts, validated automatic BP devices are becoming readily available for more accurate non-invasive measurement of brachial and central BP. This should improve the care of patients with cardiovascular disease and contribute to better outcomes.

It should also be recognised that most outcome studies included elderly patients, in whom brachial and central pressures are usually similar [38, 39]. No outcome studies have been conducted in younger patients, in whom a much greater difference between brachial and central pressures is expected. Different antihypertensive treatments can differentially reduce central and brachial BP as described in the literature [40]. However, prospective studies investigating whether the benefits of targeting CAP rather than BAP in patients with hypertension will improve clinical outcomes are lacking and need to be pursued [8].

Non-invasive CAP measurement is currently being used primarily as a research tool, while the most widely used method in the community is office brachial BP measurement. Even self-measured home BP [41] and ambulatory BP measurements [42] are not being used optimally, despite their well-described advantages. Additional clinical trial data are needed to determine whether lowering both brachial and central BP to targets can improve the outcomes in patients with hypertension. This information is needed before the widespread use of CAP in routine clinical practice can be recommended.

11.9 Conclusion and Take-Home Message

Currently available devices for measuring CAP are based on the pulse wave amplification taken from brachial artery recordings and converted to estimated CAP values, but these methods need to be standardised before being widely accepted and applicable. Despite this, measurement of CAP has benefits over BAP measurement in terms of diagnosing and managing hypertension. Vascular outcomes and prevention of TOD appear to be better when central pressures are targeted rather than brachial pressures. Different pharmacological antihypertensive treatments have differing effects on CAP and BAP. However, more evidence is needed to support the use of central BP in diagnosing and managing hypertension.

References

1. Pauca AL, Wallenhaupt SL, Kon ND, Tucker WY. Does radial artery pressure accurately reflect aortic pressure? Chest. 1992;102(4):1193–8.
2. Avolio AP, Van Bortel LM, Boutouyrie P, Cockcroft JR, McEniery CM, Protogerou AD, et al. Role of pulse pressure amplification in arterial hypertension: experts' opinion and review of the data. Hypertension. 2009;54(2):375–83.
3. Agabiti-Rosei E, Mancia G, O'Rourke MF, Roman MJ, Safar ME, Smulyan H, et al. Central blood pressure measurements and antihypertensive therapy: a consensus document. Hypertension. 2007;50(1):154–60.
4. Williams B, Lacy PS, Thom SM, Cruickshank K, Stanton A, Collier D, et al. Differential impact of blood pressure-lowering drugs on central aortic pressure and clinical outcomes: principal results of the Conduit Artery Function Evaluation (CAFE) study. Circulation. 2006;113(9):1213–25.
5. O'Rourke MF, Pauca A, Jiang XJ. Pulse wave analysis. Br J Clin Pharmacol. 2001;51(6): 507–22.
6. Safar ME, O'Rourke MF. Handbook of hypertension: arterial stiffness in hypertension. Edinburg: Elsevier; 2006.
7. Townsend RR, Black HR, Chirinos JA, Feig PU, Ferdinand KC, Germain M, et al. Clinical use of pulse wave analysis: proceedings from a symposium sponsored by North American Artery. J Clin Hypertens (Greenwich). 2015;17(7):503–13.
8. Yao Y, Wang L, Hao L, Xu L, Zhou S, Liu W. The noninvasive measurement of central aortic blood pressure waveform. In: Blood pressure-from bench to bed; 2018. IntechOpen.
9. Cheng HM, Chuang SY, Wang TD, Kario K, Buranakitjaroen P, Chia YC, et al. Central blood pressure for the management of hypertension: is it a practical clinical tool in current practice? J Clin Hypertens (Greenwich). 2020;22(3):391–406.
10. Benetos A, Tsoucaris-Kupfer D, Favereau X, Corcos T, Safar M. Carotid artery tonometry: an accurate non-invasive method for central aortic pulse pressure evaluation. J Hypertens Suppl. 1991;9(6):S144–5.
11. Karamanoglu M, O'Rourke MF, Avolio AP, Kelly RP. An analysis of the relationship between central aortic and peripheral upper limb pressure waves in man. Eur Heart J. 1993;14(2):160–7.
12. Williams B, Lacy PS, Yan P, Hwee CN, Liang C, Ting CM. Development and validation of a novel method to derive central aortic systolic pressure from the radial pressure waveform using an n-point moving average method. J Am Coll Cardiol. 2011;57(8):951–61.
13. Cheng HM, Wang KL, Chen YH, Lin SJ, Chen LC, Sung SH, et al. Estimation of central systolic blood pressure using an oscillometric blood pressure monitor. Hypertens Res. 2010;33(6):592–9.

14. Weber T, Wassertheurer S, Rammer M, Maurer E, Hametner B, Mayer CC, et al. Validation of a brachial cuff-based method for estimating central systolic blood pressure. Hypertension. 2011;58(5):825–32.
15. Millasseau SC, Guigui FG, Kelly RP, Prasad K, Cockcroft JR, Ritter JM, et al. Noninvasive assessment of the digital volume pulse. Comparison with the peripheral pressure pulse. Hypertension. 2000;36(6):952–6.
16. Roman MJ, Devereux RB, Kizer JR, Lee ET, Galloway JM, Ali T, et al. Central pressure more strongly relates to vascular disease and outcome than does brachial pressure: the Strong Heart Study. Hypertension. 2007;50(1):197–203.
17. Zhang Y, Agnoletti D, Protogerou AD, Wang JG, Topouchian J, Salvi P, et al. Radial late-SBP as a surrogate for central SBP. J Hypertens. 2011;29(4):676–81.
18. Pauca AL, Kon ND, O'Rourke MF. The second peak of the radial artery pressure wave represents aortic systolic pressure in hypertensive and elderly patients. Br J Anaesth. 2004;92(5): 651–7.
19. Chen CH, Nevo E, Fetics B, Pak PH, Yin FC, Maughan WL, et al. Estimation of central aortic pressure waveform by mathematical transformation of radial tonometry pressure. Validation of generalized transfer function. Circulation. 1997;95(7):1827–36.
20. McEniery CM, Cockcroft JR, Roman MJ, Franklin SS, Wilkinson IB. Central blood pressure: current evidence and clinical importance. Eur Heart J. 2014;35(26):1719–25.
21. Shih YT, Cheng HM, Sung SH, Hu WC, Chen CH. Application of the N-point moving average method for brachial pressure waveform-derived estimation of central aortic systolic pressure. Hypertension. 2014;63(4):865–70.
22. Sharman J, Stowasser M, Fassett R, Marwick T, Franklin S. Central blood pressure measurement may improve risk stratification. J Hum Hypertens. 2008;22(12):838–44.
23. Sharman JE, Marwick TH, Gilroy D, Otahal P, Abhayaratna WP, Stowasser M. Randomized trial of guiding hypertension management using central aortic blood pressure compared with best-practice care: principal findings of the BP GUIDE study. Hypertension. 2013;62(6):1138–45.
24. Camacho F, Avolio A, Lovell NH. Estimation of pressure pulse amplification between aorta and brachial artery using stepwise multiple regression models. Physiol Meas. 2004;25(4):879–89.
25. Wilkinson IB, Mohammad NH, Tyrrell S, Hall IR, Webb DJ, Paul VE, et al. Heart rate dependency of pulse pressure amplification and arterial stiffness. Am J Hypertens. 2002;15:24–30.
26. Albaladejo P, Copie X, Boutouyrie P, Laloux B, Déclère AD, Smulyan H, et al. Heart rate, arterial stiffness, and wave reflections in paced patients. Hypertension. 2001;38(4):949–52.
27. Kollias A, Lagou S, Zeniodi ME, Boubouchairopoulou N, Stergiou GS. Association of central versus brachial blood pressure with target-organ damage: systematic review and meta-analysis. Hypertension. 2016;67(1):183–90.
28. McEniery CM, Yasmin, McDonnell B, Munnery M, Wallace SM, Rowe CV, et al. Central pressure: variability and impact of cardiovascular risk factors: the Anglo-Cardiff Collaborative Trial II. Hypertension. 2008;51(6):1476–82.
29. Chi C, Yu X, Auckle R, Lu Y, Fan X, Yu S, et al. Hypertensive target organ damage is better associated with central than brachial blood pressure: the Northern Shanghai Study. J Clin Hypertens (Greenwich). 2017;19(12):1269–75.
30. Vlachopoulos C, Aznaouridis K, O'Rourke MF, Safar ME, Baou K, Stefanadis C. Prediction of cardiovascular events and all-cause mortality with central haemodynamics: a systematic review and meta-analysis. Eur Heart J. 2010;31(15):1865–71.
31. Ferdinand KC, Pool J, Weitzman R, Purkayastha D, Townsend R. Peripheral and central blood pressure responses of combination aliskiren/hydrochlorothiazide and amlodipine monotherapy in African American patients with stage 2 hypertension: the ATLAAST trial. J Clin Hypertens (Greenwich). 2011;13(5):366–75.
32. Matsui Y, Eguchi K, O'Rourke MF, Ishikawa J, Miyashita H, Shimada K, et al. Differential effects between a calcium channel blocker and a diuretic when used in combination with angiotensin II receptor blocker on central aortic pressure in hypertensive patients. Hypertension. 2009;54(4):716–23.

33. Borlaug BA, Olson TP, Abdelmoneim SS, Melenovsky V, Sorrell VL, Noonan K, et al. A randomized pilot study of aortic waveform guided therapy in chronic heart failure. J Am Heart Assoc. 2014;3(2):e000745.
34. Benetos A, Thomas F, Joly L, Blacher J, Pannier B, Labat C, et al. Pulse pressure amplification a mechanical biomarker of cardiovascular risk. J Am Coll Cardiol. 2010;55(10):1032–7.
35. Smulyan H, Safar ME. Blood pressure measurement: retrospective and prospective views. Am J Hypertens. 2011;24(6):628–34.
36. Chuang SY, Chang HY, Cheng HM, Pan WH, Chen CH. Prevalence of hypertension defined by central blood pressure measured using a type II device in a nationally representative cohort. Am J Hypertens. 2018;31(3):346–54.
37. Campbell NR, Gelfer M, Stergiou GS, Alpert BS, Myers MG, Rakotz MK, et al. A call to regulate manufacture and marketing of blood pressure devices and cuffs: a position statement from the World Hypertension League, International Society of Hypertension and supporting hypertension organizations. J Clin Hypertens (Greenwich). 2016;18(5):378–80.
38. Laurent S, Sharman J, Boutouyrie P. Central versus peripheral blood pressure: finding a solution. J Hypertens. 2016;34(8):1497–9.
39. Laurent S, Briet M, Boutouyrie P. Arterial stiffness as surrogate end point: needed clinical trials. Hypertension. 2012;60(2):518–22.
40. Morgan T, Lauri J, Bertram D, Anderson A. Effect of different antihypertensive drug classes on central aortic pressure. Am J Hypertens. 2004;17(2):118–23.
41. Kaul U, Wander GS, Sinha N, Mohan JC, Kumar S, Dani S, et al. Self-blood pressure measurement as compared to office blood pressure measurement in a large Indian population; the India Heart Study. J Hypertens. 2020;38(7):1262–70.
42. Kaul U, Arambam P, Rao S, Kapoor S, Swahney JPS, Sharma K, et al. Usefulness of ambulatory blood pressure measurement for hypertension management in India: the India ABPM study. J Hum Hypertens. 2020;34(6):457–67.

Diabetes and Hypertension: What Is the Connection?

12

Mukundan Aswin and Viswanathan Mohan

12.1 Introduction

Diabetes mellitus, hypertension and dyslipidaemia are all risk factors for athero-sclerotic cardiovascular disease (ASCVD), which is a major cause of mortality and morbidity worldwide. The global prevalence of diabetes is increasing in epidemic proportions, as outlined in an International Diabetes Federation (IDF) report. The IDF estimated that 366 million people worldwide had diabetes mellitus in 2011 and, if current trends continue, 552 million people (or one in ten adults) will have diabetes by 2030 [1]. In contrast, over the period 1975–2015, the prevalence of hypertension decreased in high-income, and some middle-income countries and remained unchanged elsewhere [2]. Hypertension and type 2 diabetes are common comorbidities based on data from studies conducted in Western countries and in Japan [3, 4]. It was estimated that 20% of patients with hypertension had type 2 diabetes and the 50% of patients with type 2 diabetes had hypertension; having either of these conditions increased the risk of developing the other by 1.5–2.0 times [5].

Individuals with both hypertension and diabetes have a four-fold higher risk of developing cardiovascular disease (CVD) compared with age-matched normotensive nondiabetic controls [6]. Diabetes was associated with a two- to four-fold increase in the risk of myocardial infarction (MI), congestive heart failure, peripheral arterial disease, stroke and death in the Framingham Heart Study [7]. A recent analysis of the Framingham data revealed that the population with hypertension at

M. Aswin · V. Mohan (✉)
Dr. Mohan's Diabetes Specialities Centre and Madras Diabetes Research Foundation, Chennai, India
e-mail: draswin@drmohans.com; drmohans@diabetes.ind.in; http://www.drmohans.com; http://www.mdrf.in

© The Author(s), under exclusive license to Springer Nature Switzerland AG 2022   159
C. V. S. Ram et al. (eds.), *Hypertension and Cardiovascular Disease in Asia*,
Updates in Hypertension and Cardiovascular Protection,
https://doi.org/10.1007/978-3-030-95734-6_12

the time of a diabetes mellitus diagnosis had significantly higher rates of mortality from all causes (32 vs 20 events per 1000 person-years) and cardiovascular events (52 vs 31 events per 1000 person-years) compared to normotensive subjects with diabetes mellitus, suggesting that major part of this excess risk is due to coexistent HTN [8].

12.2 Pathophysiology: Converging Pathways

The complex mechanisms involved in the pathogenesis of essential hypertension act on a genetic background. Development of hypertension in genetically vulnerable people is related to increased salt intake, obesity, excess alcohol consumption, mental stress, decreased physical activity and poor sleep [9]. While the majority of patients have essential hypertension that does not have an identifiable cause, it is important to recognise secondary causes of hypertension because they may be curable. Causes of secondary hypertension include primary aldosteronism, pheochromocytoma or renal artery stenosis, among others.

Several pathophysiologic mechanisms coexist in diabetes mellitus and hypertension including oxidative stress due to reactive oxygen species (ROS), pathogenic activation of the renin-angiotensin-aldosterone system (RAAS), inflammation, blunted insulin-mediated vasodilatation, sympathetic nervous system (SNS) activation, dysfunctional innate and adaptive immune responses, and impaired renal handling of sodium [10]. In addition, major pathogenic factors underlying the coexistence of both diabetes mellitus and hypertension are obesity and increased visceral adiposity. Tissue RAAS activation is caused by chronic low-grade inflammation and oxidative stress in adipose tissue, which leads to increased production of angiotensinogen (AGT) and angiotensin II (Ang II) [8]. Ultimately, overexpression of AGT in adipose tissue results in elevated blood pressure (BP) [11]. Thus, AGT and Ang II have local and systemic effects on BP regulation. One of the many deleterious effects of Ang II is mediated via activation of the Ang II type 1 receptor (AT1R) [12]. The production of ROS, reduced insulin metabolic signalling, and proliferative and inflammatory vascular responses culminating in endothelial dysfunction, insulin resistance and hypertension are caused by the activation of AT1R in non-adrenal tissues [12]. Figure 12.1 shows the RAAS pathway and its association with chronic low-grade inflammation in diabetes.

12.3 Coexistence of Hypertension and Diabetes Mellitus

There is a lot of evidence for an increased prevalence of hypertension in people with versus without diabetes mellitus [13]. In a paper published on American Indian and Alaska Native communities to study the prevalence of clinical hypertension and look at its coexistence with diabetes mellitus, 37% of individuals with

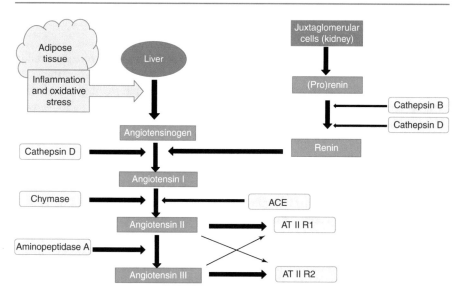

Fig. 12.1 The RAAS pathway and its association with chronic low-grade inflammation in diabetes. *ACE* angiotensin-converting enzyme, *AT II R1* angiotensin II receptor type 1, *AT II R2* angiotensin II receptor type 2

diabetes were diagnosed with hypertension [14]. In addition, the relative risk of hypertension in those with versus without diabetes varied from 4.7 to 7.7 [11]. Individuals from ethnic minorities in the United Kingdom who were aged 40–59 years and had hypertension were more likely to have diabetes than those without hypertension, and people with diabetes were at much higher risk of having hypertension [15]. Data from a meta-analysis suggest that each 20-mmHg increase in systolic BP increases the risk of new-onset type 2 diabetes mellitus by 77% [16].

The affected organs are similar in both essential hypertension and diabetes mellitus, and the vascular tree is the common target. As a result, coexisting diabetes mellitus and hypertension significantly increase the risk of developing renal failure, retinopathy, coronary heart disease [17], left ventricular hypertrophy [18], congestive heart failure [19] and stroke [20] compared with either condition alone. Rates of microvascular complications, retinopathy and nephropathy are highest in patients with hypertension and diabetes [21, 22]. Lowering BP is therefore of critical importance in patients with diabetes [23, 24]. However, how low the BP target should be is the subject of debate. The "rule of halves" in diabetes and hypertension states that more than 50% of people with either of these conditions remain undiagnosed, 50% of those in whom the disease is detected are untreated treatment, and 50% of those receiving treatment do not achieve disease control. Thus, the overall proportion of patients with diabetes mellitus or hypertension who are adequately controlled is small [25, 26].

12.4   Diagnosis of Hypertension in Diabetes Mellitus

BP should be measured routinely at each clinical visit, and should be measured in both arms to detect any differences. For patients with elevated office BP (>140/90 mmHg), this should be confirmed using multiple readings, including measurements on a separate day, to diagnose hypertension. All patients with hypertension and diabetes are advised to monitor their BP at home. Orthostatic measurement of BP should be performed at the first visit and periodically during follow-up [27, 28].

BP needs to be measured by a trained person [29] with the individual in the seated position, with feet on the floor and arm supported at heart level. Cuff size should be appropriate for the upper arm circumference [30]. To reduce within-patient variability, BP should be measured after 5 min of rest, and 2–3 readings should ideally be taken 1–2 min apart, and BP measurements should be averaged [31].

Autonomic neuropathy can be assessed by checking for postural changes in BP and heart rate, which would require adjustment of BP targets [32]. Home BP monitoring and 24-h ambulatory BP monitoring can provide evidence of white coat hypertension, masked hypertension, or other discrepancies between office and "true" BP [33].

12.5   Blood Pressure Target in DM

BP targets for all patients with diabetes and hypertension must be individualised, taking into account patient preference, cardiovascular risk, and the potential beneficial and adverse effects of antihypertensive medications. A BP target of <130/80 mmHg is appropriate for individuals with diabetes and hypertension at higher cardiovascular risk (existing or 10-year ASCVD risk ≥15%) [34]. BP should be maintained at <140/90 mmHg in individuals with diabetes and hypertension who are at lower risk for CVD (10-year ASCVD risk <15%) [34].

12.6   Glycaemic Goals

Optimal management of diabetes requires identification and optimisation of the "ABCDEs" of diabetes: A1C (glycosylated haemoglobin; HbA1c), BP, cholesterol (i.e., dyslipidaemia), diet and exercise. Glucose goals should be established on an individual basis for each patient, taking both clinical characteristics and the patient's psycho-socioeconomic circumstances into consideration [35, 36].

The American Diabetes Association (ADA) also recommends individualising glycaemic targets based on patient attitude and expected treatment efforts, risks potentially associated with hypoglycaemia and other adverse events, disease duration, life expectancy, important comorbidities, established vascular complications, resources and support system [37].

Glycaemic recommendations for many non-pregnant adults with diabetes are summarised as follows [37]:

1. HbA1c: <7.0%.
2. Preprandial blood glucose: 80–130 mg/dL.
3. Postprandial blood glucose: <180 mg/dL.
4. Time in range as measured by continuous glucose monitoring should be >70%.

12.7   Prevention

Treating diabetes and hypertension in developing countries such as India presents a real challenge because prevention is a major goal before starting to treat. Effective management is essential to lessen the mortality and morbidity related to both diabetes and hypertension, and there are now better treatment tools than ever before [38–40]. There are multiple levels at which hypertension and diabetes can be prevented and treated.

Primordial prevention: This refers to avoidance and prevention of risk factors. Given the overlap between major risk factors for hypertension and diabetes mellitus, an integrated approach to the prevention and control of both can be undertaken. Primordial prevention focuses on health policies that create an environment that promotes healthy behaviours and necessary education programmes. This requires political commitment, involvement by health professionals, and efforts of community leaders and the mass media [41].

The extremely low proportion of patients with hypertension or diabetes mellitus who achieve adequate disease control [25, 26] represents a massive challenge and underlines the need to urgently raise community awareness of these conditions. It is important to detect diabetes and hypertension early, before any organ damage occurs, and provide patients with the best possible and yet affordable treatment. A research paper titled "Improving the prevention and management of chronic disease in low-income and middle-income countries: a priority for primary health care" talks about management of chronic diseases and how this is totally different from acute care, and dependent on several factors, including opportunistic case finding for risk factor assessment, early disease detection and high-risk status identification [42]. All these should be followed by a combination of social, psychological and pharmacological interventions, often in a stepped-care manner, and finally long-term follow-up with frequent monitoring and promotion of adherence to treatment [42].

Primary prevention: The concept of primary prevention refers to a stage where risk factors have already emerged and efforts are needed to prevent the condition in individuals with pre-diabetes or prehypertension. The co-ordination and collaboration of public primary care and private health systems will be required to make many of these recommendations a reality. Primary prevention is very important as a complementary and integrated strategy for several reasons [43]:

1. The immense public health burden associated with diabetes and hypertension justifies action at the population level.
2. Many of the presently available treatments, while valuable, come with the risk of undesirable side effects (e.g. hypoglycaemia, electrolyte imbalance), may have limited efficacy, and may not be accessible over the long term, especially for

individuals who have problems accessing medical care or adhering to self-care regimens [44].

3. Lifestyle modifications designed to facilitate the prevention of hypertension and diabetes are likely to have other beneficial effects (e.g. decreased lipid levels, and prevention of heart disease and certain cancers).
4. Many important determinants, such as a balanced diet, weight management and physical activity, are not amenable to implementation or influence by medical care practitioners alone, and are likely to be better addressed by public health efforts and educators in the society.
5. The integration of primary care and public health interventions is needed to address racial/ethnic and socioeconomic disparities because these are multifactorial in origin [45].
6. In addition to lifestyle modification, pharmacotherapy as part of primary prevention strategies in high-risk individuals has been shown to be effective in randomised controlled trials (RCTs) [46].

RCTs of structured lifestyle modification have found that reducing calorie intake plus physical activity leading to modest weight loss reduces the incidence of type 2 diabetes in high-risk adults by 50–70% [47]. The Diabetes Prevention Program (DPP) research trial led by the US National Institutes of Health (NIH) was performed in a US population, but is generalisable to other countries [48]. Participants were randomly assigned to one of three groups:

1. Lifestyle intervention to encourage reduction in calorie intake and increased physical activity.
2. Metformin, the most prescribed oral antidiabetic medication, which is known to improve insulin sensitivity.
3. A placebo control group.

The DPP lifestyle intervention group had an initial body weight reduction of ~6% after 12 months, decreasing to ~4% after 3 years. This group had showed an increase in self-reported physical activity (equivalent to brisk walking) from 100 to 190 min per week. This intervention had an impressive effect on the incidence of diabetes, which was decreased by 58% over 4 years compared with the placebo control group. These benefits were seen in men and women across race and ethnic groups and were even greater at older ages [48].

Major trials have also documented the efficacy of lifestyle modification to reduce BP and prevent hypertension in high-risk adults [49]. A prospective study on the contribution of physical activity and body mass index to the risk of hypertension conducted in drug-naïve adults from Finland showed that subjects with higher levels of physical activity had a lower prevalence of hypertension [50]. In combination with increased community detection programmes and compelling epidemiologic data, these trials of primary and secondary prevention led to a series of strong national recommendations for the prevention, detection and treatment of high BP [28].

In the Trials of Hypertension Prevention-phase I (TOHP-I), which enrolled 2182 participants with prehypertension, of the three lifestyle changes proposed (weight reduction, sodium reduction and stress management), weight reduction was the most powerful strategy, producing a net weight loss of 3.9 kg and a BP change of −2.3/−2.9 mmHg [51].

Secondary prevention: Secondary prevention in people with diabetes and hypertension is essential because of the significant financial burden associated with the complications of these conditions. Unless proper treatment is provided to patients with hypertension and/or diabetes, these individuals face a significant burden of complications in the future. Three landmark studies on glycaemic control in diabetes mellitus—the Diabetes Complications and Control Trial (DCCT) [52], the United Kingdom Prospective Diabetes Study (UKPDS) [53] and the Kumamoto Study [54]—have documented the beneficial effects of glycaemic control in preventing microvascular complications. In the UKPDS study there was a 16% reduction in the occurrence of MI but this did not reach statistical significance, indicating that blood sugar control alone is not sufficient to prevent MI. Thus, a multifaceted approach, including controlling glucose, BP, and serum lipid levels, is needed to prevent CVD in patients with diabetes, as shown by the STENO-2 Study [55].

Unfortunately, replicating the international consensus on the treatment and care of patients with diabetes mellitus and hypertension is a huge task in developing nations such as India, especially in rural areas. One major challenge is the availability of trained physicians to screen, identify and treat the conditions in these regions. Screening can be successfully performed by non-physicians, as shown in the Chunampet Rural Diabetes Prevention Project (CRDPP) [56]. The CRDPP is a successful model of diabetes healthcare and prevention to underserved rural areas in developing countries such as India, which provides an example for others on how to deliver quality screening and treatment tools. Large-scale screening for diabetes mellitus and hypertension is possible, but the challenge then becomes long-term follow-up and delivery of pharmacological therapy in a cost-effective manner because the diseases are lifelong.

12.8 Conclusion

The prevalence of diabetes and hypertension is rising in epidemic proportions, and these conditions represent some of the biggest challenges facing healthcare systems in developing countries like India. Diabetes increases the risk of CVD, and this risk is multiplied by coexisting hypertension. All major molecular mechanisms that contribute to the microvascular and macrovascular complications of diabetes, including oxidative stress, inflammation and fibrosis, also cause vascular remodelling and dysfunction in hypertension. Targeting diabetes at all levels of prevention including primordial prevention, primary prevention, secondary prevention and tertiary prevention along with controlling comorbidities, especially hypertension, and good strategies to promote vascular health will be important for reducing the microvascular and macrovascular complications of diabetes. There is an urgent need to improve monitoring and management of risk factors through primary care-linked programmes. Public health policies and large-scale public education have a huge role in reducing the risk of both diabetes and hypertension, and thus of CVD, in the community.

References

1. Unwin N, Whiting D, Guariguata L, Ghyoot G, Gan D, editors. Diabetes atlas. 5th ed. Brussels: International Diabetes Federation; 2011.
2. Zhou B, Bentham J, Di Cesare M, Bixby H, Danaei G, Cowan MJ, et al. Worldwide trends in blood pressure from 1975 to 2015: a pooled analysis of 1479 population-based measurement studies with 19· 1 million participants. Lancet. 2017;389(10064):37–55.
3. Kaplan NM. The deadly quartet. Upper-body obesity, glucose intolerance, hypertriglyceridemia, and hypertension. Arch Intern Med. 1989;149(7):1514–20.
4. Hayashi T, Tsumura K, Suematsu C, Endo G, Fujii S, Okada K. High normal blood pressure, hypertension, and the risk of type 2 diabetes in Japanese men. The Osaka Health Survey. Diabetes Care. 1999;22(10):1683–7.
5. Tatsumi Y, Ohkubo T. Hypertension with diabetes mellitus: significance from an epidemiological perspective for Japanese. Hypertens Res. 2017;40(9):795–806.
6. Lastra G, Syed S, Kurukulasuriya LR, Manrique C, Sowers JR. Type 2 diabetes mellitus and hypertension: an update. Endocrinol Metab Clin N Am. 2014;43(1):103–22.
7. Fox CS. Cardiovascular disease risk factors, type 2 diabetes mellitus, and the Framingham Heart Study. Trends Cardiovasc Med. 2010;20(3):90–5.
8. Chen G, McAlister FA, Walker RL, Hemmelgarn BR, Campbell NR. Cardiovascular outcomes in Framingham participants with diabetes: the importance of blood pressure. Hypertension. 2011;57(5):891–7.
9. Liang M. Epigenetic mechanisms and hypertension. Hypertension. 2018;72(6):1244–54.
10. Sowers JR, Whaley-Connell A, Hayden MR. The role of overweight and obesity in the cardiorenal syndrome. Cardiorenal Med. 2011;1(1):5–12.
11. Massiéra F, Bloch-Faure M, Ceiler D, Murakami K, Fukamizu A, Gasc JM, et al. Adipose angiotensinogen is involved in adipose tissue growth and blood pressure regulation. FASEB J. 2001;15(14):2727–9.
12. Mehta PK, Griendling KK. Angiotensin II cell signaling: physiological and pathological effects in the cardiovascular system. Am J Physiol Cell Physiol. 2007;292(1):C82–97.
13. Berraho M, El Achhab Y, Benslimane A, El Rhazi K, Chikri M, Nejjari C. Hypertension and type 2 diabetes: a cross-sectional study in Morocco (EPIDIAM Study). Pan Afr Med J. 2012;11:52.
14. Broussard BA, Valway SE, Kaufman S, Beaver S, Gohdes D. Clinical hypertension and its interaction with diabetes among American Indians and Alaska Natives. Estimated rates from ambulatory care data. Diabetes Care. 1993;16(1):292–6.
15. Cappuccio FP, Barbato A, Kerry SM. Hypertension, diabetes and cardiovascular risk in ethnic minorities in the UK. Br J Diabetes Vasc Dis. 2003;3(4):286–93.
16. Emdin CA, Anderson SG, Woodward M, Rahimi K. Usual blood pressure and risk of new-onset diabetes: evidence from 4.1 million adults and a meta-analysis of prospective studies. J Am Coll Cardiol. 2015;66(14):1552–62.
17. Assmann G, Schulte H. The Prospective Cardiovascular Munster (PROCAM) study: prevalence of hyperlipidemia in persons with hypertension and/or diabetes mellitus and the relationship to coronary heart disease. Am Heart J. 1988;116:1713–24.
18. Somaratne JB, Whalley GA, Poppe KK, ter Bals MM, Wadams G, Pearl A, et al. Screening for left ventricular hypertrophy in patients with type 2 diabetes mellitus in the community. Cardiovasc Diabetol. 2011;10:29.

19. Govind S, Saha S, Brodin LA, Ramesh SS, Arvind SR, Quintana M. Impaired myocardial functional reserve in hypertension and diabetes mellitus without coronary artery disease: searching for the possible link with congestive heart failure in the myocardial Doppler in diabetes (MYDID) study II. Am J Hypertens. 2006;19(8):851–7; discussion 858.
20. Grossman E, Messerli FH, Goldbourt U. High blood pressure and diabetes mellitus: are all antihypertensive drugs created equal? Arch Intern Med. 2000;160(16):2447–52.
21. Lea JP, Nicholas SB. Diabetes mellitus and hypertension: key risk factors for kidney disease. J Natl Med Assoc. 2002;94(8 Suppl):7S–15S.
22. Knowler WC, Bennett PH, Ballintine EJ. Increased incidence of retinopathy in diabetics with elevated blood pressure. A six-year follow-up study in Pima Indians. N Engl J Med. 1980;302(12):645–50.
23. Parving HH. Hypertension Optimal Treatment (HOT) trial. Lancet. 1998;352(9127):574–5.
24. Hansson L, Zanchetti A, Carruthers SG, Dahlöf B, Elmfeldt D, Julius S, et al. Effects of intensive blood-pressure lowering and low-dose aspirin in patients with hypertension: principal results of the Hypertension Optimal Treatment (HOT) randomised trial. HOT Study Group. Lancet. 1998;351(9118):1755–62.
25. Deepa R, Shanthirani CS, Pradeepa R, Mohan V. Is the 'rule of halves' in hypertension still valid? Evidence from the Chennai Urban Population Study. J Assoc Physicians India. 2003;51:153–7.
26. Ranjit Unnikrishnan I, Anjana RM, Mohan V. Importance of controlling diabetes early—the concept of metabolic memory, legacy effect and the case for early insulinisation. J Assoc Physicians India. 2011;59(Suppl):8–12.
27. Williams B, Mancia G, Spiering W, et al. 2018 ESC/ESH guidelines for the management of arterial hypertension. The Task Force for the management of arterial hypertension of the European Society of Cardiology (ESC) and the European Society of Hypertension (ESH). Eur Heart J. 2018;39:3021–104.
28. Whelton PK, Carey RM, Aronow WS, et al. 2017 ACC/AHA/AAPA/ABC/ACPM/AGS/APhA/vASH/ASPC/NMA/PCNA guideline for the prevention, detection, evaluation, and management of high blood pressure in adults. Hypertension. 2018;71:e13–e115.
29. Pickering TG, Hall JE, Appel LJ, Falkner BE, Graves J, Hill MN, et al.; Subcommittee of Professional and Public Education of the American Heart Association Council on High Blood Pressure Research. Recommendations for blood pressure measurement in humans and experimental animals: part 1: blood pressure measurement in humans: a statement for professionals from the Subcommittee of Professional and Public Education of the American Heart Association Council on High Blood Pressure Research. Hypertension. 2005;45(1):142–61.
30. Powers BJ, Olsen MK, Smith VA, Woolson RF, Bosworth HB, Oddone EZ. Measuring blood pressure for decision making and quality reporting: where and how many measures? Ann Intern Med. 2011;154(12):781–8.
31. Shah SN, Core Committee Members: Billimoria AB, Mukherjee S, Kamath S, Munjal YP, Maiya M, Wander GS, Mehta N. Indian guidelines on management of hypertension (I.G.H) – IV 2019. Suppl J Assoc Phys India (JAPI). 2019;67(9):1–48.
32. Shibao C, Gamboa A, Diedrich A, Biaggioni I. Management of hypertension in the setting of autonomic failure: a pathophysiological approach. Hypertension. 2005;45(4):469–76. https://doi.org/10.1161/01.HYP.0000158835.94916.0c. Epub 2005 Feb 28. PMID: 15738343.
33. de Boer IH, Bangalore S, Benetos A, Davis AM, Michos ED, Muntner P, et al. Diabetes and hypertension: a position statement by the American Diabetes Association. Diabetes Care. 2017;40(9):1273–84.
34. American Diabetes Association. 10. cardiovascular disease and risk management: standards of medical care in diabetes-2020. Diabetes Care. 2020;43(Suppl 1):S111–34.
35. Garber AJ, Abrahamson MJ, Barzilay JI, Blonde L, Bloomgarden ZT, Bush MA, et al. Consensus statement by the American Association of Clinical Endocrinologists and American College of Endocrinology on the comprehensive type 2 diabetes management algorithm—2017 executive summary. Endocr Pract. 2017;23(2):207–38.

36. American Diabetes Association. 6. Glycemic targets: standards of medical care in diabetes-2020. Diabetes Care. 2020;43(Suppl 1):S66–76.
37. Handelsman Y, Bloomgarden ZT, Grunberger G, Umpierrez G, Zimmerman RS, Bailey TS, et al. American association of clinical endocrinologists and American College of Endocrinology—clinical practice guidelines for developing a diabetes mellitus comprehensive care plan—2015. Endocr Pract. 2015;21:1–87.
38. Nathan DM, Cleary PA, Backlund JY, Genuth SM, Lachin JM, Orchard TJ, et al.; Diabetes Control and Complications Trial/Epidemiology of Diabetes Interventions and Complications (DCCT/EDIC) Study Research Group. Intensive diabetes treatment and cardiovascular disease in patients with type 1 diabetes. N Engl J Med. 2005;353(25):2643–53.
39. UK Prospective Diabetes Study (UKPDS) Group. Intensive blood-glucose control with sulphonylureas or insulin compared with conventional treatment and risk of complications in patients with type 2 diabetes (UKPDS 33). Lancet. 1998;352(9131):837–53.
40. Gaede P, Vedel P, Larsen N, Jensen GV, Parving HH, Pedersen O. Multifactorial intervention and cardiovascular disease in patients with type 2 diabetes. N Engl J Med. 2003;348(5):383–93.
41. Falkner B, Lurbe E. Primordial prevention of high blood pressure in childhood: an opportunity not to be missed. Hypertension. 2020;75(5):1142–50.
42. Beaglehole R, Epping-Jordan J, Patel V, Chopra M, Ebrahim S, Kidd M, et al. Improving the prevention and management of chronic disease in low-income and middle-income countries: a priority for primary health care. Lancet. 2008;372(9642):940–9.
43. Krousel-Wood MA, Muntner P, He J, Whelton PK. Primary prevention of essential hypertension. Med Clin North Am. 2004;88(1):223–38.
44. Whelton PK, He J, Appel LJ, Cutler JA, Havas S, Kotchen TA, et al.; National High Blood Pressure Education Program Coordinating Committee. Primary prevention of hypertension: clinical and public health advisory from the National High Blood Pressure Education Program. JAMA. 2002;288(15):1882–8.
45. National Health Care Disparities Report. 2010. https://fodh.phhp.ufl.edu/files/2011/05/AHRQ-disparities-2010.pdf.
46. Knowler WC, Barrett-Connor E, Fowler SE, Hamman RF, Lachin JM, Walker EA, et al.; Diabetes Prevention Program Research Group. Reduction in the incidence of type 2 diabetes with lifestyle intervention or metformin. N Engl J Med. 2002;346(6):393–403.
47. Klein S, Sheard NF, Pi-Sunyer X, Daly A, Wylie-Rosett J, Kulkarni K, et al.; American Diabetes Association; North American Association for the Study of Obesity; American Society for Clinical Nutrition. Weight management through lifestyle modification for the prevention and management of type 2 diabetes: rationale and strategies. A statement of the American Diabetes Association, the North American Association for the Study of Obesity, and the American Society for Clinical Nutrition. Am J Clin Nutr. 2004;80(2):257–63.
48. Knowler WC, Barrett-Connor E, Fowler SE, Hamman RF, Lachin JM, Walker EA, et al. Reduction in the incidence of type 2 diabetes with lifestyle intervention or metformin. N Engl J Med. 2002;346(6):393–403.
49. Whelton PK, Appel LJ, Espeland MA, Applegate WB, Ettinger WH Jr, Kostis JB, et al. Sodium reduction and weight loss in the treatment of hypertension in older persons: a randomized controlled trial of nonpharmacologic interventions in the elderly (TONE). TONE Collaborative Research Group. JAMA. 1998;279(11):839–46.
50. Hu G, Barengo NC, Tuomilehto J, Lakka TA, Nissinen A, Jousilahti P. Relationship of physical activity and body mass index to the risk of hypertension: a prospective study in Finland. Hypertension. 2004;43(1):25–30.
51. The Trials of Hypertension Prevention Collaborative Research Group. The effects of nonpharmacologic interventions on blood pressure of persons with high normal levels. Results of the trials of hypertension prevention, phase I. JAMA. 1992;267:1213–20.
52. Nathan DM, Genuth S, Lachin J, Cleary P, Crofford O, Davis M, et al. The effect of intensive treatment of diabetes on the development and progression of long-term complications in insulin-dependent diabetes mellitus. N Engl J Med. 1993;329(14):977–86.

53. Intensive blood-glucose control with sulphonylureas or insulin compared with conventional treatment and risk of complications in patients with type 2 diabetes (UKPDS 33). UK Prospective Diabetes Study (UKPDS) Group. Lancet. 1998;352(9131):837–53.
54. Shichiri M, Kishikawa H, Ohkubo Y, Wake N. Long-term results of the Kumamoto Study on optimal diabetes control in type 2 diabetic patients. Diabetes Care. 2000;23(Suppl 2):B21–9.
55. Gaede P, Lund-Andersen H, Parving HH, Pedersen O. Effect of a multifactorial intervention on mortality in type 2 diabetes. N Engl J Med. 2008;358(6):580–91.
56. Mohan V, Deepa M, Pradeepa R, Prathiba V, Datta M, Sethuraman R, et al. Prevention of diabetes in rural India with a telemedicine intervention. J Diabetes Sci Technol. 2012;6(6):1355–64.

Heart Failure in Different Asian Populations

13

V. K. Chopra, S. Harikrishnan, and Deepanjan Bhattacharya

13.1 Introduction

Cardiovascular diseases (CVDs) are the leading cause of mortality and morbidity all over the world. The transition from communicable to non-communicable diseases as the main contributor to overall mortality occurred in the developing countries in the mid-1990s. Heart failure (HF) is emerging as one of the main contributors to CVD burden in both the developed and the developing world [1, 2].

HF is a clinical syndrome where symptoms and signs are due to structural and/or functional cardiac abnormalities that result in elevated intracardiac pressures and/or inadequate cardiac output at rest and/or during exercise [1]. The overall prevalence of HF is 1–2%, but prevalence increases sharply with advancing age [2]. HF is categorised based on the left ventricular ejection fraction (LVEF) into HF with reduced ejection fraction (HFrEF; LVEF <40%), HF with mildly reduced ejection fraction (HFmrEF; LVEF 40–49%) and HF with preserved ejection fraction (HFpEF; LVEF >50%) [3].

HF is a disease with high morbidity and mortality. Management of HF is resource intensive and the vast majority of patients with HF require lifelong therapy. HF is also an important contributor to disability-adjusted life years (DALYs) [4]. Outcomes for patients with HF are often poor, with 1- and 5-year mortality rates of 30% and 50–60% respectively, which is worse than some common malignancies such as breast and colon cancer [5].

V. K. Chopra (✉)
Heart Failure and Research, Max Super Speciality Hospital, Saket, New Delhi, India

S. Harikrishnan · D. Bhattacharya
Department of Cardiology, Sree Chitra Tirunal Institute for Medical Sciences and Technology, Thiruvananthapuram, Kerala, India

Although there is a substantial body of data on HF in Western populations, there is comparatively less data from the developing world, especially Asian countries. Given that the Asian region is home to 60% of the world's population, it is likely to have a huge burden of HF. In addition, the Asian region is unique because of differences in ethnicity and the size and distribution of the population.

13.2   Hypertensive Heart Failure

Patients with persistent and uncontrolled hypertension initially develop left ventricular (LV) hypertrophy which progresses to HF. Significant LV hypertrophy results in diastolic dysfunction and patients initially develop HFpEF. Later, the LV starts dilating and systolic dysfunction is predominant. Degrees of heart involvement in patients with hypertension are defined in Fig. 13.1, while the different forms of cardiac involvement and progression from hypertension to HF and death are summarised in Fig. 13.2.

13.3   HF Registries from Asia

Even though there is less data on the incidence and prevalence of HF from the Asian region, data from various registries in the region is starting to emerge. This provides the scientific community with reasonably good data about HF-related morbidity and mortality, and HF management practices in the region (Table 13.1).

The Trivandrum Heart Failure Registry (THFR) was setup in 2013 and enrolled 1205 patients [8]. The most common forms of heart disease were coronary artery disease and rheumatic heart disease. Major comorbidities included hypertension (58%), diabetes mellitus (55%), chronic kidney disease (CKD; 18%) and chronic obstructive pulmonary disease (COPD; 15.4%). HFpEF was found in 19.6% of patients, HFmrEF in 18% and HFrEF in 62.4%. Patients with HFpEF were more likely to be female, have a lower prevalence of CAD, tobacco use and diabetes, a higher prevalence of atrial tachyarrhythmia, and were less likely to be receiving treatment with beta-blockers and renin-angiotensin-aldosterone system (RAAS) inhibitors.

Only 25% of patients with LV systolic dysfunction were being treated with guideline-directed medical treatment (GDMT; a combination of beta-blockers and RAAS inhibitors). The rate of hospital readmissions at 1 year was 30%, with no gender bias. Major predictors of hospital readmission were GDMT and New York Heart Association (NYHA) class IV symptoms. The median duration of hospital stay was 6 days, and rates of in-hospital and 90-day mortality were 8.5% and 18%, respectively. Mortality rates at 1, 3 and 5 years were 30.8%, 44.8% and 58.8%, respectively. Important predictors of mortality were lack of GDMT, increasing age, increasing serum creatinine, and NYHA functional class IV HF at presentation [9].

The Medanta Heart Failure Registry (MHFR) is a single centre registry from Delhi. Over the period 2014–2017 it enrolled 5590 patients with HFrEF (mean age

Fig. 13.1 Stages of hypertensive heart disease. *LV* left ventricular, *LVH* left ventricular hypertrophy

Heart disease resulting from hypertension can be divided into four degrees of severity

- Degree I   :   Isolated LV diastolic dysfunction with no LVH
- Degree II   :   LV diastolic dysfunction with concentric LVH
- Degree III :   Clinical heart failure (dyspnea and pulmonary edema with preserved ejection fraction)
- Degree IV   :   Dilated cardiomyopathy with heart failure and reduced ejection fraction

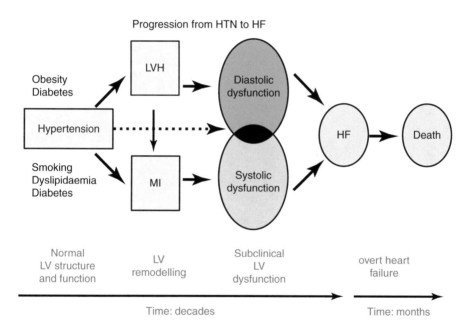

Fig. 13.2 Evolution of hypertensive heart disease in patients with hypertension and other risk factors. *HF* heart failure, *LV* left ventricular, *MI* myocardial infarction

59.1 ± 11.8 years, M:F ratio, 4.9:1). CAD was the most common aetiology for HF (in 78% of patients), while 7.8% of patients had underlying RHD. Diabetes and hypertension were present in 49% of patients. The mean estimated glomerular filtration rate (eGFR) was 76.1 ± 27.7 ml/min/m². Treatments included beta-blockers, RAAS inhibitors and diuretics in 81.8%, 65.8% and 79.4% of patients, respectively. Rates of implantable cardioverter defibrillator (ICD) and cardiac resynchronisation therapy (CRT) device implantation were 4.3% each. The 1-year mortality rate was 17.6%, and major predictors of mortality were age >50 years, higher NYHA class, LVEF <30%, renal dysfunction, anaemia and lack of GDMT [10].

Table 13.1 Baseline characteristics of patients enrolled in representative HF registries

Name [Ref]	Region/country	N	Enrolment, years	Mean age (SD), years	Female, %	Aetiology	Mortality
Heart failure registries from Asia							
Kerala HF Registry [6]	India	7512	2016–2018	64 (13)	36%	CAD: 65% HFrEF: 67.5% HFmrEF: 17.8%	In-hospital: 6.1% 90-day: 10.3%
India National HF Registry [7]	India	7500	2019–2020	60 (14)	31%	CAD: 61% HFpEF: 14.9% CAD: 65%	In-hospital: 6.1%
Trivandrum HF Registry [8, 9]	India	1205	2013	61 (14)	31%	CAD: 71% RHD: 8% HFrEF: 62%	1-year: 30.8 3-year: 40.8 5-year: 44.8
Medanta HF registry [10]	India	5590	2014–2017	59 (12)	17%	CAD: 78% RHD: 4.8%	1-year: 17.6%
ASIAN-HF [11]	North, South-East, and South Asia	6480	2012–2016	61 (13)	27%	CAD: 73% RHD: 6.3% HFpEF:13% HFrEF: 67%	1-year: 9.6%
CHART-1 [12]	Japan	1278	2001–2004	69 (13)	33%	HFrEF: 55% CAD: 25%	1-year: 7% 2-year: 16% 3-year: 22%
CHART-2 [13]	Japan	4375	2006–2010	69 (12)	32%	CAD: 53% HFrEF: 20%	
CHINA-HF [14]	China	13,687	2015	65 (15)	41%	CAD: 49.6% HFrEF: 39.6%	In-hospital: 4.1%
NCCQI-HF [15]	China	34,938	2020	67 (14)	39%		

KOR-AHF [16]	South Korea	5625	2011–2014	69 (15)	47%	CAD: 43% HFrEF: 60.5% Valvular: 14.3%	In-hospital: 7.6% 1-year 8.4%
Gulf CARE [17]	Middle East	5005	2012	59 (15)	37%	CAD: 53% HFrEF: 69%	In-hospital: 6.3% 1-year: 20.1%
Heart failure registries from the West							
ADHERE [18]	USA	105,388		72 (14)	43	HFrEF: 47% CAD: 57%	In-hospital: 4.3%
OPTIMIZE [19] HF	USA	41,267		73 (14)	38	HFrEF: 48.8% CAD: 22.9%	
EHFS II [20]	Europe	3580		70 (12)	43	HFrEF: 65.7% CAD: 53.6%	In-hospital: 6.7%

CAD coronary artery disease, *HFpEF* heart failure with preserved ejection fraction, *HFrEF* heart failure with reduced ejection fraction, *RHD* rheumatic heart disease

The **Kerala Acute Heart Failure Registry (KAHFR)** enrolled 7512 patients with acute HF from 2016 to 2018. Mean age was 64.3 ± 12.9 years with a M:F ratio of 1.7:1. More than two-thirds of patients (67.5%) had HFrEF, while 17.8% had HFmrEF and 14.9% had HFmrEF and HFpEF. The aetiology of HF was CAD accounted in 65% of patients. Rates of GDMT were low, being 27.9% in patients with HFrEF and 20.2% in those with HFmrEF. In-hospital mortality was 6.1%, and 90-day mortality was 10.3%. Mortality rates per 100 person-days were 14.1, 10.7 and 10.9 in patients with HFrEF, HFmrEF or HFpEF, respectively. Key predictors of mortality were lack of GDMT, older age, CKD, stroke, CAD, atrial fibrillation (AF) and anaemia [6].

The **National Heart Failure Registry** was established in 2019 and has enrolled more than 10,500 patients from 51 centres all over India. Interim data from 7500 patients are available (Table 13.1). Mean age at enrolment was 60.3 ± 13.5 years, with a M:F ratio of 2.2:1. HFrEF was the most common type of HF (61% of patients), followed by HFmrEF (23%) and HFpEF (13%). The underlying HF aetiology was CAD in 73% and RHD in 6.3%. Major comorbidities included hypertension (49%), diabetes mellitus (43%), CKD (9%) and atrial arrhythmia (10%). Only 43% of patients were receiving GDMT. The in-hospital mortality rate was 6.1%. Patients with HFrEF had a higher prevalence of CAD and diabetes, with lower blood pressure and higher mortality. Patients with HFpEF had a higher prevalence of atrial arrhythmias and COPD. Use of GDMT was higher in patients with HFrEF [7].

The **ASIAN HF registry** recruited 6480 patients from three Asian regions from 2012 to 2015 [11]. South-East Asia included Thailand, Malaysia, Philippines, Indonesia and Singapore, North-East Asia included South Korea, Japan, Taiwan, Hong Kong and China, while South Asia included India. Mean age was 61.6 ± 13.3 years, and M:F ratio was 2.7:1. Ischaemic aetiology accounted for 43.8% of cases. HFrEF was common (81% of patients). Major comorbidities included hypertension (55%), diabetes mellitus (41%), CKD (35.6%) and atrial arrhythmias (12.8%). ICD use was seen in 4.3% of patients, and biventricular pacing in 1.4%. Drug therapy included diuretics in 76.5% of patients, beta-blockers in 72.2%, RAAS inhibitors in 70.4% and mineralocorticoid receptor antagonists (MRAs) in 49.7%. The 1-year mortality rate was 9.6% (10.6% in HFrEF and 5.4% in HFpEF), with major predictors being LV systolic dysfunction, advanced age, HF re-hospitalisation in the previous 6 months, obesity, NYHA class III/IV symptoms, lower systolic blood pressure (SBP) at admission, AF, and renal dysfunction. Patients from South-East Asia were younger, had a higher prevalence of comorbidities, and a higher 1-year mortality rate compared with patients from other Asian regions.

The **CHART 1 registry** from Japan included 1278 patients from 2000 to 2005 (mean age 68.3 ± 13.4 years, M:F ratio 2.1:1, 55% with HFrEF) [12]. One quarter had an ischaemic HF aetiology and 26.4% had valvular pathology. Prevalence rates of hypertension, diabetes mellitus and atrial arrhythmia were 47%, 19% and 41%, respectively. Mortality at 1, 2 and 3 years was 7%, 16% and 22%, respectively. The main predictors of mortality were older age, diabetes mellitus, ventricular tachycardia, higher levels of B-type natriuretic peptide, LVEF <30% and higher NYHA functional class.

The CHART 2 registry included 10,219 patients from 2006 to 2010, with a mean age of 68.2 ± 12.3 years, M:F ratio of 2.3:1, 80% HFpEF [13]. HF has an ischaemic aetiology in 53% of patients and a valvular pathology in 19%. Major comorbidities included hypertension (76%), diabetes mellitus (23%), CKD (39%) and AF (23%). RAAS inhibitors were used by 63% of patients, beta-blockers by 40%, MRAs by 15%, and loop diuretics by 30%. At a median follow-up of 3.1 years, the mortality rate was about 50 per 1000 patient years.

The INTER CHF study enrolled 5813 patients worldwide from 2012 to 2014, of which 2661 (46%) were from Asia (M:F ratio 1.4:1) [21]. The Asian cohort had a mean age of 60.0 ± 0.3 years, older than the African population and younger than those from South America. HF aetiology was CAD in 48% of patients and RHD in 10%. Major comorbidities included hypertension (59%), diabetes mellitus (27.9%) and CKD (7.1%). HF therapy included diuretics in 62.1% of patients, RAAS inhibitors in 67.9%, beta-blockers in 60.8%, MRAs in 44% and digoxin in 27.6%. One-year mortality rates were highest in India (23%), followed by South-East Asia (15%), the Middle East (9%) and China (7%). Predictors of mortality included older age, lower body mass index, valvular heart disease, HF hospitalisation, lower SBP, LV dysfunction, COPD, illiteracy, and lack of treatment with beta-blockers and RAAS inhibitors [22].

The Gulf CARE Registry enrolled 5005 patients from 47 hospitals in seven Middle East countries (Bahrain, Kuwait, Oman, Qatar, Saudi Arabia, United Arab Emirates and Yemen) between February and November 2012 [17]. Mean age of the population was 59 ± 15 years, with M:F ratio of 2.3:1. HFrEF was seen in 69%, and median LVEF was 35%. HF aetiology was CAD in 53% and valvular heart disease in 13%. Major comorbidities included hypertension (61%), diabetes mellitus (50%), atrial arrhythmia (14%) and CKD (15%). Diuretics were being used by 94% of patients, beta-blockers by 71%, RAAS inhibitors by 78% and MRAs by 43%; 2.1% had an implanted ICD and 1.1% had undergone CRT device implantation. In-hospital mortality was 6.3%, and re-hospitalisation rates at 3 and 12 months were 18% and 40%. Cumulative mortality rates at 3 and 12 months were 13% and 20.1%.

The HEARTS registry enrolled 1090 patients from Saudi Arabia from 2009 to 2011 (mean age 60.6 ± 15.3 years, M:F ratio 1.45:1) [23]. HFrEF was seen in 63.3% of patients, and 70.6% had severe LV dysfunction. Major comorbidities included hypertension (70.9%), diabetes mellitus (57.9%), CKD (29.8%) and atrial arrhythmias (15.7%). The most commonly used HF therapies were beta-blockers (95% of patients), diuretics (94%), RAAS inhibitors (86%) and MRAs (53%). ICD and CRD devices were implanted in 29.1% and 8.8% of patients, respectively. In-hospital, 30-day and 1-year mortality rates were 5.3%, 7.5% and 9%, respectively.

The Korean Acute Heart Failure registry (KorAHF) included 5625 patients from 2011 to 2014 (mean age 68.5 ± 14.5 years, M:F ratio 1.1:1) [16]. The most common form of HF was HFrEF (60.5% of patients). Aetiology was CAD in 42.9% of patients and valvular heart disease in 14.3%. RAAS inhibitors were used by 68.8% of patients, beta-blockers by 43% and MRAs by 37.1%. Comorbidities included hypertension (59.1%), diabetes mellitus (35.3%), CKD (14.3%) and AF (34.9%). In-hospital mortality was 7.6% and 1-year mortality was 8.4%. Predictors of mortality were older age, renal dysfunction, SBP <100 mmHg and lack of GDMT.

The China-HF registry enrolled 13,687 patients from 132 centres from 2012 to 2015 [14]. The mean age of the study population was 65 ± 15 years and the M:F ratio was 1.44. CAD was the underlying HF aetiology in 49.6% of cases, and 39.6% had HFrEF. Common comorbidities included hypertension (50.9%), CKD (46.7%), diabetes (21%) and AF (24.4%). The proportion of patients taking diuretics, angiotensin-converting enzyme inhibitors or angiotensin receptor blockers, and beta-blockers was 30.1%, 27.0%, and 25.6%, respectively. The median length of hospital stay was 10 (range 7–15) days, and in-hospital mortality was 4.1 ± 0.3%. Predictors of mortality included low SBP, acute MI, infection, right bundle branch block, and elevated total bilirubin and blood urea nitrogen levels.

13.4 Special Features in Asian Population

The data above indicate that, compared with Western populations, the Asian population of HF patients is a decade younger in terms of mean age, with male predominance. The proportion of patients with HFrEF is significantly higher and the proportion with HFpEF much lower. RHD aetiology was only specifically mentioned in studies from India. The prevalence of CAD was highest in the Indian cohorts, but similar to Western data in other groups [18–20]. Asian cohorts also showed a high in-hospital mortality rate. Some registries reported long-term mortality. One of these was the THFR, which reported 1-year, 3-year and 5-year mortality rates of 30%, 45% and nearly 60%, respectively [24].

Rates of optimal GDMT usage and cardiac device implantation of devices were much lower in Asian cohorts compared with Western data. This could be due to issues relating to availability, accessibility and affordability. Late referral in patients with acute coronary syndromes and not getting evidence-based therapy (e.g., timely revascularisation) may be another reason underlying the high mortality rates.

As a comparison, the THESEUS registry reported data from Africa (2007–2010; $n = 1006$; mean age 52.3 ± 18.3 years; M:F ratio 0.97:1) [25]. The cause of HF was most commonly hypertension (45.4% of patients), followed by RHD (14.3%) and CAD (7.7%). Renal dysfunction, diabetes mellitus and AF were present in 7.7%, 11.4% and 18.3% of patients, respectively. The prevalence of HIV was quite high (13%), while the median duration of hospitalisation was 7 days. In-hospital and

6-months mortality rates were 4.2% and 17.8%, respectively. The patients in this cohort were a decade younger than the Asian patients, with a similar proportion of males and females. Surprisingly, the prevalence of CAD was significantly lower, with similar in-hospital mortality but higher 6-month mortality.

13.5 Intra-continental Variation

Comparing data between Asian regions, age distribution was similar but the proportion of male patients was slightly higher in India and Gulf nations compared with South-East Asia. The prevalence of CAD was highest in patients from India, while CKD was most prevalent in those from China, despite this group having the lowest prevalence of diabetes mellitus. Rates of GDMT prescription were lowest in patients with HF from China, while this country also had the lowest in-hospital mortality rate. Data on the contribution of RHD to HF is lacking from China and the Middle East, while this aetiology was present in one-fifth of patients from India, Korea and Saudi Arabia. Despite the high prevalence of HFrEF in these countries, the use of MRAs was low. Implantation of devices (CRT and ICD) was very low in all Asian countries, but higher in patients from Saudi Arabia. HF patients from Japan tended to be older, with a higher prevalence of HFpEF, renal dysfunction and hypertension, although medication usage was similar to that in the rest of the Asian cohort.

13.6 What Lies in the Future?

These HF registries are useful to describe the real-world clinical situation in different countries and regions. With an increasing proportion of elderly in many populations, lack of physical activity, low intake of fruits and vegetables, rising prevalence of diabetes mellitus, hypertension and CAD, the prevalence of HF in Asia is likely to rise over coming decades [26].

Treatment of HF is resource intensive and is not affordable for the majority of patients in Asia. Therefore, there needs to be a focus on the prevention of HF. These preventive measures can be primordial (e.g., cultivating healthy habits in the population at a young age), or primary (early identification and control of risk factors such as diabetes and hypertension). Secondary prevention is also important in HF by providing continued follow-up, risk factor control and evidence-based care. For HFrEF, this includes treatment with low-cost agents such as beta-blockers and RAAS inhibitors. With the advent of newer drug classes such as the angiotensin receptor-neprilysin inhibitors and sodium-glucose cotransport 2 inhibitors, there is even more scope for improving the quality of life and outcomes in patients with HF.

Lack of adequate manpower and overcrowded public health facilities are other important issues in Asian countries. HF nurse-based follow-up is a very well accepted practice in the West, which can be adopted as a cost-effective solution.

In conclusion, the prevalence of HF in most Asian countries is high, and will likely increase due to factors such as greater longevity, a sedentary lifestyle and increasing prevalence of CAD. It is important to recognise differences in HF risk factors and outcomes in Asian populations, and design preventive and therapeutic protocols accordingly. Hypertension is a major risk factor for HF in the Asian population, meaning that good control of blood pressure will help to reduce the prevalence of HF.

References

1. McDonagh TA, Metra M, Adamo M, Gardner RS, Baumbach A, Böhm M, et al. 2021 ESC guidelines for the diagnosis and treatment of acute and chronic heart failure. Eur Heart J. 2021;42(36):3599–726.
2. Savarese G, Lund LH. Global public health burden of heart failure. Card Fail Rev. 2017;3(1):7–11.
3. Bozkurt B, Coats AJS, Tsutsui H, Abdelhamid CM, Adamopoulos S, Albert N, et al. Universal definition and classification of heart failure: a report of the Heart Failure Society of America, Heart Failure Association of the European Society of Cardiology, Japanese Heart Failure Society and Writing Committee of the Universal Definition of Heart Failure: endorsed by the Canadian Heart Failure Society, Heart Failure Association of India, Cardiac Society of Australia and New Zealand, and Chinese Heart Failure Association. Eur J Heart Fail. 2021;23(3):352–80.
4. Bragazzi NL, Zhong W, Shu J, Abu Much A, Lotan D, Grupper A, et al. Burden of heart failure and underlying causes in 195 countries and territories from 1990 to 2017. Eur J Prev Cardiol. 2021;28(15):1682–90.
5. Taylor CJ, Ordóñez-Mena JM, Roalfe AK, Lay-Flurrie S, Jones NR, Marshall T, et al. Trends in survival after a diagnosis of heart failure in the United Kingdom 2000-2017: population based cohort study. BMJ. 2019;364:l223.
6. Stigi J, Jabir A, Sanjay G, Panniyammakal J, Anwar CV, Harikrishnan S. Kerala acute heart failure registry-rationale, design and methods. Indian Heart J. 2018;70 Suppl 1(Suppl 1):S118–20.
7. Harikrishnan S, Bahl A, Roy A, Mishra A, Prajapati J, Nanjappa MC, et al. National Heart Failure Registry, India: design and methods. Indian Heart J. 2019;71(6):488–91.
8. Harikrishnan S, Sanjay G, Anees T, Viswanathan S, Vijayaraghavan G, Bahuleyan CG, et al. Clinical presentation, management, in-hospital and 90-day outcomes of heart failure patients in Trivandrum, Kerala, India: the Trivandrum Heart Failure Registry. Eur J Heart Fail. 2015;17(8):794–800.
9. Sanjay G, Jeemon P, Agarwal A, Viswanathan S, Sreedharan M, Vijayaraghavan G, et al. In-hospital and three-year outcomes of heart failure patients in South India: the Trivandrum Heart Failure Registry. J Card Fail. 2018;24(12):842–8.
10. Chopra VK, Mittal S, Bansal M, Singh B, Trehan N. Clinical profile and one-year survival of patients with heart failure with reduced ejection fraction: the largest report from India. Indian Heart J. 2019;71(3):242–8.
11. MacDonald MR, Tay WT, Teng T-HK, Anand I, Ling LH, Yap J, et al. Regional variation of mortality in heart failure with reduced and preserved ejection fraction across Asia: outcomes in the ASIAN-HF Registry. J Am Heart Assoc. 2020;9(1):e012199.
12. Shiba N, Takahashi J, Matsuki M. The CHART study (Japanese). Naika. 2007;99:410–4.
13. Shiba N, Nochioka K, Miura M, Kohno H, Shimokawa H. Trend of westernization of etiology and clinical characteristics of heart failure patients in Japan—first report from the CHART-2 study. Circ J. 2011;75(4):823–33.

14. Zhang Y, Zhang J, Butler J, Yang X, Xie P, Guo D, et al. Contemporary epidemiology, management, and outcomes of patients hospitalized for heart failure in China: results from the China Heart Failure (China-HF) Registry. J Card Fail. 2017;23(12):868–75.
15. Wang H, Chai K, Du M, Wang S, Cai JP, Li Y, et al. Prevalence and incidence of heart failure among urban patients in China: a national population-based analysis. Circ Heart Fail. 2021;14(10):e008406.
16. Lee SE, Lee H-Y, Cho H-J, Choe W-S, Kim H, Choi JO, et al. Clinical characteristics and outcome of acute heart failure in Korea: results from the Korean Acute Heart Failure Registry (KorAHF). Korean Circ J. 2017;47(3):341–53.
17. Hassan M. Gulf CARE: heart failure in the Middle East. Glob Cardiol Sci Pract. 2015;2015(3):34.
18. Adams KFJ, Fonarow GC, Emerman CL, LeJemtel TH, Costanzo MR, Abraham WT, et al. Characteristics and outcomes of patients hospitalized for heart failure in the United States: rationale, design, and preliminary observations from the first 100,000 cases in the Acute Decompensated Heart Failure National Registry (ADHERE). Am Heart J. 2005;149(2):209–16.
19. Abraham WT, Fonarow GC, Albert NM, Stough WG, Gheorghiade M, Greenberg BH, et al. Predictors of in-hospital mortality in patients hospitalized for heart failure: insights from the Organized Program to Initiate Lifesaving Treatment in Hospitalized Patients with Heart Failure (OPTIMIZE-HF). J Am Coll Cardiol. 2008;52(5):347–56.
20. Nieminen MS, Brutsaert D, Dickstein K, Drexler H, Follath F, Harjola V-P, et al. EuroHeart Failure Survey II (EHFS II): a survey on hospitalized acute heart failure patients: description of population. Eur Heart J. 2006;27(22):2725–36.
21. Dokainish H, Teo K, Zhu J, Roy A, AlHabib KF, ElSayed A, et al. Heart failure in Africa, Asia, the Middle East and South America: the INTER-CHF study. Int J Cardiol. 2016;204:133–41.
22. Dokainish H, Teo K, Zhu J, Roy A, AlHabib KF, ElSayed A, et al. Global mortality variations in patients with heart failure: results from the International Congestive Heart Failure (INTER-CHF) prospective cohort study. Lancet Glob Health. 2017;5(7):e665–72.
23. Alhabeeb W, Elasfar A, AlBackr H, AlShaer F, Almasood A, Alfaleh H, et al. Clinical characteristics, management and outcomes of patients with chronic heart failure: results from the Heart Function Assessment Registry Trial in Saudi Arabia (HEARTS-chronic). Int J Cardiol. 2017;235:94–9.
24. Harikrishnan S, Jeemon P, Ganapathi S, Agarwal A, Viswanathan S, Sreedharan M, et al. Five-year mortality and readmission rates in patients with heart failure in India: results from the Trivandrum Heart Failure Registry. Int J Cardiol. 2021;326:139–43.
25. Damasceno A, Mayosi BM, Sani M, Ogah OS, Mondo C, Ojji D, et al. The causes, treatment, and outcome of acute heart failure in 1006 Africans from 9 countries. Arch Intern Med. 2012;172(18):1386–94.
26. Martinez-Amezcua P, Haque W, Khera R, Kanaya AM, Sattar N, Lam CSP, et al. The upcoming epidemic of heart failure in South Asia. Circ Heart Fail. 2020;13(10):e007218.

Hypertension and Chronic Kidney Disease in Asians

14

Gek Cher Chan and Philip Kam-Tao Li

14.1 Introduction

Chronic kidney disease (CKD) and hypertension can cause, and are known for their associations with one another. In adults with hypertension and CKD, the KDIGO 2021 Clinical Practice Guideline for the Management of Blood Pressure in Chronic Kidney Disease made several new recommendations including method of blood pressure (BP) measurement, lifestyle interventions, BP targets (SBP <120 mmHg when tolerated), and choice of antihypertensive drugs [1]. As such, blood pressure control is imperative to retard CKD progression and to avoid complications associated with hypertension. For this chapter, focusing on countries in Asia, we review the following: (1) epidemiology of hypertension and CKD; (2) relationship of hypertension, and CKD, with clinical outcomes; (3) genetics associated with hypertension and CKD; (4) impact of sodium and dietary patterns; and (5) antihypertensives prescription patterns.

G. C. Chan (✉)
Division of Nephrology, Department of Medicine, National University Hospital, Singapore, Singapore

Department of Medicine, Yong Loo Lin School of Medicine, National University of Singapore, Singapore, Singapore
e-mail: gek_cher_chan@nuhs.edu.sg

P. K.-T. Li
Department of Medicine and Therapeutics, Carol and Richard Yu Peritoneal Dialysis Research Centre, Prince of Wales Hospital, The Chinese University of Hong Kong, Shatin, Hong Kong, SAR, China
e-mail: philipli@cuhk.edu.hk

© The Author(s), under exclusive license to Springer Nature Switzerland AG 2022
C. V. S. Ram et al. (eds.), *Hypertension and Cardiovascular Disease in Asia*,
Updates in Hypertension and Cardiovascular Protection,
https://doi.org/10.1007/978-3-030-95734-6_14

14.2   Epidemiology

The prevalence of CKD in hypertensive patients, hypertension among CKD patients, and end-stage kidney disease (ESKD) caused by hypertension of various Asian countries are presented in Tables 14.1 and 14.2 [2]. The types and proportions of ethnic groups differ among countries, which in turn may contribute to differences in dietary habits, cultural beliefs and practices as well as socioeconomic status. These will be discussed in the later part of the chapter. As the sampling methodologies differ with each country or study, it is difficult and therefore, not recommended to perform direct comparisons with one another.

Knowledge on various international clinical practices in BP control remains limited. Alencar de Pinho et al. analyzed data from independent CKD cohort studies (internal Network of Chronic Kidney Disease, iNET-CKD) to compare the prevalence of uncontrolled BP in adults with CKD, as well as to illustrate prescription patterns of antihypertensives. The study observed lower prevalence ratios of uncontrolled BP in cohorts from North America and high-income Asian countries like Japan and Korea, while higher prevalence ratios in cohorts from European countries, India, and Uruguay [22].

Table 14.1 Prevalence of hypertension and CKD [2]

Area	Population	Hypertension (%)	CKD in hypertensive patients (%)	Hypertension in CKD patients (%)
China	General	35.4 [3]	61.2 [3]	79.8 [4]
Hong Kong	General	27.7 [5]	–	–
India	Opportunistic	43.1 [6]	–	64.5 [7]
Japan	General	60.0 (Male) [8] 41.0 (Female)	–	–
Malaysia	CKD (subgroup)	30.3 [9]	–	38.4 [10]
Singapore	General	23.5 [9]	7.6 [11]	–
South Korea	General	29.1 [12]	19.6 [13]	–
Taiwan	General	24.5 [14]	–	–

Table 14.2 Prevalence of ESKD caused by hypertension [2]

Area	ESKD due to hypertension (%)
China	10.5 [15]
Hong Kong	9.6 [16]
India	12.8 [17]
Japan	9.9 [18]
Singapore	5 [19]
South Korea	20 [20]
Taiwan	8.3 [21]

The prevalence of hypertension and CKD can differ among ethnic groups in countries with multi-ethnic populations, like Singapore [23]. In a cross-sectional study, varying patterns of association between individual BP components and CKD were found among Chinese, Malay, and Indians in Singapore [24]. Among Malays, higher prevalence of hypertension and CKD, and higher levels of SBP and DBP were observed [24]. Similar findings of association between Malays for hypertension and CKD were identified in another two cross-sectional studies conducted in Singapore [25, 26].

14.3 Relationship Between Hypertension and CKD

Association between hypertension and increased risk of developing CKD and cardiovascular disease have been established by several studies [27, 28]. A post-hoc analysis of the China Stroke Primary Prevention Trial (CSPPT) by Jiang et al. observed a U-shaped association between serum albumin levels and the risk of developing CKD among hypertensive patients with normal renal function [29]. In a Japanese prospective cohort study, incident hypertension was also associated with a higher risk of new-onset proteinuria [30].

Associations between hypertension and CKD have been demonstrated in studies performed in South Korea, Malaysia, Singapore, and Thailand [25, 31–34]. In the C-STRIDE study of hypertensive CKD Chinese patients, uncontrolled hypertension was associated with decreased estimated GFR, albuminuria, and higher cardiovascular risk [35, 36]. Several studies in India have also demonstrated hypertension to be a risk factor of CKD including the Screening and Early Evaluation of Kidney Disease (SEEK) [7, 37]. There are also studies evaluating the relationship of hypertension with progression of CKD. Inaguma et al. showed that elevated SBP and increased albuminuria were risk factors of CKD progression in Japan [38].

Visit-to-visit BP variability has been evaluated against various clinical outcomes. The post-hoc analysis of CSPPT study by Li et al. demonstrated SBP variability was associated with development of CKD in hypertensive patients with normal renal function [39]. Li et al. also showed significant association between visit-to-visit BP variability with the risk of subsequent first stroke in hypertensive patients with mild-to-moderate CKD [40]. In Japan, SBP variability was demonstrated to be associated with eGFR decline, particularly in those with proteinuria [41]. Among Malaysian hypertensive patients, visit-to-visit variability was shown to be a determinant of renal function decline [42].

In addition to BP variability, associations of non-dipping, reverse-dipping BP patterns were observed in hypertensive Korean patients with CKD [43]. Within-visit BP variability has also been evaluated. Azushima et al. demonstrated among Japanese CKD patients with hypertension, within-BP visit variability to be a risk factor of cardiovascular disease and use of RAAS inhibitors or alpha-blockers improved variability [44].

14.4   Genetics Associated with Hypertension and CKD

Studies have identified several genetic loci associated with blood pressure and kidney diseases through genome-wide association studies based on various types of ancestries [45, 46]. Within Asia, limited studies have been conducted but several gene polymorphisms have been identified to be associated with hypertension in patients with CKD.

Non-muscle myosin heavy chain 9 (MYH9) is a major susceptibility gene for end-stage renal disease, of which its single nucleotide polymorphisms (SNPs) are associated with hypertension and end-stage renal disease [47]. In a China case-control study by Liu et al., rs3752462 TT genotype of MYH9 was associated with higher risk of concurrent high SBP ≥140 mmHg in patients with CKD [47].

Angiotensin converting enzyme-1 (ACE) is a key player in the renin-angiotensin system that maintains BP and fluid homeostasis. The insertion (I) or absence (D, or deletion) polymorphism of ACE gene is associated with hypertensive renal damage [48]. Another case-control study evaluating ACE gene polymorphism in north Indians, demonstrated strong association of ACE DD genotype with hypertensive state and end-stage renal disease [49].

Apart from ACE, genes of other components of the renin-angiotensin system have been explored. ACE, AGT, and AGTR1 gene polymorphisms were examined by Su et al. on their associations with CKD susceptibility among Chinese in Taiwan [50]. *ACE*-A2350G AA genotype and *AGTR1*-C573T CT genotype were found to be risk factors for CKD [50].

Blood pressure increase after sodium loading is known as salt sensitivity and has been regarded as a potential blood pressure control target. The impact of sodium on hypertension and CKD will be discussed in the next section. Many SNPs have been identified to be associated with salt sensitivity in Asian populations including Chinese, Koreans, and Japanese, that are known to have high-sodium dietary habits [51–57]. As such, sodium restricted diets is a strategy to alleviate hypertension in patients with these polymorphisms.

14.5   Impact of Sodium and Dietary Patterns

In adults with pre-dialysis CKD, a recommended limit of sodium intake of <100 mmol/day or <2.3 g/day (by KDOQI Clinical Practice Guideline for Nutrition in CKD 2020 Update), and <90 mmol/day (<2 g/day) or <5 g of sodium chloride/day (by KDIGO 2021 Clinical Practice Guideline for Management of Blood Pressure in Chronic Kidney Disease) can improve volume control, and reduce blood pressure control or proteinuria [58, 59]. The renin-angiotensin-aldosterone system achieves fluid homeostasis through balancing of sodium excretion, extracellular fluid volume, and arterial BP [60]. In CKD, insufficient sodium excretion affects fluid and BP balance [61]. Through various mechanisms such as oxidative stress, inflammation, and endothelial dysfunction, excessive sodium intake has adverse effects on blood vessels, heart, kidneys and sympathetic nervous systems [62].

Table 14.3 Sodium intake by country and region

Country	Sodium intake, g/day (95% uncertainty intervals)[a] [69, 70]
China	4.83 (4.62–5.05)
India	3.72 (3.63–3.82)
Indonesia	3.36 (3.02–3.76)
Japan	4.89 (4.71–5.08)
Korea	5.21 (4.98–5.48)
Malaysia	3.57 (3.25–3.93)
Philippines	4.29 (3.65–5.10)
Singapore	5.14 (4.36–6.02)
Taiwan	3.92 (3.66–4.17)
Thailand	5.31 (4.88–5.75)

[a] Age-standardized estimated sodium intakes (g/day) in 2010

There are several major studies including Intersalt, DASH, ONTARGET, LowSALT CKD and HEMO, evaluating dietary sodium intake and its effects on various outcomes such as BP, proteinuria, cardiovascular events, and mortality [63].

A summary of sodium intake in several Asian countries is shown in Table 14.3. Quantity of salt intake can vary according to countries, regions, and ethnicity, as a result of different socio-economic-cultural factors. The INTERMAP study by China used 24-h urinary sodium excretion method and demonstrated an average sodium intake of 3990 mg/day, while the 2002 Chinese Nutrition and Health Condition Survey used 3-day dietary record and showed that the nationwide salt intake was 10.7 g/day [64–66]. A more recent 2010 China Health and Nutrition Survey revealed a decline in sodium consumption from 6.8 g/day in 1991 to 4.3 g/day in 2011 [66, 67]. In a Singapore National Nutrition Survey, the mean urinary sodium excretion was 142.2 mmol/24 h and estimated mean salt intake was 8.3 g/day [68]. Powles et al. analyzed sodium intake in adults, revealing a global mean of sodium intake of 3.95 g/day. Countries within Asia had the highest mean sodium intakes, with 4.8 g/day in East Asia countries, 5 g/day in Asia Pacific High-Income countries (Japan and South Korea), and 5.51 g/day in Central Asia [69].

He et al. showed salt reduction to 100 mmol/day led to SBP (5.8 mmHg) and DBP (2.82 mmHg) reduction in the general population [71]. For CKD, the evidence of sodium intake reductions is derived from short-term randomized controlled trials or crossover studies, and observational studies for clinical endpoints such as CKD progression, mortality, and cardiovascular events [58]. The evidence demonstrating benefits of sodium reduction on BP reduction in CKD were derived from short-term randomized trials that were mostly conducted in the United Kingdom, Netherlands, Australia, Japan, and USA [72–78].

Within Asia, the effects of lowering blood pressure and proteinuria via sodium restriction in Japanese patients with IgA nephropathy were demonstrated by Konishi et al. [73]. In China, Xu et al. demonstrated association of elevated BP and risk of hypertension with higher sodium intakes in adults in Shandong and Jiangsu [66, 79]. Another cluster-randomized trial conducted in China demonstrated a sodium reduction program involving health education and access to potassium-based salt substitute resulted in lower urinary albumin-creatinine ratio and lower odds of albuminuria [80]. A case-control study on rural population in southern India showed high serum sodium-potassium ratio excretion. Hypertension was associated with lower serum potassium levels and a lower intake of vegetables was reported by participants with hypertension [81]. Loh et al. evaluated Malaysian patients with CKD to undergo 1-month salt restricting diet prospectively. Salt intake was estimated using 24-h urinary sodium and potassium levels. With salt reduction, improvement in BP and proteinuria were observed [82].

Differences in socio-cultural beliefs and practices contribute to different dietary habits too. A cross-sectional study on adults of Sado city in Japan demonstrated associations between the intake of miso soup and Japanese pickles, with daily salt intake of 9.4 g/day based on estimated 24-h urine sodium excretion [83]. Cultural differences within Indonesia with high-fat and high-salt dietary habits in several Indonesian provinces were shown to contribute to the difference in prevalence of hypertension among the provinces [84]. Teo et al. evaluated 24-h urinary sodium excretion in Singaporean patients with CKD and found dietary sodium intake was high among those in earlier stages of CKD, of which many had declined dietician counselling [85].

Twenty-four-hour urinary excretion has been established as a method to estimate daily sodium intake. Amano et al. measured daily sodium excretion using 24-h urine collection and compared with estimated sodium excretion from a spot urine sample in Japanese patients with chronic kidney disease, using a formula by Tanaka et al. [86, 87]. The study showed a significant difference in readings when compared. In a multi-ethnic Singaporean cohort, a 5-variable equation, which included spot urine sodium, age, gender, ethnicity, and weight, was formed to predict 24-h urine sodium excretion [11].

Apart from sodium, emphasis has been placed on dietary potassium and Dietary Approaches to Stop Hypertension (DASH) diet in patients with hypertension and CKD. Mun et al. studied dietary potassium patterns in adults with stage 2 CKD in Korean rural populations and observed high dietary potassium was associated with slower CKD progression in those with hypertension [88]. Furthermore, the HEXA study found association between low potassium intake with increased odds of advanced stage CKD in hypertensive participants in South Korea [89]. The DASH diet comprises of vegetables, fruits, low-fat dairy products, whole grain, fish, poultry and nuts, with low proportions of red meat, and sugary food. Two studies revealed DASH diet when combined with low sodium intake, effectively lowered BP [90, 91]. Only one study by Lee et al. had observed low odds of developing CKD in elderly adults with greater adherence to DASH-style diet [92]. The risk of incident CKD and components of DASH diet have been evaluated in several Asian studies. Firstly, fruits and vegetable-rich diets with lower dietary acid load, were associated with lower risk of incident CKD and proteinuria in a South Korean prospective study [93]. Secondly, red meat consumption was also associated with end-stage kidney disease in a prospective study in Singapore Chinese population, and not consumption of fish, eggs, or poultry [94]. While there has been no other studies evaluating DASH diet with progression or complications of CKD, studies on effects of components of DASH diets in Asian patients with CKD have been conducted but these are beyond the scope of this chapter [95].

14.6 Prescribing Patterns of Antihypertensives in CKD

The prevalence of uncontrolled hypertension differed by regions and countries; hence, approaches to managing hypertension were likely to be heterogeneous too [22]. The study conducted by Alencar de Pinho et al. analyzed prescribing patterns of antihypertensives. The highest number of antihypertensive drug classes (3 or more) was observed in cohorts from North America while lowest number was observed in the Chinese cohort study (only 1) [22]. Across the cohorts, renin-angiotensin-aldosterone system (RAAS) inhibitors was mostly prescribed and the preferred agent for monotherapy too. In comparison to cohorts from Brazil, Europe, and North America, diuretics were less frequently prescribed in Asian cohorts, in which calcium channel blockers (CCB) were most frequently prescribed [22].

The C-STRIDE cohort study in China observed that more than half of hypertensive CKD patients were on 2 or more antihypertensives. Among them, RAAS inhibitors and CCB were prescribed the most [35]. Similar findings were demonstrated by another study in China by Zhang et al. [96]. A prospective study on patients with CKD was conducted in Singapore by Teo et al. The study described the prescribing pattern of antihypertensives within a tertiary hospital. RAAS inhibitors were observed to be commonest antihypertensive prescribed, particularly in patients with diabetes mellitus [97]. Higher frequency of diuretics, beta-blockers, or dihydropyridine CCB were prescribed at higher stages of CKD and in patients aged >65 years [97]. On the other hand, a cross-sectional study evaluating CKD patients in Pakistan, among which 74.4% had hypertension, revealed that only 48.7% of them were on antihypertensives [98]. Only 17% of patients were on RAAS inhibitors as monotherapy or in combination with other antihypertensives. Among those on monotherapy, beta-blockers were mostly commonly prescribed [98]. In the CKD-JAC study, 91.9% of Japanese CKD patients had hypertension among which RAAS inhibitors were most frequently used, followed by CCB. Between two RAAS inhibitors, angiotensin receptor blockers were commonly prescribed than ACE-inhibitors [99].

14.7 Conclusion

Among different Asian populations, this chapter has (1) illustrated the epidemiology of hypertension and CKD, (2) explored the relationship of hypertension and CKD against renal outcomes, and (3) discussed how factors including genetics, dietary choices, and varying pharmacological treatments can play powerful roles in management of blood pressure and curbing CKD progression. Taken together, it remains important to review prevalence of uncontrolled hypertension and BP management practices, to identify gaps in knowledge and implement tailored national strategies to optimize BP control. And it would add to the overall armamentarium in the prevention of chronic kidney diseases in Asia and worldwide [100].

References

1. Kidney Disease: Improving Global Outcomes Blood Pressure Work G. KDIGO 2021 clinical practice guideline for the management of blood pressure in chronic kidney disease. Kidney Int. 2021;99(3S):S1–S87. https://doi.org/10.1016/j.kint.2020.11.003.
2. Teo BW, Chan GC, Leo CCH, Tay JC, Chia YC, Siddique S, et al. Hypertension and chronic kidney disease in Asian populations. J Clin Hypertens (Greenwich). 2021;23(3):475–80. https://doi.org/10.1111/jch.14188.
3. Zhang L, Wang F, Wang L, Wang W, Liu B, Liu J, et al. Prevalence of chronic kidney disease in China: a cross-sectional survey. Lancet. 2012;379(9818):815–22. https://doi.org/10.1016/S0140-6736(12)60033-6.
4. Zheng Y, Tang L, Zhang W, Zhao D, Zhang D, Zhang L, et al. Applying the new intensive blood pressure categories to a nondialysis chronic kidney disease population: the Prevalence, Awareness and Treatment Rates in Chronic Kidney Disease Patients with Hypertension in China survey. Nephrol Dial Transplant. 2020;35(1):155–61. https://doi.org/10.1093/ndt/gfy301.
5. Surveilance and Epidemiology Branch CfHP, Department of Health, Hong Kong Special Administrative Region Government. Report of population survey 2014/2015. 2017.
6. Anchala R, Kannuri NK, Pant H, Khan H, Franco OH, Di Angelantonio E, et al. Hypertension in India: a systematic review and meta-analysis of prevalence, awareness, and control of hypertension. J Hypertens. 2014;32(6):1170–7. https://doi.org/10.1097/HJH.0000000000000146.
7. Singh AK, Farag YM, Mittal BV, Subramanian KK, Reddy SR, Acharya VN, et al. Epidemiology and risk factors of chronic kidney disease in India—results from the SEEK (Screening and Early Evaluation of Kidney Disease) study. BMC Nephrol. 2013;14:114. https://doi.org/10.1186/1471-2369-14-114.
8. Umemura S, Arima H, Arima S, Asayama K, Dohi Y, Hirooka Y, et al. The Japanese Society of Hypertension guidelines for the management of hypertension (JSH 2019). Hypertens Res. 2019;42(9):1235–481. https://doi.org/10.1038/s41440-019-0284-9.
9. Chia YC, Kario K, Turana Y, Nailes J, Tay JC, Siddique S, et al. Target blood pressure and control status in Asia. J Clin Hypertens (Greenwich). 2020;22(3):344–50. https://doi.org/10.1111/jch.13714.
10. Hooi LS, Ong LM, Ahmad G, Bavanandan S, Ahmad NA, Naidu BM, et al. A population-based study measuring the prevalence of chronic kidney disease among adults in West Malaysia. Kidney Int. 2013;84(5):1034–40. https://doi.org/10.1038/ki.2013.220.
11. Subramanian S, Teo BW, Toh QC, Koh YY, Li J, Sethi S, et al. Spot urine tests in predicting 24-hour urine sodium excretion in Asian patients. J Ren Nutr. 2013;23(6):450–5. https://doi.org/10.1053/j.jrn.2012.12.004.
12. Korean Society H, Hypertension Epidemiology Research Working G, Kim HC, Cho MC. Korea hypertension fact sheet 2018. Clin Hypertens. 2018;24:13. https://doi.org/10.1186/s40885-018-0098-0.
13. Park JI, Baek H, Jung HH. Prevalence of chronic kidney disease in Korea: the Korean National Health and Nutritional Examination Survey 2011-2013. J Korean Med Sci. 2016;31(6):915–23. https://doi.org/10.3346/jkms.2016.31.6.915.
14. Health Promotion Administration, Ministry of Health and Welfare, Taiwan. Nutrition and Health Survey in Taiwan (NAHSIT) Report 2013–2016.
15. Cai G. Etiology, comorbidity and factors associated with renal function decline in Chinese chronic kidney disease patients. J Am Soc Nephrol. 2011;22:183A–4A.
16. Ho YW, Chau K, Choy BY, et al. Hong Kong renal registry report 2010. Hong Kong J Nephrol. 2010;12(2):81–98.
17. Jha V. Current status of end-stage renal disease care in India and Pakistan. Kidney Int Suppl. 2013;3(2):157–60.
18. Masakane I, Taniguchi M, Nakai S, et al. Annual dialysis data report 2016, JSDT renal data registry. Ren Replace Ther. 2018;4:45.

19. Registry SR. First report of the Singapore renal registry 1997. 1998.
20. Kim YS, Jin DC. Global dialysis perspective: Korea. Kidney360. 2020;1:52–7.
21. Hwang SJ, Tsai JC, Chen HC. Epidemiology, impact and preventive care of chronic kidney disease in Taiwan. Nephrology (Carlton). 2010;15(Suppl 2):3–9. https://doi.org/10.1111/j.1440-1797.2010.01304.x.
22. Alencar de Pinho N, Levin A, Fukagawa M, Hoy WE, Pecoits-Filho R, Reichel H, et al. Considerable international variation exists in blood pressure control and antihypertensive prescription patterns in chronic kidney disease. Kidney Int. 2019;96(4):983–94. https://doi.org/10.1016/j.kint.2019.04.032.
23. Sabanayagam C, Lim SC, Wong TY, Lee J, Shankar A, Tai ES. Ethnic disparities in prevalence and impact of risk factors of chronic kidney disease. Nephrol Dial Transplant. 2010;25(8):2564–70. https://doi.org/10.1093/ndt/gfq084.
24. Sabanayagam C, Teo BW, Tai ES, Jafar TH, Wong TY. Ethnic differences in the association between blood pressure components and chronic kidney disease in middle aged and older Asian adults. BMC Nephrol. 2013;14:86. https://doi.org/10.1186/1471-2369-14-86.
25. Lew QLJ, Allen JC, Nguyen F, Tan NC, Jafar TH. Factors associated with chronic kidney disease and their clinical utility in primary care clinics in a multi-ethnic Southeast Asian population. Nephron. 2018;138(3):202–13. https://doi.org/10.1159/000485110.
26. Sabanayagam C, Shankar A, Lim SC, Tai ES, Wong TY. Hypertension, hypertension control, and chronic kidney disease in a Malay population in Singapore. Asia Pac J Public Health. 2011;23(6):936–45. https://doi.org/10.1177/1010539510361637.
27. Wan EYF, Yu EYT, Chin WY, Fong DYT, Choi EPH, Lam CLK. Association of blood pressure and risk of cardiovascular and chronic kidney disease in Hong Kong hypertensive patients. Hypertension. 2019;74(2):331–40. https://doi.org/10.1161/HYPERTENSIONAHA.119.13123.
28. Domrongkitchaiporn S, Sritara P, Kitiyakara C, Stitchantrakul W, Krittaphol V, Lolekha P, et al. Risk factors for development of decreased kidney function in a Southeast Asian population: a 12-year cohort study. J Am Soc Nephrol. 2005;16(3):791–9. https://doi.org/10.1681/ASN.2004030208.
29. Jiang C, Wang B, Li Y, Xie L, Zhang X, Wang J, et al. U-shaped association between serum albumin and development of chronic kidney disease in general hypertensive patients. Clin Nutr. 2020;39(1):258–64. https://doi.org/10.1016/j.clnu.2019.02.002.
30. Yano Y, Fujimoto S, Sato Y, Konta T, Iseki K, Iseki C, et al. New-onset hypertension and risk for chronic kidney disease in the Japanese general population. J Hypertens. 2014;32(12):2371–7.; ; discussion 7. https://doi.org/10.1097/HJH.0000000000000344.
31. Kang YU, Bae EH, Ma SK, Kim SW. Determinants and burden of chronic kidney disease in a high-risk population in Korea: results from a cross-sectional study. Korean J Intern Med. 2016;31(5):920–9. https://doi.org/10.3904/kjim.2014.243.
32. Chang TI, Lim H, Park CH, Rhee CM, Moradi H, Kalantar-Zadeh K, et al. Associations of systolic blood pressure with incident CKD G3-G5: a cohort study of South Korean adults. Am J Kidney Dis. 2020;76(2):224–32. https://doi.org/10.1053/j.ajkd.2020.01.013.
33. Saminathan TA, Hooi LS, Mohd Yusoff MF, Ong LM, Bavanandan S, Rodzlan Hasani WS, et al. Prevalence of chronic kidney disease and its associated factors in Malaysia; findings from a nationwide population-based cross-sectional study. BMC Nephrol. 2020;21(1):344. https://doi.org/10.1186/s12882-020-01966-8.
34. Ingsathit A, Thakkinstian A, Chaiprasert A, Sangthawan P, Gojaseni P, Kiattisunthorn K, et al. Prevalence and risk factors of chronic kidney disease in the Thai adult population: Thai SEEK study. Nephrol Dial Transplant. 2010;25(5):1567–75. https://doi.org/10.1093/ndt/gfp669.
35. Yan Z, Wang Y, Li S, Wang J, Zhang L, Tan H, et al. Hypertension control in adults with CKD in China: baseline results from the Chinese cohort study of chronic kidney disease (C-STRIDE). Am J Hypertens. 2018;31(4):486–94. https://doi.org/10.1093/ajh/hpx222.
36. Yuan J, Zou XR, Han SP, Cheng H, Wang L, Wang JW, et al. Prevalence and risk factors for cardiovascular disease among chronic kidney disease patients: results from the Chinese

cohort study of chronic kidney disease (C-STRIDE). BMC Nephrol. 2017;18(1):23. https:// doi.org/10.1186/s12882-017-0441-9.

37. Anupama YJ, Hegde SN, Uma G, Patil M. Hypertension is an important risk determinant for chronic kidney disease: results from a cross-sectional, observational study from a rural population in South India. J Hum Hypertens. 2017;31(5):327–32. https://doi.org/10.1038/ jhh.2016.81.

38. Inaguma D, Imai E, Takeuchi A, Ohashi Y, Watanabe T, Nitta K, et al. Risk factors for CKD progression in Japanese patients: findings from the Chronic Kidney Disease Japan Cohort (CKD-JAC) study. Clin Exp Nephrol. 2017;21(3):446–56. https://doi.org/10.1007/ s10157-016-1309-1.

39. Li Y, Li D, Song Y, Gao L, Fan F, Wang B, et al. Visit-to-visit variability in blood pressure and the development of chronic kidney disease in treated general hypertensive patients. Nephrol Dial Transplant. 2020;35(10):1739–46. https://doi.org/10.1093/ndt/gfz093.

40. Li Y, Zhou H, Liu M, Liang M, Wang G, Wang B, et al. Association of visit-to-visit variability in blood pressure and first stroke risk in hypertensive patients with chronic kidney disease. J Hypertens. 2020;38(4):610–7. https://doi.org/10.1097/HJH.0000000000002306.

41. Hirayama A, Konta T, Kamei K, Suzuki K, Ichikawa K, Fujimoto S, et al. Blood pressure, proteinuria, and renal function decline: associations in a large community-based population. Am J Hypertens. 2015;28(9):1150–6. https://doi.org/10.1093/ajh/hpv003.

42. Chia YC, Lim HM, Ching SM. Long-term visit-to-visit blood pressure variability and renal function decline in patients with hypertension over 15 years. J Am Heart Assoc. 2016;5(11):e003825. https://doi.org/10.1161/JAHA.116.003825.

43. Cha RH, Kim S, Ae Yoon S, Ryu DR, Eun OJ, Han SY, et al. Association between blood pressure and target organ damage in patients with chronic kidney disease and hypertension: results of the APrODiTe study. Hypertens Res. 2014;37(2):172–8. https://doi.org/10.1038/ hr.2013.127.

44. Azushima K, Wakui H, Uneda K, Haku S, Kobayashi R, Ohki K, et al. Within-visit blood pressure variability and cardiovascular risk factors in hypertensive patients with non-dialysis chronic kidney disease. Clin Exp Hypertens. 2017;39(7):665–71. https://doi.org/10.108 0/10641963.2017.1313850.

45. Wain LV, Vaez A, Jansen R, Joehanes R, van der Most PJ, Erzurumluoglu AM, et al. Novel blood pressure locus and gene discovery using genome-wide association study and expression data sets from blood and the kidney. Hypertension. 2017. https://doi.org/10.1161/ HYPERTENSIONAHA.117.09438.

46. Wuttke M, Li Y, Li M, Sieber KB, Feitosa MF, Gorski M, et al. A catalog of genetic loci associated with kidney function from analyses of a million individuals. Nat Genet. 2019;51(6):957–72. https://doi.org/10.1038/s41588-019-0407-x.

47. Liu L, Wang C, Mi Y, Liu D, Li L, Fan J, et al. Association of MYH9 polymorphisms with hypertension in patients with chronic kidney disease in China. Kidney Blood Press Res. 2016;41(6):956–65. https://doi.org/10.1159/000452597.

48. Redon J, Chaves FJ, Liao Y, Pascual JM, Rovira E, Armengod ME, et al. Influence of the I/D polymorphism of the angiotensin-converting enzyme gene on the outcome of micro-albuminuria in essential hypertension. Hypertension. 2000;35(1 Pt 2):490–5. https://doi. org/10.1161/01.hyp.35.1.490.

49. Tripathi G, Dharmani P, Khan F, Sharma RK, Pandirikkal V, Agrawal S. High prevalence of ACE DD genotype among north Indian end stage renal disease patients. BMC Nephrol. 2006;7:15. https://doi.org/10.1186/1471-2369-7-15.

50. Su SL, Lu KC, Lin YF, Hsu YJ, Lee PY, Yang HY, et al. Gene polymorphisms of angiotensin-converting enzyme and angiotensin II type 1 receptor among chronic kidney disease patients in a Chinese population. J Renin-Angiotensin-Aldosterone Syst. 2012;13(1):148–54. https:// doi.org/10.1177/1470320311430989.

51. Katsuya T, Ishikawa K, Sugimoto K, Rakugi H, Ogihara T. Salt sensitivity of Japanese from the viewpoint of gene polymorphism. Hypertens Res. 2003;26(7):521–5. https://doi. org/10.1291/hypres.26.521.

52. Lin DS, Wang TD, Buranakitjaroen P, Chen CH, Cheng HM, Chia YC, et al. Angiotensin receptor neprilysin inhibitor as a novel antihypertensive drug: evidence from Asia and around the globe. J Clin Hypertens (Greenwich). 2021;23(3):556–67. https://doi.org/10.1111/jch.14120.
53. Liu Y, Shi M, Dolan J, He J. Sodium sensitivity of blood pressure in Chinese populations. J Hum Hypertens. 2020;34(2):94–107. https://doi.org/10.1038/s41371-018-0152-0.
54. Liu Z, Qi H, Liu B, Liu K, Wu J, Cao H, et al. Genetic susceptibility to salt-sensitive hypertension in a Han Chinese population: a validation study of candidate genes. Hypertens Res. 2017;40(10):876–84. https://doi.org/10.1038/hr.2017.57.
55. Nierenberg JL, Li C, He J, Gu D, Chen J, Lu X, et al. Blood pressure genetic risk score predicts blood pressure responses to dietary sodium and potassium: The GenSalt Study (Genetic Epidemiology Network of Salt Sensitivity). Hypertension. 2017;70(6):1106–12. https://doi.org/10.1161/HYPERTENSIONAHA.117.10108.
56. Qi H, Liu B, Guo C, Liu Z, Cao H, Liu K, et al. Effects of environmental and genetic risk factors for salt sensitivity on blood pressure in northern China: the systemic epidemiology of salt sensitivity (EpiSS) cohort study. BMJ Open. 2018;8(12):e023042. https://doi.org/10.1136/bmjopen-2018-023042.
57. Rhee MY, Yang SJ, Oh SW, Park Y, Kim CI, Park HK, et al. Novel genetic variations associated with salt sensitivity in the Korean population. Hypertens Res. 2011;34(5):606–11. https://doi.org/10.1038/hr.2010.278.
58. Ikizler TA, Burrowes JD, Byham-Gray LD, Campbell KL, Carrero JJ, Chan W, et al. KDOQI clinical practice guideline for nutrition in CKD: 2020 update. Am J Kidney Dis. 2020;76(3 Suppl 1):S1–S107. https://doi.org/10.1053/j.ajkd.2020.05.006.
59. Cheung AK, Chang TI, Cushman WC, Furth SL, Hou FF, Ix JH, et al. Executive summary of the KDIGO 2021 clinical practice guideline for the management of blood pressure in chronic kidney disease. Kidney Int. 2021;99(3):559–69. https://doi.org/10.1016/j.kint.2020.10.026.
60. Schweda F. Salt feedback on the renin-angiotensin-aldosterone system. Pflugers Arch. 2015;467(3):565–76. https://doi.org/10.1007/s00424-014-1668-y.
61. Kotchen TA, Cowley AW Jr, Frohlich ED. Salt in health and disease—a delicate balance. N Engl J Med. 2013;368(26):2531–2. https://doi.org/10.1056/NEJMc1305326.
62. Dinh QN, Drummond GR, Sobey CG, Chrissobolis S. Roles of inflammation, oxidative stress, and vascular dysfunction in hypertension. Biomed Res Int. 2014;2014:406960. https://doi.org/10.1155/2014/406960.
63. Sanghavi S, Vassalotti JA. Dietary sodium: a therapeutic target in the treatment of hypertension and CKD. J Ren Nutr. 2013;23(3):223–7. https://doi.org/10.1053/j.jrn.2013.01.027.
64. Zhao L, Stamler J, Yan LL, Zhou B, Wu Y, Liu K, et al. Blood pressure differences between northern and southern Chinese: role of dietary factors: the International Study on Macronutrients and Blood Pressure. Hypertension. 2004;43(6):1332–7. https://doi.org/10.1161/01.HYP.0000128243.06502.bc.
65. Ma G. The salt consumption of resident in China. Chinese J Prev Control Chronic Dis. 2008;16(4):214–7.
66. Shao S, Hua Y, Yang Y, Liu X, Fan J, Zhang A, et al. Salt reduction in China: a state-of-the-art review. Risk Manag Healthc Policy. 2017;10:17–28. https://doi.org/10.2147/RMHP.S75918.
67. Zhai FY, Du SF, Wang ZH, Zhang JG, Du WW, Popkin BM. Dynamics of the Chinese diet and the role of urbanicity, 1991-2011. Obes Rev. 2014;15(Suppl 1):16–26. https://doi.org/10.1111/obr.12124.
68. Health Promotion Board GoS. National Nutrition Survey 2010. 2010.
69. Powles J, Fahimi S, Micha R, Khatibzadeh S, Shi P, Ezzati M, et al. Global, regional and national sodium intakes in 1990 and 2010: a systematic analysis of 24 h urinary sodium excretion and dietary surveys worldwide. BMJ Open. 2013;3(12):e003733. https://doi.org/10.1136/bmjopen-2013-003733.
70. Chan GC, Teo BW, Tay JC, Chen CH, Cheng HM, Wang TD, et al. Hypertension in a multi-ethnic Asian population of Singapore. J Clin Hypertens (Greenwich). 2021;23(3):522–8. https://doi.org/10.1111/jch.14140.

71. He FJ, Li J, Macgregor GA. Effect of longer-term modest salt reduction on blood pressure. Cochrane Database Syst Rev. 2013;(4):CD004937. https://doi.org/10.1002/14651858. CD004937.pub2.
72. de Brito-Ashurst I, Perry L, Sanders TA, Thomas JE, Dobbie H, Varagunam M, et al. The role of salt intake and salt sensitivity in the management of hypertension in South Asian people with chronic kidney disease: a randomised controlled trial. Heart. 2013;99(17):1256–60. https://doi.org/10.1136/heartjnl-2013-303688.
73. Konishi Y, Okada N, Okamura M, Morikawa T, Okumura M, Yoshioka K, et al. Sodium sensitivity of blood pressure appearing before hypertension and related to histological damage in immunoglobulin a nephropathy. Hypertension. 2001;38(1):81–5. https://doi.org/10.1161/01. hyp.38.1.81.
74. McMahon EJ, Bauer JD, Hawley CM, Isbel NM, Stowasser M, Johnson DW, et al. A randomized trial of dietary sodium restriction in CKD. J Am Soc Nephrol. 2013;24(12):2096–103. https://doi.org/10.1681/ASN.2013030285.
75. Meuleman Y, Hoekstra T, Dekker FW, Navis G, Vogt L, van der Boog PJM, et al. Sodium restriction in patients with CKD: a randomized controlled trial of self-management support. Am J Kidney Dis. 2017;69(5):576–86. https://doi.org/10.1053/j.ajkd.2016.08.042.
76. Saran R, Padilla RL, Gillespie BW, Heung M, Hummel SL, Derebail VK, et al. A randomized crossover trial of dietary sodium restriction in stage 3-4 CKD. Clin J Am Soc Nephrol. 2017;12(3):399–407. https://doi.org/10.2215/CJN.01120216.
77. Slagman MC, Waanders F, Hemmelder MH, Woittiez AJ, Janssen WM, Lambers Heerspink HJ, et al. Moderate dietary sodium restriction added to angiotensin converting enzyme inhibition compared with dual blockade in lowering proteinuria and blood pressure: randomised controlled trial. BMJ. 2011;343:d4366. https://doi.org/10.1136/bmj.d4366.
78. Vogt L, Waanders F, Boomsma F, de Zeeuw D, Navis G. Effects of dietary sodium and hydrochlorothiazide on the antiproteinuric efficacy of losartan. J Am Soc Nephrol. 2008;19(5):999–1007. https://doi.org/10.1681/ASN.2007060693.
79. Xu J, Wang M, Chen Y, Zhen B, Li J, Luan W, et al. Estimation of salt intake by 24-hour urinary sodium excretion: a cross-sectional study in Yantai, China. BMC Public Health. 2014;14:136. https://doi.org/10.1186/1471-2458-14-136.
80. Jardine MJ, Li N, Ninomiya T, Feng X, Zhang J, Shi J, et al. Dietary sodium reduction reduces albuminuria: a cluster randomized trial. J Ren Nutr. 2019;29(4):276–84. https://doi.org/10.1053/j.jrn.2018.10.009.
81. Evans RG, Subasinghe AK, Busingye D, Srikanth VK, Kartik K, Kalyanram K, et al. Renal and dietary factors associated with hypertension in a setting of disadvantage in rural India. J Hum Hypertens. 2021. https://doi.org/10.1038/s41371-020-00473-5.
82. Koh KH, Wei-Soon LH, Jun L, Lui-Sian LN, Hui-Hong CT. Study of low salt diet in hypertensive patients with chronic kidney disease. Med J Malaysia. 2018;73(6):376–81.
83. Wakasugi M, James Kazama J, Narita I. Associations between the intake of miso soup and Japanese pickles and the estimated 24-hour urinary sodium excretion: a population-based cross-sectional study. Intern Med. 2015;54(8):903–10. https://doi.org/10.2169/internalmedicine.54.4336.
84. Health Research and Development Agency MoHRoI. Basic Health Research Report. 2018.
85. Teo BW, Bagchi S, Xu H, Toh QC, Li J, Lee EJ. Dietary sodium intake in a multiethnic Asian population of healthy participants and chronic kidney disease patients. Singap Med J. 2014;55(12):652–5. https://doi.org/10.11622/smedj.2014180.
86. Tanaka T, Okamura T, Miura K, Kadowaki T, Ueshima H, Nakagawa H, et al. A simple method to estimate populational 24-h urinary sodium and potassium excretion using a casual urine specimen. J Hum Hypertens. 2002;16(2):97–103. https://doi.org/10.1038/sj.jhh.1001307.
87. Amano H, Kobayashi S, Terawaki H, Ogura M, Kawaguchi Y, Yokoo T. Measurement of daily sodium excretion in patients with chronic kidney disease; special reference to the difference between the amount measured from 24 h collected urine sample and the estimated amount from a spot urine. Ren Fail. 2018;40(1):238–42. https://doi.org/10.1080/0886022X.2018.1456452.

88. Mun KH, Yu GI, Choi BY, Kim MK, Shin MH, Shin DH. Association of dietary potassium intake with the development of chronic kidney disease and renal function in patients with mildly decreased kidney function: the Korean Multi-Rural Communities Cohort Study. Med Sci Monit. 2019;25:1061–70. https://doi.org/10.12659/MSM.913504.
89. Kim J, Lee J, Kim KN, Oh KH, Ahn C, Lee J, et al. Association between dietary mineral intake and chronic kidney disease: the Health Examinees (HEXA) Study. Int J Environ Res Public Health. 2018;15(6):1070. https://doi.org/10.3390/ijerph15061070.
90. Sacks FM, Obarzanek E, Windhauser MM, Svetkey LP, Vollmer WM, McCullough M, et al. Rationale and design of the Dietary Approaches to Stop Hypertension trial (DASH). A multicenter controlled-feeding study of dietary patterns to lower blood pressure. Ann Epidemiol. 1995;5(2):108–18. https://doi.org/10.1016/1047-2797(94)00055-x.
91. Svetkey LP, Sacks FM, Obarzanek E, Vollmer WM, Appel LJ, Lin PH, et al. The DASH Diet, Sodium Intake and Blood Pressure Trial (DASH-sodium): rationale and design. DASH-Sodium Collaborative Research Group. J Am Diet Assoc. 1999;99(8 Suppl):S96–104. https://doi.org/10.1016/s0002-8223(99)00423-x.
92. Lee HS, Lee KB, Hyun YY, Chang Y, Ryu S, Choi Y. DASH dietary pattern and chronic kidney disease in elderly Korean adults. Eur J Clin Nutr. 2017;71(6):755–61. https://doi.org/10.1038/ejcn.2016.240.
93. Jhee JH, Kee YK, Park JT, Chang TI, Kang EW, Yoo TH, et al. A diet rich in vegetables and fruit and incident CKD: a community-based prospective cohort study. Am J Kidney Dis. 2019;74(4):491–500. https://doi.org/10.1053/j.ajkd.2019.02.023.
94. Lew QJ, Jafar TH, Koh HW, Jin A, Chow KY, Yuan JM, et al. Red meat intake and risk of ESRD. J Am Soc Nephrol. 2017;28(1):304–12. https://doi.org/10.1681/ASN.2016030248.
95. Song Y, Lobene AJ, Wang Y, Hill Gallant KM. The DASH diet and cardiometabolic health and chronic kidney disease: a narrative review of the evidence in East Asian countries. Nutrients. 2021;13(3):984. https://doi.org/10.3390/nu13030984.
96. Zhang W, Shi W, Liu Z, Gu Y, Chen Q, Yuan W, et al. A nationwide cross-sectional survey on prevalence, management and pharmacoepidemiology patterns on hypertension in Chinese patients with chronic kidney disease. Sci Rep. 2016;6:38768. https://doi.org/10.1038/srep38768.
97. Teo BW, Chua HR, Wong WK, Haroon S, Subramanian S, Loh PT, et al. Blood pressure and antihypertensive medication profile in a multiethnic Asian population of stable chronic kidney disease patients. Singap Med J. 2016;57(5):267–73. https://doi.org/10.11622/smedj.2016089.
98. Jessani S, Bux R, Jafar TH. Prevalence, determinants, and management of chronic kidney disease in Karachi, Pakistan—a community based cross-sectional study. BMC Nephrol. 2014;15:90. https://doi.org/10.1186/1471-2369-15-90.
99. Imai E, Matsuo S, Makino H, Watanabe T, Akizawa T, Nitta K, et al. Chronic Kidney Disease Japan Cohort study: baseline characteristics and factors associated with causative diseases and renal function. Clin Exp Nephrol. 2010;14(6):558–70. https://doi.org/10.1007/s10157-010-0328-6.
100. Li PK, Garcia-Garcia G, Lui SF, Andreoli S, Fung WW, Hradsky A, et al. Kidney health for everyone everywhere-from prevention to detection and equitable access to care. Kidney Int. 2020;97(2):226–32. https://doi.org/10.1016/j.kint.2019.12.002.

Secondary Causes of Hypertension: An Overview

15

Meifen Zhang, Hang Siang Wong, Roy Debajyoti Malakar, and Troy H Puar

15.1 Introduction

While most patients with hypertension have essential (idiopathic) hypertension, a significant proportion of patients may have an underlying treatable condition or secondary hypertension. Asia has a high prevalence of hypertension, also known as the silent killer, which is a major contributor of cardiovascular death and healthcare burden. Cardiovascular Disease (CVD), the leading cause of deaths worldwide, contributes to about 30% of all deaths, with half of the cases of CVD is estimated to be in Asia; and hypertension, a major modifiable risk factor for CVD, results in more deaths than any other CV risk factors in the Asian regions [1]. Hypertension is also the most prevalent risk factor for stroke in Asia [2].

In a study of the prevalence of hypertension in Singapore, among 10, 215 participants (47.2% Chinese, 26.0% Malay, and 26.8% Indian), hypertension prevalence was estimated to be 31.1% [3]. In a study from China, among Chinese adults aged 35–75 years, nearly 50% have hypertension, fewer than a third are being treated, and fewer than 1 in 12 have controlled blood pressure [4]. Secondary hypertension is estimated to affect 10–20% of patients with

M. Zhang (✉) · T. H. Puar
Department of Endocrinology, Changi General Hospital, Singapore, Singapore
e-mail: zhang.meifen@singhealth.com.sg; troy.puar.h.k@singhealth.com.sg

H. S. Wong
Department of Respiratory and Critical Care Medicine, Changi General Hospital, Singapore, Singapore
e-mail: wong.hang.siang@singhealth.com.sg

R. D. Malakar
Department of Renal Medicine, Changi General Hospital, Singapore, Singapore
e-mail: roy.debajyoti.malakar@singhealth.com.sg

© The Author(s), under exclusive license to Springer Nature Switzerland AG 2022
C. V. S. Ram et al. (eds.), *Hypertension and Cardiovascular Disease in Asia*,
Updates in Hypertension and Cardiovascular Protection,
https://doi.org/10.1007/978-3-030-95734-6_15

hypertension in a US study [5]. However, the prevalence of secondary hypertension across Asia is not known. One study from Japan estimated a prevalence of secondary hypertension of 9.1% among 1020 hypertensive patients [6]. Screening for secondary hypertension is recommended by major guidelines from Europe [7], America [8], and China [9]. The European Societies of Cardiology and Hypertension (ESC/ESH) guidelines suggest that the screening for secondary hypertension should be restricted to patients with features, such as younger age (<40 years old), acute worsening of hypertension in patients with previously stable normotension, severe (grade 3) hypertension, or drug resistance, or presence of extensive hypertension-mediated organ damage [7]. The Chinese guideline also mentions the need to consider screening. Hence, it is important that clinicians managing patients with hypertension consider the possibility of secondary hypertension.

The causes of secondary hypertension include renal parenchymal and renovascular hypertension, aortic coarctation, obstructive sleep apnea, primary aldosteronism and other adrenal hypertension, drug-induced hypertension, and various forms of monogenic hypertension [9].

One of the clinical features that may point the clinician to consider secondary causes of hypertension is resistant hypertension. Resistant hypertension was initially defined as failure to keep blood pressure below 140/90 mmHg despite three antihypertensive agents at maximal tolerated doses including a diuretic [10], with recent updates by both the ESC/ESH (included the use of 24 h ambulatory blood pressure monitoring to exclude white coat hypertension) and AHA guideline (lowering of cut-off to less than 130/80 mmHg). Prevalence of resistant hypertension is estimated at 10% amongst patients with hypertension [11], and these patients are also at higher risk of cardiovascular complications. In a study by Kvapil and colleagues [12], secondary hypertension was much more frequent (31%) amongst patients with resistant hypertension than in non-selected hypertensive population (5–15%). Additionally, patients with secondary hypertension also had more advanced target organ damage [12].

Another clinical feature to suspect secondary hypertension is the presence of hypertension and hypokalemia (both spontaneous or diuretic-induced) which is suggestive of primary aldosteronism (PA). It must be highlighted that the majority of patients with PA have normokalemia. The prevalence of PA among all patients with hypertension is about 5–10% and increases to 28.1% in patients with hypokalemia. The prevalence of PA increased with decreasing serum potassium concentrations, going up to 88.5% amongst patients with spontaneous hypokalemia and serum potassium concentrations below 2.5 mmol/L [13]. Other symptoms and signs that may raise suspicion for the clinician to screen for secondary hypertension are summarized in Table 15.1 [14, 15].

Table 15.1 Overview of the most common secondary causes

Cause	Prevalence in hypertensive	Prevalence in resistant hypertension	History	Physical findings	Screening
Renal causes					
Renal parenchymal diseases	1.6–8.0%	2–10%	Loss of good BP control; diabetes; smoking; generalized atherosclerosis; previous renal failure; nocturia	Peripheral edema; pallor; loss of muscle mass	Serum creatinine, urinalysis; ultrasound of the kidneys
Renal artery stenosis	1.0–8.0%	2.5–20%	Generalized atherosclerosis; diabetes; smoking; recurrent flash pulmonary edema; worsening of kidney function >30% after use of angiotensin-converting enzyme inhibitor or angiotensin II receptor blocker	Abdominal bruit; peripheral arterial disease (PAD)	Duplex or CT or MRI or angiography of the renal artery
Cardiovascular					
Coarctation of the aorta	<1%	<1%	Headache; nose bleeding; leg weakness or claudication	Different BP (≥20/10 mmHg) between upper-lower extremities; and/or between right-left arm; reduced and delayed femoral pulsation; interscapular ejection murmur; rib notching on chest X-ray	Transthoracic echocardiography
Endocrine					
Primary aldosteronism	1.4–10%	6–23%	Fatigue; constipation; polyuria; polydipsia; mostly asymptomatic	Muscle weakness	Aldosterone-Renin ratio (ARR)

(continued)

Table 15.1 (continued)

Cause	Prevalence in hypertensive	Prevalence in resistant hypertension	History	Physical findings	Screening
Thyroid disease	1–2%	1–3%	Hyperthyroidism: Palpitations; weight loss; anxiety; heat intolerance Hypothyroidism: Weight gain; fatigue; obstipation	Hyperthyroidism: Tachycardia; AF; accentuated heart sounds; exophthalmos Hypothyroidism: Bradycardia; muscle weakness; myxedema	Free T4/thyroid stimulating hormone (TSH)
Cushing's syndrome	0.5%	<1.0%	Weight gain; impotence; fatigue; psychological changes; polydipsia; polyuria	Obesity; hirsutism; skin atrophy; striae rubrae; muscle weakness; osteoporosis	24 h urinary cortisol; dexamethasone suppression test
Pheochromocytoma	0.2–0.5%	<1%	Headaches; palpitations; flushing; anxiety; labile BP; hypertensive episodes triggered by medications such as D2-antagonisms (metoclopramide), beta-blockers, sympathomimetics, opioids, tricyclic antidepressants	5Ps: Paroxysmal hypertension; pounding headache; perspiration; palpitations; pallor	Plasma or 24-h urine metanephrines
Others					
Obstructive sleep apnea	>5–15%	>30%	Snoring; daytime sleepiness; morning headaches; irritability	Increased neck circumference; obesity; peripheral edema	Screening questionnaire; polysomnography
Medications			Oral contraceptives pills Non-steroidal anti-inflammatory drugs Steroids Sympathomimetic drugs (decongestants, diet pills) Illicit drugs (cocaine, amphetamines, ecstasy/3,4-methylenedioxymethamphetamine)		

15.2 Renal Causes of Secondary Hypertension

Renal causes can be categorized into renal parenchymal diseases and renovascular diseases. Renal parenchymal diseases include polycystic kidney disease which is usually inherited in an autosomal dominant pattern. Other causes are glomerulonephritis, chronic tubulointerstitial disease, diabetic nephropathy, and obstructive uropathy. As chronic hypertension can lead to chronic kidney disease (CKD), these two conditions commonly co-exist. CKD as a cause for secondary hypertension can be difficult to establish. Some clinical clues may be helpful to the thoughtful clinician. For example, if abnormal findings are seen on urinalysis, or renal dysfunction has appeared before onset of hypertension, or if the presence of hypertension, proteinuria, or renal dysfunction from an early phase of pregnancy (superimposed preeclampsia) is confirmed, hypertension is likely to be caused by CKD, if hypertension is mild relative to abnormal urinary findings or kidney damage or lack of concurrent cardiovascular complications [16]. Baseline investigations for the hypertension patient should include urinalysis and serum creatinine concentration measurement and imaging of the kidneys. If there are significant abnormalities, then referral to a nephrologist is warranted.

However, not all patients may have full assessment of secondary causes of hypertension at diagnosis. If during the initiation of angiotensin-converting enzyme (ACE) inhibitors or angiotensin II receptor blockers (ARB), there is a significant worsening of creatinine to more than 30% of baseline, recurrent flash pulmonary edema in a patient with normal left ventricular ejection fraction, or if baseline imaging sizes in the kidneys differ more than 1.5 cm, then a search for renal artery stenosis is warranted. For renovascular diseases, fibromuscular dysplasia tends to occur in the young while atherosclerotic renal artery stenosis tends to occur in older patients. Renal artery stenosis-related hyperactivity of the renin–angiotensin (RA) system contributes to an increase in blood pressure. The initial imaging modality recommended is duplex Doppler ultrasonography of the renal arteries [17]. However, this test is not as accurate in patients who are obese, and its accuracy is operator-dependent. Magnetic resonance imaging or computed tomographic angiography can also be considered depending on kidney function. Should non-invasive imaging be non-diagnostic and clinical suspicion remains high, the gold standard for diagnosis is renal artery angiography [18].

The mainstay of treatment of renal artery stenosis is controlling blood pressure. ACE-I and ARBs can be used to achieve blood pressure targets in patients with both unilateral stenosis and bilateral stenosis. Generally for atherosclerotic renal artery stenosis, there is less evidence for angioplasty and medical therapy to achieve blood pressure targets and percutaneous angioplasty revascularization only considered should medical therapy fail [19]. Renin–angiotensin blockade is well tolerated even in patient with severe bilateral renal artery stenosis, and reduces mortality in a large group of atherosclerotic renal artery stenosis patients; and should be considered unless absolute contraindications exist [20]. In the CORAL study comparing stenting and medical therapy for atherosclerotic renal artery stenosis, the authors concluded that stenting renal artery stenosis did not confer a significant benefit with regard to the prevention of clinical events when added to comprehensive, multifactorial medical therapy [19]. In fibromuscular dysplasia which typically occurs in the younger patients and tends to affect the mid and distal part of the renal artery, percutaneous angioplasty is associated with better outcomes and should be considered. Other factors for percutaneous intervention include short duration of blood pressure elevation prior to diagnosis of RAS, recurrent flash pulmonary edema, rapid worsening of blood pressure, resistant hypertension, and progression of severe hypertension [14].

15.3 Endocrine Causes of Secondary Hypertension

15.3.1 Primary Aldosteronism (PA)

PA was initially thought to be uncommon when first described in 1950 by Jerome Conn. However, the diagnosis of PA prevalence has increased worldwide due to various factors such as more sensitive laboratory assays, use of aldosterone-to-renin ratio (ARR) for screening, increased use of advanced imaging and resultant detection of adrenal incidentalomas. It is now well recognized that hypokalemia is only present in the minority of patients [21] and most patients with PA are normokalemic. PA is associated with increased cardiovascular morbidity and mortality than age-, sex-, and BP-matched patients with essential hypertension [22]. It is important to diagnose this condition early, as treatment can reverse end organ damage [23]. PA can be caused by unilateral adenoma (coined Conn's Syndrome), bilateral hyperplasia, or unilateral hyperplasia. Patients with unilateral disease who wish for surgical treatment are generally recommended to undergo unilateral laparoscopic adrenalectomy. Surgery improves blood pressure control in all patients and hypertension is reported to be cured in about 50% of patients [21]. Patients with bilateral hyperplasia are treated with mineralocorticoid receptor antagonists such as spironolactone or eplerenone.

Patients with PA have an age of onset of hypertension of about 40–70 years old, are mostly asymptomatic [24], and a minority (~30%) have hypokalemia. In 2016, Endocrine Society [21] recommended for screening of PA in (1) patients with blood pressures repeatedly >150/100 mmHg; (2) resistant hypertension; (3) controlled blood pressure (<140/90 mmHg) on ≥4 antihypertensive medications; (4) hypertension with hypokalemia (spontaneous or diuretics-induced). However, with the developments in PA, expert opinions have called for screening for PA at least once on diagnosis of hypertension.

Some important pointers to bear in mind before screening for PA include correction of hypokalemia as hypokalemia may falsely lower aldosterone and therefore give false negative ARR results. Spironolactone/eplerenone/amiloride should be stopped for at least 6 weeks before performing ARR. Other interfering medications such as beta-blockers, angiotensin-converting enzyme (ACE) inhibitors, angiotensin receptor blockers (ARBs), and diuretics should ideally be switched to non-interfering medications such as verapamil, hydralazine and prazosin/doxazosin as these have minimal effects on ARR. However, in situations where it may not be safe or feasible to switch antihypertensives prior to testing, screening can proceed while on these medications with results interpreted accordingly. Aldosterone, renin, and potassium are performed at 8–10 am with patients in the seated position. Patients with PA have an elevated aldosterone concentration (often >10 ng/dL), suppressed plasma renin activity (<1 ng/mL/h), leading to an elevated ARR (>20), and positive screening cases should be referred to an endocrinologist.

15.3.2 Cushing's Syndrome (CS)

CS is a less common endocrine cause of secondary hypertension, affecting <0.1% of the general population [25]. Important clinical signs to suggest CS include typical body habitus with central obesity, facial plethora, skin atrophy, easy bruisability, violaceous striae, hirsutism, and buffalo hump [26]. Hypertension is common and affects about 80% of patients with CS. Other suggestive clinical cues include sudden worsening of metabolic control, unexplained osteoporosis and fractures, psychiatric overlay such as depression and psychosis. Exogenous CS is by far the most common cause of CS, and it is important to exclude exogenous CS from over-the-counter medication use or from potent steroids taken for other co-morbidities. Screening for endogenous CS should only proceed after exogenous CS have been excluded from the history. Screening tests include 24-h urinary cortisol, overnight dexamethasone suppression tests, salivary cortisol, and cortisol day curve. If suspicion is high, referral to an endocrinologist for further workup is warranted.

15.3.3 Hyper-/Hypothyroidism

Both hyper- and hypothyroidism can be associated with hypertension. Diastolic hypertension is associated with hypothyroidism since low cardiac output is compensated by peripheral vasoconstriction to maintain adequate tissue perfusion [14]. Hyperthyroidism is associated with increased cardiac output and hence elevated systolic hypertension is typically described. Concomitant symptoms to assess thyroid functional state clinically should be assessed (Table 15.1). Biochemical assessment of the thyroid state can be evaluated using free thyroid hormone (FT4) and thyroid stimulating hormone (TSH).

15.3.4 Pheochromocytoma

This is a rare but important cause of secondary hypertension to consider with a prevalence of about 0.2% in unselected hypertensive patients. However, it is one common cause for patients who present with hypertensive emergencies. The clinical course can be stormy with fluctuating blood pressures: severe hypertension (200/100 mmHg) alternating with episodes of hypotension [27], and may lead to multiorgan failure. Failure to recognize this can result in a catastrophic outcome for the patient [28]. Clinical features are due to paroxysmal increase in plasma catecholamines and are characterized by the 5 "P"s [29]: paroxysmal hypertensions, palpitations, perspiration, pallor, and pounding headache.

Screening for pheochromocytoma is done using 24-h urinary fractionated metanephrines or plasma metanephrines. Screening is considered in patients who have symptoms; resistant hypertension; family history of pheochromocytoma (MEN 2; von Hippel Lindau, SDH mutations, neurofibromatosis); and adrenal mass with characteristics suggestive of pheochromocytoma. If screening is positive, further imaging and referral to an endocrinologist are made.

15.4   Coarctation of the Aorta

In children and young adults, coarctation of the aorta is the second most common cause of hypertension [30]. There is usually constriction of the lumen of the aorta near the ligamentum arteriosum [14]. Frequent symptoms are headache, cold feet, and pain in the legs during exercise. Clinical clues are hypertension in the presence of weak femoral pulses. A systolic murmur in the front/or back of the chest may be heard, and posterior rib notches on chest radiographs suggest collateral circulation. Transthoracic echocardiography is the screening method of choice and other imaging modalities such as CT or MRI can also be used. Early surgical repair or percutaneous balloon angioplasty appears equally effective [31].

15.5   Obstructive Sleep Apnea (OSA)

OSA is one of the most common causes of secondary hypertension [32]. It is increasingly recognized to be associated with resistant hypertension [33]. One study showed that OSA was prevalent in 65% of women and 95% of men with resistant hypertension [34].

In a study in Singapore, the prevalence of moderate to severe OSA is estimated to be 30.5% [35]. OSA is characterized by repeated partial or complete upper airway obstruction during sleep, leading to oxygen desaturation and arousals. OSA is diagnosed when AHI (apnea-hypopnea index) is ≥ 5. AHI is defined as the number of apnea plus hypopnea per hour of sleep. OSA can be classified into mild (AHI: 5–15), moderate (AHI: 16–30), and severe (AHI: >30) [36]. Associated risk factors for OSA include male gender, obesity, and middle age [37]. Typical symptoms are fatigue, daytime somnolence, snoring, morning headaches, inability to concentrate, and irritability. On physical examination, patients are often of obese habitus, and may have an enlarged neck circumference from significant soft tissue around the neck or a big tongue.

Various tools are used to screen patients for OSA including: STOP, STOP-BANG (SB), Epworth Sleepiness Scale (ESS), and the 4-Variable screening tool (4-V) [38]. These tools have different ease of administration, and sensitivity and specificity.

The most sensitive and simple screening tool is the STOP-BANG questionnaire. Using this screening tool would improve identification of patients at high risk for sleep apnea [38]. The STOP-BANG questionnaire consists of eight binary (yes/no) items related to the clinical features of sleep apnea [39]. The total score ranges from 0 to 8. Patients are classified for the risk of OSA based on their respective scores. The sensitivity of STOP-Bang score ≥ 3 to detect moderate to severe OSA (AHI > 15) and severe OSA (AHI > 30) is 93% and 100%, respectively [39].

Polysomnography is the standard diagnostic sleep test for OSA. In recent years, home sleep test has also been increasingly used for diagnosis of OSA. Home sleep test is recommended for use in uncomplicated patients with high pre-test probability of OSA. In patients with typical symptoms, suggestive physical examination, known risk factors or positive screening questionnaire should be referred to a sleep specialist.

Treatment for OSA is usually continuous positive airway pressure (CPAP), primarily to improve daytime sleepiness and quality of life [40]. CPAP or Mandibular Advancement Devices (MAD) should be used in conjunction with antihypertensive medications, and dietary and lifestyle modifications [41]. Since many patients are undiagnosed and unaware of this condition, clinicians should maintain high index of suspicion and screen the likely patients especially in the context of resistant hypertension.

15.6   Conclusion

Secondary causes of hypertension are common with prevalence reported to be 10–20% of hypertensive patients and may lead to difficult to control hypertension as well as increased end organ damages. Earlier recognition and treatment are important as it can reduce the burden of care and can lead to better patient outcomes.

References

1. An overview of hypertension and cardiac involvement in Asia: focus on heart failure. [cited 2021 Jun 11]. https://onlinelibrary.wiley.com/doi/10.1111/jch.13753.
2. Venketasubramanian N, Yoon BW, Pandian J, Navarro JC. Stroke epidemiology in South, East, and South-East Asia: a review. J Stroke. 2017;19(3):286–94.
3. Liew SJ, Lee JT, Tan CS, Koh CHG, Van Dam R, Müller-Riemenschneider F. Sociodemographic factors in relation to hypertension prevalence, awareness, treatment and control in a multi-ethnic Asian population: a cross-sectional study. BMJ Open. 2019;9(5):e025869.
4. Lu J, Lu Y, Wang X, Li X, Linderman GC, Wu C, et al. Prevalence, awareness, treatment, and control of hypertension in China: data from 1·7 million adults in a population-based screening study (China PEACE Million Persons Project). Lancet. 2017;390(10112):2549–58.
5. Hirsch JS, Hong S. The demystification of secondary hypertension: diagnostic strategies and treatment algorithms. Curr Treat Options Cardiovasc Med. 2019;21(12):90.
6. Omura M, Saito J, Yamaguchi K, Kakuta Y, Nishikawa T. Prospective study on the prevalence of secondary hypertension among hypertensive patients visiting a general outpatient clinic in Japan. Hypertens Res. 2004;27(3):193–202.

7. Mancia G, Rosei EA, Azizi M, Burnier M, Clement DL, Coca A, et al. 2018 ESC/ESH guidelines for the management of arterial hypertension. 98.
8. 2017 ACC/AHA/AAPA/ABC/ACPM/AGS/APhA/ASH/ASPC/NMA/PCNA guideline for the prevention, detection, evaluation, and management of high blood pressure in adults: a report of the American College of Cardiology/American Heart Association Task Force on Clinical Practice Guidelines [Internet]. [cited 2021 Jun 20]. https://www.ahajournals.org/doi/epub/10.1161/HYP.0000000000000065.
9. 2018 Chinese guidelines for prevention and treatment of hypertension—a report of the Revision Committee of Chinese guidelines for prevention and treatment of hypertension. J Geriatr Cardiol. 2019;16(3):182–241.
10. Resistant hypertension: diagnosis, evaluation, and treatment [Internet]. [cited 2021 Jun 20]. https://www.ahajournals.org/doi/epub/10.1161/CIRCULATIONAHA.108.189141.
11. Guillaume L, Mathieu A, Mélanie D, Nicolas B-C, Nantes University Hospital Working Group on Hypertension. Resistant hypertension: novel insights. Curr Hypertens Rev. 2020;16(1):61–72.
12. Kvapil T, Vaclavik J, Benesova K, Jarkovsky J, Kocianova E, Kamasova M, et al. Prevalence of secondary hypertension in patients with resistant arterial hypertension. J Hypertens. 2021;39:e357.
13. Burrello J, Monticone S, Losano I, Cavaglia' G, Buffolo F, Tetti M, et al. Prevalence of primary aldosteronism and hypokalemia in 5,100 patients referred to a tertiary hypertension unit. J Hypertens. 2021;39:e61.
14. Rimoldi SF, Scherrer U, Messerli FH. Secondary arterial hypertension: when, who, and how to screen? Eur Heart J. 2014;35(19):1245–54.
15. Puar T, Mok Y, Debajyoti R, Khoo J, How C, Ng A. Secondary hypertension in adults. Singapore Med J. 2016;57(05):228–32.
16. Chapter 13. Secondary hypertension. Hypertens Res. 2014;37(4):349–61.
17. Williams GJ, Macaskill P, Chan SF, Karplus TE, Yung W, Hodson EM, et al. Comparative accuracy of renal duplex sonographic parameters in the diagnosis of renal artery stenosis: paired and unpaired analysis. AJR Am J Roentgenol. 2007;188(3):798–811.
18. Vasbinder GBC, Nelemans PJ, Kessels AGH, Kroon AA, Maki JH, Leiner T, et al. Accuracy of computed tomographic angiography and magnetic resonance angiography for diagnosing renal artery stenosis. Ann Intern Med. 2004;141(9):674–82; discussion 682.
19. Stenting and medical therapy for atherosclerotic renal-artery stenosis I NEJM [Internet]. [cited 2021 Jun 15]. https://www.nejm.org/doi/full/10.1056/nejmoa1310753.
20. Chrysochou C, Foley RN, Young JF, Khavandi K, Cheung CM, Kalra PA. Dispelling the myth: the use of renin-angiotensin blockade in atheromatous renovascular disease. Nephrol Dial Transplant. 2012;27(4):1403–9.
21. Funder JW, Carey RM, Mantero F, Murad MH, Reincke M, Shibata H, et al. The management of primary aldosteronism: case detection, diagnosis, and treatment: an Endocrine Society clinical practice guideline. J Clin Endocrinol Metabol. 2016;101(5):1889–916.
22. Milliez P, Girerd X, Plouin P-F, Blacher J, Safar ME, Mourad J-J. Evidence for an increased rate of cardiovascular events in patients with primary aldosteronism. J Am Coll Cardiol. 2005;45(8):1243–8.
23. Catena C, Colussi G, Lapenna R, Nadalini E, Chiuch A, Gianfagna P, et al. Long-term cardiac effects of adrenalectomy or mineralocorticoid antagonists in patients with primary aldosteronism. Hypertension. 2007;50(5):911–8.
24. Mulatero P, Stowasser M, Loh K-C, Fardella CE, Gordon RD, Mosso L, et al. Increased diagnosis of primary aldosteronism, including surgically correctable forms, in centers from five continents. J Clin Endocrinol Metabol. 2004;89(3):1045–50.
25. Newell-Price J, Bertagna X, Grossman AB, Nieman LK. Cushing's syndrome. Lancet. 2006;367(9522):1605–17.
26. Diagnosis of Cushing's syndrome: an Endocrine Society clinical practice guideline. J Clin Endocrinol Metab. Oxford Academic [Internet]. [cited 2021 Jun 15]. https://academic.oup.com/jcem/article/93/5/1526/2598096.

27. Meifen Z, Zainudin SB, Ling CC. Pheochromocytoma crisis triggered by extra-corporal membrane oxygenation explanation. J Adv Med Med Res. 2016:1–5.
28. Lo CY, Lam KY, Wat MS, Lam KS. Adrenal pheochromocytoma remains a frequently overlooked diagnosis. Am J Surg. 2000;179(3):212–5.
29. Young WF. Adrenal causes of hypertension: pheochromocytoma and primary aldosteronism. Rev Endocr Metab Disord. 2007;8(4):309–20.
30. Arar MY, Hogg RJ, Arant BS, Seikaly MG. Etiology of sustained hypertension in children in the southwestern United States. Pediatr Nephrol. 1994;8(2):186–9.
31. Weber HS, Cyran SE. Endovascular stenting for native coarctation of the aorta is an effective alternative to surgical intervention in older children. Congenit Heart Dis. 2008;3(1):54–9.
32. Pedrosa RP, Drager LF, Gonzaga CC, Sousa MG, de Paula LKG, Amaro ACS, et al. Obstructive sleep apnea: the most common secondary cause of hypertension associated with resistant hypertension. Hypertension. 2011;58(5):811–7.
33. Hou H, Zhao Y, Yu W, Dong H, Xue X, Ding J, et al. Association of obstructive sleep apnea with hypertension: a systematic review and meta-analysis. J Glob Health. 2018;8(1):010405.
34. Logan AG, Perlikowski SM, Mente A, Tisler A, Tkacova R, Niroumand M, et al. High prevalence of unrecognized sleep apnoea in drug-resistant hypertension. J Hypertens. 2001;19(12):2271–7.
35. Tan A, Cheung YY, Yin J, Lim W-Y, Tan LWL, Lee C-H. Prevalence of sleep-disordered breathing in a multiethnic Asian population in Singapore: a community-based study. Respirology. 2016;21(5):943–50.
36. The-AASM-Manual-for-Scoring-of-Sleep-and-Associated-Events-2007-.pdf [Internet]. [cited 2021 Jun 16]. https://www.sleep.pitt.edu/wp-content/uploads/2020/03/The-AASM-Manual-for-Scoring-of-Sleep-and-Associated-Events-2007-.pdf.
37. Young T, Skatrud J, Peppard PE. Risk factors for obstructive sleep apnea in adults. JAMA. 2004;291(16):2013–6.
38. Screening tools for the obstructive sleep apnea for the cardiovascular clinician [Internet]. American College of Cardiology. [cited 2021 Jun 20]. https://www.acc.org/latest-in-cardiology/articles/2015/07/14/11/04/http%3a%2f%2fwww.acc.org%2flatest-in-cardiology%2farticles%2f2015%2f07%2f14%2f11%2f04%2fscreeing-tools-for-the-obstructive-sleep-apnea-for-the-cardiovascular-clinician.
39. Chung F, Abdullah HR, Liao P. STOP-Bang Questionnaire: a practical approach to screen for obstructive sleep apnea. Chest. 2016;149(3):631–8.
40. Giles TL, Lasserson TJ, Smith BJ, White J, Wright J, Cates CJ. Continuous positive airways pressure for obstructive sleep apnoea in adults. Cochrane Database Syst Rev. 2006;(1):CD001106.
41. Bratton DJ, Gaisl T, Wons AM, Kohler M. CPAP vs mandibular advancement devices and blood pressure in patients with obstructive sleep apnea: a systematic review and meta-analysis. JAMA. 2015;314(21):2280–93.

Hypertension in Children

16

Subhankar Sarkar and Isaac Liu

16.1 Introduction

Hypertension, though relatively less common in children and adolescents than in adults, is an important issue due to a number of reasons. First, early onset hypertension merits searching for a secondary cause of hypertension in children. Second, onset in childhood of uncontrolled blood pressure results in a longer runway for the accruement of end-organ damage in adulthood. Third, the practitioner should be mindful of the unique challenges and considerations of blood pressure measurements in young children.

16.2 Epidemiology

The prevalence of pediatric hypertension varies with ethnic group, geographic location, blood pressure (BP) measurement technique, and definition criteria. Overall data shows that prevalence of hypertension has been increasing in last few decades mainly due to rising trends of obesity [1]. The estimated prevalence of pediatric hypertension ranges from 1 to 5% in developed countries but is as high as 20% in

S. Sarkar
Pediatric Nephrology, Institute of Child Health, Kolkata, India

Peerless Hospital, Kolkata, India

I. Liu (✉)
Raffles Hospital, Singapore, Singapore

Cardiovascular and Metabolic Diseases Program, Duke-NUS Medical School, Singapore, Singapore

Department of Medicine, Yong Loo Lin School of Medicine, NUS, Singapore, Singapore
e-mail: liu_isaac@rafflesmedical.com; mdcldi@nus.edu.sg

© The Author(s), under exclusive license to Springer Nature Switzerland AG 2022
C. V. S. Ram et al. (eds.), *Hypertension and Cardiovascular Disease in Asia*,
Updates in Hypertension and Cardiovascular Protection,
https://doi.org/10.1007/978-3-030-95734-6_16

developing countries. Furthermore, it is higher in boys (15–19%) than girls (7–12%) [2–7]. Initial data were based on a single BP measurement. After using repeated measurement, the prevalence of confirmed hypertension is about 3.5% [4, 8]. In a recent systematic review and meta-analysis, Song et al. reported that pooled global prevalence of childhood hypertension is 4% [9]. The prevalence of hypertension is much higher among children and adolescents in presence of risk factors such as obesity (4–14%) [3, 10], type 1 diabetes (4–16%) [11], type 2 diabetes (12–31%) [12], sleep apnea syndrome (3.6–14%) [13], neurofibromatosis type 1 (6.1%) [14], chronic renal disease (50%), and end-stage renal disease (48–70%) [15, 16].

16.3   Definition and Classification

In 2004, the "Fourth Report on the Diagnosis, Evaluation, and Treatment of High Blood Pressure in Children and Adolescents" (Fourth Report) defined "normal blood pressure" if the systolic blood pressure (SBP) and diastolic blood pressure (DBP) values were <90th percentile [17]. "Prehypertension" was defined as SBP and/or DBP ≥90th percentile to <95th percentile or BP ≥120/80 mmHg to <95th percentile in adolescents, whichever was lower. Hypertension was defined as SBP and/or DBP ≥95th percentile [17]. In 2017, the American Academy of Pediatrics (AAP 2017) guideline updated the definitions and classification of hypertension (Table 16.1), which was subsequently endorsed by the American Heart Association (AHA) and American College of Cardiology (ACC) in 2017 [1, 16]. According to the AAP 2017, for the diagnosis of hypertension, the mean of two consecutive BP measurements taken by auscultation technique repeated at three different visits should be used. Important modifications of the definition from the Fourth report are: (1) for children and adolescent ≥13 years—BP thresholds used for defining hypertension transit seamlessly into the adult hypertension guideline by AHA/ACC, (2) the terminology "prehypertension" has changed to "elevated blood pressure." Both the "European Society of Hypertension (ESH)" and AAP 2017 guideline defined "normal blood pressure" as BP <90th percentile by age, sex, and hypertension is defined if SBP and/or DBP ≥95th percentile [1, 18]. The AHA/ACC guidelines thus become applicable in children and adolescents age ≥ 13 years by AAP 2017 guideline and ≥16 years by ESH guideline. It must be noted, however, that the normative

Table 16.1 Definition and classification of hypertension

Criteria	Age < 13 years	Age ≥ 13 years
Diagnosis or stage	SBP or DBP (percentile)	SBP/DBP (mmHg)
Normal	<90th	<120/80
Elevated BP	≥90th to <95th or 120/80 mmHg (whichever is lower)	120–129/<80
Stage 1 hypertension	≥95th to <(95th +12 mmHg) or 130/80 to 139/89 mmHg (whichever is lower)	130–139/80–89
Stage 2 hypertension	≥95th + 12 mmHg or ≥140/90 mmHg (whichever is lower)	≥140/90

BP tables in the AAP 2017 guideline have deliberately excluded children and adolescents with overweight and obesity, the norms are now several mmHg lower than the Fourth Report.

16.4   Etiology

16.4.1 Primary or Essential Hypertension

Essential or primary or idiopathic childhood hypertension is a common cause of hypertension among older children and adolescents, particularly as the prevalence of overweight and obesity is increasing. Features of primary hypertension are a positive family history of hypertension, presence of overweight or obesity in children ≥6 years, and mainly systolic hypertension. In children with these risk factors, the AAP 2017 recommends against an extensive over-investigation into causes for secondary hypertension if they do not have history or physical examination findings suggesting the latter.

16.4.2 Secondary Hypertension

Secondary hypertension is common among younger children. Renal parenchymal disease and renovascular abnormalities are the most common cause of secondary hypertension. Among children with secondary hypertension, approximately 34–79% patients have renal parenchymal disorders and 5–25% have renovascular abnormalities [19–21]. Secondary hypertension is suspected especially when children present at an early age with severely elevated hypertension. Secondary causes of hypertension are listed in Table 16.2.

16.4.2.1   Renal Parenchymal Disease
Renal parenchymal damage can lead to hypertension. Both acute and chronic glomerulonephritis (GN) are important causes of secondary hypertension in children. Post-infectious GN, especially post-streptococcal GN, is a major cause of GN and related hypertension particularly in developing countries in Asia. Volume expansion due to fluid and sodium retention and activation of the renin-angiotensin-aldosterone axis are common mechanisms of hypertension in acute GN. Chronic renal scarring due to reflux nephropathy, pyelonephritis or congenital anomaly of kidney with compensatory hypertrophy can also cause hypertension. Hypertension in chronic kidney disease (CKD) and kidney transplant recipients can further be exacerbated by volume expansion and adverse effects of medicines.

16.4.2.2   Renovascular Hypertension
Renovascular hypertension (RVH) is the second most common correctable hypertension in children caused by abnormal hormonal response secondary to the impaired blood flow to a part or all, of one or both kidneys. Significant renal artery

Table 16.2 Secondary causes of hypertension

System	Cause
Renal parenchymal disease	Reflux nephropathy
	Obstructive uropathy
	Renal dysplasia/hypoplasia
	Acute and chronic glomerulonephritis
	Nephrotic syndrome
	Hemolytic uremic syndrome
	Chronic kidney disease
	End-stage renal disease
	Polycystic kidney disease
Renovascular disease	Takayasu aortoarteritis
	Renal artery stenosis
	Renal artery thrombosis
Cardiovascular	Coarctation of aorta
	Patent ductus arteriosus
	Arteriovenous fistula
	Renal vein thrombosis
Endocrine	Pheochromocytoma
	Cushing syndrome
	Primary hyperaldosteronism
	Primary hyperparathyroidism
	Hypercalcemia
	Congenital adrenal hyperplasia
	Neuroblastoma
Neurological	Mental stress
	Anxiety
	Guillain-Barré syndrome
	Raised intracranial pressure
Respiratory	Obstructive sleep apnea
Monogenic hypertension	Liddle's syndrome
	Syndrome of apparent mineralocorticoid excess
	Familial hyperaldosteronism type 1 or glucocorticoid remediable aldosteronism
	Pseudohypoaldosteronism type 2 (Gordon syndrome)
	Overt mineralocorticoid excess
	Familial glucocorticoid resistance
	Mineralocorticoid receptor activating mutation
Medications or drugs	Glucocorticoids
	Non-steroidal anti-inflammatory drugs
	Oral contraceptives
	Sympathomimetics
	Erythropoietin
	Calcineurin inhibitors (ciclosporin, and tacrolimus)
	Cocaine
	Anabolic steroids
	Pseudoephedrine
	Phencyclidine
	Caffeine
	Nicotine
Heavy metal toxicity	Lead, cadmium, mercury, phthalates
Others	Neurofibromatosis
	Tuberous sclerosis
	Collagen vascular disease

Table 16.3 Etiologies of renovascular hypertension

Fibrous and fibromuscular dysplasia	• Medial fibroplasia • Perimedial fibroplasia • Intimal fibroplasia • Medial hyperplasia
Vasculitis	• Takayasu disease • Kawasaki disease • Moyamoya disease • Other systemic vasculitis
Vascular malformations	• Thromboembolism (e.g., prior history of umbilical catheterization, post-angiography, post-trauma) • Arteriovenous malformation • Renal artery aneurysm
Inherited conditions	• Neurofibromatosis type 1 • Tuberous sclerosis • Williams' syndrome • Marfan syndrome • Feuerstein-Mimms syndrome • Tuberous sclerosis
Extrinsic compression	• Neuroblastoma • Wilms' tumour/other tumours • Congenital fibrous bands • Post-traumatic • Retroperitoneal fibrosis
Miscellaneous	• Radiation • Congenital rubella syndrome • Transplant renal artery stenosis

stenosis (RAS) is usually defined as stenosis of the renal artery greater than 75% of the vessel lumen or at least 50% with post-stenotic dilation, and invariably results in severe hypertension. RVH is a heterogeneous disease process that can be secondary to various etiologies (Table 16.3) including intrinsic lesions of the renal arteries, extrinsic compression by masses, and intraluminal occlusion by thrombosis. Fibromuscular dysplasia is the most prevalent cause of RVH in the Western World, whereas in countries like India and South Africa, Takayasu's arteritis results in a high proportion of RAS (73–89%) [22–25]. Activation of renin-angiotensin-aldosterone system (RAAS) and presence of normal or abnormal contralateral kidney influence the RVH. Renovascular hypertension can also be caused by renal vein thrombosis.

16.4.2.3 Endocrine Cause of Hypertension

Secondary hypertension due to endocrine causes are rare, with prevalence ranging from 0.5 to 6% but they respond dramatically with medical and surgical treatment, hence the importance of achieving a precise diagnosis [1]. Common causes of hypertension due to catecholamine excess are pheochromocytoma, neuroblastoma, and sympathomimetic drugs. Pheochromocytoma is a rare disease occurring more often in adults than in children, and is thought to be responsible for approximately 1% of childhood hypertension. Traditional teaching suggested that 10% of pheochromocytomas are familial but recent advances in molecular genetics however have revealed an identifiable germline mutation in up to 59% of apparently sporadic pheochromocytomas presenting 18 years or younger and in 70% of those presenting before 10 years of age. Eighty-five percent of pheochromocytomas are located in the adrenal glands; the rest develop in the extra-adrenal parasympathetic and sympathetic paraganglia. Most tumors are less than 5 cm in size, and 25–33% are bilateral. Approximately 10% of intra-adrenal and 40% of extra-adrenal pheochromocytomas are malignant. Children often have sustained severe hypertension as their primary sign and are less likely than adults to present with the classic triad of tachycardia, headache, and diaphoresis.

Corticosteroid excess may lead to hypertension. A common cause of excess corticosteroid is exogenous administration, endogenous causes are Cushing syndrome, congenital adrenal hyperplasia (CAH), aldosterone secreting tumor, and glucocorticoid-remediable aldosteronism (GRA). Rarely, primary hyperparathyroidism and thyrotoxicosis can cause hypertension.

16.4.2.4 Sleep Apnea Syndrome

Obstructive sleep apnea (OSA) is an established risk factor for hypertension in adults. Although the meta-analysis failed to establish OSA as a risk of hypertension in children, high blood pressure along with change in diurnal variations of BP (particularly, diminished nocturnal dipping) may coexist with OSA [26, 27]. The probable mechanism of hypertension in OSA is decreased sensitivity of the baroreflex system, which is supported by evidence of differences in diurnal variation of acute phase reactants and pro-inflammatory cytokines compared to healthy children [28].

16.4.2.5 Monogenic Hypertension

Monogenic hypertension is due to mutations of single gene function which leads to early onset refractory hypertension, and, in mutations which impact the renin-angiotensin aldosterone axis, disturbances in potassium. Several genes associated with monogenic hypertension are listed in Table 16.4. Monogenic hypertension should be suspected in suppressed plasma renin activity or elevated aldosterone to renin ratio along with positive family of early onset hypertension [29]. The mechanisms of hypertension includes volume expansion due to excess sodium reabsorption, deficiency of enzymes involved in steroid metabolism in the adrenal gland, and aldosterone excess [29]. Conditions associated with increased sodium transport may be mediated by mineralocorticoid (apparent mineralocorticoid excess (AME) syndromes, glucocorticoid-remediable aldosteronism, and congenital adrenal hyperplasia due to 11β-hydroxylase or 17α-hydroxylase deficiency) or independent of mineralocorticoids (Liddle and Gordon syndromes) [30]. Apparent mineralocorticoid excess (AME) is a rare autosomal recessive condition due to deficiency of enzyme 11 beta-hydroxysteroid dehydrogenase II (HSD11B2) which converts active cortisol to inactive cortisone. Deficiency of HSD11B2 results in persistence of high cortisol level which binds to mineralocorticoid receptors resulting in sodium retention and urinary loss of potassium. The ultimate consequence is hypokalemia, metabolic alkalosis, and suppressed plasma renin and aldosterone level.

Table 16.4 Overview of monogenic hypertension

Condition	Inheritance	Genetic mutation	Encoded protein	Pathophysiology	Clinical feature	Management
Liddle syndrome	AD	*SCNN1B* *SCNN1G* *SCNN1A*	ENaC	Overactive ENaC leads to excess reabsorbs sodium, resulting in volume expansion and hypertension	Early onset hypertension, hypokalemia and metabolic acidosis	ENaC inhibitory agents like amiloride, triamterene might help
Syndrome of apparent mineralocorticoid excess	AR	*HSD11B2*	11β-HSD2	HSD11B2 deficiency leads to excess cortisol stimulation at MR	Early onset hypertension, low birth weight and developmental delay	MRA helpful
Congenital adrenal hyperplasia	AR	*CYP11B1* *CYP17A1*	11β-OH 17α-OH	Deficiency of enzyme for steroid synthesis leads to accumulation of intermediate metabolites which acts on MR	Hyperandrogenemia, defective sexual development	Glucocorticoid supplementation but treatment targets to reduce sexual dysfunction
Familial hyperaldosteronism type I (GRA)	AD	*CYP11B1/CYP11B2* hybrid gene	ADS	Hybride gene produced a chimeric product that is ACTH-sensitive and produces aldosterone	Early onset hypertension, remediable	Glucocorticoids and MRA
Familial hyperaldosteronism type II	AD	*CLCN2*	ClC-2	Excess aldosterone production	Refractory hypertension, not response to glucocorticoids	MRA and surgical resection
Familial hyperaldosteronism type III	AD	*KCNJ5*	GIRK-4	Gain-of-function mutations in potassium channels leads to cortical cells to depolarize and subsequently activate aldosterone synthase	Severe hypertension	MRA and surgical resection

(continued)

Table 16.4 (continued)

Condition	Inheritance	Genetic mutation	Encoded protein	Pathophysiology	Clinical feature	Management
Familial hyperaldosteronism type IV	AD	*CACNA1H*	Cav3.2	Gain-of-function mutations in calcium channels allowing enhancing aldosterone synthase activity		MRA and surgical resection
Geller syndrome	AD	*NR3C2*	MR	Altered MR allowing atypical stimulation by other steroids, especially progesterone	Hypertension in early adult life specially during in pregnancy	Delivery of the child and subsequent monitoring
Gordon syndrome (pseudohypoaldosteronism type II)	AD	*WNK1, WNK4, KLHL3, CUL3*		Altered NCC channel allow electrolyte and fluid over absorption		Thiazide diuretic

AD autosomal dominant, *AR* autosomal recessive, *MRA* mineralocorticoid antagonist

Glucocorticoid-Remediable Aldosteronism (GRA) is a rare autosomal dominant disorder, also known as familial hyperaldosteronism type I, characterized by an increased expression of a chimeric gene. This gene is stimulated by adrenocorticotropic hormone (ACTH) and encodes a hybrid protein that stimulates aldosterone production, independent of renin. Hypokalemia is common but serum potassium may be normal.

Liddle syndrome (pseudo-hyperaldosteronism type I) is a rare autosomal dominant disorder due to overactivity of the epithelial sodium channel (ENaC). Increased sodium reabsorption in the distal nephron leads to volume expansion and hypertension. There is metabolic acidosis and hypokalemia with suppressed plasma renin and aldosterone activity. Patients respond to triamterene and amiloride, but spironolactone is ineffective.

Gordon syndrome (pseudo-hypoaldosteronism type II) is an autosomal dominant disorder caused by gain-of-function of the Na-K-Cl cotransporter activity in the distal convoluted tubules of the kidney. Overexpression of cotransporters leads to increased sodium reabsorption resulting in hypertension, hyperkalemia, and hyperchloremic metabolic acidosis. Serum aldosterone concentrations are low and plasma renin activity is suppressed. Electrolyte abnormalities and elevated blood pressure can be managed with thiazides.

16.5 Blood Pressure (BP) Measurement

16.5.1 Steps of Clinic BP Measurement [1, 31]

Standardization of clinic BP measurement is particularly important due to special considerations in children such as cuff size.

(a) The child should have avoided stimulant drugs or foods, have been sitting quietly for 3–5 min, seated with back supported, feet uncrossed on the floor and right arm supported, cubital fossa at heart level.

(b) The right arm is preferred for consistency and comparison with standard tables and because of the possibility of coarctation of the aorta, which might lead to falsely low readings in the left arm. While measuring BP, both the patient and observer should not speak.

(c) Appropriate cuff size—a cuff with an inflatable bladder width that is at least 40% of the arm circumference with a point midway between the olecranon and the acromion, with bladder length covering 80–100% of the arm circumference should be used.

(d) Auscultatory BP measurement using a sphygmomanometer is the gold standard for diagnosis of hypertension in young patients. For this technique, the bell of stethoscope should be placed over the brachial artery 2–3 cm above the antecubital fossa. The cuff should be inflated 20–30 mmHg above the point at which the radial pulse disappears and cuff should be deflated at the rate of 2–3 mmHg per second. SBP is determined by the onset of the tapping Korotkoff sounds

(K1) and DBP is determined by the disappearance of the Korotkoff sounds (K5). If the Korotkoff sounds can be heard to zero mmHg, then BP measurement should be repeated with less pressure on the head of the stethoscope and only if the very low K5 persists should K4 (muffling of the sounds) be recorded as the DBP.

(e) To measure BP in the legs by the auscultatory method, the patient should be in the prone position. A correct sized cuff should be placed mid-thigh and the stethoscope placed over the popliteal artery. SBP in the legs is usually 10–20% higher than the brachial artery pressure.

(f) In infants, the auscultatory method may be difficult, and hence BP obtained by oscillometric device may be used instead.

16.5.2 BP Measurement Frequency

According to the AAP, children aged $\geq$3 years should receive blood pressure measurements yearly. Children with risk factors such as obesity, renal disease, diabetes, aortic arch abnormalities, and medicines that may elevate BP should receive BP measurements at each follow-up visit. Children younger than 3 years should receive BP measurements at every well-child care visit if they have risk factors for developing hypertension such as history of prematurity, known renal or urological abnormalities, family history of congenital renal disease, recurrent urinary tract infection, hematuria, proteinuria, diabetes, aortic coarctation, congenital heart disease, medications that predispose to hypertension, malignancy, bone marrow or solid organ transplant, evidence of raised intracranial pressure and any other systemic illness.

16.5.3 Ambulatory Blood Pressure Monitoring (ABPM)

ABPM is being increasingly recognized as a valuable tool in the investigation of pediatric hypertension [1]. ABPM involves repetitive non-invasive blood pressure measurements using portable devices in outpatient settings over an entire day. A much better approximation of true blood pressure is achieved by multiple readings as compared to single measurements. The detection and monitoring of hypertension has significantly improved with the use of ABPM, which allows for a more accurate classification of hypertension, confirming the diagnosis of hypertension, and differentiating between true and "white coat" hypertension. It also helps in detecting masked hypertension, and assessment of the severity and persistence of blood pressure elevation. It is possible to determine other parameters such as the dipping status in patients at high risk for end-organ damage and also to evaluate the effectiveness of antihypertensive drug therapy. Where feasible, ABPM is a useful tool for assessing the presence of ambulatory hypertension and for guiding therapy for retarding end-organ damage. ABPM also permits an assessment of the circadian blood pressure profile. By more accurately and reliably measuring blood pressure, especially circadian changes, ABPM has been shown to predict cardiovascular morbidity and mortality and end-organ damage.

16.5.3.1 Circadian Patterns of Blood Pressure Variation [1, 32]

Blood pressure typically follows a circadian pattern. In normotensive individuals, a peak in blood pressure usually occurs in the early morning, followed by a gradual decline to lower blood pressure levels during the evening and even lower levels during the night. Blood pressure is generally lowest between 2 am and 4 am, decreasing by approximately 13 and 17% of daytime levels for systolic and diastolic blood pressure, respectively. The daytime surge in BP may trigger cardiovascular events; therefore, it is ideal to provide 24-h blood pressure control to cover this surge. This is especially critical for agents that are administered once daily in the morning, as the morning surge comes at the end of the previous day's dosing interval. The use of ABPM enables blood pressure to be assessed throughout the circadian cycle to determine fluctuations in blood pressure and the efficacy of antihypertensive therapy.

16.5.3.2 Blood Pressure Load

On ABPM, blood pressure load is the percentage of readings above the ambulatory 95th percentile for that age, gender, and height, based on smoothed data of Wuhl [32, 33]. Patients showing BP load in excess of 25–30% are typically considered elevated [32]. Load in excess of 50% was predictive of left ventricular hypertrophy in one study [32]. Patients with normal mean BP may have elevated BP load, these patients may not be picked up with office BP alone though they may be at higher risk for end-organ damage [32].

16.5.3.3 Dipping

Dipping is calculated by percentage drop from mean daytime to mean night time BP in ABPM [32]. This refers to the physiological decline in systolic blood pressure and diastolic blood pressure seen at night. Normal dipping is generally defined as a >10% decline in mean systolic and diastolic ambulatory BP levels from day to night. Non-dipper status is defined as <10% decline in mean ambulatory BP from day to night [32].

16.5.3.4 White Coat Hypertension

White coat hypertension (WCH) is generally characterized by BP more than the 95th percentile in clinic or hospital setting, but below 90th percentile outside of the doctor's office. By using ABPM WCH can be diagnosed when mean SBP and DBP are <95th percentile and SBP and DBP load are <25% while the clinic BP is ≥95th percentile [32]. The proportion of children with WCH varies between 22 and 88% [32]. Data in children are sparse, but young adolescents with white coat hypertension have higher left ventricular mass index, suggesting the need for close follow-up [32].

16.5.3.5   Masked Hypertension

Masked hypertension (MH) is defined as normal clinic BP but mean SBP and DBP are >95th percentile and SBP and DBP load are ≥25%. MH can only be diagnosed by ABPM. The prevalence of MH ranges from 5.7 to 9.4% [32]. Such patients might show sustained clinic hypertension and higher left ventricular mass [34, 35].

16.5.3.6   Procedure [32]

ABPM is a fully automated oscillometric BP machine which is set to measure multiple blood pressures at regular intervals (usually every 15–30 min) over a 24–48-h period, providing a continuous blood pressure record during normal daily activities. A patient diary should be kept in order to record corresponding physical activity, sleep, and drug intake. The criteria for a successful ABPM recording include: at least 40–50 readings for a full 24-h recording, minimum of 1 reading every hour, and 65–75% of all possible BP readings for a partial day report. After a successful recording, data is analyzed by software which is based on age and gender specific normative data and BP is classified according to Table 16.5. BP index is measured in the following manner: average ABP value divided by BP value of 95th percentile of the normative data. BP index >1.0 indicates presence of hypertension.

16.5.3.7   Indications for ABPM

ABP should be performed for the confirmation of the diagnosis of hypertension in children and adolescents with elevated BP for more than 1 year, or stage 1 hypertension over three clinic visits. Additionally, it is recommended in children with high risk conditions to assess severity of hypertension and if abnormal circadian BP patterns are present which may indicate increased risk of target organ damage [1]. Table 16.6 summarizes conditions where ABPM is useful for monitoring of BP.

Table 16.5 Suggested schema for staging of ABPM levels in children

Classification	Clinic blood pressure[a]	Mean ambulatory systolic blood pressure[b]	Systolic blood pressure load (%)
Normal blood pressure	<95th percentile	<95th percentile	<25
White coat hypertension	>95th percentile	<95th percentile	<25
Masked hypertension	<95th percentile	>95th percentile	>25
Elevated blood pressure	>95th percentile	<95th percentile	25–50
Ambulatory hypertension	>95th percentile	>95th percentile	25–50
Severe ambulatory hypertension	>95th percentile	>95th percentile	>50

Modified from Lurbe et al.
[a] Based on National High Blood Pressure Education Program Task Force Standard
[b] Based on the ABPM values of smoothed values of Wuhl

Table 16.6 Value of ABP monitoring in high risk patients

- Apparent drug resistant hypertension
- Determining the efficacy of drug treatment over 24 h
- Feature of target organ damage despite normal clinic BPs
- Hypotensive symptoms with antihypertensive drugs
- Unusual BP variability or episodic hypertension
- Obesity
- Diabetes mellitus
- Chronic kidney disease
- Obstructive sleep apnoea
- Solid organ transplantation
- Genetic syndromes such as neurofibromatosis, Turner syndrome, Williams syndrome
- Repaired coarctation of the aorta
- History of prematurity

16.5.4 Normative Data for Hypertension

AAP 2017 published new normative BP table for boys and girls according to their age and height percentile [1]. This new BP table is derived from normal body weight children; overweight and obese individuals were excluded from analysis. A simplified BP table established for screening of BP which is based on BP of 95th percentile and fifth percentile of height [1]. Currently there is no normative data available for neonates. Dionne et al. established a neonatal BP table with BP value of 95th and 99th percentile from 26 to 44 weeks postmenstrual age [36]. For children between 0 and 1 years, no updated data is available, thus the AAP 2017 guideline suggests using the "Report of the Second Task Force on Blood Pressure Control in Children" [37].

16.6   Evaluation of Hypertension

After diagnosis, the causes and complications of hypertension requires investigation. Detailed history and comprehensive examination are necessary. History taking includes perinatal history, prematurity, low birth weight status and family history of hypertension or kidney disease, nutritional, psychosocial, physical activity, and medications. Perinatal risk factors such as maternal hypertension, low birth weight, and prematurity are risk factors for childhood hypertension [38, 39]. Prematurity and low birth weight are associated with a low nephron endowment which is a risk factor of hypertension in children. Other important perinatal history in ex-premature infants is bronchopulmonary dysplasia and umbilical line catheterization. Nutritional history is important to assess dietary contribution for hypertension. Sodium intake is linked to hypertension and left ventricular hypertrophy [40]. High sodium intake leads to twofold increased risk of hypertension and this risk was compounded (threefold) in obese patients [41]. High dietary fat intake and central adiposity were associated with high BP [42–44]. Lack of fruits and vegetables, and decreased physical activity are also linked to obesity and hypertension. Questions regarding physical activity include duration of screen time, and hours of exercise per week. Stress, anxiety, maltreatment, bullying, and body perception disorder are risk factors for the development of future hypertension [45–47]. Older children and adolescents should be interrogated for history of smoking, alcohol, and other substance abuse. Family history includes hypertension, obesity, diabetes, dyslipidemia, and premature cardiovascular or cerebrovascular or renal disorders.

16.6.1 Physical Examination

Weight and height are measured and body mass index calculated to identify overweight and malnutrition children. All features of malnutrition should be looked for, as it could be a sole feature of chronic illness. Four-limb blood pressure measurements and all peripheral pulses should be examined for inequality. A detailed physical examination focusing on secondary causes of hypertension and neurological and cardiovascular examinations looking for signs of end-organ damage are necessary.

16.6.2 Laboratory Evaluation

Children with suspected primary hypertension do not require extensive evaluation for secondary causes. However, diagnostic tests should be performed for etiology and complications in children at risk. Table 16.7 summarized basic diagnostic test for evaluation of hypertension. Specific tests are selected on the basis of history and physical examination finding.

16.6.3 Electrocardiography

Electrocardiography (ECG) is easily available with economy of cost. It has high specificity but poor sensitivity and extremely low positive predictive value [48–50].

Table 16.7 Basic diagnostic workup

All patients	Obese patients	Based on clue from history and physical examination	For end-organ damage
Blood urea nitrogen	Hemoglobin A1c	Fasting serum glucose	Retinal examination
Electrolytes	Aspartate transaminase	Thyroid-stimulating hormone	Urine spot protein to creatinine ratio
Creatinine	Alanine transaminase	Drug screen	Chest X-ray, ECG, echocardiography
Lipid profile	Fasting lipid panel	Sleep study	
Urinalysis		Complete blood count	
Renal sonography if age < 6 years or abnormal urine or renal function			

The current AAP 2017 guideline not recommended ECG for evaluating left ventricular hypertrophy, but can be conveniently used in in low-resource areas.

16.6.4 Echocardiography

Echocardiography was mentioned in the Fourth Report as a tool to evaluate hypertension-related left ventricular (LV) changes. The AAP 2017 guideline recommends echocardiography for the evaluation of hypertension-related left ventricular injury [1]. Left ventricular hypertrophy is defined as LV mass > 51 g/m$^{2.7}$ or LV mass > 115 g or > 95 g per body surface area for boys and girls, respectively [51]. Concentric hypertrophy is defined as LV relative wall thickness > 0.42 cm. LV ejection fraction <53% is considered to be decreased. Echocardiography should be performed at initiation of treatment and repeated at 6- to 12-month intervals in persistent hypertension despite treatment, concentric hypertrophy, or reduced LV ejection fraction. If initially normal, echocardiography should be repeated at yearly intervals in secondary hypertension, stage 2 or chronic stage 1 hypertension that is incompletely treated.

16.6.4.1 Investigation for Renovascular Hypertension (RVH)

Although non-invasive tests are initially preferred over invasive tests, the limitations of various non-invasive tests should be kept in mind particularly in reference to the challenges posed in pediatric practice. Computed tomography (CT) or magnetic resonance (MR) angiogram can be done in younger children but they are more reliable for proximal RAS and may miss distal or intrarenal RAS. Digital subtraction angiography (DSA) remains the reference standard particularly when one is suspicious of RAS. In addition, therapy for RVH such as balloon angioplasty may be performed in the same sitting. Both DSA and CT angiography carry the risk of contrast nephropathy and appropriate precautions should be followed.

Duplex Doppler ultrasonography is non-invasive, relatively inexpensive and suitable for serial monitoring to determine the disease progression. Resistive index (RI) >80% is associated with greater degree of intrinsic kidney damage or irreversible intrarenal vascular disease and less favorable outcome after revascularization. Most of these data are from adult studies and pediatric literature is lacking. These studies, however, are operator-dependent, time consuming and challenging in young children as well as in obese patients. Thus, the AAP 2017 guideline suggests for doppler renal ultrasonography to be done in centers with appropriate expertise and to be restricted to normal-weight children and adolescents ≥8 years of age who are suspected of having RVH and who will cooperate with the procedure [1].

Computed tomographic angiography (CTA) is a useful non-invasive test with adult studies reporting sensitivity and specificity for detecting RAS at 94% and 93%, respectively [1, 52]. Its usefulness in children where FMD is more common needs further evaluation. The yield of multi-detector CTA in diagnosing intrarenal stenosis particularly in children needs to be properly evaluated. The major disadvantage of CTA remains radiation exposure and risk of contrast nephropathy in CKD patients.

Magnetic resonance angiography (MRA) is a highly sensitive test for detecting proximal RAS. The sensitivity and specificity for the detection of RAS is 90% and 94% respectively for pediatric and up to 100% sensitivity has been reported in adult studies [1, 52]. However, MRA may miss accessory renal arteries and significant lesions in the intrarenal portion of renal arteries. The main concern for MRA is gadolinium-associated systemic fibrosis among those with reduced renal function. With improvements such as imaging without gadolinium-based contrast or using breath-holding MRA with paramagnetic contrast such as gadopentetate dimeglumine, this modality may be increasingly used.

Renal scintigraphy either with 99m-technetium-dimercaptosuccinic acid (DMSA) or 99m-Tc-mercaptoacetyltriglycine (MAG3) along with angiotensin-converting enzyme inhibitor (ACEI) has been evaluated as a potentially useful non-invasive test. Among adults the sensitivities and specificities for detecting RVH have been reported as 68–93% and 70–93% [53]. Unfortunately, pediatric studies have not been promising with reports of missing potentially treatable renovascular lesions [54]. The AAP guideline does not recommend its use [1].

Renal vein renin sampling of renin concentrations may be useful for localizing the site of significant lesion particularly when RAS is bilateral, segmental, or both. Renin samples are taken from the infra-renal inferior vena cava and the main renal veins and their larger intrarenal tributaries through a femoral vein approach with the idea that the site with significant RAS will have higher renin activity [55, 56]. This technique helps in localizing the kidney with RAS and may also localize the segment of the kidney being perfused by any stenosed tributaries of the main renal arteries. This guides segmental renal embolization for cases refractory to pharmacotherapy or not amenable to angioplasty [57, 58]. The main limitation of renal vein renin sampling is obviously its invasiveness which has resulted in it not being a routine pediatric practice.

Catheter-based digital subtraction angiography (DSA) remains the most accurate technique for assessment of suspected renovascular disease in children. In the presence of amenable stenosis, intervention can be carried out in the same setting. Among carefully selected cohorts it has been reported to have a diagnostic yield of 40% [59, 60].

16.6.4.2 Evaluation of Suspected Aldosterone Excess States

Features of primary hyperaldosteronism are hypokalemia, metabolic alkalosis, increased 24-h urinary potassium excretion, decreased 24-h urinary sodium excretion, and hypertension which is usually moderate-to-severe and refractory to treatment. Before evaluation, the patient needs to be optimized for evaluation for hyperaldosteronism. Drugs that alter aldosterone or renin secretion like beta-blockers or spironolactone should be avoided if possible. If the screening test is performed while on ACE inhibitors, angiotensin-receptor blockers, calcium channel antagonists, or alpha-blockers, and aldosterone levels remain frankly elevated in the setting of suppressed renin activity, the likelihood of primary aldosteronism remains high. Hypokalemia should be corrected as it directly inhibits aldosterone release. The patient should not be dehydrated as this stimulates the renin-angiotensin-aldosterone axis.

Saline suppression test: Acute intravascular volume expansion with isotonic saline normally suppresses the renin-angiotensin system. Isotonic saline is infused over 4–6 h. This test should not be performed in patients with compromised cardiac function due to the risk of pulmonary edema. In normal subjects, plasma aldosterone (PA) concentrations normally decrease below 166 pmol/L (6 ng/dL) at the end of the saline infusion. PA concentrations >277 pmol/L (10 ng/dL) are diagnostic of autonomous aldosterone production and values between 166 and 277 pmol/L (6–10 ng/dL) are considered indeterminate but are highly suspicious for this disorder.

Oral salt loading test: Oral salt loading for 3 days results in extravascular and intravascular volume expansion and renin-angiotensin-aldosterone axis suppression in normal individuals. Patients are instructed to consume a high (200 mmol) sodium diet for 3 days, or take four 500 mg sodium chloride tablets (1 tablet = 8.2 mmol sodium) with each meal (3 meals) for 3 days. On the third day of the high sodium diet, a 24-h urine collection for aldosterone excretion, creatinine, and sodium is collected. Aldosterone excretion greater than 39 nmol/day, in the presence of a urinary sodium excretion greater than 200 mmol per 24 h, is 96% sensitive and 93% specific for the diagnosis of primary hyperaldosteronism. Oral salt loading can be performed on an outpatient basis. However, BP and potassium levels should be monitored during the testing to avoid acceleration of hypertension or the precipitation of severe hypokalemia.

Posture study: Adrenal adenomas producing aldosterone are sensitive to adrenocorticotrophic hormone (ACTH), whereas bilateral idiopathic adrenal hyperplasia is sensitive to angiotensin II. Patients have blood tests at 8 am for aldosterone, cortisol, renin, and 18-hydroxycorticosterone after overnight recumbency. After 4 h of being upright and walking around (which stimulates angiotensin II production due to blood pressure changes, not affecting ACTH), the blood tests are repeated. Patients should be well-hydrated for this test with urine sodium >20 mmol/L. Adrenal adenomas will show very little change in aldosterone levels, whereas the benign adrenal hyperplasia cases will show increased aldosterone (due to stimulation by increased angiotensin II.).

16.6.4.3 Evaluation of Suspected Glucocorticoid Excess States

Hypercortisolism: Features of Cushing syndrome are usually apparent. Screening tests are 24-h urine free cortisol, overnight standard (low-dose) dexamethasone suppression test (1 mg of dexamethasone to be taken at 11 pm). Check serum cortisol level at 8 am and 8 pm for diurnal rhythm. The diagnostic test for hypercortisolism is overnight high-dose dexamethasone suppression test which differentiates Cushing's disease (pituitary ACTH hypersecretion) from other causes of Cushing's syndrome (ectopic ACTH or adrenal tumors).

Congenital adrenal hyperplasia: To evaluate this condition, investigations to be considered are: aldosterone, cortisol, deoxycorticosterone (DOC), 11-deoxycortisol, 17-hydroxyprogesterone, and dehydroepiandrosterone (DHEAS).

16.6.4.4 Evaluation of Catecholamine Excess States

Catecholamine secreting tumors: Conditions are pheochromocytoma (80%), paraganglioma (20%), Von Hippel–Landau syndrome, neurofibromatosis type 1, multiple endocrine neoplasia type 2A or 2B (MEN2), familial paraganglioma syndrome. Screenings test include: (a) 24-h urine fractionated metanephrines (metanephrine and normetanephrine) and catecholamines (adrenaline, noradrenaline, dopamine) for 3 days (3 samples), (b) 24-h urine vanillylmandelic acid for 3 days (3 samples), (c) plasma fractionated metanephrines (drawn from indwelling cannula) following 30 min of supine rest: if high index of suspicion. Imaging techniques used to localize the tumor are adrenal/abdominal CT scan or scans specific for adrenaline-producing tumor substances such as *meta*-iodo-benzylguanidine (MIBG) scan or In-III pentetreotide scan. Drugs which inhibit uptake of MIBG should be stopped before doing the scan include: propranolol, labetalol, calcium channel blockers, phenylephrine, pseudoephedrine, phenothiazines, tricyclic antidepressants, and monoamine oxidase inhibitors. Other scans which help to localize the tumor and potential spread are positron emission tomography (PET)-CT scan, gallium (68Ga)-1,4,7,10-tetraazacyclododecane-1,4,7,10-tetraacetic acid (DOTA)-octreotate (DOTATATE), fluoro (18F)-3,4-dihydroxyphenylalanine (DOPA), and fluoro (18F)-2-deoxy-glucose (FDG).18F-FDG PET-CT has the greatest sensitivity for identifying paragangliomas and recurrent tumors.

16.7 Management

The goal of BP reduction is to prevent end-organ damage and lower the risk of cardiovascular morbidity in adulthood. The previous recommendation was to control SBP and DBP below the 95th percentile in the absence of CKD or DM. Emerging evidence suggests that LVH develops even if BP is <95th percentile but >90th percentile or >120/80 mmHg [61–63]. Thus, the target of BP reduction has been set lower. According to AAP 2017, the target BP is <130/80 mmHg or below 90th percentile SBP and DBP whichever is lower.

16.7.1 Lifestyle Modification

Evidence supports that lifestyle changes reduce BP and are recommended in all children with hypertension [62].

16.7.2 Weight Reduction

Reduction in body weight lowers the incidence of cardiovascular risk factors such as dyslipidemia and insulin resistance, and maintaining ideal body weight controls BP. A 10% lowering of BMI has been reported to lower SBP by as much as 8–12 mmHg [17]. Ideal body weight can be achieved by dietary modification and regular exercise.

16.7.3 Diet Modification

Dietary interventions which help to lower BP include low dietary sodium [62], consumption of olive oil [63], and high intake of fruits, vegetables, and legumes [64]. The "Dietary Approach to Stop Hypertension" (DASH) diet recommends high intake of fruit, vegetables, low-fat milk products, whole grain products, less chicken or fish or lean meat, and lower salt intake.

16.7.4 Physical Activity

Any type of exercise is beneficial for reduction of BP [65]. Studies show that moderate to vigorous aerobic exercise on average 40 min for 3–5 days per week reduce SBP by 6.6 mmHg and lower vascular dysfunction [66]. AAP 2017 recommends that after diagnosis of hypertension along with DASH diet, moderate to vigorous physical activity for 30–60 min per day for 3–5 days per week should be advised [1].

16.7.5 Motivational Interviewing and Stress Reduction

Motivational interviewing in combination with other behavioral techniques helps to deal with overweight and obesity. It also improves adherence to antihypertensive medications. Stress reduction techniques like breathing awareness meditation and yoga may also be helpful for reduction of BP.

16.7.6 Antihypertensive Drug Therapy

Symptomatic patients with stage 2 hypertension require immediate evaluation and treatment. Patients with stage 2 hypertension may need more prompt evaluation and pharmacologic therapy. Stage 1 hypertension allows time for evaluation before initiating treatment unless patient is symptomatic.

16.7.6.1 Indication of Drugs

Children who remain hypertensive despite a trial of lifestyle modifications or who have symptomatic hypertension, stage 2 hypertension without a clearly modifiable factor (such as obesity), or any stage of hypertension associated with CKD or diabetes mellitus therapy should be started on a single medication at the low end of the dosing range. Depending on repeated BP measurements, the dose of the initial medication can be increased every 2–4 weeks until BP is controlled, the maximal dose is reached, or adverse effects occur. Although the dose can be titrated every 2–4 weeks using home BP measurements, the patient should be seen every

4–6 weeks until BP has normalized. If BP is not controlled with a single agent, a second agent can be added and titrated. Because of the salt and water retention in many hypertensive conditions, a thiazide diuretic is often the preferred second agent.

16.7.6.2   Follow-Up Management on Basis of Clinic (Office) BP

Child with normal BP or whose BP normalized on follow-up will need to have their BP re-measured in the next well-child visit. Once the child has been diagnosed with elevated BP, lifestyle modification should be started and BP measurement repeated after 6 months. If the child's BP still remains elevated after 6 months, four-limb BP should be measured and a repeat BP measured after 6 months. ABPM and diagnostic evaluation is indicated if child remains in elevated BP category even after lifestyle modification for more than 12 months. Asymptomatic child with stage 1 hypertension requires lifestyle modification and repeat BP measurement after 1–2 weeks. If they remain at stage 1 hypertension, the blood pressure should be re-measured in 3 months. ABPM and diagnostic evaluation is indicated if the child is displaying stage 1 hypertension even after consequent three clinic visits. For symptomatic child with stage 2 hypertension, four-limb BP should be measured, and lifestyle modification initiated. Treatment should be initiated after ABPM and diagnostic evaluation, if the child remains at stage 2 hypertension after 1 week of follow-up. Symptomatic stage 2 hypertension or BP >30 mmHg beyond 95th percentile or >180/120 mmHg needs urgent emergency treatment.

16.7.6.3   Principles of Pharmacotherapy

Acceptable drug classes in children include angiotensin-converting enzyme (ACE) inhibitors, angiotensin-receptor blockers (ARBs), beta-blockers, calcium channel blockers, and diuretics. The choice of initial antihypertensive therapy depends on the preference of the physician. Thiazide diuretics and beta-blockers have a long history of safety and efficacy. Newer drugs like ACE inhibitors, ARBs, and calcium channel blockers have also been proven to be safe and efficacious in children. For all antihypertensive drugs, start at the lowest recommended dose, and then increase till desired blood pressure (BP) is reached. Once the highest recommended dose is reached, or child experiences side effects, then a second drug from a different class should be added. Drugs with a longer duration of action are preferred due to better compliance. It is better to avoid frequent dose adjustment. Certain combinations of drugs have complementary mechanisms of actions; for example, ACE inhibitors with diuretics, vasodilator with diuretic or beta-blocker. Routine use of combination therapy is not recommended, but can be considered once BP control has been achieved to reduce the cost and improve compliance. Specific classes of antihypertensive drugs should be used in specific clinical circumstances. These are ACE inhibitors or ARBs in those with diabetes and microalbuminuria, or proteinuric renal diseases, and beta-blockers or calcium channel blockers in those with migraine. Pure beta-blockers should not be used alone without prior adequate alpha-adrenergic blockade in children with hyperadrenergic states such as pheochromocytoma and methamphetamine overdose. Monitor electrolytes and serum creatinine periodically in those treated with ACE inhibitors or diuretics. ABPM can be used to assess effective control of BP during follow-up.

16.7.6.4 Step-Down of Drug Therapy

The ultimate aim is to completely stop the drug in children with uncomplicated hypertension, especially overweight ones who have successfully lost weight. After an extended course of good BP control, dose is gradually reduced to a stop. BP monitoring must continue after cessation of drug.

16.7.7 Resistant Hypertension

In adults, resistant hypertension is defined as persistently elevated BP despite treatment with three or more antihypertensive agents of different classes all at maximally effective doses. Management includes dietary sodium restriction, elimination of substances known to elevate BP, identification of previously undiagnosed secondary causes of hypertension, optimization of current therapy; additional agent such as an aldosterone receptor antagonist may be considered as it helps address volume excess as well as untreated hyperaldosteronism [1].

16.8 Malignant Hypertension

The term malignant hypertension is usually restricted for extremely high blood pressure (beyond stage II hypertension) that has developed rapidly and has resulted in some type of organ damage. The current AAP guideline advocates that clinicians should be highly concerned about acute target organ damage for any blood pressure which exceeds 30 mmHg above the 95th percentile blood pressure target for the child's sex, age, and height [1]. The primary reason for end-organ damage is acute change in mean arterial pressure well above the tolerated limits. Children with chronic hypertension may also have very high BP but are often asymptomatic as body has had time to adjust to the change in blood pressure. Common causes of severe acute hypertension are summarized in Table 16.8.

16.8.1 Definition

Severe acute hypertension can be classified as hypertensive urgency or emergency.

Hypertensive emergency: Hypertensive emergency is defined when patients with stage 2 hypertension present with acute potentially life-threatening symptoms

Table 16.8 Some common cause of malignant hypertension

Age group	Common causes
Neonates	Renovascular disease such as renal artery thrombosis, congenital anomalies (such as autosomal recessive polycystic kidney disease), coarctation of aorta, renal parenchymal disease bronchopulmonary dysplasia, congenital adrenal hyperplasia
Children	Glomerulonephritis, endocrine disease, renovascular causes
Adolescents	Glomerulonephritis, chronic kidney disease, drugs intoxication

or features of end-organ damage involving the central nervous system (encephalopathy, cerebral infarction, cerebral hemorrhage, seizures), heart (pulmonary edema), kidneys (acute kidney injury), or eyes (papilledema, retinal hemorrhage). Among these the most common presentation is hypertensive encephalopathy. Acute rise of blood pressure leads to loss of cerebrovascular autoregulation and endothelial damage causing vasogenic cerebral edema. Children with hypertensive emergencies need rapid assessment and immediate initiation of hypertension management in a controlled setup.

Hypertensive urgency: These are patients with stage 2 hypertension but without any evidence of end-organ damage. Symptoms are less and mild (example, headache and/or vomiting). These children if not managed urgently are at risk for progression to hypertensive emergencies.

It has to be emphasized that in defining hypertensive emergency/urgency presence or absence of life-threatening symptoms and/or end-organ damage are more important than the absolute blood pressure level.

16.8.2 Pathophysiology and Clinical Feature of Malignant Hypertension

Clinical features depend on etiology and age. Neonates may present with poor feeding irritability, vomiting, apnea, and cyanosis [67, 68]. Older children present according to end-organ damage.

16.8.2.1 Neurological Manifestations

Cerebral blood flow is relatively constant over a wide range of systemic blood pressures because of autoregulation. Beyond a certain level of perfusion pressure autoregulatory vasodilatation or vasoconstriction fails to work and may result in cerebral ischemia. This is particularly true for any acute significant change in blood pressure.

Hence due to acute rise of blood pressure, cerebral autoregulation is lost leading to disruption of the blood–brain barrier and endothelial dysfunction resulting in imbalance in oxygen delivery, cerebral edema formation, and micro-hemorrhages [66]. Headache is the most common presentation along with other neurological signs and symptoms including seizures (25%), altered mental status or encephalopathy (25%), facial nerve palsy (12%), hemiplegia (8%), vomiting, and signs of raised intracranial pressure and/or focal neurological deficits.

Children with hypertensive encephalopathy often have features of posterior reversible encephalopathy syndrome (PRES) wherein brain imaging shows lesions predominantly in the occipitoparietal white matter with occasional spread to basal ganglia, cerebellum, and brainstem. Other than hypertension, post-chemotherapy, post-transplant, post-infectious, and autoimmune conditions may also cause PRES. PRES lesions as suggested by its nomenclature are usually reversible but can have permanent sequelae.

Cardiovascular manifestations: Cardiac involvement may vary in the form of left ventricular hypertrophy (LVH), left ventricular failure, or left ventricular ischemia, depending on the duration and persistence of hypertension [69–71]. The incidence of LVH due to hypertension is up to 41% particularly in children with high Body Mass Index (BMI) [71]. Sudden acute increase in blood pressure may precipitate left ventricular failure and the child may present with increased work of breathing, shortness of breath, chest pain, palpitations, decreased urine output, and poor appetite.

Renal manifestations: Hypertension leads to loss of autoregulation of renal blood flow and results in kidney injury. Clinically child may present with hematuria, flank pain, and oliguria. Renal histopathology may demonstrate fibrinoid necrosis with thrombosis of intrarenal arteries [72, 73].

Ophthalmological manifestations: Severe hypertension may lead to retinal bleeds, papilledema, loss of visual acuity, and cortical blindness [74]. The prevalence of hypertensive retinopathy in children with hypertension varies from 8.9% (assessed by direct fundoscopy) to 50% (assessed by retinal photographs) [75, 76]. Moderate-to-severe grades of retinopathy are relatively rare in children.

16.8.3 Evaluation

While controlling blood pressure to a safe level is an emergency priority, accurate diagnosis and evaluation of etiology of malignant hypertension is also important. Proper history and a thorough physical examination are essential. This will help in identifying conditions which may require deviation from the usual algorithm of hypertensive emergency management. These include clinical conditions such as intracranial lesions, coarctation of aorta, pheochromocytoma, hypertension secondary to scorpion stings or intake of recreational drugs. An extensive workup should be undertaken particularly in a young child who has extremely high pressure to rule out secondary causes of hypertension.

16.8.4 Management

Severe hypertension has the potential to cause end-organ damage, hence prompt recognition and treatment is important. A child with hypertensive emergency should preferably be admitted to the intensive care unit for close monitoring and prompt initiation of appropriate intravenous antihypertensive therapy. Initial management involves stabilization as per standard emergency guidelines. Oral antihypertensive agents may be tried in hypertensive urgencies provided the child is not vomiting and is tolerating oral intake. Seizures should be controlled with lorazepam. Blood pressure should be monitored continuously by an intra-arterial line. Although intra-arterial blood pressure monitoring is desirable, treatment should not be delayed for the lack of intra-arterial access. Treatment should be initiated based on results of intermittent blood pressure measurement by non-invasive methods which need to be done frequently, i.e., at least every 10–15 min. Target reduction of blood pressure to below 95th percentile for sex, age, and height (<130/80 for adolescent) over initial 24–48 h and thereafter target below 90th percentile [1]. Blood pressure should not be lowered to more than 25% of the planned reduction over the first 8 h with aim to achieve 95th percentile target within 24–48 h. In patients with malignant hypertension and hypertensive encephalopathy it is vital to lower the blood pressure, but it is equally important to do it in a controlled way, and much effort should be devoted to avoid precipitous fall in blood pressure in the early phase of treatment. When the blood pressure is lowered acutely it may drop below the lower limit of autoregulation, with a consequent fall in cerebral blood flow. Focal ischemia may be provoked if a significant stenosis is present on a larger cerebral resistance vessel [66]. Table 16.9 summarizes some special scenarios and standard approach to emergency blood pressure control.

Table 16.9 Special clinical scenarios and standard approach to management

Cause of malignant hypertension	Mechanism	Management
Pheochromocytoma	Excessive release of catecholamines	Non-competitive α2-blocker like phenoxybenzamine is preferred Laparoscopic cortical-sparing adrenalectomy is the preferred surgery in children for localized tumors. Treatment options for invasive and metastatic tumors include embolization, systemic therapy with MIBG, or chemotherapy
Increased intra cranial pressure	Neurogenic	Priority should be given to managing the increased intracranial pressure. Significant reduction of blood pressure before controlling intracranial pressure might lead to reduced cerebral perfusion
Advanced chronic kidney disease (CKD)	Volume overload	Maintaining dry weight by strict fluid intake and perform dialysis if indicated Dietary sodium restriction Angiotensin converting enzyme inhibitor (ACEi) and angiotensin receptor blocker (ARB) are preferred in for its nephro-protection. In advanced CKD caution is advocated and avoid its use when glomerular filtration rate is <30 mL/min/1.73 m² Calcium channel blocker and beta-blockers are other options
Hypertension in thrombotic microangiopathies	Volume overload and endothelial damage	Dialysis is often required particularly in oliguric kidney failure. Hypertension in hemolytic uremic syndrome and thrombotic thrombocytopenic purpura may require multiple antihypertensive agents

Table 16.9 (continued)

Cause of malignant hypertension	Mechanism	Management
Kidney transplant recipients	Preexisting hypertension	Calcium-channel blockers decrease the vasoconstriction induced by the calcineurin inhibitors
	Native kidney disease	Thiazide and loop diuretics, ACEi/ARB are also used
	Medications like steroids, calcineurin inhibitors	Minoxidil may be considered in patients with tacrolimus-induced alopecia
	Graft dysfunction	
	Renal artery stenosis	
	Thrombotic microangiopathy	
	Post biopsy arteriovenous fistula	
Medications or drugs	Sudden cessation of opiates, benzodiazepines, and clonidine, Cocaine toxicity	For cocaine toxicity IV alpha-blocker like phentolamine is treatment of choice and beta-blockers may be added later
	Monoamine oxidase inhibitors interact with food containing tyramine and other medications like dextromethorphan, methylene blue, selective serotonin reuptake inhibitors, and linezolid	Alpha blocker or sodium nitroprusside
	Amphetamine toxicity	Decontamination, cooling, sedation, intravenous alpha-blockers, or sodium nitroprusside is the treatment of choice Beta-blockers alone are absolutely contraindicated in all these toxidromes as they will worsen the hypertensive crisis due to unopposed action on alpha receptors

16.9 Comparison Between the Guidelines

Differences between the American Academy of Pediatrics and European Society for Hypertension (ESH) guidelines have been discussed and summarized by Brady [1, 18]. Firstly, normative values used by the latest AAP guidelines are more conservative, specifically excluding data from children with overweight or obesity, and hence result in BP norms (for 95th percentile BP) that are several mmHg lower compared to the corresponding ESH norms, which reference the 2004 Fourth Report, which include children with a BMI > 85th percentile.

In terms of measurement frequency, the AAP guidelines generally recommend a more frequent schedule than the ESH. In patients with elevated blood pressure at first encounter, a re-measurement is recommended in 6 months by the AAP as opposed to 1 year by the ESH; in children with normotensive BP at initial measurement, the AAP recommends re-measurement in 1 year (2 years by ESH). Furthermore, the AAP recommends BP measurement at each health encounter for children with comorbidities such as obesity. In terms of workup and investigation, the AAP explicitly recommends against over-investigation for secondary causes in children whom primary hypertension is suspected, whereas the ESH guidelines maintain an extensive evaluation. Finally, the ESH guidelines, different to the AAP guidelines, recommend home BP monitoring, practical guidance for such measurement as well as advocate for the use of ambulatory BP monitoring (ABPM) when considering starting BP medications. In terms of treatment goals, the ESH guidelines recommend BP below the 75th percentile in children without proteinuria, and below the 50th percentile in cases of proteinuria, contrasting to the AAP guidelines which recommend the 50th percentile for both cases.

References

1. Flynn JT, Kaelber DC, Baker-Smith CM, et al. Clinical practice guideline for screening and management of high blood pressure in children and adolescents. Pediatrics. 2017;140(3):e20171904.
2. Adrogue HE, Sinaiko AR. Prevalence of hypertension in junior high school-aged children: effect of new recommendations in the 1996 Updated Task Force Report. Am J Hypertens. 2001;14:412–4.
3. Sorof JM, Lai D, Turner J, Poffenbarger T, et al. Overweight, ethnicity, and the prevalence of hypertension in school-aged children. Pediatrics. 2004;113:475–82.
4. McNiece KL, Poffenbarger TS, Turner JL, Franco KD, et al. Prevalence of hypertension and pre-hypertension among adolescents. J Pediatr. 2007;150:640–4, 644.e641.
5. Antal M, Regoly-Merei A, Nagy K, Greiner E, et al. Representative study for the evaluation of age- and gender-specific anthropometric parameters and blood pressure in an adolescent Hungarian population. Ann Nutr Metab. 2004;48:307–13.
6. Genovesi S, Giussani M, Pieruzzi F, Vigorita F, et al. Results of blood pressure screening in a population of school-aged children in the province of Milan: role of overweight. J Hypertens. 2005;23:493–7.
7. Jafar TH, Islam M, Poulter N, Hatcher J, et al. Children in South Asia have higher body mass-adjusted blood pressure levels than white children in the United States: a comparative study. Circulation. 2005;111:1291–7.
8. Hansen ML, Gunn PW, Kaelber DC. Underdiagnosis of hypertension in children and adolescents. JAMA. 2007;298(8):874–9.
9. Song P, Zhang Y, Yu J, Zha M, Zhu Y, Rahimi K, Rudan I. Global prevalence of hypertension in children: a systematic review and meta-analysis. JAMA Pediatr. 2019;173:1–10.
10. Freedman DS, Dietz WH, Srinivasan SR, Berenson GS. The relation of overweight to cardiovascular risk factors among children and adolescents: the Bogalusa Heart Study. Pediatrics. 1999;103:1175–82.
11. Orchard TJ, Forrest KY, Kuller LH, Becker DJ, Pittsburgh Epidemiology of Diabetes Complications Study. Lipid and blood pressure treatment goals for type 1 diabetes: 10-year incidence data from the Pittsburgh Epidemiology of Diabetes Complications Study. Diabetes Care. 2001;24:1053–9.
12. Copeland KC, Zeitler P, Geffner M, et al. Characteristics of adolescents and youth with recent-onset type 2 diabetes: the TODAY cohort at baseline. J Clin Endocrinol Metab. 2011;96:159–67.
13. Li AM, Au CT, Ng C, Lam HS, Ho CKW, Wing YK. A 4-year prospective follow-up study of childhood OSA and its association with BP. Chest. 2014;145:1255–63.

14. Dubov T, Toledano-Alhadef H, Chernin G, Constantini S, Cleper R, Ben-Shachar S. High prevalence of elevated blood pressure among children with neurofibromatosis type 1. Pediatr Nephrol. 2016;31:131–6.
15. Chavers BM, Solid CA, Daniels FX, et al. Hypertension in pediatric long-term hemodialysis patients in the United States. Clin J Am Soc Nephrol. 2009;4:1363–9.
16. Whelton PK, Carey RM, Aronow WS, et al. 2017 ACC/AHA/AAPA/ABC/ACPM/AGS/ APhA/ASH/ASPC/NMA/PCNA Guideline for the Prevention, Detection, Evaluation, and Management of High Blood Pressure in Adults: a report of the American College of Cardiology/ American Heart Association Task Force on Clinical Practice Guidelines [published correction appears in Hypertension]. Hypertension. 2018;71:e13–115.
17. National High Blood Pressure Education Program Working Group on High Blood Pressure in Children and Adolescents. The fourth report on the diagnosis, evaluation, and treatment of high blood pressure in children and adolescents. Pediatrics. 2004;114(2, suppl 4th Report):555–76.
18. Lurbe E, Cifkova R, Cruickshank JK, et al. Management of high blood pressure in children and adolescents: recommendations of the European Society of Hypertension. J Hypertens. 2009;27:1719–42.
19. Gupta-Malhotra M, Banker A, Shete S, et al. Essential hypertension vs. secondary hypertension among children. Am J Hypertens. 2015;28(1):73–80.
20. Baracco R, Kapur G, Mattoo T, et al. Prediction of primary vs secondary hypertension in children. J Clin Hypertens (Greenwich). 2012;14(5):316–21.
21. Silverstein DM, Champoux E, Aviles DH, Vehaskari VM. Treatment of primary and secondary hypertension in children. Pediatr Nephrol. 2006;21(6):820–7.
22. Hari P, Bagga A, Srivastava RN. Sustained hypertension in children. Indian Pediatr. 2000;37:268–74.
23. Kumar P, Arora P, Kher V, et al. Malignant hypertension in children in India. Nephrol Dial Transplant. 1996;11(7):1261–6.
24. McCulloch M, Andronikou S, Goddard E, et al. Angiographic features of 26 children with Takayasu arteritis. Pediatr Radiol. 2003;33(4):230–5.
25. Tyagi S, Kaul UA, Satsangi DK, Arora R. Percutaneous transluminal angioplasty for renovascular hypertension in children: initial and long-term results. Pediatrics. 1997;99(1):44–9.
26. Zintzaras E, Kaditis AG. Sleep-disordered breathing and blood pressure in children: a meta-analysis. Arch Pediatr Adolesc Med. 2007;161:172.
27. Vlahandonis A, Yiallourou SR, Sands SA, et al. Long-term changes in blood pressure control in elementary school-aged children with sleep-disordered breathing. Sleep Med. 2014;15:83.
28. Brown OE, Manning SC, Ridenour B. Cor pulmonale secondary to tonsillar and adenoidal hypertrophy: management considerations. Int J Pediatr Otorhinolaryngol. 1988;16:131.
29. Precone V, Krasi G, Guerri G, et al. Monogenic hypertension. Acta Biomed. 2019;90(10-S):50–2.
30. Lu YT, Fan P, Zhang D, Zhang Y, Meng X, Zhang QY, Zhao L, Yang KQ, Zhou XL. Overview of monogenic forms of hypertension combined with hypokalemia. Front Pediatr. 2021;8:543309.
31. Pickering TG, Hall JE, Appel LJ, et al. Recommendations for blood pressure measurement in humans and experimental animals: part 1: blood pressure measurement in humans: a statement for professionals from the Subcommittee of Professional and Public Education of the American Heart Association Council on High Blood Pressure Research. Circulation. 2005;111(5):697–716.
32. Urbina E, Alpert B, Flynn J, Hayman L, Harshfield GA, Jacobson M, Mahoney L, McCrindle B, Mietus-Snyder M, Steinberger J, Daniels S. Ambulatory blood pressure monitoring in children and adolescents: recommendations for standard assessment: a scientific statement from the American Heart Association Atherosclerosis, Hypertension, and Obesity in Youth Committee of the Council on Cardiovascular Disease in the Young and the Council for High Blood Pressure Research. Hypertension. 2008;52:433–51.
33. Flynn JT, Daniels SR, Hayman LL, Maahs DM, McCrindle BW, Mitsnefes M, Zachariah JP, Urbina EM, American Heart Association Atherosclerosis, Hypertension and Obesity in Youth Committee of the Council on Cardiovascular Disease in the Young. Update: ambulatory blood pressure monitoring in children and adolescents: a scientific statement from the American Heart Association. Hypertension. 2014;63(5):1116–35.

34. Stabouli S, Kotsis V, Toumanidis S, Papamichael C, Constantopoulos A, Zakopoulos N. White-coat and masked hypertension in children: association with target-organ damage. Pediatr Nephrol. 2005;20(8):1151–5.
35. Mitsnefes M, Flynn J, Cohn S, et al. CKiD Study Group. Masked hypertension associates with left ventricular hypertrophy in children with CKD. J Am Soc Nephrol. 2010;21(1):137–44.
36. Dionne JM, Abitbol CL, Flynn JT. Hypertension in infancy: diagnosis, management and outcome [published correction appears in Pediatr Nephrol. 2012;27(1):159-60]. Pediatr Nephrol. 2012;27(1):17–32.
37. Report of the second task force on blood pressure control in children–1987. Task force on blood pressure control in children. National Heart, Lung, and Blood Institute, Bethesda, Maryland. Pediatrics.1987;79(1):1–25.
38. Edvardsson VO, Steinthorsdottir SD, Eliasdottir SB, Indridason OS, Palsson R. Birth weight and childhood blood pressure. Curr Hypertens Rep. 2012;14(6):596–602.
39. Staley JR, Bradley J, Silverwood RJ, et al. Associations of blood pressure in pregnancy with offspring blood pressure trajectories during childhood and adolescence: findings from a prospective study. J Am Heart Assoc. 2015;4(5):e001422.
40. Daniels SD, Meyer RA, Loggie JM. Determinants of cardiac involvement in children and adolescents with essential hypertension. Circulation. 1990;82(4):1243–8.
41. Yang Q, Zhang Z, Kuklina EV, et al. Sodium intake and blood pressure among US children and adolescents. Pediatrics. 2012;130(4):611–9.
42. Aeberli I, Spinas GA, Lehmann R, Allemand D, Molinari L, Zimmermann MB. Diet determines features of the metabolic syndrome in 6- to 14-yearold children. Int J Vitam Nutr Res. 2009;79(1):14–23.
43. Colin-Ramirez E, Castillo-Martinez L, Orea-Tejeda A, Villa Romero AR, Vergara Castaneda A, Asensio LE. Waist circumference and fat intake are associated with high blood pressure in Mexican children aged 8 to 10 years. J Am Diet Assoc. 2009;109(6):996–1003.
44. Niinikoski H, Jula A, Viikari J, et al. Blood pressure is lower in children and adolescents with a low saturated-fat diet since infancy: the special turku coronary risk factor intervention project. Hypertension. 2009;53(6):918–24.
45. Stein DJ, Scott K, Haro Abad JM, et al. Early childhood adversity and later hypertension: data from the World Mental Health Survey. Ann Clin Psychiatry. 2010;22(1):19–28.
46. Halonen JI, Stenholm S, Pentti J, et al. Childhood psychosocial adversity and adult neighborhood disadvantage as predictors of cardiovascular disease: a cohort study. Circulation. 2015;132(5):371–9.
47. Maggio AB, Martin XE, Saunders Gasser C, et al. Medical and nonmedical complications among children and adolescents with excessive body weight. BMC Pediatr. 2014;14:232.
48. Killian L, Simpson JM, Savis A, Rawlins D, Sinha MD. Electrocardiography is a poor screening test to detect left ventricular hypertrophy in children. Arch Dis Child. 2010;95(10):832–6.
49. Ramaswamy P, Patel E, Fahey M, Mahgerefteh J, Lytrivi ID, Kupferman JC. Electrocardiographic predictors of left ventricular hypertrophy in pediatric hypertension. J Pediatr. 2009;154(1):106–10.
50. Rijnbeek PR, van Herpen G, Kapusta L, Ten Harkel AD, Witsenburg M, Kors JA. Electrocardiographic criteria for left ventricular hypertrophy in children. Pediatr Cardiol. 2008;29(5):923–8.
51. Lang RM, Badano LP, Mor-Avi V, et al. Recommendations for cardiac chamber quantification by echocardiography in adults: an update from the American Society of Echocardiography and the European Association of Cardiovascular Imaging. J Am Soc Echocardiogr. 2015;28(1):1–39.e14.
52. Rountas C, Vlychou M, Vassiou K, et al. Imaging modalities for renal artery stenosis in suspected renovascular hypertension: prospective intraindividual comparison of color Doppler US, CT angiography, GD-enhanced MR angiography, and digital substraction angiography. Ren Fail. 2007;29(3):295–302.
53. Fommei E, Ghione S, Hilson AJ, et al. Captopril radionuclide test in renovascular hypertension: a European multicentre study. Eur J Nucl Med. 1993;20:617–23.

54. Arora P, Kher V, Singhal MK, et al. Renal artery stenosis in aortoarteritis: spectrum of disease in children and adult. Kidney Blood Press Res. 1997;20:285–9.
55. Dillon MJ, Ryness JM. Plasma renin activity and aldosterone concentration in children. Br Med J. 1975;4(5992):316–9.
56. Goonasekera CD, Shah V, Wade AM, Dillon MJ. The usefulness of renal vein renin studies in hypertensive children: a 25-year experience. Pediatr Nephrol. 2002;17(11):943–9.
57. Teigen CL, Mitchell SE, Venbrux AC, Christenson MJ, McLean RH. Segmental renal artery embolization for treatment of pediatric renovascular hypertension. J Vasc Interv Radiol. 1992;3:111–7.
58. Roebuck DJ. Interventional radiology in children. Imaging. 2001;13:302–20.
59. Vo NJ, Hammelman BD, Racadio JM, Strife CF, Johnson ND, Racadio JM. Anatomic distribution of renal artery stenosis in children: implications for imaging. Pediatr Radiol. 2006;36(10):1032–6.
60. Shahdadpuri J, Frank R, Gauthier BG, Siegel DN, Trachtman H. Yield of renal arteriography in the evaluation of pediatric hypertension. Pediatr Nephrol. 2000;14:816–9.
61. Urbina EM, Khoury PR, McCoy C, Daniels SR, Kimball TR, Dolan LM. Cardiac and vascular consequences of pre-hypertension in youth. J Clin Hypertens (Greenwich). 2011;13(5):332–42.
62. Adler AJ, Taylor F, Martin N, Gottlieb S, Taylor RS, Ebrahim S. Reduced dietary salt for the prevention of cardiovascular disease. Cochrane Database Syst Rev. 2014;(12):CD009217.
63. Moreno-Luna R, Munoz-Hernandez R, Miranda ML, et al. Olive oil polyphenols decrease blood pressure and improve endothelial function in young women with mild hypertension. Am J Hypertens. 2012;25(12):1299–304.
64. Damasceno MM, de Araujo MF, de Freitas RW, de Almeida PC, Zanetti ML. The association between blood pressure in adolescents and the consumption of fruits, vegetables and fruit juice--an exploratory study. J Clin Nurs. 2011;20(11–12):1553–60.
65. Monzavi R, Dreimane D, Geffner ME, et al. Improvement in risk factors for metabolic syndrome and insulin resistance in overweight youth who are treated with lifestyle intervention. Pediatrics. 2006;117(6). www.pediatrics.org/cgi/content/full/117/6/e1111.
66. MacKenzie ET, Strandgaard S, Graham DI, Jones JV, Harper AM, Farrar JK. Effects of acutely induced hypertension in cats on pial arteriolar caliber, local cerebral blood flow, and the blood-brain barrier. Circ Res. 1976;39(1):33–41.
67. Flynn JT. Evaluation and management of hypertension in childhood. Prog Pediatr Cardiol. 2001;12(2):177–88.
68. Chandar J, Zilleruelo G. Hypertensive crisis in children. Pediatr Nephrol. 2012;27(5):741–51.
69. Chaversm BM, Herzog CA. The spectrum of cardiovascular disease in children with predialysis chronic kidney disease. Adv Chronic Kidney Dis. 2004;11(3):319–27.
70. Wilson AC, Mitsnefes MM. Cardiovascular disease in CKD in children: update on risk factors, risk assessment and management. Am J Kidney Dis. 2009;54(2):345–60.
71. Hanevold C, Waller J, Daniels S, Portman R, Sorof J. The effects of obesity, gender, and ethnic group on left ventricular hypertrophy and geometry in hypertensive children: a collaborative study of the International Pediatric Hypertension Association. Pediatrics. 2004;113(2):328–33.
72. Shavit L, Reinus C, Slotki I. Severe renal failure and microangiopathic hemolysis induced by malignant hypertension—case series and review of literature. Clin Nephrol. 2010;73(2):147–52.
73. Otsuka Y, Abe K, Sato Y. Malignant hypertension and microangiopathic hemolyticanemia. Jpn Heart J. 1976;17(2):258–64.
74. Logan P, Eustace P, Robinson R. Hypertensive retinopathy: a cause of decreased visual acuity in children. J Pediatr Ophthalmol Strabismus. 1992;29(5):287–9.
75. Daniels SR, Lipman MJ, Burke MJ, Loggie JMH. The prevalence of retinal vascular abnormalities in children and adolescents with essential hypertension. Am J Ophthalmol. 1991;111(2):205–8.
76. Foster BJ, Ali H, Mamber S, Polomeno RC, MacKie AS. Prevalence and severity of hypertensive retinopathy in children. Clin Pediatr. 2009;48(9):926–30.

Hypertension in the Elderly: Pathophysiology and Clinical Significance

17

Yook-Chin Chia

17.1 The Aging Population in Asia

The definitions of elderly vary somewhat in Asia where some countries use age ≥ 60 to define "elderly" while others adopt age ≥ 65 years as a cut-off point to define the elderly group. Notwithstanding these arbitrary age definitions of elderly, the number of people age ≥ 60 is increasing rapidly. It is estimated that between 2015 and 2050, the proportion of the world's population over 60 years will nearly double from 12 to 22% [1]. In Asia, with falling fertility rates and increased life expectancy, the population age 65 and older has been rising since 2000 (Fig. 17.1). The proportion of the population age ≥ 65 years in Asia was 4.8% in 2019 and is predicted to rise to 17.8% in 2050, with an increase in absolute numbers from 400 million to 900 million [2]. Due to this rapid rise, several countries in Asia will have between 15% to as high as 34% of the total their population being elderly (Figs. 17.2 and 17.3). In India and China with the largest populations in the world, the proportion of the population aged ≥ 65 years in India was 3.8% in 1990 and is predicted to rise to 13.7% in 2050. In China, the proportion of the population aged ≥ 65 years was 5.6% in 1990 and is predicted to rise to 26% in 2050 [2]. Japan is a country with the longest life expectancy in the world, the proportion of the Japanese population aged ≥ 65 years was 11.8% in 1990 and was predicted to rise to 37.7% in 2050. Japan may represent the possible future age structure in many Asian countries. As a result of this rapid rise in the aging population, Asia is on track in the next few decades to become one of the oldest in the world [3] by the middle of this century.

Y.-C. Chia (✉)
Department of Medical Sciences, School of Medical and Life Sciences, Sunway University, Bandar Sunway, Selangor Darul Ehsan, Malaysia

Faculty of Medicine, Department of Primary Care Medicine, University of Malays, Kuala Lumpur, Malaysia
e-mail: ycchia@sunway.edu.my; chiayc@um.edu.my

© The Author(s), under exclusive license to Springer Nature Switzerland AG 2022
C. V. S. Ram et al. (eds.), *Hypertension and Cardiovascular Disease in Asia*,
Updates in Hypertension and Cardiovascular Protection,
https://doi.org/10.1007/978-3-030-95734-6_17

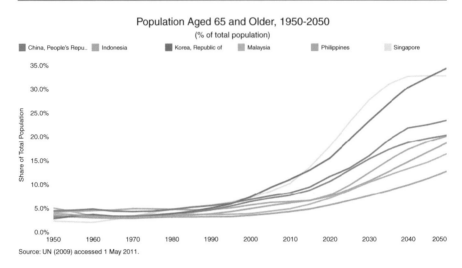

Fig. 17.1 Population aged 65 and older in Asia 1950–2050. (Source: Asia Development Bank. https://data.adb.org/dataset/population-and-aging-asia-and-pacific (accessed 9 Aug 2021))

17.2 Prevalence of Hypertension in Asia

Globally, the prevalence of hypertension is high where in 2015, 1 in 4 men and 1 in 5 women had hypertension resulting in an estimated 1.13 billion people worldwide suffering from hypertension. Most of these people live in Asia which is the world's most populous continent and most (two-thirds) living in low- and middle-income countries [4].

While it is recognized that high-income countries like the United States of America (USA) has a high prevalence of 47.1% of adults age 20 and older in 2017–2018 having hypertension [5], this also occurs in high-income countries in Asia where 60% and 32.9% of adult men in Japan and Korea, respectively, are hypertensive [6]. Furthermore it is equally high in many low- and middle-income economies in Asia where prevalence of hypertension is 42.3% in adults age 30 years or older in Malaysia and 50.3% in Pakistan [6–9] and notably, hypertension is more prevalent in people of East and South Asian ancestry [10, 11].

The prevalence of hypertension increases with age. In the USA, among those aged 60 years and older, the prevalence is 74.5% compared to lower prevalence of 54.5% among the 40–59 age group [5]. This higher prevalence in older adults is also observed in the low- and middle-income countries in Asia where in countries like Malaysia, the prevalence of hypertension in older adults is much higher than in younger adults, reaching a prevalence of 75.4% in those age between 74 and 75 years [7] (Fig. 17.3). Many governments in Asia are generally poorly prepared for this change. There will be implications not only on the health care system but also social and economic consequences [12].

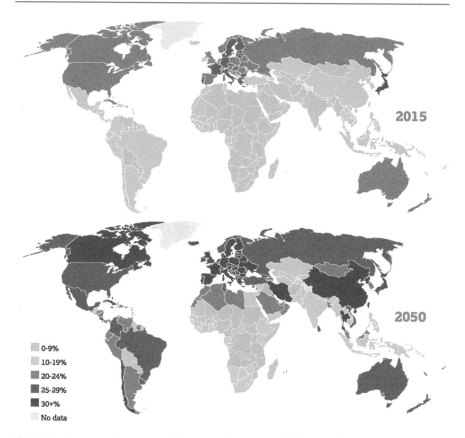

Fig. 17.2 The proportion of population aged 60 or over in 2015 and 2050. (Source: Global Age Watch https://www.helpage.org/global-agewatch/about/about-global-agewatch/(accessed 9 Aug 2021). Map: https://www.helpage.org/global-agewatch/population-ageing-data/population-ageing-map/ (accessed 9 August 2021))

17.3 Pathophysiology of Hypertension in the Elderly

Multiple systems like the cardiac, renal, sympathetic nervous system, and other CV control mechanisms interact and can affect cardiac and vascular homeostasis resulting in changes in blood pressure (BP).

17.3.1 The Renin-Angiotensin-Aldosterone System in the Elderly

The kidneys are an important regulator role of blood pressure [13, 14]. Parenchymal damage to the kidneys or any impairment of renal function almost invariably leads to the development of hypertension [15]. The kidneys control blood pressure via natriuresis and the renin-angiotensin-aldosterone system (RAAS) [13, 14, 16]. BP

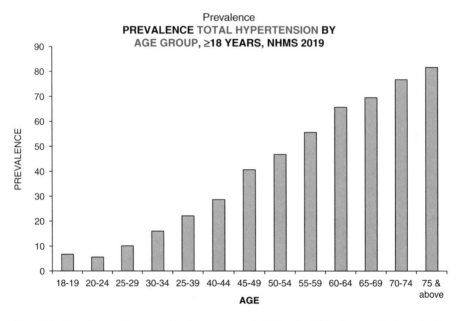

Fig. 17.3 Prevalence of hypertension by age group in Malaysia 2019. (Source: National health and Morbidity Survey 2019 Volume ll Report on Non-communicable Diseases http://iku.moh.gov. my/images/IKU/Document/REPORT/NHMS2019/Report_NHMS2019-NCD_v2.pdf)

and salt excretion is regulated by the RAAS system. The RAAS system is activated by a low salt intake stimulating renal sodium reabsorption and preserving intravascular volume to maintain BP. Conversely, a high salt intake dampens the RAAS, facilitating natriuresis, excreting more sodium to reduce intravascular volume and BP. Many therapeutic agents targeting the RAAS had been developed over the past few decades, starting with captopril, the first angiotensin converting enzyme (ACE) inhibitor in 1981, followed by losartan, the first AT-ll angiotensin receptor blocker (ARB) in 1990 and the direct renin inhibitor, aliskiren in 2005 [17, 18]. Spironolactone has been available since the 1970s and used as a potassium-sparing diuretic initially but is now used in resistant hypertension as demonstrated in recent clinical trials [19–21], particularly in hyperaldosteronism.

In the elderly, plasma renin activity declines with age and this has been attributed to the effect of age-associated nephrosclerosis [22, 23]. A study reported a reduced plasma renin activity in older adults age 65 years and older compared to younger individuals who were matched for mean arterial pressure, race, sex, height, and weight [24]. Nevertheless, a randomized control study comparing the efficacy of BP lowering of aliskiren to ramipril in the elderly found aliskiren lowered systolic and diastolic BP by 14 and 5.1 mmHg respectively, relative to baseline BP, while ramipril lowered systolic and diastolic BP by 11.76 and 3.6 mmHg respectively [25], indicating that aliskiren is able to lower BP in the elderly despite renin activity being reduced. However, to date no specific clinical outcome trials have been done on the efficacy and role of direct renin inhibitors in reducing cardiovascular mortality and morbidity in the hypertensive elderly.

Similarly, although plasma renin activity is decreased in the elderly, the fall in BP after ACE inhibitors is at least as great as in the younger age groups. One reason is due in part to a sustained ACE-inhibitor concentration as a consequence of slower clearance of the drug due to the poorer kidney function that is associated with aging.

Aging is also associated with a decline in renal function which may be amplified by intrinsic renal disease; for example, that caused by diabetes or other renovascular abnormalities. Hence, changes in tubular reabsorption of sodium will affect the ability and capacity of the kidney to adjust to fluctuations in normal sodium intake [26]. It has also been shown that compared to younger normal healthy individuals and normal older healthy individuals, the half-life for renal sodium excretion was prolonged. The elderly is less able to handle a high sodium load well, rendering them more salt-sensitive [27]. Herein suggests the role of diuretics for the treatment of hypertension in the elderly and is reflected in most guidelines for the management of hypertension.

17.3.2 Vascular Mechanism in the Elderly

Total peripheral resistance is one of the components determining BP. Structural and functional damage to the micro- or macrocirculation results in increased arterial stiffening [28] and total peripheral resistance causing a rise in BP. Microcirculatory damage could be a cause or a result of hypertension. Peripheral vascular resistance is controlled mainly at the arterioles and small arteries. Vascular tone is regulated by various factors including the sympathetic nervous system, humoral factors (the RAAS and endothelium), and auto-regulation. Endothelial damage and the resulting inflammation caused by low-density lipoprotein (LDL)-cholesterol, or hyperglycemia, stimulates the production of factors like the reactive oxygen species (ROS) and endothelin, leading to reduced nitric oxide and changes in the vasculature, and eventually raising BP.

Aging is associated with remodeling of large arteries leading to structural changes such as increased collagen deposition and rupture of elastin fibers [29] leading to a loss of elasticity, hypertrophy and sclerosis of muscular arteries and arterioles, and increasing arterial stiffness [30]. The loss of this elasticity with aging promotes the early return of reflected waves from the peripheral arterial circulation. Early wave reflection amplifies the systolic pressure wave generated with each heartbeat, leading to an increase in systolic pressure and a fall in diastolic pressure. These changes are responsible for the physiological changes in BP with aging, which is seen as a decline in diastolic BP and a rise in systolic BP, often presenting as isolated systolic hypertension (ISH), seen almost always in the elderly only. Abundant data has already confirmed that ISH is associated with at least twice the hazards ratio of increased CV mortality and morbidity compared to older individuals without ISH [31, 32]. Randomized clinical outcome trials have also demonstrated that treating ISH reduces CV mortality and morbidity [33, 34].

White coat hypertension (WCH) is a situation where in an untreated individual the BP measured in the clinic is $\geq 140/90$ mmHg, while the day BP using ambulatory BP measurements or home BP measurements is $<135/85$ mmHg. WCH is common and is reported to be more common in the elderly [35, 36]. Although the term WCH was originally defined for individuals not on hypertensive medications, it is now being used to describe discrepancies between office and out-of-office BP in patients already on treatment for hypertension, with the term white coat uncontrolled hypertension (WUCH) or as white coat effect (WCE) [37]. WCE is reserved more for patients already on anti-hypertensive drugs, where similar to the WCH situation, the office BP is high but home is low [36, 38, 39]. These two terms are frequently used interchangeably.

Studies have found a correlation between WCE and arterial stiffness [36, 40, 41] and WCH/WCE is not uncommon. In a study involving many countries in Asia, WCE was as high as 40% in one country [42]. WCH WCE is more common in the elderly and was as high as 50% in the very old [35, 37, 38]. This has clinical implications especially in the elderly, as not identifying WCH/WCE may mean unnecessary anti-hypertensive therapy in an elderly who is often already burdened by many other co-morbidities or up-titrating hypertensive drugs giving rise to hypotension and/or postural hypotension leading to falls.

Masked hypertension (MH) is the reverse of WCH/WCE and is also referred to as isolated uncontrolled home hypertension. In this category, the clinic/office BP is normal, that is, <140/80 mmHg but the out-of-office home BP is ≥135/85 mmHg. MH is associated with increased risk of CV events and has been shown to be more common in the elderly [43–47]. In the Japanese J-HOME study, among a group of treated older hypertensive adults with mean age of 66 years, 23% had uncontrolled home hypertension [48]. As almost one in four treated older hypertensives adults have MH, it is important to identify such patients and modify management to reduce their cardiovascular risk.

17.3.3 The Nervous System and Hypertension in the Elderly

Both short- and long-term control of BP are regulated by the sympathetic system. Studies have shown a significant decline in sympathetic and para-sympathetic response with aging [49, 50]. Postural hypotension occurs due to age-related impairment in baroreflex-mediated vasoconstriction, cardiac chronotropic responses, and deterioration of the diastolic filling of the heart [51, 52]. Postural hypotension increases with age where, in community dwelling individuals ≥65 years of age, its prevalence is approximately 20%; in those ≥75 years of age it is as high as 30% and in frail elderly individuals living in nursing homes, the prevalence is up to 50% [51, 53].

Patients with apparent and true resistant hypertension (RHT) often have increased sympathetic activity [54, 55]. Based on these findings, technology was developed to treat RHT by suppressing sympathetic activity with electrical stimulation of the carotid baroreflex and catheter-based renal denervation (RDN). Initial studies of renal nerve denervation (RDN) showed good decreases in BP with minimal adverse effects [56–58]. However, in sham-control trials, RDN was found to be not better at lowering BP compared with use of anti-hypertensive drugs alone [59]. One of the reasons for the failure of this study was that RDN was not standardized and many operators of RDN did not have any experience before the study.

17.3.4 Hypertension, Left Ventricular Hypertrophy, and Heart Failure in the Elderly

The prevalence of left ventricular hypertrophy (LVH) based on echocardiographic studies in hypertensive patients ranged from 36 to 41% [60]. In Asian hypertensive patients attending a primary care clinic with a mean age of 59.2 ± 7.7 years, 1 in 4 had echocardiography evidence of LVH [61]. The concentric LVH seen in hypertension makes the myocardium less compliant and results in diastolic dysfunction. This is more common in women, the elderly, and patents with diabetes [62, 63]. Also it is reported that isolated systolic hypertension, almost exclusively seen in the elderly, is associated more with concentric LVH [64, 65]. Hypertension with LVH and diastolic dysfunction when left untreated or not controlled well, leads to heart failure with preserved ejection fraction (HFpEF) [66].

Hypertensive heart failure (HF) is common in patients with hypertension in general and it is more common in the elderly than younger adults [67, 68]. The incidence of HF is strongly dependent on age, with an estimated incidence of 1% at age 65 that approximately doubles with each decade of age thereafter [69]. The prevalence of HF was about ten times higher in those aged ≥ 80 years (12.8% in men and 12% in women), compared to those aged 40–59 years (1.2% in men and 1.7% in women) [68]. Mortality and hospitalization rates due to HF is also higher in those aged ≥ 65 years than those aged <65 years [68, 69].

Several large HF registries showed that hypertension was the primary cause of HF in 11–23% of the patients [70, 71]. In an Asian HF registry, one-third of the patients with HF had hypertensive HF [72, 73]. A more recent study hypertensive HF accounted for 4% of HF, suggesting better control of hypertension [74]. The diagnosis and management of HF in the elderly is more challenging than for younger adults due to multi-comorbid diseases, polypharmacy, cognitive impairment, functional ability, and frailty. Studies have shown that 60% of elderly with HF usually have three or more other co-morbid conditions and 17% had cognitive impairment [75–77].

Many clinical trials on HF did not include patients over 75 years. HFpEF in the elderly remains without definitive treatment as most clinical trials were performed in HF patients with reduced ejection fraction (HFrEF). As hypertension is often the pre-disposing cause of HF in the elderly, control of hypertension is critical. Reduction in salt consumption is also important as the elderly are more salt-sensitive and even more so in elderly Asians with HF as Asians have been shown to not only have a higher salt intake but are also more salt-sensitive than their European counter-parts (see Sec. 17.4.3 on Salt Intake and Hypertension in the Elderly) (see Chap. 13 on HF for more details).

17.4   Atrial Fibrillation

The prevalence of atrial fibrillation (AF) increases with age, and is the most common arrhythmia in adults age 65 years and older [78]. The prevalence in individuals aged ≥75 years is 10% [79–81]. As high as 70% of individuals with AF were between 65 and 85 of age [78, 82]. AF often co-exists with hypertension and hypertension increases the risk of AF, with 60–80% of patients with AF also having hypertension [83, 84].

Although the epidemiological association between hypertension and AF is well established, the pathogenic mechanisms explaining the higher propensity of hypertensive patients to develop AF are still incompletely known. Experimental animal studies seem to suggest that it is due to a remodeling of the left atrium caused by hypertension. Although the entire mechanism is not well understood, it stands to reason that hypertension should be treated early and controlled well to prevent the progression to left atrium dilatation [84].

17.4.1 Genetic Factors and Hypertension in the Elderly

Hypertension develops because of an interaction between genetic and multiple environmental factors. Familial clustering of hypertension is well recognized and a family history of hypertension in an individual is a determinant of not only future hypertension but an earlier onset in the said individual. This implies that genetic factors are at play in the causation of hypertension.

Although gene analyses have identified many gene variants that predispose an individual to the development of hypertension, they are of modest effect. It has been estimated that the combination of different gene variants associated with the risk of hypertension contribute only around 3.5% of the trait variance [85]. Besides genetic predisposition to future risk of hypertension, there is the interaction between genes with epigenetics where environmental and life-style factors may influence the actual and potential onset of hypertension. This has been evaluated in a recent genome-wide association (GWAS) and replication study of BP phenotypes in 320,251 East and South Asian and European ancestry that suggested an interaction of genomic with epigenomic in the regulation of BP [86]. One of the suggested use of these GWAS studies is that it could generate a genetic risk score to predict future cardiovascular risk [87].

17.4.2 Environmental and Life-Style Factors and Hypertension in the Elderly

The prevalence of obesity is increasing rapidly worldwide [88]. Asia is not spared from this "epidemic" in spite of several countries being in the low-middle income group, with prevalence of obesity along with the metabolic syndrome, being just as high as in developed countries [89, 90]. Risk factors for CV disease like diabetes and even actual coronary heart disease occur at lower body mass index in Asians than in Europeans [11, 90, 91]. Obesity predisposes to an increased risk of diabetes mellitus and sleep apnea, which are both associated with higher prevalence of hypertension, and is independently associated with risk of CV and cerebrovascular events [92].

This increase in prevalence of obesity is also seen in the elderly, for example in the USA, to more than 30% in men and women aged 60 years and over [93, 94]. Similarly an Asian country also reports obesity rate of 30.2% in those aged 60 and above [95].

17.4.3 Salt Consumption and Hypertension in the Elderly

The daily average intake of salt globally is 9.8 g (equivalent to 3.95 g of sodium) [96]. This is much higher than the recommended daily intake of 5 g of salt by the World Health Organization (WHO) [97]. Asians consume more salt than populations in the USA and the United Kingdom [98] and studies have shown that higher salt consumption is associated with higher BP [99–103] and that different individuals have different salt sensitivity [100, 104–106].

Salt sensitivity is associated with reduced blood pressure dipping at night [107, 108], insulin resistance [109, 110], and increased sympathetic nerve activity [109, 111]. In the elderly, there is less nocturnal BP decline [112, 113] and this could be related to salt sensitivity and/or amount of salt intake. That salt sensitivity and/or high salt load in the elderly has been studied and when salt intake was reduced, there was a shift from non-dipping to a dipping state [114, 115]. Furthermore salt sensitivity is seen more commonly in hypertensives, blacks, women [116, 117], the elderly [116, 118], the obese [117, 119], and in Asians. Genetic studies done in Asia identified higher prevalence of salt sensitivity among the Chinese [120] and Japanese [104]. Due to the high intake of salt in Asia coupled with higher salt sensitivity in Asians and the elderly, and the evidence that salt reduction does indeed reduce BP, it would be beneficial to advise reduction in salt intake in elderly Asians.

17.4.4 Clinical Significance of the Pathophysiological Differences of Hypertension in the Elderly in Asia

The elderly is a vulnerable group with a high burden of diseases [75, 76, 121, 122] necessitating multiple drugs. Hence, it is not uncommon to see polypharmacy in the elderly. In Korea, 86.4% of elderly age 65 years and older were on multiple drugs, and similarly, in Japan the number of drugs ranged from 4.8 to 5.6 [123–125]. Because of multi-morbidity and differences in the pathophysiology of hypertension in the elderly, management of hypertension has to be adjusted accordingly when compared to younger individuals.

The differences between younger and older hypertensive patients can be summarized:

- Isolated systolic hypertension is more common in the elderly and in the very old adults aged 75 years and older [31, 32].
- The elderly is more salt-sensitive [104, 107, 110, 119].
- Heart failure, HFpEF, is more common in the elderly [66, 69, 72, 73].
- Elderly patients have more atrial fibrillation [79–81, 83, 84].
- Renin concentrations are lower in the elderly [24].
- White coat effect (isolated office hypertension) is more common in the elderly [35–38].
- Masked hypertension is more common in the elderly [43, 47, 48].
- Postural hypotension is more common in the elderly [45, 51–53].
- Nocturnal dipping is commonly absent in elderly patients [112, 113].
- Multi-comorbidities, more functional and cognitive impairments occur in the elderly [75, 76, 121, 122].
- Polypharmacy is high in the elderly [123–125].

The clinical diagnosis of hypertension in the elderly is universally the same as for younger adults, ie a BP ≥140/90 mmHg [126–128] except for the recommendations by the USA where in 2017, their cut-off point for diagnosis was lowered by 10 mmHg to 130/80 irrespective of age [129]. Similarly, the diagnosis of isolated systolic hypertension is based on a systolic BP of ≥140 and diastolic BP of <90 mmHg while the USA uses systolic BP ≥130 and diastolic BP <80 mmHg.

Hypertension guidelines recommend assessing total CV risk rather than relying on the BP alone to decide on treatment. This is useful as often several mildly raised CV risk factors in someone with BP in the range of SBP of 140–150 mmHg may appear to be at low risk and hence dismissed as not needing treatment when in fact a formal calculation shows the patient actually to be at medium to high risk, thus necessitating treatment [130]. Most assessments in older patients for total/global CV risk invariably will be of medium or high risk, and a formal calculation is usually not needed. However, it is useful to do a global CV risk as it helps to decide how far to treat and the target BP. It is also helpful when there is uncertainty as to whether one really need to put a patient on anti-hypertensive therapy especially if patient is frail, has many other co-morbid conditions and a limited life span.

The elderly with hypertension of various forms will benefit from treatment [33, 34], and even the very elderly aged 80 and older also benefit [131, 132]. Several of the Asian guidelines do not differentiate between the "old" (≥65–74 years) and the "very old" (≥75 years) elderly. Most of the Asian guidelines for the management of hypertension in the elderly recommend a target that is similar to younger adults, <140/90 mmHg but not going below systolic BP of 130 mmHg for the old, and a target of <150/90 mmHg for the very old [127] (Table 17.1).

The recommended drug of choice for treatment of hypertension in the elderly is based on the mechanistic and pathophysiologic differences between younger and

Table 17.1 Comparison of blood pressure thresholds and targets for the elderly between US, European, International, and Asian Guidelines

	ACC/AHA2017	ESC/ESH 2018	ISH 2020	Asian guidelines
Definition of older patients	≥65 years	Elderly 65–79 years Very old ≥80 years	≥65 years	≥65 years
BP threshold for initiation of pharmacotherapy	≥130/80 mmHg	Elderly ≥140/90 mmHg Very old ≥160/90 mmHg	≥140/90 mmHg	≥140/90 mmHg
Blood pressure target	<130/80 mmHg	SBP 130–139 mmHg DBP 70–79 mmHg	<140/80 mmHg	Most Asian countries <140/90 mmHg Japan: 65–74 years <130/80 mmHg ≥75 years <140/90 mmHg

ISH International Society of Hypertension

older adults and on clinical outcome trials. In Asia, all the guidelines recommend diuretics and calcium blockers for the elderly while a few countries included the RAS blockers [127, 133–138].

Many commonly used classes of anti-hypertensive drugs are generic, and more affordable to treat hypertension, even in the low middle-income countries in Asia. Despite this, BP control rates in Asia are all <50%, except for countries like Korea, Taiwan, and Singapore [6, 9].

In summary, hypertension in the elderly should be treated. However, the management of hypertension in the elderly needs to be individualized, where in healthier, ambulant elderly who are community dwelling, greater efforts could be made to achieve target BP. Rather than base treatment on chronological age, we should be using biological age as our guide on the management of hypertension in the elderly.

References

1. WHO. Ageing and health fact sheet. WHO Publication. 2018. https://www.who.int/news-room/fact-sheets/detail/ageing-and-health. Accessed 9 Aug 2021.
2. Department of Economic and Social Affairs, Population Division, United Nations. World Population Prospects 2019. https://population.un.org/wpp/. Accessed 9 Aug 2021.
3. Population and aging in Asia: the growing elderly population. Asia Development Bank. 2018. https://data.adb.org/dataset/population-and-aging-asia-and-pacific. Accessed 9 Aug 2021.
4. WHO. Hypertension fact sheet. WHO Publication. 2021. https://www.who.int/news-room/fact-sheets/detail/hypertension. Accessed 9 Aug 2021.
5. CDC. Prevalence of select measures among adults aged 20 and over: United States, 1999–2000 through 2017–2018. NHANES Interactive Data Visualizations. 2018. https://www.cdc.gov/nchs/nhanes/visualization/. Accessed 9 Aug 2021.

6. Chia Y-C, Buranakitjaroen P, Chen C-H, Divinagracia R, Hoshide S, Park S, et al. Current status of home blood pressure monitoring in Asia: statement from the HOPE Asia Network. J Clin Hypertens. 2017;19(11):1192–201.
7. Malaysia National Health and Morbidity Survey 2015. Non-communicable diseases, risk factors and other health related problems, Vol. II. Ministry of Health Malaysia. 2015. http:// www.moh.gov.my/moh/resources/nhmsreport2015vol2.pdf. Accessed 9 Aug 2021.
8. Pakistan Health Research Council. Non-communicable Diseases Risk Factors Survey—Pakistan. Islamabad: World Health Organization; 2016. ISBN 978-969-499-008-8.
9. Kario K, Wang J-G. Could 130/80 mmHg be adopted as the diagnostic threshold and management goal of hypertension in consideration of the characteristics of Asian populations? Hypertension. 2018;71(6):979–84.
10. Ueshima H, Sekikawa A, Miura K, Turin TC, Takashima N, Kita Y, et al. Cardiovascular disease and risk factors in Asia: a selected review. Circulation. 2008;118(25):2702–9.
11. Forouhi NG, Sattar N, Tillin T, McKeigue PM, Chaturvedi N. Do known risk factors explain the higher coronary heart disease mortality in South Asian compared with European men? Prospective follow-up of the Southall and Brent studies, UK. Diabetologia. 2006;49(11):2580–8.
12. Dou L, Liu X, Zhang T, Wu Y. Health care utilization in older people with cardiovascular disease in China. Int J Equity Health. 2015;14(1):59.
13. Coffman TM. The inextricable role of the kidney in hypertension. J Clin Invest. 2014;124(6):2341–7.
14. Coffman TM, Crowley SD. Kidney in hypertension. Hypertension. 2008;51(4):811–6.
15. Bidani AK, Griffin KA. Pathophysiology of hypertensive renal damage. Hypertension. 2004;44(5):595–601.
16. Hall JE, Granger JP, do Carmo JM, da Silva AA, Dubinion J, George E, et al. Hypertension: physiology and pathophysiology. Compr Physiol. 2012;2(4):2393–442.
17. Kotchen TA. Historical trends and milestones in hypertension research. Hypertension. 2011;58(4):522–38.
18. Sever PS, Messerli FH. Hypertension management 2011: optimal combination therapy. Eur Heart J. 2011;32(20):2499–506.
19. Zhao D, Liu H, Dong P, Zhao J. A meta-analysis of add-on use of spironolactone in patients with resistant hypertension. Int J Cardiol. 2017;233:113–7.
20. Wang C, Xiong B, Huang J. Efficacy and safety of spironolactone in patients with resistant hypertension: a meta-analysis of randomised controlled trials. Heart Lung Circ. 2016;25(10):1021–30.
21. Williams B, MacDonald TM, Morant SV, Webb DJ, Sever P, McInnes GT, et al. Endocrine and haemodynamic changes in resistant hypertension, and blood pressure responses to spironolactone or amiloride: the PATHWAY-2 mechanisms substudies. Lancet Diabet Endocrinol. 2018;6(6):464–75.
22. Weidmann P, De Myttenaere-Bursztein S, Maxwell MH, de Lima J. Effect on aging on plasma renin and aldosterone in normal man. Kidney Int. 1975;8(5):325–33.
23. Bauer JH. Age-related changes in the renin-aldosterone system. Physiological effects and clinical implications. Drugs Aging. 1993;3(3):238–45.
24. Messerli FH, Sundgaard-Riise K, Ventura HO, Dunn FG, Glade LB, Frohlich ED. Essential hypertension in the elderly: haemodynamics, intravascular volume, plasma renin activity, and circulating catecholamine levels. Lancet. 1983;2(8357):983–6.
25. Duprez DA, Munger MA, Botha J, Keefe DL, Charney AN. Aliskiren for geriatric lowering of systolic hypertension: a randomized controlled trial. J Hum Hypertens. 2010;24(9):600–8.
26. Mimran A, Ribstein J, Jover B. Aging and sodium homeostasis. Kidney Int Suppl. 1992;37:S107–13.
27. Epstein M, Hollenberg NK. Age as a determinant of renal sodium conservation in normal man. J Lab Clin Med. 1976;87(3):411–7.
28. Laurent S, Boutouyrie P. Arterial stiffness and hypertension in the elderly. Front Cardiovasc Med. 2020;7(202):544302.

29. Safar ME, Levy BI, Struijker-Boudier H. Current perspectives on arterial stiffness and pulse pressure in hypertension and cardiovascular diseases. Circulation. 2003;107(22):2864–9.
30. Deakin CD, Nolan JP, Soar J, Sunde K, Koster RW, Smith GB, et al. European Resuscitation Council guidelines for resuscitation 2010 section 4. Adult advanced life support. Resuscitation. 2010;81(10):1305–52.
31. Kannel WB, Dawber TR, McGee DL. Perspectives on systolic hypertension. The Framingham study. Circulation. 1980;61(6):1179–82.
32. Himmelmann A, Hedner T, Hansson L, O'Donnell CJ, Levy D. Isolated systolic hypertension: an important cardiovascular risk factor. Blood Press. 1998;7(4):197–207.
33. Staessen JA, Fagard R, Thijs L, Celis H, Arabidze GG, Birkenhäger WH, et al. Randomised double-blind comparison of placebo and active treatment for older patients with isolated systolic hypertension. Lancet. 1997;350(9080):757–64.
34. Prevention of stroke by antihypertensive drug treatment in older persons with isolated systolic hypertension. Final results of the Systolic Hypertension in the Elderly Program (SHEP). SHEP Cooperative Research Group. JAMA. 1991;265(24):3255–64.
35. Tanner RM, Shimbo D, Seals SR, Reynolds K, Bowling CB, Ogedegbe G, et al. White-coat effect among older adults: data from the Jackson heart study. J Clin Hypertens (Greenwich). 2016;18(2):139–45.
36. Pioli MR, Ritter AM, de Faria AP, Modolo R. White coat syndrome and its variations: differences and clinical impact. Integr Blood Press Control. 2018;11:73–9.
37. Williams B, Mancia G, Spiering W. ESC/ESH guidelines for the management of arterial hypertension. Eur Heart J. 2018;2018:39.
38. Helvaci MR, Seyhanli M. What a high prevalence of white coat hypertension in society! Intern Med. 2006;45(10):671–4.
39. Reddy AK, Jogendra MR, Rosendorff C. Blood pressure measurement in the geriatric population. Blood Press Monit. 2014;19(2):59–63.
40. de Simone G, Schillaci G, Chinali M, Angeli F, Reboldi GP, Verdecchia P. Estimate of white-coat effect and arterial stiffness. J Hypertens. 2007;25(4):827–31.
41. Cai P, Peng Y, Wang Y, Wang X. Effect of white-coat hypertension on arterial stiffness: a meta-analysis. Medicine. 2018;97(42):e12888.
42. Kario K, Tomitani N, Buranakitjaroen P, Chia YC, Park S, Chen CH, et al. Home blood pressure control status in 2017-2018 for hypertension specialist centers in Asia: results of the Asia BP@Home study. J Clin Hypertens. 2018;20(12):1686–95.
43. Cacciolati C, Hanon O, Alpérovitch A, Dufouil C, Tzourio C. Masked hypertension in the elderly: cross-sectional analysis of a population-based sample. Am J Hypertens. 2011;24(6):674–80.
44. Fagard RH, Cornelissen VA. Incidence of cardiovascular events in white-coat, masked and sustained hypertension versus true normotension: a meta-analysis. J Hypertens. 2007;25(11):2193–8.
45. Pierdomenico SD, Cuccurullo F. Prognostic value of white-coat and masked hypertension diagnosed by ambulatory monitoring in initially untreated subjects: an updated meta analysis. Am J Hypertens. 2011;24(1):52–8.
46. Mallion J-M, Clerson P, Bobrie G, Genes N, Vaisse B, Chatellier G. Predictive factors for masked hypertension within a population of controlled hypertensives. J Hypertens. 2006;24(12):2365–70.
47. Gorostidi M, Vinyoles E, Banegas JR, de la Sierra A. Prevalence of white-coat and masked hypertension in national and international registries. Hypertens Res. 2015;38(1):1–7.
48. Obara T, Ohkubo T, Funahashi J, Kikuya M, Asayama K, Metoki H, et al. Isolated uncontrolled hypertension at home and in the office among treated hypertensive patients from the J-HOME study. J Hypertens. 2005;23(9):1653–60.
49. Parashar R, Amir M, Pakhare A, Rathi P, Chaudhary L. Age related changes in autonomic functions. J Clin Diagn Res. 2016;10(3):CC11–C5.
50. Baker SE, Limberg JK, Dillon GA, Curry TB, Joyner MJ, Nicholson WT. Aging alters the relative contributions of the sympathetic and parasympathetic nervous system to blood pressure control in women. Hypertension. 2018;72(5):1236–42.

51. Gupta V, Lipsitz LA. Orthostatic hypotension in the elderly: diagnosis and treatment. Am J Med. 2007;120(10):841–7.
52. Brignole M. Progressive orthostatic hypotension in the elderly. e-journal of the ESC Council for Cardiology Practice 2006;5(10). https://www.escardio.org/Journals/E-Journal-of-Cardiology-Practice/Volume-5/Progressive-orthostatic-hypotension-in-the-elderly-Title-Progressive-orthosta. Accessed 9 Aug 2021.
53. Saedon NI, Pin Tan M, Frith J. The prevalence of orthostatic hypotension: a systematic review and meta-analysis. J Gerontol A Biol Sci Med Sci. 2020;75(1):117–22.
54. Grassi G, Mark A, Esler M. The sympathetic nervous system alterations in human hypertension. Circ Res. 2015;116(6):976–90.
55. Esler M. The sympathetic nervous system in hypertension: back to the future? Curr Hypertens Rep. 2015;17(2):11.
56. Catheter-based renal sympathetic denervation for resistant hypertension: Durability of blood pressure reduction out to 24 months. Hypertension 2011;57(5):911–7.
57. Esler MD, Krum H, Sobotka PA, Schlaich MP, Schmieder RE, Böhm M. Renal sympathetic denervation in patients with treatment-resistant hypertension (The Symplicity HTN-2 Trial): a randomised controlled trial. Lancet. 2010;376(9756):1903–9.
58. Courand PY, Feugier P, Workineh S, Harbaoui B, Bricca G, Lantelme P. Baroreceptor stimulation for resistant hypertension: first implantation in France and literature review. Arch Cardiovasc Dis. 2014;107(12):690–6.
59. Bhatt DL, Kandzari DE, O'Neill WW, D'Agostino R, Flack JM, Katzen BT, et al. A controlled trial of renal denervation for resistant hypertension. N Engl J Med. 2014;370(15):1393–401.
60. Cuspidi C, Sala C, Negri F, Mancia G, Morganti A. Prevalence of left-ventricular hypertrophy in hypertension: an updated review of echocardiographic studies. J Hum Hypertens. 2012;26(6):343–9.
61. Ching S, Chia Y, Azman WW. Prevalence and determinants of left ventricular hypertrophy in hypertensive patients at a primary care clinic. Malays Fam Physician. 2012;7(2–3):2.
62. Krumholz HM, Larson M, Levy D. Sex differences in cardiac adaptation to isolated systolic hypertension. Am J Cardiol. 1993;72(3):310–3.
63. Chahal NS, Lim TK, Jain P, Chambers JC, Kooner JS, Senior R. New insights into the relationship of left ventricular geometry and left ventricular mass with cardiac function: a population study of hypertensive subjects. Eur Heart J. 2010;31(5):588–94.
64. Ganau A, Devereux RB, Roman MJ, de Simone G, Pickering TG, Saba PS, et al. Patterns of left ventricular hypertrophy and geometric remodeling in essential hypertension. J Am Coll Cardiol. 1992;19(7):1550–8.
65. Devereux RB, James GD, Pickering TG. What is normal blood pressure? Comparison of ambulatory pressure level and variability in patients with normal or abnormal left ventricular geometry. Am J Hypertens. 1993;6(6 Pt 2):211s–5s.
66. Sweitzer NK, Lopatin M, Yancy CW, Mills RM, Stevenson LW. Comparison of clinical features and outcomes of patients hospitalized with heart failure and normal ejection fraction (> or =55%) versus those with mildly reduced (40% to 55%) and moderately to severely reduced (<40%) fractions. Am J Cardiol. 2008;101(8):1151–6.
67. Goldberg RJ, Spencer FA, Farmer C, Meyer TE, Pezzella S. Incidence and hospital death rates associated with heart failure: a community-wide perspective. Am J Med. 2005;118(7):728–34.
68. Virani SS, Alonso A, Benjamin EJ, Bittencourt MS, Callaway CW, Carson AP, et al. Heart disease and stroke statistics 2020 update: a report from the American Heart Association. Circulation. 2020;141(9):e139–596.
69. Butrous H, Hummel SL. Heart failure in older adults. Can J Cardiol. 2016;32(9):1140–7.
70. Nieminen MS, Brutsaert D, Dickstein K, Drexler H, Follath F, Harjola VP, et al. EuroHeart Failure Survey II (EHFS II): a survey on hospitalized acute heart failure patients: description of population. Eur Heart J. 2006;27(22):2725–36.
71. Gheorghiade M, Abraham WT, Albert NM, Greenberg BH, O'Connor CM, She L, et al. Systolic blood pressure at admission, clinical characteristics, and outcomes in patients hospitalized with acute heart failure. JAMA. 2006;296(18):2217–26.

72. Oh GC, Cho H-J. Blood pressure and heart failure. Clin Hypertens. 2020;26(1):1.
73. Choi DJ, Han S, Jeon ES, Cho MC, Kim JJ, Yoo BS, et al. Characteristics, outcomes and predictors of long-term mortality for patients hospitalized for acute heart failure: a report from the Korean Heart Failure Registry. Korean Circ J. 2011;41(7):363–71.
74. Lee SE, Lee HY, Cho HJ, Choe WS, Kim H, Choi JO, et al. Clinical characteristics and outcome of acute heart failure in Korea: results from the Korean Acute Heart Failure Registry (KorAHF). Korean Circ J. 2017;47(3):341–53.
75. Wolff JL, Starfield B, Anderson G. Prevalence, expenditures, and complications of multiple chronic conditions in the elderly. Arch Intern Med. 2002;162(20):2269–76.
76. Murad K, Goff DC Jr, Morgan TM, Burke GL, Bartz TM, Kizer JR, et al. Burden of comorbidities and functional and cognitive impairments in elderly patients at the initial diagnosis of heart failure and their impact on total mortality: the cardiovascular health study. JACC Heart Fail. 2015;3(7):542–50.
77. Marengoni A, Tazzeo C, Calderón-Larrañaga A, Roso-Llorach A, Onder G, Zucchelli A, et al. Multimorbidity patterns and 6-year risk of institutionalization in older persons: the role of social formal and informal care. J Am Med Dir Assoc. 2021;22:2184.
78. Karamichalakis N, Letsas KP, Vlachos K, Georgopoulos S, Bakalakos A, Efremidis M, et al. Managing atrial fibrillation in the very elderly patient: challenges and solutions. Vasc Health Risk Manag. 2015;11:555–62.
79. Go AS, Hylek EM, Phillips KA, Chang Y, Henault LE, Selby JV, et al. Prevalence of diagnosed atrial fibrillation in adults: national implications for rhythm management and stroke prevention: the AnTicoagulation and Risk Factors in Atrial Fibrillation (ATRIA) Study. JAMA. 2001;285(18):2370–5.
80. Krijthe BP, Kunst A, Benjamin EJ, Lip GY, Franco OH, Hofman A, et al. Projections on the number of individuals with atrial fibrillation in the European Union, from 2000 to 2060. Eur Heart J. 2013;34(35):2746–51.
81. Chugh SS, Havmoeller R, Narayanan K, Singh D, Rienstra M, Benjamin EJ, et al. Worldwide epidemiology of atrial fibrillation. Circulation. 2014;129(8):837–47.
82. Kistler PM, Sanders P, Fynn SP, Stevenson IH, Spence SJ, Vohra JK, et al. Electrophysiologic and electroanatomic changes in the human atrium associated with age. J Am Coll Cardiol. 2004;44(1):109–16.
83. Nabauer M, Gerth A, Limbourg T, Schneider S, Oeff M, Kirchhof P, et al. The Registry of the German competence NETwork on atrial fibrillation: patient characteristics and initial management. Europace. 2009;11(4):423–34.
84. Verdecchia P, Angeli F, Reboldi G. Hypertension and atrial fibrillation. Circ Res. 2018;122(2):352–68.
85. Dominiczak A, Delles C, Padmanabhan S. Genomics and precision medicine for clinicians and scientists in hypertension. Hypertension. 2017;69(4):e10–e3.
86. Kato N, Loh M, Takeuchi F, Verweij N, Wang X, Zhang W, et al. Trans-ancestry genome-wide association study identifies 12 genetic loci influencing blood pressure and implicates a role for DNA methylation. Nat Genet. 2015;47(11):1282–93.
87. Mattson DL, Liang M. Hypertension: from GWAS to functional genomics-based precision medicine. Nat Rev Nephrol. 2017;13(4):195–6.
88. Blüher M. Obesity: global epidemiology and pathogenesis. Nat Rev Endocrinol. 2019;15(5):288–98.
89. Cameron AJ, Shaw JE, Zimmet PZ. The metabolic syndrome: prevalence in worldwide populations. Endocrinol Metab Clin N Am. 2004;33(2):351–75.
90. Ramachandran A, Chamukuttan S, Shetty SA, Arun N, Susairaj P. Obesity in Asia – is it different from rest of the world. Diabetes Metab Res Rev. 2012;28(s2):47–51.
91. McKeigue PM, Shah B, Marmot MG. Relation of central obesity and insulin resistance with high diabetes prevalence and cardiovascular risk in South Asians. Lancet. 1991;337(8738):382–6.
92. Larsson SC, Bäck M, Rees JMB, Mason AM, Burgess S. Body mass index and body composition in relation to 14 cardiovascular conditions in UK Biobank: a Mendelian randomization study. Eur Heart J. 2020;41(2):221–6.

93. Zamboni M, Mazzali G. Obesity in the elderly: an emerging health issue. Int J Obes. 2012;36(9):1151–2.
94. Flegal KM, Carroll MD, Kit BK, Ogden CL. Prevalence of obesity and trends in the distribution of body mass index among US adults, 1999-2010. JAMA. 2012;307(5):491–7.
95. Ariaratnam S, Rodzlan Hasani WS, Krishnapillai AD, Abd Hamid HA, Jane Ling MY, Ho BK, et al. Prevalence of obesity and its associated risk factors among the elderly in Malaysia: findings from The National Health and Morbidity Survey (NHMS) 2015. PLoS One. 2020;15(9):e0238566.
96. Mozaffarian D, Fahimi S, Singh GM, Micha R, Khatibzadeh S, Engell RE, et al. Global sodium consumption and death from cardiovascular causes. N Engl J Med. 2014;371(7):624–34.
97. WHO. Fact sheet on salt reduction. WHO Publication. 2020. https://www.who.int/newsroom/fact-sheets/detail/salt-reduction. Accessed 9 Aug 2021.
98. INTERSALT study an international co-operative study on the relation of blood pressure to electrolyte excretion in populations. I. Design and methods. The INTERSALT Co-operative Research Group. J Hypertens. 1986;4(6):781–7.
99. Dahl. Possible role of salt intake in the development of essential hypertension. In: Cottier P, Bock D, editors. Essential hypertension: an international symposium. Berlin: Springer; 1960. p. 52–65.
100. Stamler J. The INTERSALT Study: background, methods, findings, and implications. Am J Clin Nutr. 1997;65(2 Suppl):626s–42s.
101. Intersalt: an international study of electrolyte excretion and blood pressure. Results for 24 hour urinary sodium and potassium excretion. Intersalt Cooperative Research Group. BMJ. 1988;297(6644):319–28.
102. Sacks FM, Svetkey LP, Vollmer WM, Appel LJ, Bray GA, Harsha D, et al. Effects on blood pressure of reduced dietary sodium and the dietary approaches to stop hypertension (DASH) diet. N Engl J Med. 2001;344(1):3–10.
103. Zhou BF, Stamler J, Dennis B, Moag-Stahlberg A, Okuda N, Robertson C, et al. Nutrient intakes of middle-aged men and women in China, Japan, United Kingdom, and United States in the late 1990s: the INTERMAP study. J Hum Hypertens. 2003;17(9):623–30.
104. Katsuya T, Ishikawa K, Sugimoto K, Rakugi H, Ogihara T. Salt sensitivity of Japanese from the viewpoint of gene polymorphism. Hypertens Res. 2003;26(7):521–5.
105. Obarzanek E, Proschan MA, Vollmer WM, Moore TJ, Sacks FM, Appel LJ, et al. Individual blood pressure responses to changes in salt intake. Hypertension. 2003;42(4):459–67.
106. Stamler J, Rose G, Stamler R, Elliott P, Dyer A, Marmot M. INTERSALT study findings. Public health and medical care implications. Hypertension. 1989;14(5):570–7.
107. Higashi Y, Oshima T, Ozono R, Nakano Y, Matsuura H, Kambe M, et al. Nocturnal decline in blood pressure is attenuated by NaCl loading in salt-sensitive patients with essential hypertension: noninvasive 24-hour ambulatory blood pressure monitoring. Hypertension. 1997;30(2 Pt 1):163–7.
108. Takakuwa H, Shimizu K, Izumiya Y, Kato T, Nakaya I, Yokoyama H, et al. Dietary sodium restriction restores nocturnal reduction of blood pressure in patients with primary aldosteronism. Hypertens Res. 2002;25(5):737–42.
109. Yatabe MS, Yatabe J, Yoneda M, Watanabe T, Otsuki M, Felder RA, et al. Salt sensitivity is associated with insulin resistance, sympathetic overactivity, and decreased suppression of circulating renin activity in lean patients with essential hypertension. Am J Clin Nutr. 2010;92(1):77–82.
110. Melander O, Groop L, Hulthén UL. Effect of salt on insulin sensitivity differs according to gender and degree of salt sensitivity. Hypertension. 2000;35(3):827–31.
111. Ebata H, Hojo Y, Ikeda U, Ishida H, Natsume T, Shimada K. Differential effects of an alpha 1-blocker (doxazosin) on diurnal blood pressure variation in dipper and non-dipper type hypertension. Hypertens Res. 1995;18(2):125–30.
112. O'Sullivan C, Duggan J, Atkins N, O'Brien E. Twenty-four-hour ambulatory blood pressure in community-dwelling elderly men and women, aged 60-102 years. J Hypertens. 2003;21(9):1641–7.

113. Imai Y, Nagai K, Sakuma M, Sakuma H, Nakatsuka H, Satoh H, et al. Ambulatory blood pressure of adults in Ohasama, Japan. Hypertension. 1993;22(6):900–12.
114. Uzu T, Ishikawa K, Fujii T, Nakamura S, Inenaga T, Kimura G. Sodium restriction shifts circadian rhythm of blood pressure from nondipper to dipper in essential hypertension. Circulation. 1997;96(6):1859–62.
115. Uzu T, Kimura G, Yamauchi A, Kanasaki M, Isshiki K, Araki SI. Enhanced sodium sensitivity and disturbed circadian rhythm of blood pressure in essential hypertension. J Hypertens. 2006;24:1627.
116. He J, Gu D, Chen J, Jaquish CE, Rao DC, Hixson JE, et al. Gender difference in blood pressure responses to dietary sodium intervention in the GenSalt study. J Hypertens. 2009;27(1):48–54.
117. Vollmer WM, Sacks FM, Ard J, Appel LJ, Bray GA, Simons-Morton DG, et al. Effects of diet and sodium intake on blood pressure: subgroup analysis of the DASH-sodium trial. Ann Intern Med. 2001;135(12):1019–28.
118. Weinberger MH, Fineberg NS. Sodium and volume sensitivity of blood pressure. Age and pressure change over time. Hypertension. 1991;18(1):67–71.
119. Elijovich F, Weinberger MH, Anderson CA, Appel LJ, Bursztyn M, Cook NR, et al. Salt sensitivity of blood pressure: a scientific statement from the American Heart Association. Hypertension. 2016;68(3):e7–e46.
120. Liu Y, Shi M, Dolan J, He J. Sodium sensitivity of blood pressure in Chinese populations. J Hum Hypertens. 2020;34(2):94–107.
121. Mitsutake SIT, Teramoto C, Shimizu S, Ito H. Patterns of co-occurrence of chronic disease among older adults in Tokyo, Japan. Prev Chronic Dis. 2019;16(180170) https://doi.org/10.5888/pcd16.180170. Accessed 9 Aug 2021.
122. Jang I-Y, Lee HY, Lee E. Th anniversary Committee of Korean Geriatrics S. geriatrics fact sheet in Korea 2018 from National Statistics. Ann Geriatr Med Res. 2019;23(2):50–3.
123. Kim H-A, Shin J-Y, Kim M-H, Park B-J. Prevalence and predictors of polypharmacy among Korean elderly. PLoS One. 2014;9(6):e98043.
124. Mabuchi T, Hosomi K, Yokoyama S, Takada M. Polypharmacy in elderly patients in Japan: analysis of Japanese real-world databases. J Clin Pharm Ther. 2020;45(5):991–6.
125. Morin L, Johnell K, Laroche M-L, Fastbom J, Wastesson JW. The epidemiology of polypharmacy in older adults: register-based prospective cohort study. Clin Epidemiol. 2018;10:289–98.
126. Williams B, Mancia G, Spiering W, Agabiti Rosei E, Azizi M, Burnier M, et al. 2018 ESC/ESH guidelines for the management of arterial hypertension: the task force for the management of arterial hypertension of the European Society of Cardiology (ESC) and the European Society of Hypertension (ESH). Eur Heart J. 2018;39(33):3021–104.
127. Chia Y-C, Turana Y, Sukonthasarn A, Zhang Y, Shin J, Cheng H-M, et al. Comparison of guidelines for the management of hypertension: similarities and differences between international and Asian countries; perspectives from HOPE-Asia Network. J Clin Hypertens. 2021;23(3):422–34.
128. Unger T, Borghi C, Charchar F, Khan NA, Poulter NR, Prabhakaran D, et al. 2020 International Society of Hypertension global hypertension practice guidelines. Hypertension. 2020;75(6):1334–57.
129. Whelton PK, Williams B. The 2018 European Society of Cardiology/European Society of Hypertension and 2017 American College of Cardiology/American Heart Association blood pressure guidelines: more similar than DifferentComparison of the 2018 ESC/ESH and 2017 ACC/AHA hypertension GuidelinesComparison of the 2018 ESC/ESH and 2017 ACC/AHA hypertension guidelines. JAMA. 2018;320(17):1749–50.
130. Chia Y. Review of tools of cardiovascular disease risk stratification: interpretation, customisation and application in clinical practice. Singap Med J. 2011;52(2):116–23.
131. Bulpitt C, Fletcher A, Beckett N, Coope J, Gil-Extremera B, Forette F, et al. Hypertension in the very elderly trial (HYVET): protocol for the main trial. Drugs Aging. 2001;18(3):151–64.
132. Beckett NS, Peters R, Fletcher AE, Staessen JA, Liu L, Dumitrascu D, et al. Treatment of hypertension in patients 80 years of age or older. N Engl J Med. 2008;358(18):1887–98.

133. Liu J. Highlights of the 2018 Chinese hypertension guidelines. Clin Hypertens. 2020;26(1):8.
134. Kim HC, Ihm S-H, Kim G-H, Kim JH, Kim K-i, Lee H-Y, et al. 2018 Korean Society of Hypertension guidelines for the management of hypertension: part I-epidemiology of hypertension. Clin Hyperten. 2019;25(1):16.
135. Umemura S, Arima H, Arima S, Asayama K, Dohi Y, Hirooka Y, et al. The Japanese Society of Hypertension guidelines for the management of hypertension (JSH 2019). Hypertens Res. 2019;42(9):1235–481.
136. Philippines clinical practice guideline for adult hypertension-prevention, screening, counseling and management. 2018. www.mahealthcare.com/pdf/practice_guidelines.
137. 2018 Vietnam National Heart Association/Vietnam Society of Hypertension guidelines on diagnosis and treatment of arterial hypertension in adults. 2018 VNHA/VSH guidelines for diagnosis and treatment of hypertension … slideshare.net. Accessed 9 Aug 2021.
138. 2018 Pakistan Hypertension League 3rd National Guideline for the prevention, detection, evaluation and Managment of hypertension 2018. PHL 2018 Magazine Final 1. phlpk.org. Accessed 9 Aug 2021.

Hypertensive Disorders in Pregnancy

18

Marjorie I. Santos, Carmela Madrigal-Dy, and Deborah Ignacia D. Ona

18.1 Introduction

Hypertensive disorders of pregnancy constitute one of the major causes of maternal and perinatal morbidity and mortality worldwide. It has been estimated that pre-eclampsia complicates 2–8% of all pregnancies globally [1]. In the Philippines, HDP account for 36.7% of all maternal deaths [2] and remains as the second leading cause of maternal mortality from 1991 to 2006 according to the data culled from the Department of Health [2] which is much higher than the worldwide rate of 18% according to the WHO [3].

Hypertensive disorders of pregnancy can be sub-classified into four groups: chronic hypertension, gestational hypertension, preeclampsia, and superimposed preeclampsia in the setting of chronic hypertension as presented in the American College of Obstetricians and Gynecologists (ACOG) Guideline in 2013 [4]. The International Society for the Study of Hypertension in Pregnancy (ISSHP) has published guidelines on diagnosis to establish global unity of definition in referring to

M. I. Santos
Section of Fetal Maternal Medicine, Department of Obstetrics and Gynecology, MCU- Filemon D. Tanchoco Medical Foundation Hospital, Manila, Philippines

C. Madrigal-Dy
Section of Fetal Maternal Medicine, Department of Obstetrics and Gynecology, Cardinal Santos Medical Center, Manila, Philippines

D. I. D. Ona (✉)
Division of Hypertension, Department of Medicine, University of the Philippines-Philippine General Hospital, Manila, Philippines

Department of Medicine, St. Luke's Medical Center, Manila, Philippines
e-mail: dadavidona@up.edu.ph; daona@stlukes.com.ph

© The Author(s), under exclusive license to Springer Nature Switzerland AG 2022
C. V. S. Ram et al. (eds.), *Hypertension and Cardiovascular Disease in Asia*,
Updates in Hypertension and Cardiovascular Protection,
https://doi.org/10.1007/978-3-030-95734-6_18

the various hypertensive disorders seen in pregnancy, with the most recent guidelines released in 2018 which include the category of "white coat hypertension" and "masked hypertension" [5].

Although each condition increases the risk of maternal and neonatal morbidity, the greatest risks are associated with a diagnosis of preeclampsia, either de novo or in the setting of chronic hypertension [6, 7]. Women with a history of preeclampsia have an elevated risk of cardiovascular disease in subsequent years. Preeclampsia has been linked by several systematic reviews and meta-analyses to the development of cardiovascular disease (hypertension, myocardial infarction, congestive heart failure), cerebrovascular events (stroke), peripheral vascular disease and cardiovascular mortality later in life [8, 9]. It is postulated that endothelial dysfunction, which has been linked to atherosclerosis, persists in women with a history of preeclampsia, years after an affected pregnancy [8].

The diagnostic criteria for these disorders vary somewhat among published international guidelines, the terms and definitions used in the Classification of Hypertensive Disorders of Pregnancy in the CPG on Hypertension in Pregnancy by the Philippine Obstetrical and Gynecological Society (POGS) in 2015 will be presented here. The classification system adapted by the POGS was based on the Task Force Report on Hypertension in Pregnancy by the ACOG published in November 2013. HDP has four categories as follows:

1. Preeclampsia-Eclampsia.
2. Chronic hypertension (of whatever cause).
3. Chronic hypertension with superimposed preeclampsia.
4. Gestational hypertension.

This classification system is very basic, precise, and practical and probably the most appropriate classification scheme for HDP that the obstetrician and general practitioner can use (Table 18.1).

Hypertension is diagnosed empirically when appropriately taken blood pressure (BP) exceeds 140 mmHg systolic or 90 mmHg diastolic. Korotkoff phase V is used to define diastolic BP.

Previously, incremental increases of 30 mmHg systolic or 15 mmHg diastolic above BP values taken mid-pregnancy were used as diagnostic criteria, even when absolute values were <140/90 mmHg. These incremental changes are no longer used to define hypertension, but it is recommended that such women be observed more closely [10].

Abnormal protein excretion during pregnancy is empirically defined as 300 mg/dL of protein or more in a 24-h urine collection [11, 12] or a protein-to-creatinine ratio of 0.30 or more [13]. When quantitative methods are not available or rapid decisions are required, a urine protein dipstick reading can be used. If urinalysis is the only available means of assessing proteinuria, overall accuracy is better when using 2+ as the discriminant value [14, 15].

Table 18.1 Classification of hypertension of pregnancy

Hypertensive disorder of pregnancy (HDP)	Criteria
Preeclampsia	Blood pressure • Systolic BP of 140 mmHg or more or diastolic BP of 90 mmHg or more on two occasions at least 4 h apart after 20 weeks of gestation in a woman with a previously normal blood pressure • Systolic BP of 160 mmHg or more diastolic BP of 110 mmHg or more (severe hypertension can be confirmed within a short interval (minutes) to facilitate timely antihypertensive therapy) And *proteinuria* • 300 mg or more per 24 h urine collection (or this amount extrapolated from a timed collection) Or • Protein/creatinine ratio of 0.3 mg/dL or more or • Dipstick reading of 2+ (used only if other quantitative methods not available) **OR** in the absence of proteinuria, new-onset hypertension with the new-onset of any of the following: • Thrombocytopenia: Platelet count less than $100,000 \times 10^9$/L • Renal insufficiency: Serum creatinine concentrations greater than 1.1 mg/dL or a doubling of the serum creatinine concentration in the absence of other renal disease • Impaired liver function: Elevated blood concentrations of liver transaminases to twice normal concentration • Pulmonary edema • New-onset headache unresponsive to medication and not accounted for by alternative diagnoses or visual symptoms
Eclampsia	• New-onset tonic-clonic, focal, or multifocal seizures in the absence of other causative conditions such as epilepsy, cerebral arterial ischemia and infarction, intracranial hemorrhage, or drug use
Chronic hypertension	• Hypertension of any cause that predates pregnancy. BP $\geq$ 140/90 mmHg before pregnancy or before 20 weeks gestation or both
Superimposed preeclampsia	• Chronic hypertension in association with preeclampsia. Others define it as worsening baseline hypertension accompanied by new-onset proteinuria or other findings supportive of preeclampsia
Gestational hypertension	• Systolic BP 140 mmHg or more or a diastolic BP of 90 mmHg or more, or both, on two occasions at least 4 h apart after 20 weeks of gestation, in a woman with a previously normal BP • Hypertension without proteinuria or severe features develops after 20 weeks of gestation and BP levels return to normal in the postpartum period (12 weeks postpartum)

18.2 Pathogenesis of Preeclampsia

Any hypertensive disorder of pregnancy can result in preeclampsia. It occurs in 35% of women with gestational hypertension and in 25% of women with chronic hypertension [16].

The underlying pathophysiology that results in the transition to or superimposition of preeclampsia is not well understood and is probably related to a mechanism of reduced placental perfusion inducing systemic vascular endothelial dysfunction.

There are thought to be two stages to cytotrophoblast invasion: the first involves invasion of the decidual segments of the spiral arteries, at around 10–12 weeks gestation; the second involves invasion of the myometrial segments at 15–16 weeks [17]. In preeclampsia, cytotrophoblast invasion of the myometrial segments is impaired: the effect is the spiral arteries remain narrow which restricts the blood supply to the fetus. The effect on the fetus becomes more significant as pregnancy progresses. Placental ischemia is thought to develop as a result of the abnormal cytotrophoblast invasion which leads to the release of placental factors and an imbalance of angiogenic factors thereby causing widespread endothelial dysfunction which characterizes preeclampsia [17].

18.3   Prevention Strategies

Women with any of the *high-risk factors* for preeclampsia (previous pregnancy with preeclampsia, multifetal gestation, renal disease, autoimmune disease, type 1 or type 2 diabetes mellitus and chronic hypertension) and those with more than one of the *moderate-risk factors* (first pregnancy, maternal age of 35 years or older, a body mass index >30, family history of preeclampsia, sociodemographic characteristics, and personal history factors) should receive 150 mg/day aspirin best given at bedtime, for preeclampsia prophylaxis initiated between 12 and 28 weeks of gestation (optimally before 16 weeks of gestation) and continuing until delivery (Table 18.2) [18].

Table 18.2   Clinical risk factors and aspirin use [18]

Level of risk	Risk factors	Recommendation
High	• History of preeclampsia, especially when accompanied by an adverse outcome • Multifetal gestation • Chronic hypertension • Type 1 or 2 diabetes • Autoimmune disease (i.e., systemic lupus erythematosus, antiphospholipid syndrome)	Recommend low-dose aspirin if the patient has one or more of these high-risk factors
Moderate	• Nulliparity • Obesity (BMI > 30) • Family history of preeclampsia (mother or sister) • Sociodemographic characteristics (African American race, low socioeconomic status) • Age 35 years or old • Personal history factors (e.g., low birth weight or small for gestational age, previous adverse pregnancy outcome, more than 10-year pregnancy interval)	Consider low-dose aspirin if the patient has more than one of these moderate-risk factors
Low	• Previous uncomplicated full term delivery	Do not recommend low-dose aspirin

The use of high dose calcium supplementation of 1500–2000 mg/day is recommended for patients with adequate and inadequate calcium intake based on the meta-analysis by Hofmeyr et al. in 2014, there was a statistically significant reduction in hypertension with or without proteinuria in pregnancy and the risk for severe preeclampsia in calcium deficient patients was likewise decreased with high dose calcium intake [19].

18.4   Approach to Management

The goal of management of hypertension in pregnancy is to prevent significant cerebrovascular and cardiovascular events in the mother without compromising fetal well-being. After initial evaluation of the mother and the fetus, the BP measurement will dictate the need for urgent care or conservative management. Figure 18.1 shows the BP values with the corresponding management.

18.5   Severe Hypertension

Acute-onset severe hypertension (systolic BP of 160 mmHg or more or diastolic BP of 110 mmHg or more, or both) can occur in the prenatal, intrapartum, and postpartum period. It is accurately measured using standard techniques and is persistent for 15 min or more. The objectives of treating severe hypertension are to prevent

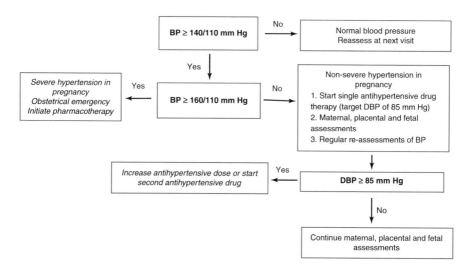

Fig. 18.1 Management of hypertension in pregnancy. (From Butalia, S et al. Hypertension Canada's 2018 Guidelines for the Management of Hypertension in Pregnancy. Canadian Journal of Cardiology 34 (2018) 526–531)

congestive heart failure, myocardial ischemia, renal injury or failure, and ischemic or hemorrhagic stroke [20]. The available literature suggests that antihypertensive agents should be administered within 30–60 min. However, it is recommended to administer antihypertensive therapy as soon as reasonably possible after the criteria for acute-onset severe hypertension are met.

The goal of treatment is not to normalize BP, but to achieve a range of 140–15-/90–100 mmHg in order to prevent repeated, prolonged exposure to severe systolic hypertension, with subsequent loss of cerebral vasculature autoregulation. Maternal stabilization should be done before delivery. When acute-onset, severe hypertension is diagnosed in the office setting, the patient should be expeditiously sent to the hospital for treatment.

The first line of treatment is intravenous (IV) hydralazine and labetalol. If labetalol alone is ineffective, switching to hydralazine is recommended. Extended-release oral nifedipine also may be considered as a first-line therapy, particularly when IV access is not available. Use of these drugs does not require cardiac monitoring (Table 18.3) [20].

A recent Cochrane systematic review that involved 3573 women found no significant differences regarding either efficacy or safety between hydralazine and labetalol or between hydralazine and calcium channel blockers. Thus, any of these agents can be used to treat acute severe hypertension in pregnancy [21]. Although parenteral antihypertensive therapy may be needed initially for acute control of BP, oral medications can be used as expectant management is continued. One approach is to begin an initial regimen of labetalol at 200 mg orally every 12 h and increase the dose up to 800 mg orally every 8–12 h as needed (maximum total 2400 mg/day). If the maximum dose is inadequate to achieve the desired BP goal, or the dosage is limited by adverse effect, then short-acting oral nifedipine can be added gradually. The extended-release nifedipine 30 mg oral tablet is an effective antihypertensive agent that is less likely to result in a rapid and severe fall in BP than the immediate-release capsule and provides antihypertensive effects over several hours [21].

Options for second-line therapy includes nicardipine by infusion pump. A review of studies of intravenous nicardipine for treatment of severe hypertension in pregnancy found that target BP was reached within 23 min in 70% of pregnant patients with severe hypertension and 91% reached target BP within 130 min, with no severe maternal or fetal side effects [22].

Treatment of severe hypertension (BP of $\geq$160/100 mmHg) is always recommended as it prevents serious maternal and fetal complications to set in. Initiating therapy in non-severe disease, however, is a subject of controversy. The NICE, ISSHP, and SOGC recommend therapy when the BP remains above 140/90 mmHg but SOGC suggests a lower threshold in patients with other comorbidities [21, 23–25]. The ACOG recommends conservative management of non-severe disease but stressed on the importance of control in the severe type [20].

Table 18.3 Antihypertensive agents used for urgent blood pressure control in pregnancy

Drug	Initial dose	Follow-up dose

Labetalol	20 mg IV gradually over 2 min	Repeat BP measurement at 10-min intervals:
		  If BP remains above target level at 10 min, give 40 mg IV over 2 min
		  If BP remains above target level at 20 min, give 80 mg IV over 2 min
		  If BP remains above target level at 30 min, give 80 mg IV over 2 min
		  If BP remains above target level at 40 min, give 80 mg IV over 2 min
		Cumulative maximum dose is 300 mg. If target BP is not achieved, switch to another class of agent.

	A continuous IV infusion of 1–2 mg/min can be used instead of intermittent therapy or started after 20 mg IV dose	Adjust dose within this range to achieve target BP
	Requires use of programmable infusion pump and continuous noninvasive monitoring of BP and heart rate	Cumulative maximum dose is 300 mg. If target BP is not achieved, switch to another class of agent

Hydralazine	5 mg IV gradually over 1–2 min	Repeat BP measurement at 20-min intervals:
		  If BP remains above target level at 20 min, give 5 or 10 mg IV over 2 min, depending on the initial response
		  If BP remains above target level at 40 min, give 10 mg IV over 2 min, depending on the previous response
		  Cumulative maximum dose is 30 mg. If target BP is not achieved, switch to another class of agent

	Adequate reduction of BP is less predictable than with IV labetalol	
Nicardipine (parenteral)	The initial dose is 5 mg/h IV by infusion pump and can be increased to a maximum of 15 mg/h	Adjust dose within this range to achieve target BP
	Onset of action is delayed by 5–15 min; in general, rapid titration is avoided to minimize risk of overshooting dose	
	Requires use of a programmable infusion pump and continuous noninvasive monitoring of blood pressure and heart rate	
Nifedipine extended release	30 mg orally	If target BP is not achieved in 1–2 h, another dose can be administered If target BP is not achieved, switch to another class of agent *Maximum daily dose is 180 mg*

(continued)

Table 18.3 (continued)

Drug	Initial dose	Follow-up dose
Nifedipine immediate release[a]	10 mg orally	Repeat BP measurement at 20-min intervals:
	May be associated with precipitous drops in BP in some women, with associated FHR decelerations for which emergency cesarean delivery may be indicated. As such, this regimen is not typically used as a first-line option and is usually reserved only for women without IV access. If used, FHR should be monitored while administering short-acting nifedipine	If BP remains above target at 20 min, give 10 or 20 mg orally, depending on the initial response If BP remains above target at 40 min, give 10 or 20 mg orally, depending on the previous response If target BP is not achieved, switch to another class of agent *Maximum daily dose is 180 mg*

Adapted from: American College of Obstetricians and Gynecologists Committee on Obstetric Practice. Committee Opinion No. 767: Emergent therapy for acute-onset, severe hypertension during pregnancy and the postpartum period. Obstet Gynecol 2019

Bernstein PS, Martin JN Jr, Barton JR, et al. National Partnership for Maternal Safety: Consensus Bundle on Severe Hypertension During Pregnancy and the Postpartum Period. Obstet Gynecol 2017; 130:347

[a]We caution against use of immediate-release oral nifedipine, although some obstetric guidelines have endorsed its use as a first-line option for emergency treatment of acute, severe hypertension in pregnancy or postpartum (other options were labetalol and hydralazine), particularly when IV access is not in place. In most cases, use of immediate-release oral nifedipine will be safe and well tolerated; however, there is a risk of an acute, precipitous fall in BP, which may result in a reduction in uteroplacental perfusion. The immediate-release preparations are also associated with a higher incidence of headache and tachycardia. In non-pregnant adults, the package insert states that "nifedipine capsules should not be used for the acute reduction of BP"

18.6 Mild to Moderate Hypertension

A judicious approach to the treatment of mild (BP of 140–150 mmHg/90–100 mmHg) to moderate (BP of 150–159 mmHg/100–109 mmHg) hypertension is made with consideration on the patient's comorbidities and symptoms (e.g., headaches, visual disturbances).

The goal of antihypertensive therapy is not to normalize the BP, but to achieve a value that would prevent repeated prolonged exposure to severe systolic hypertension, with subsequent loss of cerebral vasculature autoregulation. It is important to avoid hypotension because the degree by which placental blood flow is autoregulated is not established, and aggressive lowering may cause fetal distress [9]. The Canadian guidelines recommend 130–150/90–105 mmHg in the absence of comorbid conditions. The NICE guidelines recommend aiming for 135/85 mmHg or less [25]. The ISSHP endorses an approach that seeks to control BP levels to 110–140/85 mmHg [24].

18.7 Pharmacologic Management

The choice of antihypertensive drug for initial therapy should be based on the characteristics of the patient, contraindications to a particular drug, and physician and patient preferences. The first-line drugs are methyldopa, calcium channel blockers, or beta blockers. Antihypertensives may be used to keep systolic BP at 130–155 mmHg and diastolic BP at 80–105 mmHg.

In a 2014 Cochrane Collaboration systematic review, antihypertensive medication use for non-severe hypertension in pregnancy (49 randomized trials; n 1/44,723) was associated with a halving in the risk of progression to severe hypertension (relative risk, 0.49; 95% CI, 0.40–0.60). The number needed to treat was 10 (95% CI, 8–13). This finding was consistent across all HDP ranges of conditions [26].

Initial antihypertensive therapy should be monotherapy from the following first-line drugs: oral labetalol, oral methyldopa, long-acting oral nifedipine, or other oral beta blockers (acebutolol, metoprolol, pindolol, and propranolol) [27]. A diastolic BP of 85 mmHg should be targeted for pregnant women receiving antihypertensive therapy with chronic hypertension or gestational hypertension [27]. A similar target could be considered for pregnant women with preeclampsia [27] (Fig. 18.1).

Additional antihypertensive drugs should be used if target BP levels are not achieved with standard-dose monotherapy. Add-on drugs should be from a different drug class chosen from first-line or second-line options (Table 18.4) [27].

For patients with chronic hypertension, it can be difficult to differentiate worsening of the hypertension from superimposed preeclampsia. Conditions that may indicate superimposed preeclampsia, that warrants a referral to a maternal fetal medicine specialist/perinatologist, include the following:

1. Acute, severe, and persistent elevations in blood pressure.
2. Sudden increase in baseline hypertension.
3. New-onset proteinuria or sudden increase in proteinuria (above the threshold for normal or a clear change from baseline).

Angiotensin converting enzyme (ACE) inhibitors and angiotensin receptor blockers (ARBs) should NOT be given before conception and during the first

Table 18.4 Antihypertensive medications commonly used in pregnancy

First-line oral drugs (Grade C)	Second-line oral drugs (Grade D)	Medications to avoid
Labetalol	Clonidine	Angiotensin converting enzyme inhibitors[a] (Grade C)
Methyldopa	Thiazide diuretics	Angiotensin receptor blockers[a] (Grade D)
Long-acting oral nifedipine		
Other beta blockers (acebutolol, metoprolol, pindolol, and propranolol)		

From Butalia, S et al. Hypertension Canada's 2018 Guidelines for the Management of Hypertension in Pregnancy. Canadian Journal of Cardiology 34 (2018) 526–531
[a] Fetotoxicity of renal system

trimester of pregnancy because of evidence of teratogenicity (ACE inhibitor exposure increased the risk for cardiovascular and central nervous system anomalies). Likewise, they should NOT be used during the second and third trimesters of pregnancy because of evidence of fetopathy (fetal and neonatal death, renal failure, oligohydramnios, arterial hypertension, intrauterine growth restriction, respiratory distress syndrome, pulmonary hypoplasia, hypocalvaria, and limb defects).

18.8 Non-pharmacologic Management

Non-pharmacological management of women with the hypertensive disorders of pregnancy involves consideration of dietary interventions, lifestyle, and place of care [28]. Data on the role of dietary interventions and lifestyle changes have mainly focused on their role in prevention and not on women with established hypertension in pregnancy.

18.9 Dietary Interventions

Dietary interventions have been studied to curb weight gain in pregnancy, primarily among overweight and obese women [29, 30]. However, the objective was for prevention rather than treatment in hypertensive pregnant women. Actual weight loss is not recommended during pregnancy because of the potential adverse effects of catabolism and ketosis on fetal brain development.

A reduction in salt intake and the DASH diet were independently effective in lowering BP, and the effects of both were greater than the effects of either intervention alone. In a trial of sodium reduction and the DASH diet (i.e., Dietary Approaches to Stop Hypertension), both were shown to decrease BP [31]. Among non-pregnant subjects of whom 59% were women, the DASH diet lowered BP in all subjects, particularly those who were already hypertensive, and the BP reduction occurred regardless of pre-trial salt intake (that was high, intermediate, or low). Reducing the sodium intake from the high to the intermediate level reduced the SBP by 2.1 mmHg ($p < 0.001$) during the control diet and by 1.3 mmHg ($p = 0.03$) during the DASH diet. Reducing the sodium intake from the intermediate to the low level caused additional reductions of 4.6 mmHg during the control diet ($p < 0.001$) and 1.7 mmHg during the DASH diet ($p < 0.01$) [31].

18.10 Lifestyle Changes

It is common practice to recommend workload reduction or cessation, or stress management (e.g., meditation) when non-severe elevations in BP are found in association with chronic or gestational hypertension, or preeclampsia and outpatient care is continued. However, there is currently no evidence that these lifestyle changes improve pregnancy outcomes [28].

18.11 Place of Care

Outpatient care for women with hypertension in pregnancy should be reserved for women without severe disease. A full assessment of maternal and fetal well-being must be done to exclude women with severe hypertension or severe preeclampsia using Hypertension Canada's 2018 Guidelines for the Management of Hypertension in Pregnancy (Fig. 18.1).

Options for outpatient care include obstetric day units and antepartum home care that is delivered through structured antepartum home care program [28]. A woman's eligibility is dependent on the proximity of the hospital to her residence, a home environment that allows the home care team to provide the necessary maternal and fetal surveillance, a woman's likelihood of compliance, the lability of a woman's blood pressure, the absence of comorbid conditions, and no evidence of active progression of preeclampsia [28].

18.12 Management of Hypertension in Special Cases: Immediate Postpartum and Breastfeeding Periods

Blood pressure should be recorded shortly after birth and if normal again within 6 h. All women should have BP recorded and defer discharge for at least 24 h or until vital signs are normal and/or treated or referred. Any woman with an obstetric complication and/or newborn with complications should stay in the hospital until both are stable.

WHO [21] recommendations include:

1. In hospital stay for at least 24 h.
2. Checkup within 48–72 h of the birth and again 7–14 days and at 6 weeks postpartum.
3. All women should be reminded of the danger signs of preeclampsia following birth including headaches, visual disturbances, nausea, vomiting, epigastric or hypochondrial pain, feeling faint or convulsions.

Advise women with hypertension who wish to breastfeed that their treatment can be adapted to accommodate breastfeeding, and that the need to take antihypertensive medication does not prevent them from breastfeeding. Although no antihypertensive drugs are licensed for use in breastfeeding, since most evidence is based on observational studies and expert opinion, most antihypertensive medicines taken while breastfeeding only lead to very low levels in breast milk, so the amounts taken in by babies are very small and would be unlikely to have any clinical effect [22].

Most medicines are not tested in pregnant or breastfeeding women, so disclaimers in the manufacturer's information are not because of any specific safety concerns or evidence of harm. Hence, make decisions on treatment together with the woman, based on her preferences (Table 18.5).

As antihypertensive agents have the potential to transfer into breast milk, consider monitoring the BP of babies, especially those born preterm, who have

Table 18.5 Antihypertensive drugs used in pregnancy and lactation

	Atenolol	Captopril	Enalapril
Mechanism	Beta blocker	ACE inhibitor	ACE inhibitor
Pregnancy	Avoid in first and second trimester—associated with fetal growth restriction and bradycardia. Reduces uteroplacental blood flow	No—associated with severe fetal anomaly, fetal nephropathy, and intrauterine death	No—associated with severe fetal anomaly, fetal nephropathy, and intrauterine death
Breastfeeding	No known evidence of harm (NICE)	Manufacturers advise to avoid	Not for preterm infants
	Second line after labetalol	However recommended by SOGC	No evidence of harm (NICE)
		No known evidence of harm (NICE)	Particularly for women needing cardio/renal protection
Postnatal	Yes	Yes	Yes
Side effects	Risk of fetal growth restriction and bradycardia in pregnancy	Cough	Cough
Contraindications	Asthma		

NICE National Institute for Health and Care Excellence, *SOGC* Society of Obstetricians and Gynecologists of Canada

symptoms of low BP for the first few weeks. When discharged, advise women to monitor their babies for drowsiness, lethargy, pallor, cold peripheries, or poor feeding.

For women with hypertension in the postnatal period, if BP is not controlled with a single medicine, consider a combination of nifedipine or amlodipine and an ACE inhibitor either enalapril and captopril, which has been shown to be safe and effective in breastfeeding women. If this combination is not tolerated or is ineffective, consider either adding atenolol or labetalol to the combination treatment or swapping one of the medicines already being used for atenolol or labetalol. When treating women with antihypertensive medication during the postnatal period, use medicines that are taken once daily when possible. Where possible, avoid using diuretics or angiotensin receptor blockers to treat hypertension in women in the postnatal period who are breastfeeding or expressing milk.

For women with hypertension in the postnatal period who are not breastfeeding and who are not planning to breastfeed, treat them based on general guidelines on hypertension in adults.

References

1. Steegers EA, von Dadelszen P, Duvekot JJ, Pijnenborg R. Pre-eclampsia. Lancet. 2010;376:631–44.
2. National Epidemiology Center, Department of Health. The 2013 Philippine health statistics. 2013. http://www.dohgovph/sites/default/files/publications.

3. Khan KS, Wojdyla D, Say L, Gulmezoglu AM, Van Look PF. WHO analysis of causes of maternal death: a systematic review. Lancet. 2006;367:1066–74.
4. American College of Obstetricians and Gynecologists. Executive summary: hypertension in pregnancy. Obstet Gynecol. 2013;122:1122–31.
5. Brown MA, Magee LA, Kenny LC, et al. The hypertensive disorders of pregnancy: ISSHP classification, diagnosis and management recommendations for international practice. Pregnancy Hypertens. 2018;13:291–310.
6. Bramham K, Parnell B, Nelson-Piercy C, Seed PT, Postol L, Chappell LC. Chronic hypertension and pregnancy outcomes: systematic review and meta-analysis. BMJ. 2014;348:g2301.
7. Chappell LC, Enye S, Seed P, Briley AL, Poston L, Shennan AH. Adverse perinatal outcomes and risk factors for preeclampsia in women with chronic hypertension: a prospective study. Hypertension. 2008;51:1002–9.
8. Lee ES, Oh MJ, Jung JW, et al. The levels of circulating vascular endothelial growth factor and soluble FLT-1 in pregnancies complicated by preeclampsia. J Korean Med Sci. 2007;22:94–8. PubMed: 17297258.
9. Immunology of preeclampsia; current views and hypothesis. *In:* Kurpisz M.; Fernandez, N., *editors* Immunology of human reproduction. Sargent, I.L and Smarason, A.K. BIOS Scientific Publishers; Oxford:, 1995, pp. 355–370.
10. Clinical practice guidelines on hypertension in pregnancy. Philippine Obstetrical, and Gynecological Society; 2015.
11. Alexander JM, McIntire DD, Leveno KJ, et al. Magnesium sulfate for the prevention of preeclampsia in women with mild hypertension. Am J Obstet Gynecol. 2006;108:826.
12. Report of the National High Blood Pressure Education Program Working Group on High Blood Pressure in Pregnancy. Am J Obstet Gynecol. 2000;183:S1–22.
13. Kuo VS, Koumantakis G, Gallery ED. Proteinuria and its assessment in normal and hypertensive pregnancy. Am J Obstet Gynecol. 1992;167:723–8.
14. Morris RK, Riley RD, Doug M, Deeks JJ, Kilby MD. Diagnostic accuracy of spot urinary protein and albumin to creatinine ratios for detection of significant proteinuria or adverse pregnancy outcome in patients with suspected pre-eclampsia: systematic review and meta-analysis. BMJ. 2012;345:e4342.
15. Phelan LK, Brown MA, Davis GK, Mangos G. A prospective study of the impact of automated dipstick urinalysis on the diagnosis of preeclampsia. Hypertens Pregnancy. 2004;23:135–42.
16. North RA, Taylor RS, Schellenberg JC. Evaluation of a definition of preeclampsia. Br J Obstet Gynaecol. 1999;106:767–73.
17. Meher S, Duley L, Hunter K, Askie L. Antiplatelet therapy before or after 16 weeks' gestation for preventing preeclampsia: an individual participant data meta-analysis. Am J Obstet Gynecol. 2017;216:121.
18. LeFevre ML, U.S. Preventive Services Task. Low-dose aspirin use for the prevention of morbidity and mortality from preeclampsia: U.S. Preventive Services Task Force Recommendation Statement. Ann Intern Med. 2014;161:819–26.
19. Hofmeyr GJ, Lawrie TA, Atallah AN, Duley L, Torloni MR. Calcium supplementation during pregnancy for preventing hypertensive disorders and other related problems. Cochrane Database Syst Rev. 2014;(6):CD001059. https://doi.org/10.1002/14651858.CD001059.pub4.
20. American College of Obstetricians and Gynecologists. Committee Opinion No. 692: emergent therapy for acute-onset, severe hypertension during pregnancy and the postpartum period. Obstet Gynecol. 2017;129:e90–5.
21. American College of Obstetricians and Gynecologists. ACOG Practice Bulletin No. 202: gestational hypertension and preeclampsia. Obstet Gynecol. 2019;133:e1–25.
22. Nij Bivank SW, Duvekot JJ. Nicardipine for the treatment of severe hypertension in pregnancy; a review of the literature. Obstet Gynecol Surv. 2010;65:341.
23. Brown MA, et al. The hypertensive disorders of pregnancy: ISSHP classification, diagnosis & management recommendations for international practice. Pregnancy Hypertens. 2018;13:291–310.
24. SOGC Clinical Practice Guideline. Diagnosis, evaluation, and management of the hypertensive disorders of pregnancy: executive summary. J Obstet Gynaecol Can. 2014;36(5):416–38.

25. Behrens I, Basit S, Melbye M, Lykke JA, Wolfhart J, Bundgaard H, et al. Risk of post-pregnancy hypertension in women with a history of hypertensive disorders of pregnancy: nationwide cohort study. BMJ. 2017;358:3078.
26. Brown MC, Best KE, Pearce MS, Waugh J, Robson SC, Bell R. Cardiovascular disease risk in women with preeclampsia: systematic review and meta-analysis. Eur J Epidemiol. 2013;281:1–19.
27. Powe CE, Levine RJ, Karumachi SA. Preeclampsia, a disease of maternal endothelium: the role of antiangiogenic factors and implications for later cardiovascular disease. Circulation. 2011;123:2856–69.
28. Magee L, von Dadelszen P, Stones W, Mathai M. Diet, lifestyle and place of care. In:The FIGO Textbook of Pregnancy Hypertension: an evidence-based guide to monitoring, prevention and management. London, 2016. 456 p.
29. Thangaratinam S, Rogozinska E, Jolly K, Glinkowski S, Roseboom T, Tomlinson JW, et al. Effects of interventions in pregnancy on maternal weight and obstetric outcomes: meta-analysis of randomised evidence. BMJ. 2012;344:e2088.
30. Muktabhant B, Lawrie TA, Lumbiganon P, Laopaiboon M. Diet or exercise, or both, for preventing excessive weight gain in pregnancy. Cochrane Database Syst Rev. 2015;6:CD007145.
31. Sacks FM, Svetkey LP, Vollmer WM, Appel LJ, Bray GA, Harsha D, et al. Effects on blood pressure of reduced dietary sodium and the Dietary Approaches to Stop Hypertension (DASH) Diet. N Engl J Med. 2001;344(1):3–10.

Cerebrovascular Disease in Asia: Causative Factors

Alejandro Bimbo F. Diaz, Allan A. Belen, Anne Marie Joyce Tenorio-Javier, and Dan Neftalie A. Juangco

Cerebrovascular disease (CVD) or stroke is the second leading cause of mortality and leading cause of long-term disability worldwide [1, 2]. Globally, there was a downward trend of stroke statistics since the 1990s; however, the collective stroke burden in terms of absolute number of people affected still increases [3]. Stroke is an important health concern as about 60% of the world's population is in Asia and with many countries regarded as developing economies [4]. Notably, Asia has a higher burden related to stroke compared to coronary artery disease (CAD), while the reverse is observed in Western countries [5]. Also, Asia has a higher burden of cerebrovascular risk factors with hypertension (HTN) being the most prevalent cause of both ischemic and hemorrhagic strokes [4].

In contrast to Western countries, Asia has a higher incidence of hemorrhagic strokes [6] and acute ischemic strokes (AIS) related to cerebral small vessel disease (CSVD) [7]. Cases of large vessel occlusion (LVO) and subarachnoid hemorrhages (SAH) are variable.

A. B. F. Diaz (✉)
Department of Neurosciences and Behavioral Medicine, University of Santo Tomas Hospital, Manila, Philippines
e-mail: afdiaz@ust.edu.ph

A. A. Belen
Department of Internal Medicine, Section of Neurosciences, Community General Hospital of San Pablo City, Inc., San Pablo City, Laguna, Philippines

A. M. J. Tenorio-Javier
Department of Neuroscience, Mary Mediatrix Medical Center and Lipa Medix Medical Center, Lipa, Batangas, Philippines

D. N. A. Juangco
Institute for Neurosciences, St. Luke's Medical Center, Quezon City, Philippines

Department of Neuroscience, East Avenue Medical Center, Quezon City, Philippines

© The Author(s), under exclusive license to Springer Nature Switzerland AG 2022
C. V. S. Ram et al. (eds.), *Hypertension and Cardiovascular Disease in Asia*,
Updates in Hypertension and Cardiovascular Protection,
https://doi.org/10.1007/978-3-030-95734-6_19

This chapter highlights CVD in Asia by discussing the epidemiology, causative risk factors, pathophysiology, and outcome of these different stroke subtypes: AIS due to large vessel occlusion (LVO) and CSVD as well as ICH and SAH. The impact of HTN for each stroke subtype is explored. A section on blood pressure variability (BPV) is provided as it is currently one of the emerging factors contributing to the development of strokes including the Asian population.

19.1 Large Vessel Occlusion and Atherosclerosis

Large vessel occlusion (LVO), a subtype of ischemic stroke under The Trial of Org 10172 in Acute Stroke Treatment Classification, has been variably defined in different clinical studies. Despite this variation, LVO was generally defined as any acute-onset occlusion of the intracranial internal carotid artery (ICA), proximal posterior, middle, and anterior cerebral arteries (PCA, MCA, and ACA, respectively), intracranial vertebral artery (VA), and/or basilar artery (BA) [8]. Prior to the advent of thrombectomy, LVO studies typically defined LVO as those involving ICA and first segment of the MCA (M1). As endovascular interventions such as thrombectomy gained popularity, the definition of LVO expanded to include the occlusions in the other segments as defined above.

Large vessel occlusion has been widely studied through the years due to its huge impact in stroke outcomes in relation to infarct size, severity of neurologic deficits, mortality, and treatment availability.

In a systematic review by Lakomkin et al. identifying the proportion of patients with AIS presenting with LVO on image analysis, it was shown that the prevalence of LVO ranges from 7.3% to 60.6% with a mean prevalence of 31.1% across all included studies [9]. Rai et al., in a US population-based study, indicated that LVOs involving occlusions of the M1, ICA terminus, and BA had an estimated annual incidence of 24 per 100,000 people per year [10]. Approximately two-thirds of LVOs occur in the anterior circulation, mainly in the ICA and MCA, and the remaining occur in the posterior circulation with equal distribution among VA, BA, and PCA [11].

The epidemiology of stroke subtype distribution differs across races and ethnicity. Among the causes of stroke, cardioembolic stroke accounts for about 25–30% of cases in Western countries, while intracranial atherosclerosis accounts for up to 25–65% in Asian countries [12]. Young AIS Asian patients aged 18–49 were investigated by Kay Sin Tan et al. and their study showed that the predominant stroke subtypes were large artery atherosclerosis (29.8%) and small vessel occlusion (20.2%) [13]. The preceding statements suggest that a large proportion of Asian stroke patients with AIS are classified as having large vessel disease and as will be discussed further in the later sections, CSVD as well.

Risk factors for all AIS include HTN, diabetes, dyslipidemia, and excessive alcohol consumption. Notably, cigarette smoking is associated with large vessel disease leading to LVO [13–15].

Several mechanisms are associated with the development of LVOs, namely: occlusion at the primary site resulting from intracranial atherosclerosis; embolism from an atherosclerotic extracranial artery or plaque rupture resulting from intracranial vessel occlusion (artery-to-artery embolism); intracranial artery occlusion due to cardioembolism which is usually from atrial fibrillation and cryptogenic etiology [8, 16]. As with the other types of IS, LVOs occur when there is inadequate blood supply to the brain parenchyma, causing cellular, metabolic derangements, inflammatory mechanisms, and activation of the ischemic cascade leading to neuronal cell death. However, a more important feature of LVO is the infarct size and volume which are determined by the degree and length of hypoperfusion and the status of the collateral circulation which is significant in determining stroke severity, progression, and outcome [8, 17].

Large vessel occlusions, if not treated promptly, result in worse outcomes with increased morbidity and mortality. Patients with AIS must be immediately assessed for eligibility to undergo reperfusion therapies, such as intravenous-recombinant tissue Plasminogen Activator (IV-rtPA) and/or mechanical thrombectomy. However, there have been studies demonstrating the limitation of IV-rtPA in the treatment of LVO with successful reperfusion seen only in 13–50% of patients with intracranial ICA or MCA occlusion [16, 18, 19].

In a review by Malhotra et al., LVOs were found to contribute disproportionately to severe functional outcomes after AIS. In their review, LVOs cause about 38.7% of acutely presenting acute cerebral ischemic events, but significantly contributes to poststroke dependence or death in about 61.6%, and poststroke mortality in 95.6% [20]. Endovascular interventions made significant impact in improving the prognosis of patients with LVOs. Using literature-based projections of AIS patients with LVO treatable within 8 h of onset, in the review by Malhotra et al., endovascular thrombectomy could potentially be applied to 21.4% of all the AIS patients and improve the numbers to only 34% of poststroke dependence and death, and 52.8% poststroke mortality. In the meta-analysis by Goyal et al. assessing the efficacy of endovascular thrombectomy in proximal anterior circulation occlusions performed up to 6 h from symptom onset, thrombectomy significantly reduced 90-day disability with a number-needed-to-treat of 2.6 to reduce the modified Rankin Scale Score in one patient by at least one point [8, 21]. Though endovascular treatment revolutionized management of eligible AIS patients, there is still a large gap in access to endovascular thrombectomy between high- and low-income countries [22].

Hypertension remains to be the most important modifiable risk factor for all types of AIS and is said to worsen stroke outcome. In a study by Inoue et al., atrial fibrillation and systolic BP of 170 mmHg or lower were found to have a significant correlation with LVO [23]. Other studies have shown that patients who are known to be hypertensive tend to have less viable or salvageable brain tissue and are prone to develop larger infarctions compared to patients with normal BP levels [24–26]. In some animal studies, chronic HTN has detrimental effects on the cerebral circulation, both in the small and large arteries, as it may cause structural abnormalities such as hypertrophy and inward re-modeling. Decrease in the diameter of the vessel lumen and vasodilatory reserve may ensue leading to hypoperfusion and hemodynamic instability contributing to more cerebral damage [27–32]. Another factor that was studied in relation to large infarct size and HTN is the status of the collateral circulation especially in chronically hypertensive patients. Studies showed that poor collateral circulation, especially the pial or leptomeningeal collaterals in patients with HTN, may result in large infarctions due to a decrease in salvageable tissue or penumbra [24, 31–33].

19.2   Cerebral Small Vessel Disease

Cerebral small vessel disease (CSVD) is a disease affecting the smaller blood vessels that impairs the perfusion to the brain parenchyma. It can affect the small penetrating arteries, arterioles, venules, and capillaries. Arteriosclerosis and small vessel atherosclerosis are the most common pathologies seen related to HTN [34].

The epidemiology of CSVD varies depending on which parameters are taken into account. Lacunar infarcts account for almost 25% of all AIS [35]. Obviously, this does not include the clinically undetected silent ischemic changes that are discovered using cranial Magnetic Resonance Imaging (MRI). There is little data in the literature regarding the prevalence of CSVD as it is challenging to determine its prevalence since most of the small vessel diseases start insidiously and are clinically silent. In many cases, it will take years before they become clinically evident. Moreover, these patients may present with cognitive impairment or vascular dementia later in life.

There are few Asian prevalence studies on CSVD, including white matter hyperintensities (WMH), periventricular hyperintensities (PVHI), silent strokes (SS), and cerebral microbleeds (CMB). In a study by Akthar et al., involving participants from the Middle East and Southeast Asia who were admitted to their stroke service, a preexisting CSVD in the MRI was seen in 65% and distributed as follows: 19.6% with WMH, 33.2% with PVHI, 51.4% with SS, and 22% with CMB. Silent strokes were more common in IS, while CMBs were more common in ICH [7]. In a study by Hilal et al., 3-T brain MRI assessments were performed among participants from Singapore, Hong Kong, and Korea. The results showed that the prevalence of WMH was 36.6%, lacunes was 24.6%, and CMB was 26.9%. Also, the presence of the three small vessel disease parameters demonstrated a sharp increase with increasing age rising from 1.9% in the lowest to 46.2% in the highest 5-year age strata [36].

The pathogenesis of CSVD is still poorly understood although many pathophysiologic mechanisms are in development. Among the most commonly recognized mechanisms are small vascular atherosclerosis and arteriolosclerosis. While arteriolosclerosis increases with age, the hardening and loss of elasticity of the arterioles are further aggravated by HTN as well as diabetes mellitus. Both HTN and age are thought to be the most important risk factors in the development of small vessel disease due to narrowing of the arteriolar lumen and small penetrating arteries [37].

The effects of CSVD on the cerebral tissue are best visualized using MRI. Brain changes that are most commonly seen are WMH, enlarged perivascular spaces, small subcortical infarcts, CMB, lacunar infarcts, and atrophy. In a cohort study conducted in chronically hypertensive patients, a significant correlation was seen between the progression of periventricular white matter lesions and executive function impairment [38].

Over time, chronic HTN leads to a slow and gradual development of minute structural changes in the microvasculature of the cerebral small blood vessels. These are brought about by multiple factors such as chronic inflammation, oxidative stress, and other HTN-related changes that occur over time [37]. The vasculopathy associated with HTN, together with concurrent inflammatory changes, significantly contributes to the development of CSVD. In a paper by Rouhl et al., putative endothelial progenitor cells (cell types that play roles in the regeneration of the endothelial lining of vessels) were lower among hypertensive patients with CSVD in contrast to hypertensive patients without CSVD [39].

Small vessel disease plays a crucial role in various conditions such as aging, stroke, cognitive impairment, and other age-related disabilities including motor-gait issues and mood disorders. It can manifest with significant functional disabilities in the later stage of life. While CSVD may present abruptly such as seen in ICH and lacunar strokes, it may also present insidiously as progressive cognitive decline, among others [40]. Among the important long-term sequelae of CSVD in the Asian population is its effect on cognitive decline and dementia in the elderly. Hilal et al. demonstrated that increasing severity of small vessel disease markers showed significant association with worse cognitive performance independent of the other concomitant risk factors [36].

19.3 Primary Intracerebral Hemorrhage

Intracerebral hemorrhage is defined as any bleeding in the brain parenchyma that may extend into the ventricles and subarachnoid space [41]. Spontaneous, nontraumatic ICH may be primary or secondary. Primary ICH represents about 78–88% of all cases and is usually due to spontaneous rupture of small vessels as complications of chronic HTN or cerebral amyloid angiopathy (CAA) [41], while secondary ICH is due to other known bleeding etiologies such as arteriovenous malformations (AVM), tumors, or cerebral aneurysms.

Intracerebral hemorrhage may be categorized as *lobar* or *deep* depending on its location within the brain parenchyma. *Lobar* ICH refers to hemorrhages which are in the cortical-subcortical brain regions involving one or more lobes of the brain and this accounts for about one-third of all ICH cases. Conversely, *deep* ICH refers to hemorrhages which are in the deeper subcortical regions such as the basal ganglia and internal capsule, brainstem, and cerebellum, and accounts for the remaining two-thirds [42, 43].

The worldwide incidence of ICH is 10–20 cases per 100,000 individuals [41] but it varies substantially across different countries and race. In a systematic review by van Asch et al., the incidence rate of ICH per 100,000 person-years was 51.8 in Asians, 24.2 in Whites, 22.9 in Blacks, and 19.6 in Hispanics [44]. While it accounts for approximately 10–20% of all strokes, Western countries such as the USA and UK have lower incidence with about 8–15% compared to Asian countries such as Japan, China, and Korea with about 18–24% [6, 45].

A review of various Asian studies explored epidemiological data and highlighted the temporal trends. Studies from Japan, China, Korea, Taiwan, Malaysia, Singapore, India, Bangladesh, Pakistan, Iran, and Israel were included to ensure adequate representation from almost all regions in Asia. The frequency of ICH was higher compared to Western countries. Intracerebral hemorrhage accounts for about 20–39% of cases in these countries except for Iran, Thailand, and Israel with 17%, 13%, and 10%, respectively. The review showed an increase in the absolute number of stroke cases except for Japan where there was a significant decline in ICH incidence probably due to better HTN control [46]. Cases of ICH accounted for more than one-third of stroke cases in Asian countries especially in East and South Asia but notably decreased over time [47].

In the summary of findings from the Global Burden of Disease 2010 Study by Krishnamurthi et al. [48], the absolute number of hemorrhagic strokes (combined ICH and SAH) worldwide between 1990 and 2010 showed a 47% increase. Low- and middle-income countries such as Sub-Saharan Africa, Central Asia, and Southeast Asia showed the largest proportion of ICH incidence of about 80%. Comparing high-income and low- and middle-income countries, the age-adjusted incidence rate of hemorrhagic stroke is reduced by 8% in the former while increased by 22% in the latter. A systematic analysis of the Global Burden of Disease Study 2017 showed higher primary ICH rates in Oceania, East, Central, and Southeast Asia. There was a trend towards greater burden of primary ICH in high-income countries which may be associated with better evaluation by neuroimaging of acute strokes, increased prevalence of diabetes mellitus, and overweight. There was a slightly reduced burden in low- and middle-income countries which may be attributed to epidemiologic transitions in these countries [49].

Hypertension is considered as the most important risk factor for spontaneous, non-traumatic ICH and is seen more commonly in patients with *deep* ICH than *lobar* ICH [6]. The second most common risk factor is CAA which is predominantly seen in the elderly population and is regarded as a significant etiology of *lobar* ICH. Other risk factors include advancing age, anticoagulation, use of antiplatelet agents, leukoaraiosis, prior stroke, hematologic abnormalities, chronic kidney disease, excessive alcohol consumption, and use of sympathomimetic drugs [6, 50].

As mentioned, the primary etiologies leading to primary ICH and vasculopathy are HTN and CAA. Chronic HTN leads to lipohyalinoses of the small penetrating arteries which eventually rupture leading to deep hemorrhages frequently extending to the ventricles [41]. Deep hemorrhages have been attributed to the degeneration of media and smooth muscles of the vessel wall in association with long-standing HTN, advancing age, diabetes, and other vascular risk factors. Most bleeding occurs at or near the bifurcation of the affected arterioles [41]. The underlying mechanism in CAA-associated ICH is a combination of deposition of amyloid-β peptide at capillaries, arterioles, and arteries in the cerebral cortex, leptomeninges, and cerebellum [51] and breakdown of the vessel wall. The amyloid deposits lead to vasculopathic changes such as microaneurysms, concentric splitting of the vessel wall, fibrinoid necrosis, and chronic inflammation of the perivascular space [50]. Lipohyalinoses, which is more prominently associated with chronic HTN, are mostly seen in *deep* ICH, while CAA is more associated with *lobar* ICH [6].

The outcome of ICH is variable but comparing with IS, ICH leads to higher mortality and more severe disability. The case fatality rate is approximately 40% at 30 days and about 54% at 1 year [6]. In the acute setting, general predictors of 30-day mortality include size of the hematoma and its expansion, age, presence of coma, intraventricular extension, infratentorial location, and other co-morbid cardiovascular conditions [52]. In two Asian studies, identified independent predictors of poor outcome after primary ICH include fever, low initial Glasgow Coma Scale, large hematoma, intraventricular hemorrhage, and concomitant diabetes [53, 54]. Almost 50% of mortality occurs during the first 2–3 days and are usually associated with neurological complications such as mass effect, increased intracranial pressure, and/or herniation. Deaths occurring after 7 days of hospitalization are usually due to medical complications including pneumonia, aspiration, respiratory distress/failure, or sepsis [55].

The quality of life of ICH survivors are expected to be worse compared to the general population. Caucasians and Asians differ in their views about family values, cultural attitudes, and care preferences which greatly affect the stroke outcome. Despite Asians having worse function, they are more likely to be discharged early and cared for at home, while Caucasians are less likely to be discharged to home despite better functional status [56].

Many studies have identified HTN as a major risk factor for developing primary ICH. There seems to be a stronger association between BP and hemorrhagic stroke compared to that for IS [57] although the relationship between the patterns of changes in BP over time and the risks of ischemic versus hemorrhagic stroke still remains unclear. Elevated BP is associated with an increased risk of stroke incidence and mortality. In the Asia-Pacific Cohort Studies Collaboration study [58], the risk of ischemic and hemorrhagic stroke incidence increased with elevations in BP levels in a dose-dependent manner from the systolic BP level of 115 mmHg. The INTERSTROKE study [59] identified that self-reported HTN of BP ≥ 140/90 mmHg conferred a population-attributable risk of about 56.4% for hemorrhagic stroke. The Hisayama study in Japan [60] showed a considerable decline in stroke incidence. The age-adjusted ICH incidence declined in men and women particularly with a significant decline of 61% from the first cohort (enrolled in 1961) to the second cohort (enrolled in 1974) in men and sustained in both sexes in the third cohort (enrolled in 1988). During this period, the prevalence of severe HTN was significantly decreased and the use of antihypertensive medication increased. With all of these studies it is clear without any doubt that better control of HTN leads to better protection against hemorrhagic strokes.

19.4   Subarachnoid Hemorrhage

Subarachnoid hemorrhage means blood between the arachnoid layer and pia mater. Bleeding into this space is caused by a number of factors, most commonly trauma (tearing of blood vessels) and ruptured intracranial aneurysms. Other causes include presence of an AVM, blood dyscrasia, anticoagulation, and intracranial hemorrhage. This section focuses on aneurysmal SAH (aSAH) since HTN is an independent risk factor for the development, growth, and rupture of an intracranial aneurysm.

The worldwide incidence of aSAH is 7.9 per 100,000 person-years [61]. Looking at temporal trends, the incidence of SAH declined from 10.2 per 100,000 person-years in 1980 to 6.1 per 100,000 person-years in 2010. The incidence widely varies across geographic regions [62]. In 2010, the incidence per 100,000 person-years are as follows: 6.9 in North America, 5.1 in South and Central America, 5.1 in Switzerland, 7.4 in Australia/New Zealand, and 3.7 in Asia with the exception of Japan, 28 [61]. In Asia, registries from different countries showed the incidence of SAH as follows: 6.2 in China, 3.3 in Iran, 4.5 in India, 14.2 in Israel, 0.5 in Kuwait, while 13.7–27.9 in Japan [63]. There is a 2.0% decline annually in Asian countries except Japan, where an increase of 1.6% annually was noted [61].

Known risk factors for aneurysm formation, growth, and rupture include HTN, cigarette smoking, family history, and connective tissue disease. Particularly in the Asia-Pacific region, cigarette smoking and elevated systolic BP each doubles the risk of aSAH [63].

Intracranial aneurysms are pathologic protrusions from the intracranial arteries. Compared to the normal vessels, the walls of aneurysms are composed of a very thin or absent tunica media, and an absent or fragmented internal elastic lamina [64]. Most intracranial aneurysms are acquired and not congenital. Several factors contribute to aneurysm formation [65]. Turbulent and hyperdynamic blood flow produces changes in the vessel wall and subsequent aneurysm development. Hemodynamic stress produces excessive wear and tear in the vessels and eventual breakdown of the internal elastic lamina. Turbulence further produces vibrations resulting in structural fatigue of the vessel wall. The above-mentioned risk factors further play contributory roles in aneurysm development and growth.

However, the rupture rate of aneurysms varies and is influenced by several risk factors: size, growth rate, location, prior hemorrhage, family history, and race. Some aneurysms do not rupture at all and are only found incidentally or during autopsy series. Two of the largest prospective studies on unruptured intracranial aneurysms (ISUIA [66] and UCAS [67]) have identified that smaller aneurysms have lower rates of rupture than larger ones. Low risk of rupture for both studies was seen with aneurysms <7 mm in diameter, and for which growth rate is also slower. However, this rupture risk is modified by the location: with posterior circulation aneurysms generally having the highest rate of rupture, followed by anterior circulation, and lowest with cavernous carotid artery aneurysms [66, 67].

Aneurysmal SAH is associated with significant neurologic morbidity and mortality especially if left untreated. The risk of rebleeding is highest in the first few hours from rupture and this re-rupture is associated with an even higher mortality. Aneurysm treatment by surgical or endovascular means is the only way to prevent rebleeding.

Around 18% of patients with aSAH die even before reaching the hospital [68]. Even among patients who get evaluated in the hospital, rates of early mortality are still high due to rebleeding, vasospasm and cerebral ischemia, hydrocephalus, increased intracranial pressure, seizures, and cardiac complications. Due to medical advances and more research, the case fatality rate has significantly decreased from 50% to about 30% if treated in a timely manner [69]. Among survivors, long-term neurologic complications include cognitive dysfunction, epilepsy, and focal deficits.

As mentioned above, HTN is a major risk factor for aneurysm formation, growth, and eventual rupture, causing SAH. In a systematic review, the presence of HTN has a statistically significant relative risk of 2.5% (95% CI 2.0–3.1) in longitudinal studies and an odds ratio of 2.6% (95% CI 2.0–3.1) in case-control studies [3]. There is a corresponding decrease in the incidence of SAH per unit decrease of BP: 7.1% decrease per mmHg of systolic BP decrease, while 11.5% for diastolic BP [61]. A global decline in the incidence of aSAH is noted and this is partly due to a parallel decline in prevalence of HTN. In China, it was found that higher average levels of BP were linearly and positively associated with SAH, and elevated BP accounted for about 23% of all SAH cases [70].

After rupture of an intracranial aneurysm, BP control is of paramount importance to avoid further morbidity. Although optimal BP therapy in SAH is not clear, hypotension should definitely be avoided. While lowering the BP may decrease the risk of rebleeding of an unsecured aneurysm, it may also present a risk of cerebral infarction if too low. Some guidelines recommend to have a goal of maintaining SBP <160 mmHg, while others use the mean arterial pressure (MAP) of <110 mmHg. It may be reasonable to use the patient's premorbid baseline BP control to define targets of therapy [71, 72]. Intravenous agents, such as labetalol or nicardipine, are preferred for BP control because they can easily be titrated.

19.5   Blood Pressure Variability and Its Impact in Stroke and Vascular Dementia

Normally, there are some short-term fluctuations in BP during the 24-h period which are largely influenced by the circadian modulations. In about 10–20% of the population, there is a physiological reduction of nocturnal BP. Blood pressure variability can be classified as very-short-term (beat-to-beat), short-term (during 24 h), mid-term (day-by-day), long-term (less than 5 years), and very-long-term (more than 5 years) [73]. This is associated with hypertension-mediated-organ-damage or target-organ-damage, and in some cases, it triggers significant vascular events during transient rises in the BP.

Blood pressure variability is a novel factor in the development of cardiac events, stroke, and cognitive impairment. Emerging literatures are showing consequences of increased BPV among certain populations including people with morning surge HTN, night sleep HTN, non-dippers, reverse dippers, and dippers with morning surges of BP. The wide fluctuations of BP in these instances are thought to increase the cerebral microcirculatory changes and cumulative lesions in the brain—be it ischemic (as seen in ischemic white matter changes) or hemorrhagic (as observed in CMBs).

Increasing BPV seems to negatively impact stroke outcome in the acute phase and probably, even in the subacute phase. This is seen in both the acute ischemic and hemorrhagic stroke types. Why this is so is not yet fully understood. One mechanism that possibly explains the negative effect of BPV is the impaired cerebral autoregulation in the region affected with stroke. For instance, in AIS, a drop in the cerebral perfusion secondary to BPV can worsen the condition of the ischemic core and convert the ischemic penumbra into an enlarging infarct core [74]. Chung et al. considered BPV as an independent factor that is associated with early neurological deterioration [75]. Increased BPV can also negatively impact hemorrhagic strokes. Very high BP can increase hematoma size and brain edema. A significant drop in systolic BP can also be detrimental as it potentially can increase perihematomal ischemia and cause acute renal hypoperfusion leading to acute kidney injury [76].

Aside from ischemic and hemorrhagic strokes, large BPV is also associated with cognitive impairment and dementia. There is mounting data linking marked BPV to higher risk of heart diseases, strokes, and dementia, even beyond the effect of the BP per se [77]. In a Japanese elderly population study by Oishi et al. on the relationship of BPV and incident dementia, the results suggested that increased day-to-day BPV is independent of average home BP and is a significant risk factor for the development of all-cause dementia, vascular dementia, and Alzheimer's disease [78].

In a systematic review and meta-analysis of prospective cohort studies that looked into the association of BPV with the presence or progression of CSVD markers such as WMH, lacunes, and CMBs in MRI, the analysis showed that the association of systolic BPV with the presence of lacunes and CMBs were not statistically significant. However, increased BPV was associated with increased odds of presence or progression of WMHs [79].

A study by Liu et al. conducted in China among the geriatric community dwelling areas looked at the effect of systolic BP variability through self-measured BP at home for 7 consecutive days. Mini Mental State Examination (MMSE) scores as well as brain MRI WMH at baseline and at final follow-up visits were taken. After an average of 2.3 years, a decline in MMSE score and an increase in brain MRI WMHs were noted. The results of the study suggested that excessive variability in self-measured systolic BP exacerbates the progression of cognitive impairment and brain white matter lesions in the oldest-old geriatric population [80].

References

1. Feigin VL. Stroke in developing countries: can the epidemic be stopped and outcomes improved? Lancet Neurol. 2007;6(2):94–7.
2. Strong K, Mathers C, Bonita R. Preventing stroke: saving lives around the world. Lancet Neurol. 2007;6(2):182–7.
3. Feigin VL, Norrving B, Mensah GA. Global burden of stroke. Circ Res. 2017;120:439–48.
4. Venketasubramanian N, Yoon BW, Pandian J, Navarro JC. Stroke epidemiology in south, east, and south-East Asia: a review. J Stroke. 2017;19:286–94.
5. Ueshima H, Sekikawa A, Miura K, Turin TC, Takashima N, Kita Y, et al. Cardiovascular disease and risk factors in Asia: a selected review. Circulation. 2008;118:2702–9.

6. An SJ, Kim TJ, Yoon BW. Epidemiology, risk factors, and clinical features of intracerebral hemorrhage: an update. J Stroke. 2017;19:3–10.
7. Akhtar N, Salam A, Kamran S, D'Souza A, Imam Y, Own A, et al. Pre-existing small vessel disease in patients with acute stroke from the Middle East, Southeast Asia, and Philippines. Transl Stroke Res. 2018;9:274–82.
8. Rennert RC, Wali AR, Steinberg JA, Santiago-Dieppa DR, Olson SE, Pannell JS, et al. Epidemiology, natural history, and clinical presentation of large vessel ischemic stroke. Clin Neurosurg. 2019;85:S4–8.
9. Lakomkin N, Dhamoon M, Carroll K, Singh IP, Tuhrim S, Lee J, et al. Prevalence of large vessel occlusion in patients presenting with acute ischemic stroke: a 10-year systematic review of the literature. J Neurointerv Surg. 2019;11(3):241–5.
10. Rai AT, Seldon AE, Boo S, Link PS, Domico JR, Tarabishy AR, et al. A population-based incidence of acute large vessel occlusions and thrombectomy eligible patients indicates significant potential for growth of endovascular stroke therapy in the USA. J Neurointerv Surg. 2017;9:722–6.
11. Smith WS, Lev MH, English JD, Camargo EC, Chou M, Johnston SC, et al. Significance of large vessel intracranial occlusion causing acute ischemic stroke and tia. Stroke. 2009;40:3834–40.
12. Bang OY. Considerations when subtyping ischemic stroke in Asian patients. J Clin Neurol. 2016;12(2):129–36.
13. Tan KS, Navarro JC, Wong KS, Huang YN, Chiu HC, Poungvarin N, et al. Clinical profile, risk factors and aetiology of young ischaemic stroke patients in Asia: a prospective, multicentre, observational, hospital-based study in eight cities. Neurol Asia. 2014;19:117–27.
14. Ihle-Hansen H, Thommessen B, Wyller TB, Engedal K, Fure B. Risk factors for and incidence of subtypes of ischemic stroke. Funct Neurol. 2012;27:35–40.
15. Bejot Y, Caillier M, Ben Salem D, Couvreur G, Rouaud O, Osseby GV, et al. Ischaemic stroke subtypes and associated risk factors: a French population based study. J Neurol Neurosurg Psychiatry. 2008;79:1344–8.
16. Al Kasab S, Holmstedt CA, Jauch EC, Schrock J. Acute ischemic stroke due to large vessel occlusion. Emerg Med Rep. 2018;39:13–22.
17. Brouns R, De Deyn PP. The complexity of neurobiological processes in acute ischemic stroke. Clin Neurol Neurosurg. 2009;111(6):483–95.
18. De Silva DA, Brekenfeld C, Ebinger M, Christensen S, Barber PA, Butcher KS, et al. The benefits of intravenous thrombolysis relate to the site of baseline arterial occlusion in the echo-planar imaging thrombolytic evaluation trial (EPITHET). Stroke. 2010;41:295–9.
19. Jansen O, Von Kummer R, Forsting M, Hacke W, Sartor K. Thrombolytic therapy in acute occlusion of the intracranial internal carotid artery bifurcation. Am J Neuroradiol. 1995;16:1977–86.
20. Malhotra K, Gornbein J, Saver JL. Ischemic strokes due to large-vessel occlusions contribute disproportionately to stroke-related dependence and death: a review. Front Neurol. 2017;8:651.
21. Goyal M, Menon BK, Van Zwam WH, Dippel DWJ, Mitchell PJ, Demchuk AM, et al. Endovascular thrombectomy after large-vessel ischaemic stroke: a meta-analysis of individual patient data from five randomised trials. Lancet. 2016;387:1723–31.
22. Tsang ACO, Yang IH, Orru E, Nguyen QA, Pamatmat RV, Medhi G, et al. Overview of endovascular thrombectomy accessibility gap for acute ischemic stroke in Asia: a multi-national survey. Int J Stroke. 2020;15:516–20.
23. Inoue M, Noda R, Yamaguchi S, Tamai Y, Miyahara M, Yanagisawa S, et al. Specific factors to predict large-vessel occlusion in acute stroke patients. J Stroke Cerebrovasc Dis. 2018;27:886–91.
24. Cipolla MJ, Liebeskind DS, Chan SL. The importance of comorbidities in ischemic stroke: impact of hypertension on the cerebral circulation. J Cereb Blood Flow Metab. 2018;38(12):2129–49.
25. Lima FO, Furie KL, Silva GS, Lev MH, Camargo ÉCS, Singhal AB, et al. The pattern of leptomeningeal collaterals on CT angiography is a strong predictor of long-term functional outcome in stroke patients with large vessel intracranial occlusion. Stroke. 2010;41:2316–22.

26. Menon BK, Smith EE, Coutts SB, Welsh DG, Faber JE, Goyal M, et al. Leptomeningeal collaterals are associated with modifiable metabolic risk factors. Ann Neurol. 2013;74:241–8.
27. Ahmed N, Wahlgren N, Brainin M, Castillo J, Ford GA, Kaste M, et al. Relationship of blood pressure, antihypertensive therapy, and outcome in ischemic stroke treated with intravenous thrombolysis: retrospective analysis from safe implementation of thrombolysis in stroke-international stroke thrombolysis register (SITS-ISTR). Stroke. 2009;40:2442–9.
28. Leonardi-Bee J, Bath PMW, Phillips SJ, Sandercock PAG. Blood pressure and clinical outcomes in the International Stroke Trial. Stroke. 2002;33:1315–20.
29. Strandgaard S. Autoregulation of cerebral circulation in hypertension. Acta Neurol Scand Suppl. 1978;57:1–82.
30. Muller M, Van Der Graaf Y, Visseren FL, Mali WPTM, Geerlings MI. Hypertension and longitudinal changes in cerebral blood flow: the SMART-MR study. Ann Neurol. 2012;71:825–33.
31. Coyle P, Heistad DD. Development of collaterals in the cerebral circulation. Blood Vessels. 1991;28(1-3):183–9.
32. Shuaib A, Butcher K, Mohammad AA, Saqqur M, Liebeskind DS. Collateral blood vessels in acute ischaemic stroke: a potential therapeutic target. Lancet Neurol. 2011;10(10):909–21.
33. Zhang H, Prabhakar P, Sealock R, Faber JE. Wide genetic variation in the native pial collateral circulation is a major determinant of variation in severity of stroke. J Cereb Blood Flow Metab. 2010;30:923–34.
34. Pantoni L. Cerebral small vessel disease: from pathogenesis and clinical characteristics to therapeutic challenges. Lancet Neurol. 2010;10(10):689–701.
35. Petty GW, Brown RD, Whisnant JP, Sicks JRD, O'Fallon WM, Wiebers DO. Ischemic stroke subtypes: a population-based study of functional outcome, survival, and recurrence. Stroke. 2000;31:1062–8.
36. Hilal S, Mok V, Youn YC, Wong A, Ikram MK, Chen CLH. Prevalence, risk factors and consequences of cerebral small vessel diseases: data from three Asian countries. J Neurol Neurosurg Psychiatry. 2017;88:669–74.
37. Liu Y, Dong YH, Lyu PY, Chen WH, Li R. Hypertension-induced cerebral small vessel disease leading to cognitive impairment. Chin Med J (Engl). 2018;131:615–9.
38. Uiterwijk R, Staals J, Huijts M, De Leeuw PW, Kroon AA, Van Oostenbrugge RJ. MRI progression of cerebral small vessel disease and cognitive decline in patients with hypertension. J Hypertens. 2017;35:1263–70.
39. Rouhl RPW, Mertens AECS, Van Oostenbrugge RJ, Damoiseaux JGMC, Debrus-Palmans LL, Henskens LHG, et al. Angiogenic T-cells and putative endothelial progenitor cells in hypertension-related cerebral small vessel disease. Stroke. 2012;43:256–8.
40. Pasi M, Cordonnier C. Clinical relevance of cerebral small vessel diseases. Stroke. 2020;51:47–53.
41. Qureshi AI, Tuhrim S, Broderick JP, Batjer HH, Hondo H, Hanley DF. Spontaneous intracerebral hemorrhage. N Engl J Med. 2001;344:1450–60.
42. Flaherty ML, Woo D, Haverbusch M, Sekar P, Khoury J, Sauerbeck L, et al. Racial variations in location and risk of intracerebral hemorrhage. Stroke. 2005;36:934–7.
43. Grysiewicz RA, Thomas K, Pandey DK. Epidemiology of ischemic and hemorrhagic stroke: incidence, prevalence, mortality, and risk factors. Neurol Clin. 2008;26:871–95.
44. van Asch CJ, Luitse MJ, Rinkel GJ, van der Tweel I, Algra A, Klijn CJ. Incidence, case fatality, and functional outcome of intracerebral haemorrhage over time, according to age, sex, and ethnic origin: a systematic review and meta-analysis. Lancet Neurol. 2010;9:167–76.
45. Wang W, Jiang B, Sun H, Ru X, Sun D, Wang L, et al. Prevalence, incidence, and mortality of stroke in China: results from a nationwide population-based survey of 480687 adults. Circulation. 2017;135:759–71.
46. Toyoda K. Epidemiology and registry studies of stroke in Japan. J Stroke. 2013;15:21.
47. Mehndiratta MM, Khan M, Mehndiratta P, Wasay M. Stroke in Asia: geographical variations and temporal trends. J Neurol Neurosurg Psychiatry. 2014;85:1308–12.
48. Krishnamurthi RRV, Feigin VL, Forouzanfar MH, Mensah GA, Connor M, Bennett DA, et al. Global and regional burden of first-ever ischaemic and haemorrhagic stroke during 1990–2010: findings from the Global Burden of Disease Study 2010. Lancet Glob Health. 2013;1:e259–81.

49. Krishnamurthi RV, Ikeda T, Feigin VL. Global, regional and country-specific burden of Ischaemic stroke, intracerebral haemorrhage and subarachnoid haemorrhage: a systematic analysis of the global burden of disease study 2017. Neuroepidemiology. 2020;54:171–9.
50. Aguilar MI, Brott TG. Update in intracerebral hemorrhage. Neurohospitalist. 2011;1:148–59.
51. Rosand J, Hylek EM, O'Donnell HC, Greenberg SM. Warfarin-associated hemorrhage and cerebral amyloid angiopathy: a genetic and pathologic study. Neurology. 2000;55:947–51.
52. Safatli DA, Günther A, Schlattmann P, Schwarz F, Kalff R, Ewald C. Predictors of 30-day mortality in patients with spontaneous primary intracerebral hemorrhage. Surg Neurol Int. 2016;7:S510–7.
53. Poungvarin N, Suwanwela NC, Venketasubramanian N, Wong LKS, Navarro JC, Bitanga E, et al. Grave prognosis on spontaneous intracerebral haemorrhage: GP on stage score. J Med Assoc Thail. 2006;89:84–93.
54. Chen HS, Hsieh CF, Chau TT, Yang CD, Chen YW. Risk factors of in-hospital mortality of intracerebral hemorrhage and comparison of ICH scores in a Taiwanese population. Eur Neurol. 2011;66:59–63.
55. Hemphill JC, Greenberg SM, Anderson CS, Becker K, Bendok BR, Cushman M, et al. Guidelines for the Management of Spontaneous Intracerebral Hemorrhage: a guideline for healthcare professionals from the American Heart Association/American Stroke Association. Stroke. 2015;46:2032–60.
56. McNaughton H, Feigin V, Kerse N, Barber PA, Weatherall M, Bennett D, et al. Ethnicity and functional outcome after stroke. Stroke. 2011;42:960–4.
57. Zia E, Hedblad B, Pessah-Rasmussen H, Berglund G, Janzon L, Engström G. Blood pressure in relation to the incidence of cerebral infarction and intracerebral hemorrhage—hypertensive hemorrhage: debated nomenclature is still relevant. Stroke. 2007;38:2681–5.
58. Lawes C. Blood pressure indices and cardiovascular disease in the Asia Pacific region: a pooled analysis. Hypertension. 2003;42:69–75.
59. O'Donnell MJ, Chin SL, Rangarajan S, Xavier D, Liu L, Zhang H, et al. Global and regional effects of potentially modifiable risk factors associated with acute stroke in 32 countries (INTERSTROKE): a case-control study. Lancet. 2016;388:761–75.
60. Kubo M, Kiyohara Y, Kato I, Tanizaki Y, Arima H, Tanaka K, et al. Trends in the incidence, mortality, and survival rate of cardiovascular disease in a Japanese community: the Hisayama study. Stroke. 2003;34:2349–54.
61. Etminan N, Chang HS, Hackenberg K, De Rooij NK, Vergouwen MDI, Rinkel GJE, et al. Worldwide incidence of aneurysmal subarachnoid hemorrhage according to region, time period, blood pressure, and smoking prevalence in the population: a systematic review and meta-analysis. JAMA Neurol. 2019;76:588–97.
62. Hughes JD, Bond KM, Mekary RA, Dewan MC, Rattani A, Baticulon R, et al. Estimating the global incidence of aneurysmal subarachnoid hemorrhage: a systematic review for central nervous system vascular lesions and meta-analysis of ruptured aneurysms. World Neurosurg. 2018;115:430–447.e7.
63. Feigin V, Parag V, Lawes CMM, Rodgers A, Suh I, Woodward M, et al. Smoking and elevated blood pressure are the most important risk factors for subarachnoid hemorrhage in the Asia-Pacific region: an overview of 26 cohorts involving 306620 participants. Stroke. 2005;36:1360–5.
64. Austin G, Fisher S, Dickson D, Anderson D, Richardson S. The significance of the extracellular matrix in intracranial aneurysms. Ann Clin Lab Sci. 1993;23:97–105.
65. Wiebers DO, Piepgras DG, Meyer FB, Kallmes DF, Meissner I, Atkinson JLD, et al. Pathogenesis, natural history, and treatment of unruptured intracranial aneurysms. Neuroradiol J. 2006;19:504–15.
66. Wiebers DO. Unruptured intracranial aneurysms: natural history, clinical outcome, and risks of surgical and endovascular treatment. Lancet. 2003;362:103–10.
67. Morita A, Kirino T, Hashi K, Aoki N, Fukuhara S. The natural course of unruptured cerebral aneurysms in a Japanese cohort. N Engl J Med. 2012;366:2474–82.
68. Lindbohm JV, Kaprio J, Jousilahti P, Salomaa V, Korja M. Risk factors of sudden death from subarachnoid hemorrhage. Stroke. 2017;48:2399–404.

69. Mackey J, Khoury JC, Alwell K, Moomaw CJ, Kissela BM, Flaherty ML, et al. Stable incidence but declining case-fatality rates of subarachnoid hemorrhage in a population. Neurology. 2016;87:2192–7.
70. McGurgan IJ, Clarke R, Lacey B, Kong XL, Chen Z, Chen Y, et al. Blood pressure and risk of subarachnoid hemorrhage in China. Stroke. 2019;50:38–44.
71. Diringer MN, Bleck TP, Hemphill JC, Menon D, Shutter L, Vespa P, et al. Critical care management of patients following aneurysmal subarachnoid hemorrhage: recommendations from the neurocritical care society's multidisciplinary consensus conference. Neurocrit Care. 2011;15:211–40.
72. Connolly ES, Rabinstein AA, Carhuapoma JR, Derdeyn CP, Dion J, Higashida RT, et al. Guidelines for the management of aneurysmal subarachnoid hemorrhage: a guideline for healthcare professionals from the American Heart Association/American Stroke Association. Stroke. 2012;43(6):1711–37.
73. Rosei EA, Chiarini G, Rizzoni D. How important is blood pressure variability? Eur Heart J Suppl. 2020;22:E1–6.
74. Zhang T, Wang X, Wen C, Zhou F, Gao S, Zhang X, et al. Effect of short-term blood pressure variability on functional outcome after intra-arterial treatment in acute stroke patients with large-vessel occlusion. BMC Neurol. 2019;19(1):228.
75. Chung JW, Kim N, Kang J, Park SH, Kim WJ, Ko Y, et al. Blood pressure variability and the development of early neurological deterioration following acute ischemic stroke. J Hypertens. 2015;33:2099–106.
76. Qureshi AI, Palesch YY, Barsan WG, Hanley DF, Hsu CY, Martin RL, et al. Intensive blood-pressure lowering in patients with acute cerebral hemorrhage. N Engl J Med. 2016;375:1033–43.
77. Rothwell PM, Howard SC, Dolan E, O'Brien E, Dobson JE, Dahlöf B, et al. Prognostic significance of visit-to-visit variability, maximum systolic blood pressure, and episodic hypertension. Lancet. 2010;375:895–905.
78. Oishi E, Ohara T, Sakata S, Fukuhara M, Hata J, Yoshida D, et al. Day-to-day blood pressure variability and risk of dementia in a general Japanese elderly population: the Hisayama study. Circulation. 2017;136:516–25.
79. Ma Y, Song A, Viswanathan A, Blacker D, Vernooij MW, Hofman A, et al. Blood pressure variability and cerebral small vessel disease: a systematic review and meta-analysis of population-based cohorts. Stroke. 2020;51:82–9.
80. Liu Z, Zhao Y, Zhang H, Chai Q, Cui Y, Diao Y, et al. Excessive variability in systolic blood pressure that is self-measured at home exacerbates the progression of brain white matter lesions and cognitive impairment in the oldest old. Hypertens Res. 2016;39:245–53.

Clinical Pharmacology of Antihypertensive Drugs

20

Myeong-Chan Cho, Ki Chul Sung, Eun Joo Cho, and Jinho Shin

20.1 β-Adrenergic Receptor Blockers (β-Blockers)

The precise mechanisms of the antihypertensive effects of β-blockers remain incompletely understood. β-blockers differ in absorption and metabolism depending on the type and are usually metabolized in the liver, with a relatively short plasma half-life. β-blockers attenuate sympathetic stimulation by competing with catecholamines in β-adrenergic receptors [1]. Many tissues have both β1 and β2 receptors, and the concept of cardiac selective drugs is only relative, but β1-adrenergic receptors are found primarily in the heart, brain, and adipose tissue, and β2 receptors are widely distributed in the lungs, liver, and muscle. Cardioselective β-blockers such as metoprolol, bisoprolol, and nebivolol selectively block β1-receptors when used at the approved dose. Inhibition of β1-adrenergic receptors in the kidney can inhibit renin release. β2-blockade may blunt the antihypertensive effects of β1-blockade [2]. Certain β-blockers have antihypertensive effects that are mediated through

M.-C. Cho (✉)
Department of Internal Medicine, College of Medicine, Chungbuk National University, Cheongju, South Korea
e-mail: mccho@cbnu.ac.kr

K. C. Sung
Division of Cardiology, Department of Internal Medicine, Kangbuk Samsung Hospital, Sungkyunkwan University, Seoul, South Korea
e-mail: kcmd.sung@samsung.com

E. J. Cho
Division of Cardiology, Department of Internal Medicine, Yeouido St. Mary's Hospital, Catholic University, Seoul, South Korea

J. Shin
Division of Cardiology, Department of Internal Medicine, Hanyang University Hospital, Seoul, South Korea

© The Author(s), under exclusive license to Springer Nature Switzerland AG 2022
C. V. S. Ram et al. (eds.), *Hypertension and Cardiovascular Disease in Asia*,
Updates in Hypertension and Cardiovascular Protection,
https://doi.org/10.1007/978-3-030-95734-6_20

α1-adrenergic antagonist activity and nitric oxide-dependent vasodilator action, in addition to β-adrenergic receptor antagonism. Carvedilol and labetalol, mixed α−/β--blockers, block both β1/β2-receptors and peripheral α1-receptors. Recent guidelines do not recommend β-blockers as first-line therapy in patients with simple hypertension without complications, as β-blockers have limited effects on cardiovascular disease (CVD) prevention in several random trials, but have caused many metabolic disorders. Abrupt discontinuation of a large amount of β-blockers can cause rebound hypertension.

The side effects of β-blockers including erectile dysfunction, hyperglycemia, and dyslipidemia can be minimized by use of low to moderate doses of traditional β-blockers, vasodilating β-blockers (nebivolol), and mixed α−/β-blockers (carvedilol) than with high doses of traditional β-blockers [3]. β-blockers have a compelling indication for post-myocardial infarction and left ventricular dysfunction [4]. β-blockers could be helpful for patients with essential tremor, tachycardia, or arrhythmias.

20.2 Diuretics

Diuretics are a heterogenous and popular class of antihypertensives. Diuretics can increase renin and improve the effectiveness of angiotensin conversion enzyme inhibitors and aldosterone receptor blockers. Diuretics are effective in salt-sensitive hypertension, which is prevalent in elderly, obese, and black patients. Diuretics are also very important for the management of resistant hypertension. Excessive salt consumption or administration of nonsteroidal anti-inflammatory drugs hinders the antihypertensive effect of diuretics by preventing volume reduction and cardiac output reduction.

20.2.1 Thiazides

Diuretics such as thiazide and thiazide-like diuretics have been shown to reduce hypertension-related diseases and mortality in many controlled clinical trials [5, 6]. Diuretics combine well with most other antihypertensive drug classes [7] and are especially effective in lowering blood pressure (BP) in African-Americans and older adults. The effects of thiazide or thiazide-like diuretics on BP can be divided into three sequential phases: short-term, long-term, and chronic reduction of BP [7]. During the short-term phase, which corresponds to the first few weeks, reduction in BP is related to reduction in cardiac output and plasma volume. In contrast, the long-term and chronic antihypertensive effects of diuretics are more closely related to continuous reduction of total vascular resistance rather than to volume reduction [7]. Although chlorthalidone is a diuretic used in many groundbreaking clinical trials [5, 6] hydrochlorothiazide (HCTZ) is most commonly used. The pharmacodynamic and pharmacokinetic profiles of chlorthalidone differ markedly from those of HCTZ. Chlorthalidone is 1.5 to 2 times more potent than HCTZ and has a longer half-life (9–10 h) [7]. Also, using the recommended dose, chlorthalidone is more effective in lowering systolic BP [8]. The thiazide diuretics can cause a number of metabolic disorders, including glucose intolerance, hypokalemia, hypomagnese- mia, hyperuricemia, and hypercholesterolemia. In the early stages of use, thiazide was used in high doses (100 to 200 mg/day) and many side effects were observed, but these side effects can be minimized at the doses currently used. Patients treated with chlorthalidone in clinical trials had a higher incidence of type 2 diabetes than control groups [9]. However, elevated blood glucose levels and diabetes caused by chlorthalidone treatment did not increase risk of CVD events [10].

20.2.2 Loop Diuretics

A recent Cochrane analysis report shows that loop diuretics are less effective in lowering BP than thiazide and thiazide-like diuretics, especially in non-edematous patients [11]. Loop diuretics would be reserved for patients who require treatment for volume overload or edema in addition to lowering BP, and loop diuretics are more appropriate and effective than thiazide in patients with severe chronic kidney disease (estimated glomerular filtration rate (eGFR) <30 mL/min/1.73 m^2) [7].

20.2.3 Spironolactone and Eplerenone

Spironolactone belongs to the mineralocorticoid receptor antagonist drug family, and non-selectively antagonize androgen and progesterone receptors but competi- tively block aldosterone receptor-mediated action. Spironolactone inhibits sodium– potassium exchange at the mineralocorticoid receptor in the kidney and antagonizes aldosterone-induced vasoconstriction. The most common adverse effects are breast complaints and hyperkalemia. Specifically, men may experience gynecomastia, general feminization, and loss of libido and spironolactone is considered pregnancy category C [12]. Eplerenone, an aldosterone receptor antagonist similar to spirono- lactone, is weaker but more tolerable, more selective mineralocorticoid receptor antagonist. For patients with side effects of spironolactone, eplerenone could be the preferred choice.

20.3   Angiotensin Converting Enzyme (ACE) Inhibitors

As for blocking agents for renin–angiotensin–aldosterone system (RAS), there are direct renin inhibitor, ACE inhibitors, angiotensin II receptor blockers, and aldoste- rone receptor blocker. Among them, ACE inhibitor was firstly introduced as the clinically available antihypertensive drug. According to the ligand type for zinc ion in ACE, ACE inhibitor can be categorized as sulfhydryl ligand drug (captopril), carboxyl ligand drug (enalapril, lisinopril, perindopril, quinapril, cilazapril, moexi- pril, ramipril, trandolapril, benazepril), and phosphoryl ligand drug (fosinopril).

The first mechanism of action of ACE inhibitors is to block the formation of angiotensin II resulting in the alleviation of increased angiotensin II actions such as vasoconstriction, stimulation of aldosterone synthesis, activation of sympathetic nervous system, and tissue damage or proliferation. The second mechanism is to block kallikrein II which inactivates bradykinin resulting in the accumulation of bradykinin. Bradykinin can increase prostaglandin I2 and E2, which induce vasodilation. And increased bradykinin is suggested as one of the mechanisms of angioedema.

Clinical indications of ACE inhibitors could include regression of left ventricular hypertrophy, improvement of left ventricular function, improve congestive heart failure, renoprotection, and reduction of proteinuria, slowing the atherosclerosis, and neutral or positive effect on glucose and lipid metabolism. In classical dogma of high versus low renin hypertensions, ACE inhibitor is preferred for high renin hypertension as well as hypertension in the young, hypertension due to renal parenchymal diseases, or renovascular hypertension. For normal or low renin hypertension, ACE inhibitor is preferred in combination with diuretics. ACE inhibitor is first choice drug for heart failure, coronary artery disease, acute myocardial infarction, chronic kidney disease with or without proteinuria, and stroke. Regarding the superiority of the combinations of ACE inhibitor and calcium-channel blocker (CCB) versus ACE inhibitor and diuretics, there are controversies and different study results. Bottom line will be the combination of the long-acting drugs to cover 24-h period such as long-acting CCB or long-acting thiazide-like diuretics.

Regarding side effects of ACE inhibitors, first-dose phenomenon of hypotension can be anticipated in the elderly patient with volume depletion. Adverse reactions such as hyperkalemia, hypoglycemia, taste change, skin eruption, leukopenia were reported. Angioedema is the most serious adverse effect and it will usually be manifested within several hours after oral intake but sometimes it occurs after several days. It involves face, lips, oropharynx, and larynx and it will last up to 2–3 days. It was reported that angioedema can increase when ACE inhibitor is combined with dipeptidyl peptidase-4 inhibitor. From Asian perspectives, the most important side effect of ACE inhibitor is a dry cough which was reported up to 40–50%. It usually observed after a week but sometimes it can be observed after several months following ACE inhibitor administration. Whether there are differences among many ACE inhibitors is not clear because there are few comparative studies available. In patients with bilateral renal artery stenosis or renal artery stenosis in single kidney, ACE inhibitor can precipitate severe hypotension or shock and acute kidney injury, which requires rapid volume replacement. ACE inhibitor can elevate creatinine level especially in the patient with chronic kidney disease. Creatinine elevation up to 30% should be tolerated because of the renoprotective benefit but for higher increase, dosage reduction can be considered and it is usually stopped when eGFR decreased more than 50%. It is important to follow up the patient with potassium and creatinine level after 1 or 2 weeks when ACE inhibitor started. Additional regular monitoring is also recommended. Low dialyzability ACE inhibitors (ramipril, fosinopril) could result in hypotension and bone marrow suppression unless the dosage is adjusted. ACE inhibitor is absolutely contraindicated during second and third trimester in pregnancy but captopril and enalapril are safely used for breast feeding women because of the low secretion rate to apocrine gland.

Antihypertensive efficacy of ACE inhibitor could be influenced by the volume or salt retention status as exemplified by low renin hypertension. Poor initial response could suggest the salt overload so that diuretics could be a very effective combination drug. In general, ACE inhibitor and angiotensin II receptor blockers combination is not recommended [13].

20.4   Angiotensin II Receptor Blockers

Angiotensin II receptor blocker (ARB) blocks selectively the type I angiotensin II receptor (AT_1R) and there are the compounds such as losartan, valsartan, irbesartan, telmisartan, candesartan, eprosartan, olmesartan, azilsartan, and fimasartan. Instead of ACE inhibitor, ARB is much more frequently prescribed in Asian patients because of almost no dry cough.

Action mechanism of ARB is to block selectively AT_1R which mediates pathophysiologic effect of angiotensin II but type II angiotensin II receptor (AT_2R) generally mediates opposite effect to AT_1R. ARBs can block the activities of angiotensin II regardless of its origin from ACE or chymase system. Blocking of AT_1R results in the increase in angiotensin II concentration, which in turn competes the binding of ARBs to AT_1R or even surmount or overcome the binding of ARB to AT_1R. Surmountability means weakened action of ARBs by increased angiotensin II concentration. In contrast, when higher concentrations of angiotensin II cannot overcome the effect of an ARB, it means that the ARB is insurmountable. But AT_1R binding affinity is not directly correlated with the antihypertensive efficacy. Moreover, the impact of surmountability of ARBs on clinical outcomes has not been established.

In general, the indication or benefit as well as contraindication of ARBs is almost the same with ACE inhibitors. But ARBs are not recommended for breast feeding women. The more important aspect of the clinical application of ARBs could be related to their pharmacokinetic aspects. All ARBs increase renal resorption of lithium so that the relevant history should be checked and ARBs should be avoided. Losartan undergoes first pass metabolism via cytochrome P450 system so that the dosage should be decreased by half in severe hepatic dysfunction. Fluconazole and rifampin could reduce C_{max} or AUC of losartan significantly. And other CYP2C9 enzyme inhibitor could reduce the efficacy of losartan. Candesartan, olmesartan, and azilsartan are prodrugs to be activated in gastrointestinal tract to be active metabolites. Telmisartan will increase the plasma digoxin level and induce toxicity. Most of ARBs are not dialyzable but, at least, biliary or fecal route of elimination is more than 50% so that ARBs can be used in chronic kidney diseases and hemodialysis patients [14] .

20.5   Calcium-Channel Blockers

Calcium-channel blockers (CCBs) lower BP by blocking the opening of voltage-gated (L-type) calcium channels in cardiac myocytes and vascular smooth muscle cells. This group of drugs includes three classes of drugs: phenylalkylamines (verapamil), benzothiazepines (diltiazem), and 1,4-dihydropyridines (nifedipine-like).

They lower BP by causing peripheral arterial dilation and potency of BP lowering is dihydropyridines, benzothiazepines, phenylalkylamines in order. CCBs are generally well tolerated, do not require monitoring with blood tests, and have proved safe and effective in many large randomized clinical trials (RCTs). CCBs also have antianginal and antiarrhythmic effects and more protective against cerebrovascular accident than other antihypertensive agents. ALLHAT (Antihypertensive Lowering to Prevent Heart Attack Trial) showed that CCBs (represented by amlodipine) prevent coronary events as effectively as diuretics and RAS blockers [15].

Amlodipine, by far the best studied of the dihydropyridine CCBs, has been investigated in multiple RCTs. Advantages of amlodipine include predictable dose-dependent potency, once-daily dosing because of its long half-life, tolerability, and cost. These drugs have some diuretic action because of dilation of the afferent renal arteriole, which may reduce the efficacy of diuretics in combination therapy. ASCOT (Anglo-Scandinavian Cardiovascular Outcomes Trial) [16] and the ACCOMPLISH (Avoiding Cardiovascular Events Through Combination Therapy in Patients Living with Systolic Hypertension) [17] trial indicated that amlodipine plus ACE inhibitor is one of the most effective drug combinations for preventing cardiovascular complications of hypertension [17]. Multiple fixed-dose single-pill combinations of amlodipine with ACE inhibitors or ARBs are available; some have added a thiazide for triple combination therapy.

Dihydropyridine CCBs such as amlodipine are less renoprotective than ACE inhibitors or ARBs in patients with proteinuric chronic kidney disease, but may be useful as adjunctive therapy after initiation of ACE inhibitors or ARBs. Verapamil is weakly antihypertensive and has limited usefulness because of dose-dependent constipation. Diltiazem is intermediate in potency between verapamil and the dihydropyridines and is usually well tolerated.

The principal side effect of the dihydropyridines is dose-dependent ankle edema. The edema can be improved by concomitant therapy with ACE inhibitors or ARBs that causes balanced arterial and venous dilation. Long-acting dihydropyridine CCBs are rarely associated with flushing and headache. All CCBs can cause gingival hyperplasia, that is reversible if detected early. Verapamil and diltiazem can impair cardiac conduction especially in elderly patients who take other agents such as β-blockers.

20.6   Other Pharmacological Classes

20.6.1 Renin Inhibitors

Direct renin inhibitor inhibits plasma renin activity (PRA) by up to 80% in contrast with the other antihypertensive drugs such as diuretics, ACE inhibitors, and ARBs, which increase PRA. The characteristic role of renin inhibitors on PRA to prevent cardiovascular event has not been suggested or proven. The efficacy of direct renin inhibitor was reported in Asian hypertensive patients, in which aliskiren was significantly superior to ramipril in systolic BP lowering efficacy and comparable in diastolic BP reduction [18]. Dry cough was significantly lower than ramipril. Until a large, formal, and prospective randomized study with major endpoints with aliskiren has not been completed, renin inhibitor now is reserved for use as an alternative to ACE inhibitors or ARBs, when these are ineffective or not tolerated.

20.6.2 α-Adrenergic Receptor Blockers

α-adrenergic receptor antagonists (α-blockers), at therapeutic doses, block peripheral α-adrenergic receptors and result in arterial vasodilation. α-blockers attenuate the stimulation of peripheral α-receptors so orthostatic hypotension can occur. α-blockers may also induce compensatory renal sodium retention and volume expansion. Dizziness and headaches are other common side effects of α-blockers. To minimize the risk of first-dose phenomenon such as hypotension or syncope, an initial low dose prescription followed by dose escalation is necessary. First-dose phenomenon is less common in longer half-life agents (terazosin or doxazosin) but it occurs with all α-blockers, especially in volume contracted status. α-blockers are not recommended as a first-line antihypertensive treatment, but for patients with benign prostatic hyperplasia, α-blockers may provide symptomatic relief.

20.6.3 Centrally Acting Agents

The classic centrally acting antihypertensives such as clonidine, guanfacine, and α-methyldopa (via its active metabolite α-methyl-noradrenaline) induce peripheral sympathoinhibition and a fall in BP as a result of α2-adrenoceptor stimulation in the brain stem.

Central sympatholytics stimulate postsynaptic α2-adrenergic receptors and imidazoline receptors in the central nervous system that lowers central sympathetic outflow. Whereas stimulation of presynaptic α2-receptors causes feedback inhibition of norepinephrine release from peripheral sympathetic nerve terminals. These combined actions reduce adrenergic drive to the heart and peripheral circulation and decrease peripheral resistance.

This class of drugs is used for short-term oral treatment of hypertensive urgency when β-blockers are contraindicated and add-on therapy for very difficult hypertension. To avoid rebound hypertension between doses, short-acting clonidine must be given every 6 to 8 h or, whenever possible, discontinued through gradual tapering [19]. Rebound hypertension is less of a problem with guanfacine, a longer-acting oral central sympatholytic that is dosed at bedtime. The transdermal clonidine causes frequent dermatitis. α-methyldopa is poorly tolerated and no longer a first-line therapy for hypertension in pregnancy.

20.6.4 Direct-Acting Vasodilators

20.6.4.1 Minoxidil

Minoxidil is a potent vasodilator and its antihypertensive activity of minoxidil is mediated by its sulfate metabolite, minoxidil sulfate. About 90% of the administered drug is metabolized by hepatic conjugation with glucuronic acid. The mechanism of antihypertensive action is opening of adenosine triphosphate-sensitive potassium channels in vascular smooth muscle cells resulting in vasodilation. Minoxidil is thought to promote survival by increasing the ratio of Bcl-2/Bax, i.e., by anti-apoptotic effects on dermal papilla cells. This mechanism explains frequent side effects of hypertrichosis which is reversible within a few weeks after it is stopped.

In general, minoxidil as a direct vasodilator, it is reserved until other antihypertensive drugs cannot achieve target goal. Because half-life of minoxidil is about 4 h, BP variability may be increased. Because the duration and efficacy of BP reduction are highly variable among patients. Starting dose could be as low as 2.5 mg/day with a maintenance dose of 10–40 mg/day. And it can be given from one to three times daily according to the individual responses. Twice-daily or thrice-daily dosing regimen is preferred in patients receiving high doses of minoxidil to avoid an excessive peak hypotensive effect. BP rebound following sudden withdrawal should be cautioned.

Due to strong vasodilatory action, reflex tachycardia or sympathetic activation and salt retention are usually expected. Combination with β-blockers and/or diuretics will be helpful to avoid side effect. Minoxidil associated ST-T wave changes in electrocardiogram can be observed and it will often resolve during long-term treatment. But it will be associated with left ventricular hypertrophy or myocardial ischemia. For the prevention of left ventricular hypertrophy, combined ACE inhibitor or ARB could be useful.

20.6.4.2 Hydralazine

The potent hyperpolarizing arterial vasodilators hydralazine act by opening vascular ATP-sensitive potassium channels. Selective and rapid arterial dilation would induce profound reflex sympathetic activation and tachycardia, RAS activation, and sodium retention.

Hydralazine is useful for the treatment of preeclampsia and as rescue therapy for very difficult hypertension. A combination of hydralazine plus nitrates is useful for the treatment of heart failure.

Usually, most effective when added to a combination that includes a diuretic and β-blocker. Hydralazine is a potent direct vasodilator that has antioxidant and nitric oxide-enhancing actions. Hydralazine may induce a lupus-like syndrome. Intravenous nitroprusside can be used to treat malignant hypertension and life-threatening left ventricular heart failure associated with elevated arterial pressure.

20.7 New Class of Antihypertensive Drugs

20.7.1 Aminopeptidase A (APA) Inhibitor

20.7.1.1 Brain RAS
The brain RAS plays an important role in the control of cardiovascular function and BP regulation [20, 21]. All members of the systemic RAS, including the precursor angiotensinogen, enzymes (renin, ACE, ACE2, APA, aminopeptidase N), peptides (angiotensin I, II, III, IV, and angiotensin 1–7), and angiotensin receptors (AT$_1$R, AT$_2$R, Mas receptor) are present in the brain (Fig. 20.1) [22, 23].

Hyperactivity of the brain RAS plays an important role in pathophysiology of hypertension, and its interruption is associated with a beneficial outcome in hypertension [20, 21] and its components could thus constitute interesting targets for treatment of hypertension.

20.7.1.2 APA and Aminopeptidase N (APN) Inhibitors
Hydrolysis of Ang II and Ang III includes 2 membrane-bound zinc-metallopeptidases, APA and APN [24, 25]. Brain angiotensin III plays a role by a tonic stimulatory control over BP in hypertensive rats. Targeting angiotensin III by inhibiting brain APA is an important target in the management of hypertension. Because APA

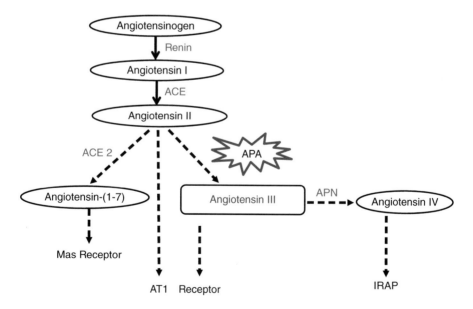

Fig. 20.1 Schematic diagram of the brain renin-angiotensin system. *ACE* angiotensin I-converting enzyme, *ACE2* angiotensin-converting enzyme type 2, *APA* aminopeptidase A, *APN* aminopeptidase N, *AT1* angiotensin type 1 receptor, *IRAP* insulin-regulated aminopeptidase. (Adapted from Hypertension 2020;75:6–15)

inhibitor EC33 does not cross the blood–brain barrier after systemic administration it has limitation in clinical use. RB150, a prodrug of EC33, was developed to be an effective oral agent [26, 27]. Orally administered RB150 crosses the gastrointestinal and blood–brain barriers and generates 2 active molecules of EC33 that block brain APA activity. This results in decreased brain angiotensin III formation and reduced BP. The mechanism of RB150-induced BP lowering effect is from a reduced vasopressin release, increases in diuresis, and reducing extracellular volume. Also, a decrease in sympathetic tone leads to a reduction of vascular resistances and the improvement of the baroreflex function. RB150 renamed as firibastat, and phase Ia/Ib clinical trials showed that firibastat is clinically well tolerated in healthy volunteers. Clinical efficacy of firibastat in hypertensive patients was demonstrated in two phase II studies. Therefore, firibastat could represent the first drug of a novel class of antihypertensive drugs targeting the brain RAS and has the potential to be groundbreaking in the management of resistant hypertension (Table 20.1).

Table 20.1 Clinical pharmacology of antihypertensive drugs

Class	Mechanism	Drug	Usual dose range in mg/day	Usual daily frequency
β-Adrenergic receptor blockers (β-blockers)				
Cardioselective (β1 selective)	β-Blockers reduce cardiac output and heart rate. They also block renin release, decrease adrenergic central nervous system effects, and reduce catecholamine release/response	Atenolol	25–100	1 or 2
		Betaxolol	5–20	1
		Bisoprolol	2.5–20	1
		Metoprolol tartrate	100–450	2
		Metoprolol succinate extended release	25–400	1
Cardio-nonselective		Nebivolol	5–40	1
		Nadolol	20–320	1
		Propranolol	40–640	2
		Propranolol, long-acting	60–640	1
		Timolol	20–60	2
Intrinsic sympathomimetic activity		Acebutolol	400–1200	2
		Penbutolol	20–80	1
		Pindolol	10–60	2
Mixed α/β blockers		Carvedilol	12.5–50	2
		Carvedilol phosphate	20–80	1
		Labetalol	200–2400	2
Diuretics				
Thiazide diuretics	Initially they cause natriuresis resulting in decreased blood volume and cardiac output. Long-term persistent effects result in decreased peripheral vascular resistance	Chlorthalidone	12.5–50	1
		Hydrochlorothiazide	12.5–50	1
		Indapamide	1.25–5	1
Loop diuretics		Furosemide	20–600	2
		Torsemide	5–10	1
Potassium-sparing diuretics		Amiloride	5–20	1
		Triamterene	37.5–75	1

Table 20.1   (continued)

Class	Mechanism	Drug	Usual dose range in mg/day	Usual daily frequency
Aldosterone receptor blockers	Blockade of aldosterone receptor results in decreased vasoconstriction and decreased sodium/water retention	Eplerenone	50–100	1 or 2
		Spironolactone	25–50	1 or 2
Angiotensin converting enzyme (ACE) inhibitors	ACE inhibition results in decreased angiotensin II production which causes decreased vasoconstriction, decreased aldosterone secretion, and sodium and water retention. ACE inhibitors also result in decreased breakdown of bradykinin and other vasoactive peptides which cause vasodilation and allergic responses	Captopril	75–450	2 or 3
		Enalapril	5–40	1 or 2
		Lisinopril	10–80	1
		Perindopril	4–16	1 or 2
		Quinapril	10–80	1 or 2
		Cilazapril	0.25–5	1 or 2
		Moexipril	7.5–30	1
		Ramipril	2.5–20	1 or 2
		Trandolapril	1–8	1 or 2
		Benazepril	10–80	1 or 2
		Fosinopril	10–80	1 or 2
Angiotensin II receptor blockers (ARBs)	Angiotensin II type I receptor blockade results in decreased angiotensin II effects which causes decreased vasoconstriction, decreased aldosterone secretion, and sodium/ water retention	Losartan	50–100	1 or 2
		Valsartan	80–320	1
		Irbesartan	150–300	1
		Telmisartan	40–80	1
		Candesartan	16–32	1
		Eprosartan	600–800	1 or 2
		Olmesartan	20–40	1
		Azilsartan	80	1
		Fimasartan	30–120	1
Dihydropyridine calcium-channel blockers (CCBs)	Blocking the opening of voltage-gated (L-type) calcium channels in cardiac myocytes and vascular smooth muscle cells results in reduced total peripheral resistance through arterial vasodilation	Amlodipine	2.5–10	1
		Felodipine	2.5–10	1
		Isradipine SR	5–20	1
		Nicardipine, sustained-release	60–120	2
		Nifedipine, long-acting	30–120	1
		Nilsoldipine	17–34	1
Nondihydropyridine calcium-channel blockers (CCBs)	Nondihydropyridine CCBs reduce the cellular calcium entry through the L-type channel which results in reduced total peripheral resistance through arterial vasodilation. They decrease myocardial contractility and heart rate	Diltiazem, sustained-release and extended release	120–540	1
		Verapamil, sustained-release	120–480	1 or 2
		Verapamil, controlled-onset, extended-release	180–480	1
		Verapamil, sustained-release, slow-onset	100–400	1

(continued)

Table 20.1 (continued)

Class	Mechanism	Drug	Usual dose range in mg/day	Usual daily frequency
Renin inhibitors	Renin is an enzyme that converts angiotensinogen to angiotensin I, which is then converted to angiotensin II by ACE. Renin inhibitors block the activity of renin and cause vasodilatation	Aliskiren	150–300	1
α-Adrenergic receptor blockers (α-blockers)	Blockade of the peripheral α-adrenergic receptors results in arterial vasodilation. α-blockers attenuate the stimulation of peripheral α-receptors which may develop orthostatic hypotension. They also induce compensatory renal sodium retention and volume expansion	Doxazosin Prazosin Terazosin	1–16 2–20 1–20	1 2 or 3 1 or 2
Centrally acting agents	α2-adrenoceptor stimulation in the brain stem results in inhibition of peripheral sympathetic nerve system and causes a fall in BP Central sympatholytics stimulate postsynaptic α2-adrenergic receptors and imidazoline receptors in the central nervous system that lowers central sympathetic outflow. Whereas stimulation of presynaptic α2-receptors causes feedback inhibition of norepinephrine release from peripheral sympathetic nerve terminals. These combined actions reduce adrenergic drive to the heart and peripheral circulation and decrease peripheral resistance	Clonidine Clonidine patch Methyldopa Guanfacine	0.2–2.4 0.1–0.6 500–3000 0.5–2	2 Once weekly 2–4 1

Table 20.1 (continued)

Class	Mechanism	Drug	Usual dose range in mg/day	Usual daily frequency
Direct-acting vasodilators	Potent direct vasodilators lower BP by exerting a peripheral vasodilating effect through direct relaxation of vascular smooth muscle	Minoxidil	2.5–100	1 or 2
		Hydralazine	20–300	2–4
New class of antihypertensive drugs				
Aminopeptidase A inhibitors (APA inhibitors)	Brain angiotensin III plays a role by a tonic stimulatory control over BP. Blocking angiotensin III by inhibiting brain APA may decrease BP	Firibastat	Phase II clinical study ongoing	

References

1. Reiter MJ. Cardiovascular drug class specificity: beta-blockers. Prog Cardiovasc Dis. 2004;47:11–33.
2. Fitzgerald JD. Do partial agonist beta-blockers have improved clinical utility? Cardiovasc Drugs Ther. 1993;7:303–10.
3. Bakris GL, Fonseca V, Katholi RE, McGill JB, Messerli FH, Phillips RA, et al. Metabolic effects of carvedilol vs metoprolol in patients with type 2 diabetes mellitus and hypertension: a randomized controlled trial. JAMA. 2004;292:2227–36.
4. Chobanian AV, Bakris GL, Black HR, Cushman WC, Green LA, Izzo JL, et al. Seventh report of the joint National Committee on prevention, detection, evaluation, and treatment of high blood pressure. Hypertension. 2003;42:1206–52.
5. Dahlof B, Lindholm LH, Hansson L, Schersten S, Ekbom T, Wester PO. Morbidity and mortality in the Swedish trial in old patients with hypertension (STOP-hypertension). Lancet. 1991;338:1281–5.
6. SHEP Cooperative Research Group. Prevention of stroke by antihypertensive drug treatment in older persons with isolated systolic hypertension: final results of the systolic hypertension in the elderly program (SHEP). JAMA. 1991;265:3255–64.
7. Ernst ME, Moser M. Use of diuretics in patients with hypertension. N Engl J Med. 2009;361:2153–64.
8. Ernst ME, Carter BL, Goerdt CJ, Steffensmeier JJG, Phillips BB, Zimmerman MB, et al. Comparative antihypertensive effects of hydrochlorothiazide and chlorthalidone on ambulatory and office blood pressure. Hypertension. 2006;47:352–8.
9. Elliott WJ, Meyer PM. Incident diabetes in clinical trials of antihypertensive drugs: a network meta-analysis. Lancet. 2007;369:201–7.
10. Black HR, Davis B, Barzilay J, Nwachuku C, Baimbridge C, Marginean H, et al. Metabolic and clinical outcomes in nondiabetic individuals with the metabolic syndrome assigned to chlorthalidone, amlodipine, or lisinopril as initial treatment for hypertension: a report from the antihypertensive and lipid-lowering treatment to prevent heart attack trial (ALLHAT). Diabetes Care. 2008;31:353–60.
11. Musini VM, Rezapour P, Wright JM, Bassett K, Jauca CD. Blood pressure-lowering efficacy of loop diuretics for primary hypertension. Cochrane Database Syst Rev. 2015;5:CD003825.

12. Charny JW, Choi JK, James WD. Spironolactone for the treatment of acne in women, a retrospective study of 110 patients. Int J Womens Dermatol. 2017;3:111–5.
13. Woo KS, Nicholls MG. High prevalence of persistent cough with angiotensin converting enzyme inhibitors in Chinese. Br J Clin Pharmacol. 1995;40:141–4.
14. Abraham HM, White CM, White WB. The comparative efficacy and safety of the angiotensin receptor blockers in the management of hypertension and other cardiovascular diseases. Drug Saf. 2015;38:33–54.
15. ALLHAT Officers and Coordinators. Major outcomes in high-risk hypertensive patients randomized to angiotensin-converting enzyme inhibitor or calcium channel blocker vs diuretic: the antihypertensive and lipid-lowering treatment to prevent heart attack trial (ALLHAT). JAMA. 2002;288:2981–97.
16. Dahlof B, Sever PS, Poulter NR, Wedel H, Beevers DG, Caulfield M, et. al. Prevention of cardiovascular events with an antihypertensive regimen of amlodipine adding perindopril as required versus atenolol adding bendroflumethiazide as required, in the Anglo-Scandinavian cardiac outcomes trial-blood pressure lowering arm (ASCOT-BPLA): a multicentre randomised controlled trial. Lancet. 2005;366:895–906.
17. Jamerson K, Weber MA, Bakris GL, Dahlöf B, Pitt B, Shi V, et. al. Benazepril plus amlodipine or hydrochlorothiazide for hypertension in high-risk patients. N Engl J Med. 2008;359:2417–28.
18. Zhu JR, Sun NL, Yang K, Hu J, Xu G, Hong H, et al. Efficacy and safety of aliskiren, a direct renin inhibitor, compared with ramipril in Asian patients with mild to moderate hypertension. Hypertens Res. 2012;35:28–33.
19. Vongpatanasin W, Kario K, Atlas SA, Victor RG. Central sympatholytic drugs. J Clin Hypertens. 2011;13:658–61.
20. Veerasingham SJ, Raizada MK. Brain renin-angiotensin system dysfunction in hypertension: recent advances and perspectives. Br J Pharmacol. 2003;139:191–202.
21. Sakai K, Sigmund CD. Molecular evidence of tissue renin-angiotensin systems: a focus on the brain. Curr Hypertens Rep. 2005;7:135–40.
22. Lenkei Z, Palkovits M, Corvol P, Llorens-Cortès C. Expression of angiotensin type-1 (AT1) and type-2 (AT2) receptor mRNAs in the adult rat brain: a functional neuroanatomical review. Front Neuroendocrinol. 1997;18:383–439.
23. Santos RAS, Sampaio WO, Alzamora AC, Motta-Santos D, Alenina N, Bader M, et al. The ACE2/angiotensin-(1-7)/MAS axis of the renin-angiotensin system: focus on angiotensin-(1-7). Physiol Rev. 2018;98:505–53.
24. Malfroy B, Kado-Fong H, Gros C, Giros B, Schwartz JC, Hellmiss R. Molecular cloning and amino acid sequence of rat kidney aminopeptidase M: a member of a super family of zinc-metallohydrolases. Biochem Biophys Res Commun. 1989;161:236–41.
25. Wu Q, Lahti JM, Air GM, Burrows PD, Cooper MD. Molecular cloning of the murine BP-1/6C3 antigen: a member of the zinc-dependent metallopeptidase family. Proc Natl Acad Sci U S A. 1990;87:993–7.
26. Fournie-Zaluski MC, Fassot C, Valentin B, Djordjijevic D, Reaux-Le Goazigo A, Corvol P, et al. Brain renin-angiotensin system blockade by systemically active aminopeptidase a inhibitors: a potential treatment of salt-dependent hypertension. Proc Natl Acad Sci U S A. 2004;101:7775–80.
27. Marc Y, Gao J, Balavoine F, Michaud A, Roques BP, Llorens-Cortes C. Central antihypertensive effects of orally active aminopeptidase a inhibitors in spontaneously hypertensive rats. Hypertension. 2012;60:411–8.

Diuretics for Hypertension in Asians

21

Ashok L. Kirpalani and Dilip A. Kirpalani

21.1   Introduction

The number of deaths due to cardiovascular or cerebrovascular disease is decreasing in high-income countries [1]. However, this trend is not evident in the lower middle-income countries of Asia [2], one of the most densely populated regions in the world. Of the 17.5 million deaths annually due to cardiovascular disease (CVD), more than 75% occur in lower middle-income countries. South Asians have been shown to experience their first myocardial infarction (MI) almost 10 years earlier than individuals from Western countries. In 1990, ischaemic heart disease was the sixth leading cause of death in India, but it is now the leading cause of mortality [3].

In India, the prevalence of hypertension is 33.8% in urban areas and 27.6% in rural areas and is much higher (64%) in individuals aged >60 years [3]. Although these figures may be overestimates, there is no doubt that the prevalence of hypertension is increasing. It is difficult to estimate what proportion of patients with hypertension are receiving antihypertensive drug therapy, and what proportion of those being treated achieve target BP values.

Almost half the world's population lives in India or China. Looking at data from China, the prevalence of hypertension was 18.8% in 2002 and only about 25% of patients with hypertension were receiving antihypertensive treatment. By 2015, the prevalence of hypertension was 30% in men and 25% in women [4]. In addition, the population of elderly patients (age > 60 years) with hypertension increased by more than ten million each year between 2010 and 2015, meaning that the number of individuals with hypertension in China in 2015 was more than 200 million [4]. This number is expected to double to 400 million by 2035 [5]. At present, more than 50% of people in China aged >60 years have hypertension. The low rates of hypertension

A. L. Kirpalani (✉) · D. A. Kirpalani
Department of Nephrology, Bombay Hospital Institute of Medical Sciences,
Mumbai, Maharashtra, India

© The Author(s), under exclusive license to Springer Nature Switzerland AG 2022   299
C. V. S. Ram et al. (eds.), *Hypertension and Cardiovascular Disease in Asia*,
Updates in Hypertension and Cardiovascular Protection,
https://doi.org/10.1007/978-3-030-95734-6_21

treatment and target blood pressure (BP) achievement in China mean that hypertension is a serious health issue.

Data from Korea indicate a stable hypertension prevalence in individuals aged ≥30 years of 29–30% [6]. According to Korea National Health and Nutrition Examination Survey (KNHANES data), hypertension awareness, treatment and control rates are generally improving [7]. In 2001, the number of treated patients with hypertension was 22% for men and 37% for women; by 2011, this had increased to 51.7% in men and 71% in women. In 2001, the number of patients achieving target BP was quite low (10% in men and 18% in women), and this had increased by 2011 (to 36.9% in men and 49.4% in women) [6]. Over this period there was a steady decrease in the mean BP in patients with hypertension. Similar prevalence data and evidence of poor BP control are available from Japan [8] and Saudi Arabia [9].

Hypertension is therefore a major source of chronic morbidity and mortality in Asian countries and regions. The prevalence of hypertension is increasing at an alarming rate because of population growth and various other factors. Good management of hypertension in individual patients using lifestyle modifications and drug therapy has been very successful in reducing the morbidity and mortality of this disease [3, 10]. Nevertheless, fewer than one-quarter of patients with hypertension from countries delivering the best healthcare to these individuals are at target BP.

This chapter provides information on diuretic agents for the treatment of hypertension and discusses the use of these agents in Asian countries.

21.2   Diuretic Agents: Use in Hypertension

Loop diuretics: These agents are mostly used for the treatment of heart failure and chronic kidney disease (CKD). Their use in the management of hypertension is limited to patients who also have CKD stage III, IV and V in association with concomitant fluid volume overload [11, 12].

Thiazide-type diuretics: These act at the distal convoluted tubular level and prevent sodium reabsorption, producing natriuresis and diuresis [13]. In India, the most popular agent in this group is hydrochlorothiazide (HCTZ). This is used at doses of 12.5, 25 and 50 mg. HCTZ is cost effective and is therefore used in most antihypertensive combinations, such as with angiotensin receptor blockers (ARBs) and other agents.

Both HCTZ and similar thiazide-like diuretics have a short duration of action (≈6 h). When using thiazide-like diuretics, the initial fall in BP associated with a reduction in intravascular volume wears off in 6 h and is followed by rebound fluid retention. When taken alone in the morning, HCTZ may result in a masked nocturnal rise of BP (masked uncontrolled hypertension; MUCH). This phenomenon would only be detected by ambulatory blood pressure monitoring (ABPM). Using thiazide-like diuretics can overcome this problem because these agents have a longer duration of action (24 h). However, statistics obtained from the Indian pharmaceutical industry reveal that HCTZ continues to be the most prescribed diuretic for hypertension in that country.

Thiazide-like diuretics: This group of agents includes chlorthalidone, indapamide and metolazone. Chlorthalidone was introduced in India in the third quarter of the twentieth century and the dosage recommended was 50 to 100 mg/day. However, it was so potent that many patients became severely hypovolaemic and had hypotension when treated with these dosages. Thereafter, it disappeared from the market and reappeared at the start of the new millennium, with recommended dosages of 6.25, 12.5 and 25 mg/day, based on data from the Western literature. Recommendations based on research conducted in Western populations are mostly adopted in Asia, meaning that chlorthalidone has been gaining popularity in Asia. In India, usage of chlorthalidone has increased over the last decade, but it remains in second place behind HCTZ in terms of popularity. Nevertheless, chlorthalidone has a 24-h duration of action and is twice as potent as HCTZ.

In Asia, as in other countries, both thiazide-type and thiazide-like diuretics have an important role to play in antihypertensive therapy when given in combination with RAS inhibitors. RAS inhibitors have a dual action (BP lowering and anti-proteinuric) due to their pleotropic properties. However, the BP-lowering effect is often blunted in patients with a high salt intake [14, 15]. Both thiazide-type and thiazide-like diuretics have additive BP-lowering effects when given with RAS inhibitors in CKD stages 1 to 4. Their mechanism of action is volume depletion, and they are also useful because they counteract the hyperkalaemic effects of RAS inhibitors. In India, most nephrologists give diuretics and RAS inhibitors as individual pills because this allows one component of the treatment regimen to be easily withdrawn or modified. However, these treatments are of little use once the glomerular filtration rate (GFR) falls below 30 mL/min (i.e., CKD stage IV and V). In contrast, physicians and diabetologist tend to use single pill combinations which contain both agents in a single formulation.

Metolazone was developed by an Indian scientist, Dr. Bola Vithal Shetty, in the USA in the 1970s, but reached India only about 20 years ago. Its action is very similar to other thiazide-like diuretics, and metolazone is most useful in CKD stage IV and V. Unlike other agents in these classes, metolazone remains effective even when GFR decreases. Hence it does not require dosage adjustment in low GFR states. Thiazide-type and thiazide-like diuretics at the apical membrane of the distal convoluted tubule (DCT) cells must first be filtered into the tubules via the glomerulus by glomerular filtration. Metolazone apparently acts directly on the DCT via the basolateral membrane of DCT cells. When metolazone is given 30 min prior to a loop diuretic (e.g., furosemide or torsemide) it blocks reabsorption of sodium in the DCT in advance. The increased amount of sodium leaving the loop of Henle due to the loop diuretic and arriving at the DCT therefore fails to get reabsorbed. Thus, in CKD, metolazone helps to enhance natriuresis and diuresis, decreasing intravascular volume and reducing BP. Metolazone is gaining popularity amongst nephrologists and cardiologists in India.

Mineralocorticoid receptor antagonists (MRAs): These are primarily used as the fourth drug for add-on therapy in patients with resistant hypertension [16]. It is important to monitor serum potassium levels in patients taking MRAs, especially when these agents are used in combination with RAS inhibitors.

Most guidelines define resistant hypertension as a failure to reach target BP during treatment with three antihypertensive agents, but only if one of these three drugs is a diuretic used at an appropriate dosage [10, 17]. In India, most patients labelled as having resistant hypertension do not actually meet the definition because a diuretic was either not used or used at an inappropriate dosage [18].

The PATHWAY-2 study compared spironolactone with doxazosin (an alpha-blocker) to try and determine the optimal treatment for drug-resistant hypertension [16]. The results showed that the alpha-blocker was less effective than spironolactone. As a result, worldwide usage of MRAs for the management of resistant hypertension has increased significantly [19]. However, in India, the most commonly used add-on therapy for resistant hypertension is either an alpha-blocker (prazosin) or a centrally acting agent such as clonidine or moxonidine [20]. There is reluctance to use spironolactone and a preference for alpha-blockers or other agents in this setting due to concerns that spironolactone may cause hyperkalaemia. In addition, alpha-blockers have the advantage of relieving urinary outlet dysfunction, particularly in males.

Using spironolactone or other MRAs as add-on therapy in patients with resistant hypertension required regular monitoring of serum potassium levels. The risk of hyperkalaemia during MRA therapy is higher in older patients, patients with diabetes, patients receiving RAS inhibitors, and those with borderline serum creatinine values indicative of early CKD stage III particularly in old age [21].

Difficulties in diagnosing and counteracting hyperkalaemia during MRA therapy are also a deterrent to their usage. Acute life-threatening hyperkalaemia would require expensive hospital, and potentially intensive care unit, admission. Most patients in India pay for their own medical care and medical insurance is only available to a small proportion of individuals. The only treatment available for chronic hyperkalaemia in most of Asia is Calcium Resonium, which cannot be used for more than a few days at a time due to unacceptable side effects.

Other potassium-sparing diuretics: Amiloride and triamterene were initially popular, mostly in combination with loop diuretics to protect against hypokalaemia, which was dangerous for patients with heart failure being treated with digitalis; however, usage of these diuretics largely stopped when digitalis treatment of heart failure became less popular.

Other agents with diuretic effects: Drugs that reduce BP by diuresis also include the recently developed sodium-glucose cotransporter 2 (SGLT2) inhibitors and tolvaptan. SGLT2 inhibitors produce diuresis and natriuresis, and trials have shown a 3- to 5-mm reduction in SBP during treatment with these agents [22]. Tolvaptan is an aquaretic that does not have a marked effect on BP [23].

21.3 Special Features in Asians

There are a number of important differences between patients with hypertension from Asia and those from Western populations [24].

Genetic factors: Some Asian populations have a higher prevalence of metabolic syndrome and diabetes [3], meaning that they are at greater risk of developing atherosclerotic cardiovascular disease (ASCVD). Individuals from South Asia have a thrifty gene, which is related to central obesity and insulin resistance [3]. The genetic risk of hypertension and target organ damage is variable in different parts of Asia [3, 25, 26].

Salt sensitivity: Salt sensitivity increases with age and is present in many Asian populations [27]. Salt sensitivity can be present in patients with normal BP as well as those with hypertension. The "Gensalt" study [28] from China estimates that about 40% of Chinese adults are salt sensitive. In a north Indian community, a salt-sensitive phenotype was found in 40.8% of normotensive individuals and 47.6% of those with hypertension, and consumption of extra salt in the diet was independently associated with enhanced salt sensitivity [29]. Acquired salt sensitivity is also seen in CKD, obesity and diabetes. When the human body is deprived of salt in the diet, it responds by activating the renin–angiotensin–aldosterone system (RAAS) to prevent a significant reduction in BP [27]. In contrast, salt-sensitive individuals exhibit a blunted response of the RAAS meaning that BP reduces to a greater extent than in non-salt sensitive individuals. BP reductions in salt-sensitive patients with hypertension have a smaller reduction in BP during treatment with RAS inhibitors compared with that during treatment with diuretics and CCBs. It has been suggested that salt-sensitive patients with hypertension who have a low renin level would be best managed with a diuretic [30].

Salt intake: Diets in most Asian countries contain a higher intake of salt compared with Western diets [31]. Dietary salt intake comes from three main sources: as part of the food consumed (e.g., processed meat and other foods, seafood, etc.); salt added during the cooking process; and table salt sprinkled on food after serving. The World Health Organization (WHO) recommends a salt intake of 5 g/day as ideal. However, recent data show that the average Indian consumes 11–15 g/day of salt, [32] while average salt intake in China is 13 g/day [33]. Within each of these two large nations, there are major regional differences in the dietary sources of salt and in the magnitude of salt intake. Mean salt consumption in the Japanese population varies between 8.3 and 23.3 g/day [34]. In this group, those with salt intake in the lowest quartile (mean 8.3 g/day) had a mean BP that was 4.5 mmHg lower than that in individuals with salt intake in the highest quartile (23 g/day) [34]. The high proportion of salt-sensitive individuals and the higher salt intake in Asian populations means that diuretics would be very effective in the management of hypertension in Asia.

Environmental factors: The climate in Asian countries has a much greater influence on health outcomes and the effects of antihypertensive therapy than in the Western world [35]. Western populations are mostly protected from extremes by artificial environments created by air conditioning, home heating, etc. For example, in the city of Chandigarh, Punjab India, the temperature may exceed 40–45 °C in summer, but be in single digits during winter. Under these circumstances, average rural workers in fields and mines are required to consume much more water than office workers in the city [36]. If these individuals are taking diuretics, they are at much higher risk of sodium and water imbalance, resulting in hypo- or hypernatre-mia and prerenal azotaemia. In extreme circumstances, these may progress to acute kidney injury, heatstroke and heat stress renal failure. Therefore, patients using diuretics are very vulnerable and need extra care and supervision.

Socioeconomic factors: Phillip et al. [37] compared treatment initiation (step 1) and modification (steps 2 and 3) for hypertension between clinical practice guide-lines in high-income countries (HIC), upper middle-income countries (UMIC), lower middle-income countries (LMIC) and low-income countries (LIC). In LICs, 57% of guidelines recommended starting with diuretic monotherapy [37]. In the latter half of the twentieth century, many countries like India, which is now classi-fied as a LMIC, followed a similar pattern. In 30% of LMICs, current recommenda-tions now suggest starting antihypertensive therapy with a combination of a CCB and a diuretic. Most Asian guidelines now recommend starting treatment with a RAS inhibitor, CCB or diuretic for patients with Grade I hypertension (SBP 140–159 mmHg) and a combination of any two of these agents in patients with Grade II hypertension (SBP >160 mmHg) [3, 5, 25].

Physician knowledge and choices: The choice of antihypertensive drug(s) also depends on physician preference and drug availability and cost. In India, pharma-ceutical industry sales data suggest that the current antihypertensive usage pattern is as follows: beta-blockers in 28% of patients, ARBs in 28%, CCBs in 25%, diuretics in 15% and angiotensin-converting enzyme inhibitors (ACEIs) in 5%. Thus, surpris-ingly, beta-blockers remain one of the most commonly used drug classes in India, despite guideline recommendations that state otherwise. However, it is difficult to determine the proportion of beta-blocker and RAS inhibitor usage that relates solely to the treatment of hypertension because these agents can be used for other cardiac conditions. In addition, diuretic data reflects the usage of these agents in cardiac and renal disease as well as hypertension. Of available diuretics, only thiazide-type and thiazide-like agents are almost exclusively used for hypertension.

In India, 70% of the total diuretic usage is as part of combination therapy and only 30% relates to diuretic monotherapy. With respect to combination therapy con-taining diuretics, the most popular is HCTZ + ARB, followed by (in decreasing usage) chlorthalidone + ARB, HCTZ + ACEI, HCTZ + CCB, HCTZ + beta-blocker, chlorthalidone + beta-blocker, chlorthalidone + CCB, indapamide + CCB, indap-amide + beta-blocker and indapamide + ARB.

Drug trials in China suggest that best antihypertensive results are obtained when using dihydropyridine CCBs [5]. Therefore, these are the most used antihyperten-sive drugs. The guidelines suggest that the poor control of high BP is due to inade-quate drug dosing and insufficient use of RASIs and diuretics.

21.4 Antihypertensive Agents Used in Asia

Physicians in Asia have a number of antihypertensive therapy options to choose from: RAS inhibitors (ACEIs and ARBs), CCBs, diuretics, beta-blockers, alpha-blockers, drugs acting on the central sympathetic nervous system, and (in some settings) MRAs. Many of these agents are not only used for hypertension but also for other indications. For example, ACEIs and ARBs are also used in cardiac diseases and CKD. CCBs are currently used primarily for hypertension because they are not a treatment of choice in patients of ischaemic heart disease (IHD). Diuretics are also used in heart failure, cirrhosis of liver and some other fluid-retaining states. In addition to hypertension, beta-blockers are used in IHD, heart failure, anxiety states, portal hypertension, liver cirrhosis and migraine, and alpha-blockers are also used in bladder outlet disease.

21.5 Use of Diuretics in Asian Countries

Diuretics have shown to reduce the incidence of stroke and CVD [38]. The effectiveness of antihypertensive therapy diuretics is likely to be higher in Asians compared with other populations for a number of reasons. These include the fact that Asians are genetically prone to higher salt sensitivity, Asians have a higher dietary salt intake making them even more salt sensitive and diuretic sensitive, a large proportion of Asians have diabetes mellitus and/or metabolic syndrome (both of which contribute to salt sensitivity), and people become more salt sensitive as they age [39]. In addition, the lower cost of diuretic therapy compared with other antihypertensive drugs is economically beneficial in many settings.

Data from China and Japan suggest that most antihypertensive prescriptions in these countries are at dosages much lower than those recommended by guidelines and the product prescribing information [5, 26]. This is also the case in India.

Diuretics are most often used in combination with other antihypertensives, and usually in single pill combinations with other agents. For example, in Japan, diuretic monotherapy only made up 5% of prescriptions, while 95% of diuretics were prescribed with other agents. There has been a recent increase in single pill diuretic use for high BP from 4.3% to 9.3%. The J-HOME study [40] showed that CCBs were the most commonly prescribed antihypertensive agents in Japan, while diuretic monotherapy was uncommon. The most frequently prescribed combinations in Japan are CCB + diuretics (60%) or RAS inhibitor + diuretics [40]. In China, there appears to be a preference for initiating antihypertensive therapy at low dosages, using long-acting drugs, using combination preparations rather than single-agent preparations, and individualised prescription.

Why still diuretics are underutilised in Asia: Overall, diuretics are underutilised as first-line antihypertensive therapy in Asia. There are a number of potential reasons for this: Firstly, there may be concern about the metabolic adverse effects of diuretics, including glucose intolerance, hypokalaemia, hyperlipidaemia/hypertriglyceridaemia, hyperuricaemia and gout, hypomagnesaemia and hyponatraemia, especially in the elderly. Although these adverse events do occur, their incidence and severity are not enough to justify non-use of diuretics, which can be prescribed and used safely with good supervision and monitoring. Secondly, changes in socioeconomic status, the adaptation of Western guidelines, and increasing affordability of RAS inhibitors and CCBs might contribute to lower diuretic usage. Finally, lack of specialised hypertension clinics staffed by qualified doctors may also contribute to diuretic underutilisation.

21.6   Conclusion

Diuretics are one of the three drug classes currently recommended for the first-line treatment of hypertension (along with RAS inhibitors and CCBs). However, diuretics remain underutilised in many Asian countries. In addition, there remains room for improvement in the use of thiazide-type diuretics rather than thiazide-like diuretics. Appropriate usage of diuretics as part of guideline-driven antihypertensive therapy has the potential to contribute to improvements in cardiovascular morbidity and mortality. All stakeholders in hypertension care in Asia, including government health authorities, the pharmaceutical industry, scientific associations of treating physicians and individual practitioners writing prescriptions for patients with hypertension should take note and facilitate better diuretic utilisation.

Acknowledgments My grateful thanks to Mr. Ivor Dsouza for transcription and script reading without whom this would not have been possible.

References

1. O'Flaherty M, Buchan I, Capewell S. Contributions of treatment and lifestyle to declining CVD mortality: why have CVD mortality rates declined so much since the 1960s? Heart. 2013;99(3):159–62.
2. Anand S, Bradshaw C, Prabhakaran D. Prevention and management of CVD in LMICs: why do ethnicity, culture, and context matter? BMC Med. 2020;18(1):7.
3. Shah SN, Munjal YP, Kamath SA, Wander GS, Mehta N, Mukherjee S, et al. Indian guidelines on hypertension-IV (2019). J Hum Hypertens. 2020;34(11):745–58.
4. Wang JG. Chinese hypertension guidelines. Pulse (Basel). 2015;3(1):14–20.
5. Liu J. Highlights of the 2018 Chinese hypertension guidelines. Clin Hypertens. 2020;1(26):8. https://doi.org/10.1186/s40885-020-00141-3.
6. Lee HY, Park JB. The Korean Society of Hypertension Guidelines for the Management of Hypertension in 2013: its essentials and key points. Pulse (Basel). 2015;3(1):21–8.
7. Kim Y. The Korea National Health and nutrition examination survey (KNHANES): current status and challenges. Epidemiol Health. 2014;30(36):e2014002. PMID: 24839580; PMCID: PMC4017741. https://doi.org/10.4178/epih/e2014002.
8. Deedwania PC, et al. Hypertension in south Asians. Chapter B87. In: Hypertension primer. 4th ed.
9. Shnaimer JA, Gosadi IM. Primary health care physicians' knowledge and adherence regarding hypertension management guidelines in southwest of Saudi Arabia. Medicine (Baltimore). 2020;99(17):e19873.
10. Williams B, Mancia G, Spiering W, et al. 2018 ESC/ESH guidelines for the management of arterial hypertension. The task force for the management of arterial hypertension of the European Society of Cardiology (ESC) and the European Society of Hypertension (ESH). Eur Heart J. 2018;39:3021–104.
11. Araoye MA, Chang MY, Khatri IM, Freis ED. Furosemide compared with hydrochlorothiazide. Long-term treatment of hypertension. JAMA. 1978;240:1863.
12. Finnerty FA Jr, Maxwell MH, Lunn J, Moser M. Long-term effects of furosemide and hydrochlorothiazide in patients with essential hypertension a two-year comparison of efficacy and safety. Angiology. 1977;28:125.
13. Ernst ME, Carter BL, Basile JN. All thiazide-like diuretics are not chlorthalidone: putting the ACCOMPLISH study into perspective. J Clin Hypertens (Greenwich). 2009 Jan;11(1):5–10.

14. Judd E, Calhoun DA. Management of hypertension in CKD: beyond the guidelines. Adv Chronic Kidney Dis. 2015;22(2):116–22.
15. Cheung AK, Chang TI, Cushman WC, Furth SL, Hou FF, Ix JH, et al. KDIGO 2021 clinical practice guideline for the management of blood pressure in chronic kidney disease. Kidney Int. 2021;99(3):S1–87.
16. Williams B, MacDonald TM, Morant S, Webb DJ, Sever P, McInnes G, et al. British Hypertension Society's PATHWAY studies group. Spironolactone versus placebo, bisoprolol, and doxazosin to determine the optimal treatment for drug-resistant hypertension (PATHWAY-2): a randomised, double-blind, crossover trial. Lancet. 2015;386(10008):2059–68.
17. Whelton PK, Carey RM, Aronow WS, et al. 2017ACC/AHA/AAPA/ABC/ACPM/AGS/APhA/vASH/ASPC/NMA/PCNA guideline for the prevention, detection, evaluation, and Management of High Blood Pressure in adults. Hypertension. 2018;71:e13–e115.
18. Gupta R, Sharma KK, Soni S, Gupta N, Khedar RS. Resistant hypertension in clinical practice in India: Jaipur heart watch. J Assoc Physicians India. 2019;67(12):14–7.
19. Pitt B. Mineralocorticoid receptor antagonists for the treatment of hypertension and the metabolic syndrome. Hypertension. 2015;65(1):41–2.
20. Tarapdher A. Alpha adrenergic blockers in the treatment of hypertension a nephrologist perspective. JAPI. 2014;62:1–5.
21. Rosano GMC, Tamargo J, Kjeldsen KP, Lainscak M, Agewall S, Anker SD, et al. Expert consensus document on the management of hyperkalaemia in patients with cardiovascular disease treated with renin angiotensin aldosterone system inhibitors: coordinated by the working group on cardiovascular pharmacotherapy of the European Society of Cardiology. Eur Heart J Cardiovasc Pharmacother. 2018;4(3):180–8.
22. Zinman B, Wanner C, Lachin JM, Fitchett D, Bluhmki E, Hantel S, et al. EMPA-REG OUTCOME investigators. Empagliflozin, cardiovascular outcomes, and mortality in type 2 diabetes. N Engl J Med. 2015;373(22):2117–28.
23. Konstam MA, Gheorghiade M, Burnett JC Jr, Grinfeld L, Maggioni AP, Swedberg K, et al. Efficacy of vasopressin antagonism in heart failure Outcome study with Tolvaptan (EVEREST) investigators. Effects of oral tolvaptan in patients hospitalized for worsening heart failure: the EVEREST Outcome trial. JAMA. 2007;297(12):1319–31.
24. Kario K, Chia YC, Sukonthasarn A, Turana Y, Shin J, Chen CH, et al. Diversity of and initiatives for hypertension management in Asia-why we need the HOPE Asia network. J Clin Hypertens (Greenwich). 2020;22(3):331–43.
25. Lee HY, Shin J, Kim GH, Park S, Ihm SH, Kim HC, et al. 2018 Korean Society of Hypertension Guidelines for the management of hypertension: part II-diagnosis and treatment of hypertension. Clin Hypertens. 2019;1(25):20.
26. Umemura S, Arima H, Arima S, Asayama K, Dohi Y, Hirooka Y, et al. The Japanese Society of Hypertension Guidelines for the Management of Hypertension (JSH 2019). Hypertens Res. 2019;42(9):1235–481.
27. de Brito-Ashurst I, Perry L, Sanders TA, Thomas JE, Dobbie H, Varagunam M, Yaqoob MM. The role of salt intake and salt sensitivity in the management of hypertension in south Asian people with chronic kidney disease: a randomised controlled trial. Heart. 2013;99(17):1256–60.
28. GenSalt Collaborative Research Group. GenSalt: rationale, design, methods and baseline characteristics of study participants. J Hum Hypertens. 2007;21(8):639–46.
29. Borah PK, Sharma M, Kalita HC, Pasha MA, Paine SK, Hazarika D, et al. Salt-sensitive phenotypes: a community-based exploratory study from northeastern India. Natl Med J India. 2018;31(3):140–5.
30. Burnier M, Bakris G, Williams B. Redefining diuretics use in hypertension: why select a thiazide-like diuretic? J Hypertens. 2019;37(8):1574–86.
31. Firestone MJ, Beasley JM, Kwon SC, Ahn J, Trinh-Shevrin C, Yi SS. Asian American dietary sources of sodium and salt behaviors compared with other racial/ethnic groups, NHANES, 2011-2012. Ethn Dis. 2017;27(3):241.
32. Johnson C, Santos JA, Sparks E, Raj TS, Mohan S, Garg V, et al. Sources of dietary salt in north and South India estimated from 24 hour dietary recall. Nutrients. 2019;11(2):318.

33. Anderson CA, Appel LJ, Okuda N, Brown IJ, Chan Q, Zhao L, et al. Dietary sources of sodium in China, Japan, the United Kingdom, and the United States, women and men aged 40 to 59 years: the INTERMAP study. J Am Diet Assoc. 2010;110(5):736–45.
34. Miura K, Okuda N, Turin TC, Takashima N, Nakagawa H, Nakamura K, et al. Dietary salt intake and blood pressure in a representative Japanese population: baseline analyses of NIPPON DATA80. J Epidemiol. 2010;20 Suppl 3(Suppl 3):S524–30.
35. de Lorenzo A, Liaño F. High temperatures and nephrology: the climate change problem. Nefrologia. 2017;37(5):492–500.
36. Grandjean A. WHO guidelines for drinking-water quality. Geneva: World Health Organization; 2004.
37. Philip R, Beaney T, Appelbaum N, Gonzalvez CR, Koldeweij C, Golestaneh AK, et al. Variation in hypertension clinical practice guidelines: a global comparison. BMC Med. 2021;19(1):117.
38. ALLHAT Officers and Coordinators for the ALLHAT Collaborative Research Group. The Antihypertensive and Lipid-Lowering Treatment to Prevent Heart Attack Trial. Major outcomes in high-risk hypertensive patients randomized to angiotensin-converting enzyme inhibitor or calcium channel blocker vs diuretic: the antihypertensive and lipid-lowering treatment to prevent heart attack trial (ALLHAT). JAMA. 2002;288(23):2981–97.
39. Mishra S, Ingole S, Jain R. Salt sensitivity and its implication in clinical practice. Indian Heart J. 2018;70(4):556–64.
40. Murai K, Obara T, Ohkubo T, Metoki H, Oikawa T, Inoue R, et al. Current usage of diuretics among hypertensive patients in Japan: the Japan home versus office blood pressure measurement evaluation (J-HOME) study. Hypertens Res. 2006;29(11):857–63.

Beta-Blockers for Hypertension in Asian Population

22

Marie Barrientos-Regala and Joan Dymphna P. Reaño

22.1 Introduction

Worldwide, approximately a quarter of the adult population has hypertension [1]. The trend is increasing among Asian countries, particularly among low-income and middle-income countries. More than half of this rise is attributable to population aging, as well as the adoption of unhealthy lifestyles [2–4]. In 1962, Sir James Black made an important contribution in the field of medicine with the development of the first beta-blocker [5]. However, since the early 1990s, the use of beta-blockers as initial therapy in hypertension management has become somewhat controversial with most of the studies reporting its unfavorable effects in hypertension therapy using mostly the second-generation drug, atenolol [6].

In 2005, a meta-analysis by Lindholm et al. found a 16% increase in relative risk of stroke with beta-blockers compared with other drugs while in 2007, an increase of new-onset diabetes in beta-blocker-treated patients was reported [7, 8]. The large hypertension trials, Losartan Intervention for End Point Reduction in Hypertension (LIFE) and the Anglo-Scandinavian Cardiac Outcomes Trial-Blood Pressure Lowering Arm (ASCOT-BPLA) studies meanwhile also demonstrated a clear superiority of newer antihypertensives, losartan, and amlodipine, respectively, compared to atenolol and the combination with diuretics [9].

Taken together, recommendations in certain hypertension management guidelines have been based on negative clinical data with some beta-blockers particularly atenolol [6]. The hesitancy from using beta-blockers is largely attributed to unfavorable outcome results of specific beta-blockers in clinical trials, mostly those in the first and second generation [10]. In contrast to these, those of the third-generation beta-blockers were found to exert considerably better metabolic and hemodynamic

M. Barrientos-Regala · J. D. P. Reaño (✉)
Manila Doctors Hospital, Manila, Philippines

profiles. An example of this is nebivolol, which has been shown to have similar or even better treatment response when compared with other available blood pressure (BP) lowering medications, with significantly better tolerability [11].

22.2 Recommendations on the Use of Beta-Blockers Based on Current Guidelines

Although the use of beta-blockers as first-line agents for hypertension is still a subject of discourse, they are consistently part of the armamentarium for controlling hypertension. The role of beta-blockers in the management of hypertension appears certain but the strength of recommendation varies across different populations.

In the 2018 European Society of Hypertension/European Society of Cardiology (ESH/ESC) guidelines on hypertension, beta-blockers are included as one of the five major drug classes recommended for the initial treatment of hypertension (angiotensin converting enzyme inhibitors, angiotensin receptor blockers, beta-blockers, calcium channel blockers, and diuretics). The benefit from their use is mainly obtained from BP lowering, thereby reducing overall major cardiovascular outcomes and mortality [12]. However, it is recognized that between drugs there are certain differences in outcomes such as less benefit in heart failure prevention with calcium channel blockers (CCBs) and less stroke prevention with beta-blockers. Beta-blockers in general appear to be relatively less effective in controlling central aortic BP, to which cerebrovascular events are sensitive [13]. However, nebivolol has shown relatively favorable effects on central BP, aortic stiffness, and endothelial dysfunction, as well as a more favorable side effect profile than classical beta-blockers [14, 15]. The ESH/ESC guidelines state that beta-blockers have shown specific usefulness for the treatment of hypertension in particular situations such as symptomatic angina, post-myocardial infarction (MI), heart failure with reduced ejection fraction (HFrEF), and as an alternative to angiotensin converting enzyme inhibitors (ACEI) or angiotensin receptor blockers (ARBs) in young hypertensive women who are planning pregnancy or of child-bearing potential.

There is relatively not much mention about the position of beta-blockers as first-line hypertensive agents in other major guidelines. In the ACC/AHA hypertension guidelines they are recommended as first-line antihypertensive agents only if the patient has ischemic heart disease or heart failure (HF) [16]. Similarly, the International Society of Hypertension (ISH) considered beta-blockers at any antihypertensive treatment step when there is a specific indication for their use (HF, angina, post-MI, AF, or younger women with or planning pregnancy) [17]. The 2014 Evidence-Based Guideline for the Management of High Blood Pressure in Adults Report From the Panel Members Appointed to the Eighth Joint National Committee (JNC 8) recommended beta-blockers as one of the options for add-on treatment if goal BP is not achieved with first-line antihypertensives [18]. In the NICE 2020 hypertension guidelines, beta-blockers should be considered as add-on treatment to ACEI/ARB, CCB, and thiazide-like diuretic for adults with resistant hypertension [19].

In Asia, hypertension guidelines have been developed and updated by countries such as Korea, Japan, Taiwan, and China, hopefully creating the foundation for the concrete establishment of Asian hypertension guidelines [20]. China, Korea, and Taiwan include beta-blockers in the five classes of first-line drugs, while only the Japanese guidelines relegate them to second line [21].

The Korean Society of Hypertension recommend beta-blockers as first-line treatment for hypertension but emphasized that the choice of antihypertensive drug should be according to the patient's combined risk factors and comorbidities. It mentioned the absolute indications for beta-blockers such as IHD and MI and the relative indications such as tachyarrhythmia [22]. Furthermore, the guidelines stressed the special precautions for the use of beta-blockers such as in the elderly and in those at risk of developing diabetes, in which concomitant use of beta-blockers with diuretics should be avoided.

The Taiwan Society of Cardiology and the Taiwan Hypertension Society suggest that beta-blockers, except atenolol, can be used as the first-line therapy for hypertension. They are especially recommended in patients with coronary heart disease (CHD), history of MI, and in patients with higher heart rates, particularly 80 beats per minute [23].

In the Chinese hypertension guidelines, beta-blockers are included in the first-line agents for hypertension but were mentioned as especially suitable for patients with tachyarrhythmia, CHD, chronic HF, increased sympathetic activation, and high-dynamic hypertension [24].

22.3 Role and Efficacy of Beta-Blockers in the Treatment of Hypertension

22.3.1 Clinical Profile of Beta-Blockers in Hypertension

First-generation beta-blockers (e.g., propranolol, pindolol) are non-selective and show no vasodilation effect. Second-generation beta-blockers (e.g., atenolol, bisoprolol) are considered β1-selective, while third-generation beta-blockers (e.g., carvedilol, nebivolol, labetalol) show additional vasodilatory properties [2]. Non-vasodilating beta-blockers are said to have suboptimal effect in lowering BP, a reduced effect on left ventricular hypertrophy, and unfavorable hemodynamics and metabolic effects. In contrast, third-generation beta-blockers do not just have vasodilatory properties but are also said to demonstrate a more favorable effect on metabolic and hemodynamic parameters, with fewer side effects [22, 25, 26].

22.3.2 Specific Groups to Highly Benefit from Beta-Blocker Treatment

Various groups of hypertensive patients have been found to benefit from beta-blocker treatment. Among these are hypertensive patients with HF, hypertensive patients with CAD, hypertensive patients with increased sympathetic activity (including obese patients), diabetic patients with hypertension, hypertensive patients with atrial fibrillation with rapid ventricular rate and among pregnant hypertensive patients [27].

22.4 Present Hypertension Burden

Noticeably in the last 10 years, the prevalence of hypertension in most Asian countries has increased and this has been attributed to acculturation to Western lifestyle, modernization, and urbanization [21]. The prevalence of hypertension among Asian populations has been observed to be variable, ranging between 13.6 and 47.9%, with a number of countries having BP ranges that are above the global average [28].

Western hypertension guidelines are relatively more established than Asian guidelines but before adopting them into clinical practice among Asian countries, the profile of Asian patients should be considered. There is increasing awareness that Asian populations have distinctive characteristics in terms of antihypertensive medication response, complications, and outcomes [29]. Stroke, especially hemorrhagic stroke, and nonischemic HF are noted to be more common in Asian than Western populations. Stroke in the Asian population is more closely related to hypertension than CAD or renal disease [30]. Furthermore, the increased prevalence of hypertension and diabetes in the Asian population may elucidate the high rate of stroke and CAD [30–32].

Hypertension is still an area of medical concern in Asia as the control of BP is usually suboptimal in clinical practice [33]. Since it is not yet completely established whether reduction in cardiovascular events may be achieved regardless of the antihypertensive class used, selecting appropriate BP-lowering drugs is a challenge in Asian populations because of the lack of existing Asia-specific data [21].

22.5 Hypertension and Ethnicity

22.5.1 Hypertension Among Different Racial Groups

Significant ethnic differences in the determinants of hypertension are observed among populations [33–36]. For instance, BP correlation to cardiovascular disease is stronger in Asians than in white subjects from Australia and New Zealand [37, 38]. In addition, Asians tend to have higher salt sensitivity and salt intake than Western populations, which may be related to the salt-sensitive gene polymorphism of the renin–angiotensin system (RAS) [39]. Vascular aging, indicated by increased arterial stiffness, may have a greater significance on hypertension outcomes of Asian patients than Western patients [40].

Asians are predisposed to increased central aortic pulse pressure because of the relatively larger diameter and thinner media at the proximal aorta that modulates the interaction between ventricular ejection and arterial load [41]. In addition, the shorter stature of Asians contributes to increased augmentation of central pulse pressure from peripheral arterial wave reflections, which is an independent determinant of incident hypertension [42, 43]. Hence, Asians may have a greater wall stress and stiffness at the proximal aorta than white populations.

Similar to Asians, blacks also have unique features that contribute to development of hypertension as follows: (1) low renin levels leading to additional sodium reabsorption by the kidneys; (2) a more active sodium-potassium-chloride cotransporter in the thick ascending limb of the kidneys; and (3) reduced bioavailability of nitric oxide in blacks [44–48].

Among the racial groups, the Asian populations are currently said to be the most technological because studies looking into new drug targets and susceptibilities are emerging. As such, clinicians should continue to assess epidemiologic, genetic, and sociologic factors among populations to achieve maximal medical therapy [49].

22.5.2 Inter-Ethnic Differences in Response to Beta-Blockers

One of the genetic polymorphisms mostly studied for beta-blockers is observed in the beta-1-adrenergic receptor (ADRB1) gene which codes for the β1 adrenergic receptors. These are important in the regulation of cardiac rate and contractility as well as renin release in the kidneys. The Arg389Gly genotype of the ADRB1 gene was associated with better response to beta-blockers in lowering BP in a study among Chinese patients [50, 51]. Another known variant associated with responsiveness to beta-blockers includes polymorphisms of the cytochrome P450 family 2D6 (CYP2D6) [18] which has considerable variability in the frequency of individual variants between populations. This variation is not only present when comparing inter-ethnic differences between Asians and Caucasians, but within the Asian population as well. The metabolism of certain beta-blockers is particularly dependent on this variability and their effects may vary based on ethnic differences [52–54]. A study including Chinese subjects demonstrated that propranolol exerted greater effects in Chinese subjects compared to Caucasians. As for metoprolol, they have found that Chinese extensive metabolizers have a reduced capacity to metabolize metoprolol when compared to their Korean or Japanese counterparts [55]. Meanwhile, in another study by Sy et al., the Klotho Variant rs36217263 was noted to be associated with poor response to cardioselective beta-blocker therapy among Filipino subjects [17].

Among different Asian ethnic groups, pharmacodynamic responses may also vary among some groups. In a study by Rasool et al. [56], healthy Malay volunteers had more pronounced bradycardic and hypotensive responses compared to those of Indian or Chinese ethnicities.

22.6 Concerns/Disadvantages Attributed with Use of Beta-Blockers

There have been several concerns over the use of beta-blockers for hypertension among the Asian population and this has been attributed to a variety of factors. Among them is the risk for upper airway sensitivity. Clinicians often refrain from prescribing them with an underlying disease of concern such as asthma. This has created some dilemma for physicians considering treatment with beta-blockers for patients in this population. However in recent years, several studies have already shown that cardioselective beta-blockers are better beta-blocker options for patients with airway reactivity [57].

Another concern is the risk of developing new-onset diabetes. Several studies in the literature have investigated the metabolic side effects of beta-blockers. In these studies, they have noted that because of a deterioration in insulin resistance, vaso-constricting beta-blockers lead to an increased risk of new-onset diabetes and a worsening of glycemic control [58]. Conventional beta-blockers are said to provide unopposed alpha-1 activity, causing vasoconstriction and decreased blood flow to skeletal muscles resulting in a decrease in insulin-stimulated glucose uptake in mus-cles, thereby causing insulin resistance. Insulin secretion from pancreatic beta-cells is also affected by intake of nonvasodilatory beta-blockers [59]. However, several newer generation beta-blockers like nebivolol are said to decrease BP without increasing the risk of incident diabetes [60].

Other concerns over its use include severe bradycardia and inadequate control of central blood pressure especially among the elderly. Inadequate BP control when used as monotherapy is also another reason for its decreasing popularity in some Asian countries [21].

22.7   Utility of Beta-Blockers in the Management of Hypertension

Although most guidelines by Asian societies on hypertension include beta-blockers as first-line antihypertensive agents and recognize their more specific benefits in certain conditions, such as CHD and HF, there is still some reluctance in prescribing beta-blockers within Asia. Clinical experts in Asia attempt to allay this concern by proposing to hold group discussions and developing simplified treatment algorithms [27]. Beta-blockers, similar on average to all other antihypertensive agents, provide effective BP lowering, which made them find their role as first-line hypertensives in specific conditions in Asian populations. Furthermore, highly beta1-selective beta-blockers or those with vasodilating effects may be preferred as initial antihyperten-sive agents.

References

1. Forouzanfar MH, Liu P, Roth GA, Ng M, Birukov S, Marczak L, et al. Global burden of hypertension and systolic blood pressure of at least 110 to 115 mmHg, 1990-2015. JAMA. 2017;317:165–82.
2. Kim CH, Abelardo N, Buranakitjaroen P, Krittayaphong R, Lim CH, Park SH, Pham NV, Rogelio G, Wong B, Low LP. Hypertension treatment in the Asia-Pacific: the role of and treatment strategies with nebivolol. Heart Asia. 2016;8(1):22–6. https://doi.org/10.1136/heartasia-2015-010656. PMID: 27326226; PMCID: PMC4898626.

3. Chiang CE, Chen CH. Hypertension in the Asia-Pacific region. J Hum Hypertens. 2008;22:441–3.
4. NCD Risk Factor Collaboration (NCD-RisC). Worldwide trends in blood pressure from 1975 to 2015: a pooled analysis 1479 population-based measurement studies with 19.1 million participants. Lancet. 2017;389:37–55.
5. Baker JG, Hill SJ, Summers RJ. Evolution of β-blockers: from anti-anginal drugs to ligand-directed signalling. Trends Pharmacol Sci. 2011;32:227–34.
6. Wiysonge CS, Opie LH. β-Blockers as initial therapy for hypertension. JAMA. 2013;310:1851–2.
7. Lindholm LH, Carlberg B, Samuelsson O. Should beta blockers remain first choice in the treatment of primary hypertension? A meta-analysis. Lancet. 2005;366:1545–53.
8. Elliott WJ, Meyer PM. Incident diabetes in clinical trials of antihypertensive drugs: a network meta-analysis. Lancet. 2007;369:201–7.
9. De Caterina AR, Leone AM. Why beta-blockers should not be used as first choice in uncomplicated hypertension. Am J Cardiol. 2010;105:1433–8.
10. Mancia G, Laurent S, Agabiti-Rosei E, et al. Reappraisal of European guidelines on hypertension management: a European Society of Hypertension Task Force document. J Hypertens. 2009;27:2121–58.
11. Chrysant SG, Chrysant GS. Current status of β-blockers for the treatment of hypertension: an update. Drugs Today (Barc). 2012;48:353–66.
12. Williams S, Mancia G, Spiering W, Rosei EA, Azizi M, Burnier M, et al. 2018 ESC/ESH guidelines for the management of arterial hypertension: the task force for the management of arterial hypertension of the European Society of Cardiology (ESC) and the European Society of Hypertension (ESH). Eur Heart J. 2018;39(33):3021–104.
13. Williams B, Lacy PS, Thom SM, Cruickshank K, Stanton A, Collier D, et al. Differential impact of blood pressure-lowering drugs on central aortic pressure and clinical outcomes: principal results of the Conduit Artery Function Evaluation (CAFÉ) study. Circulation. 2006;113(9):1213–25.
14. Bakris GL, Fonseca V, Katholi RE, McGill JB, Messerli FH, Phillips RA, et al. Metabolic effects of carvedilol vs metoprolol in patients with type 2 diabetes mellitus and hypertension: a randomized controlled trial. JAMA. 2004;292:2227–36.
15. Ayers K, Byrne LM, DeMatteo A, Brown NJ. Differential effects of nebivolol and metoprolol on insulin sensitivity and plasminogen activator inhibitor in the metabolic syndrome. Hypertension. 2012;59:893–8.
16. Whelton PK, Carey RM, Aronow WS, Casey DE, Collins KJ, Himmelfarb CD, et al. 2017 ACC/AHA/AAPA/ABC/ACPM/AGS/APhA/ASH/ASPC/NMA/PCNA guideline for the prevention, detection, evaluation, and management of high blood pressure in adults. Hypertension. 2018;71:e13–e115. https://doi.org/10.1161/HYP.0000000000000065.
17. Unger T, Borghi C, Charchar F, Khan NA, Poulter NR, Prabhakaran D, et al. 2020 International Society of Hypertension global hypertension practice guidelines. Hypertension. 2020;75:1334–57. https://doi.org/10.1161/HYPERTENSIONAHA.120.15026.
18. James PA, Oparil S, Carter BL, Cushman WC, Dennison-Himmelfarb C, Handler J, et al. Evidence-based guideline for the management of high blood pressure in adults report from the panel members appointed to the Eighth Joint National Committee (JNC 8). JAMA. 2014; https://doi.org/10.1001/jama.2013.284427.
19. National Guideline Centre (UK). Hypertension in adults: diagnosis and management. London: National Institute for Health and Care Excellence (UK); 2019.
20. Park JB. Asian guidelines on hypertension. Pulse. 2015;3(1):12–3. https://doi.org/10.1159/000381967.
21. Angeli F, Reboldi G, Verdecchia P. The 2014 hypertension guidelines: implications for patients and practitioners in Asia. Heart Asia. 2015;7:21–5. https://doi.org/10.1136/heartasia-2015-010639.
22. Lee HY, Shin J, Kim GH, Park S, Ihm S, Kim HC, et al. 2018 Korean Society of Hypertension Guidelines for the management of hypertension: part II-diagnosis and treatment of hypertension. Clin Hypertens. 2019;25:20.

23. Chiang CE, Wang T, Ueng K, Lin T, Yeh H, Chen C, et al. 2015 Guidelines of the Taiwan Society of Cardiology and the Taiwan Hypertension Society for the management of hypertension. J Chin Med Assoc. 2014; https://doi.org/10.1016/j.jcma.2014.11.005.
24. Liu L, Wu Z, Wang J, Wang W, Bao Y, Cai J, et al. 2018 Chinese guidelines for prevention and treatment of hypertension—a report of the revision Committee of Chinese Guidelines for prevention and treatment of hypertension. J Geriatr Cardiol. 2019;16:182–241. https://doi.org/10.11909/j.issn.1671-5411.2019.03.014.
25. Shin J, Choi YJ, Hong GR, Jeon DW, Kim DH, Koh YY, et al. Real-world efficacy and safety of nebivolol in Korean patients with hypertension from the BENEFIT KOREA study. J Hypertens. 2020;38(3):527–35. https://doi.org/10.1097/HJH.0000000000002296.
26. Kallistratos MS, Poulimenos LE, Manolis AJ. Vasodilator/-blockers: a different class of antihypertensive agents? Futur Cardiol. 2014;10:669–71.
27. Tomlinson B, Dalal JJ, Huang J, Low LP, Park CG, Rahman AR, Reyes EB, et al. The role of β-blockers in the management of hypertension: an Asian perspective. Curr Med Res Opin. 2011;27(5):1021–33. https://doi.org/10.1185/03007995.2011.562884.
28. Neupane D, McLachlan CS, Sharma R, et al. Prevalence of hypertension in member countries of South Asian Association for Regional Cooperation (SAARC): systematic review and meta-analysis. Medicine (Baltimore). 2014;93:e74.
29. Kearney PM, Whelton M, Reynolds K, et al. Worldwide prevalence of hypertension: a systematic review. J Hypertens. 2004;22:11–9.
30. Ueshima H. Explanation for the Japanese paradox: prevention of increase in coronary heart disease and reduction in stroke. J Atheroscler Thromb. 2007;14:278–86.
31. Ishikawa Y, Ishikawa J, Ishikawa S, et al. Prehypertension and the risk for cardiovascular disease in the Japanese general population: the Jichi Medical School Cohort study. J Hypertens. 2010;28:1630–7.
32. Woodward M, Huxley H, Lam TH, et al. Asia Pacific Cohort Studies Collaboration. A comparison of the associations between risk factors and cardiovascular disease in Asia and Australasia. Eur J Cardiovasc Prev Rehabil. 2005;12:484–91.
33. Park JB, Kario K, Wang JG. Systolic hypertension: an increasing clinical challenge in Asia. Hypertens Res. 2015;38:227–36.
34. Kario K. Evidence and perspectives on the 24-hour management of hypertension: hemodynamic biomarker-initiated 'anticipation medicine' for zero cardiovascular event. Prog Cardiovasc Dis. 2016;59:262–81. https://doi.org/10.1016/j.pcad.2016.04.001.
35. Kario K, Tomitani N, Matsumoto Y, Hamasaki H, Okawara Y, Kondo M, Nozue R, Yamagata H, Okura A, Hoshide S. Research and development of information and communication technology-based home blood pressure monitoring from morning to nocturnal hypertension. Ann Glob Health. 2016;82:254–73. https://doi.org/10.1016/j.aogh.2016.02.004.
36. Ueshima H, Sekikawa A, Miura K, Turin TC, Takashima N, Kita Y, Watanabe M, Kadota A, Okuda N, Kadowaki T, Nakamura Y, Okamura T. Cardiovascular disease and risk factors in Asia: a selected review. Circulation. 2008;118:2702–9. https://doi.org/10.1161/CIRCULATIONAHA.108.790048.
37. Lawes CM, Bennett DA, Parag V, Woodward M, Whitlock G, Lam TH, Suh I, Rodgers A, Asia Pacific Cohort Studies Collaboration. Blood pressure indices and cardiovascular disease in the Asia Pacific region: a pooled analysis. Hypertension. 2003;42:69–75. https://doi.org/10.1161/01.HYP.0000075083.04415.4B.
38. Arima H, Murakami Y, Lam TH, Kim HC, Ueshima H, Woo J, et al. Effects of prehypertension and hypertension subtype on cardiovascular disease in the Asia-Pacific region. Hypertension. 2012;59:1118–23. https://doi.org/10.1161/HYPERTENSIONAHA.111.187252.
39. Katsuya T, Ishikawa K, Sugimoto K, Rakugi H, Ogihara T. Salt sensitivity of Japanese from the viewpoint of gene polymorphism. Hypertens Res. 2003;26:521–5.
40. Kario K, Chen C, Park S, Park C, Hoshide S, Cheng H. Consensus document on improving hypertension management in Asian patients, taking into account Asian characteristics. Hypertension. 2018;71:375–82. https://doi.org/10.1161/HYPERTENSIONAHA.117.10238.

41. Avolio A. Genetic and environmental factors in the function and structure of the arterial wall. Hypertension. 1995;26:34–7.
42. Reeve JC, Abhayaratna WP, Davies JE, Sharman JE. Central hemodynamics could explain the inverse association between height and cardiovascular mortality. Am J Hypertens. 2014;27:392–400. https://doi.org/10.1093/ajh/hpt222.
43. Kaess BM, Rong J, Larson MG, Hamburg NM, Vita JA, Levy D, Benjamin EJ, Vasan RS, Mitchell GF. Aortic stiffness, blood pressure progression, and incident hypertension. JAMA. 2012;308:875–81. https://doi.org/10.1001/2012.jama.10503.
44. Ferdinand KC, Nasser SA. Understanding the importance of race/ethnicity in 462. The care of the hypertensive patient. Curr Hypertens Rep. 2015;17:15.
45. Tu W, Pratt JH. A consideration of genetic mechanisms behind the development of hypertension in blacks. Curr Hypertens Rep. 2013;15:108–13.
46. Turban S, Thompson CB, Parekh RS, Appel LJ. Effects of sodium intake and diet on racial differences in urinary potassium excretion: results from the Dietary Approaches to Stop Hypertension (DASH)-sodium trial. Am J Kidney Dis. 2013;61:88–95.
47. Tu W, Eckert GJ, Hannon TS, et al. Racial differences in sensitivity of blood pressure to aldosterone. Hypertension. 2014;63:1212–8.
48. Ozkor MA, Rahman AM, Murrow JR, et al. Differences in vascular nitric oxide and endothelium-derived hyperpolarizing factor bioavailability in blacks and whites. Arterioscler Thromb Vasc Biol. 2014;34:1320–7.
49. Bennett A, Parto P, Krim S. Hypertension and ethnicity. Curr Opin Cardiol. 2016;31:381–6. https://doi.org/10.1097/HCO.0000000000000293.
50. Peng Y, Xue H, Luo L, Yao W, Li R. Polymorphisms of the β1-adrenergic receptor gene are associated with essential hypertension in Chinese. Clin Chem Lab Med. 2009;47:1227–31.
51. RG Sy, Nevado JB Jr, Gonzales EM, Bejarin AJP, et al. A genetic polymorphism in GCKR may be associated with low high-density lipoprotein cholesterol phenotype among filipinos: A case-control study, acta medica Philippina, 2020. https://doi.org/10.47895/amp.vi0.2350.
52. Zhou SF. Polymorphism of human cytochrome P450 2D6 and its clinical significance: part I. Clin Pharmacokinet. 2009;48:689–723.
53. Nozawa T, Taguchi M, Tahara K, et al. Influence of CYP2D6 genotype on metoprolol plasma concentration and -adrenergic inhibition during longterm treatment comparison with bisoprolol. J Cardiovasc Pharmacol. 2005;46:713–20.
54. Sohn D-R, Shin S-G, Park C-W, et al. Metoprolol oxidation polymorphism in a Korean population compared with native Japanese and Chinese populations. Br J Clin Pharmacol. 1991;32:504–7.
55. Zhou HH, Koshakji RP, Silberstein DJ, et al. Altered sensitivity to and clearance of propranolol in men of Chinese descent as compared with American whites. N Engl Med. 1989;320:565–70.
56. Rasool AHG, Rahman ARA, Ismail R, et al. Ethnic differences in response to non-selective beta blockade among racial groups in Malaysia. Int J Clin Pharmacol Ther. 2000;38:260–9.
57. Huang KY, Tseng PT, Wu YC, Tu YK, Stubbs B, Su KP, et al. Do beta-adrenergic blocking agents increase asthma exacerbation? A network meta-analysis of randomized controlled trials. Sci Rep. 2021;11(1):452. https://doi.org/10.1038/s41598-020-79837-3.
58. Marketou M, Gupta Y, Jain S, Vardas P. Differential metabolic effects of beta-blockers: an updated systematic review of Nebivolol. Curr Hypertens Rep. 2017;19(3):22. https://doi.org/10.1007/s11906-017-0716-3.
59. Münzel T, Gori T. Nebivolol: the somewhat-different beta-adrenergic receptor blocker. J Am Coll Cardiol. 2009;54:1491–9.
60. Agabiti Rosei E, Rizzoni D. Metabolic profile of nebivolol, a betaadrenoceptor antagonist with unique characteristics. Drugs. 2007;67:1097–107.

Newer Angiotensin Receptor Blockers 23

S. S. Iyengar and N. Saleha

23.1 Introduction

The renin–angiotensin–aldosterone system (RAAS) plays a central role in blood pressure (BP) regulation, and fluid and electrolyte homeostasis, and regulates endothelial function, vascular reactivity, tissue remodelling, oxidative stress, and inflammation [1]. Thus, the RAAS represents an important target for the prevention and treatment of cardiovascular disease. Of the various BP-lowering drugs currently available, those targeting the RAAS are the cornerstone of hypertension management. There are four classes of RAAS blockers: angiotensin-converting enzyme (ACE) inhibitors, angiotensin receptor blockers (ARBs), mineralocorticoid receptor antagonists (MRAs), and direct renin inhibitors (DRIs). Of these, ARBs are the most widely used to treat hypertension and heart failure due to their documented efficacy and favourable tolerability profile.

The first agent used to block the angiotensin II (AT2) receptor was saralasin. However, the peptide nature of this molecule meant that it must be administered intravenously, and it has some partial agonistic and AT II-like effects at higher doses. It also has a very short duration of action [2]. The need to selectively block the AT2 receptor led to the development of the first orally active agent, losartan, followed by a variety of others, including valsartan, irbesartan, eprosartan, candesartan, and telmisartan.

S. S. Iyengar (✉)
Manipal Hospital, Bangalore, India

N. Saleha
Sri Jayadeva Institute of Cardiovascular Sciences and Research, Bangalore, India

© The Author(s), under exclusive license to Springer Nature Switzerland AG 2022
C. V. S. Ram et al. (eds.), *Hypertension and Cardiovascular Disease in Asia*,
Updates in Hypertension and Cardiovascular Protection,
https://doi.org/10.1007/978-3-030-95734-6_23

23.2   Mechanism of Action

Angiotensin II is one of the critical components of the RAAS. It mediates various actions by binding to AT1 and AT2 receptors. Angiotensin II binds to AT1 receptors, which are highly expressed in vascular tissues and the adrenal gland, causing significant vasoconstriction. It also promotes sodium and water retention and inhibits further secretion of renin [3]. The binding of angiotensin II bound to AT2 receptors produces the opposite effects, including natriuresis, a decrease in blood pressure (BP), and attenuation of vasoconstriction [3] (Table 23.1). ARBs selectively bind to AT1 receptors to directly prevent these actions of angiotensin II (Fig. 23.1). The newer ARBs azilsartan and fimasartan are highly selective for the AT1 receptor and do not cause blockade of AT2 receptors [4, 5].

23.3   Next-Generation Multifunctional ARBs

ARBs are effective in controlling BP and have shown beneficial effects on cardiac hypertrophy, heart failure, and diabetic nephropathy [1]. However, existing agents only partially reduce cardiovascular morbidity and mortality [2, 3], and there is still room for further reduction by targeting other pathways. Therefore, newer generation ARBs that are expected to provide superior control of BP and better modulate the associated cardiovascular and metabolic risk factors in patients with hypertension, diabetes, and heart failure are under development [4, 5]. The ability of researchers to design bifunctional molecules that not only block AT_1 receptors but also target other pathways involved in the regulation of BP and the pathogenesis of cardiovascular and metabolic diseases has opened a new avenue for the development of next-generation ARBs (Fig. 23.2). ARB developmental milestones are shown in Fig. 23.3.

Table 23.1 Actions mediated by angiotensin II type 1 (AT1) and type 2 (AT2) receptors [3]

AT1 receptor actions	AT2 receptor actions
• Vasoconstriction	• Vasodilatation via bradykinin and
• Aldosterone production and release	nitric oxide
• Sodium tubular reabsorption	• Antiproliferative effects
• Cardiac hypertrophy	• Embryogenic differentiation and
• Proliferation of vascular smooth muscle	development
• Catecholamine secretion and potentiation (central and periphery)	• Stimulation of apoptosis
	• Endothelial cell growth
• Vasopressin release	
• Thirst	
• Renal vasoconstriction and reduction of renal blood flow	
• Inhibition of renin release	

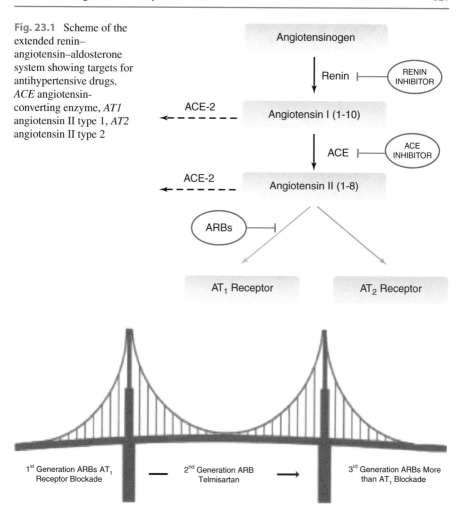

Fig. 23.1 Scheme of the extended renin–angiotensin–aldosterone system showing targets for antihypertensive drugs. *ACE* angiotensin-converting enzyme, *AT1* angiotensin II type 1, *AT2* angiotensin II type 2

Fig. 23.2 Bridge to the future for next-generation angiotensin receptor blockers (ARBs). *AT1* angiotensin II type 1

23.3.1 Nitric Oxide (NO) Donors

NO plays a major role in modulating vascular tone and has pronounced antithrombotic activity. It is cardio- and reno-protective. Dysregulation of the NO system is associated with cardiovascular disease, and reduced NO bioactivity is an important determinant of endothelial dysfunction and hypertension [6]. Several academic groups and biopharmaceutical companies are pursuing the discovery and development of NO donor hybrid drugs. Ex vivo and in vitro studies with NO-releasing

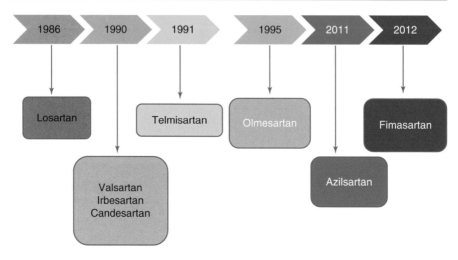

Fig. 23.3 Milestones in the development of angiotensin receptor blockers

losartan and telmisartan have shown superior cardiovascular effects compared with other ARBs [7–9]. However, there are not yet any clinical data with these agents.

23.3.2 ARBs that Block Endothelin Receptors

Sparsentan is the first and only dual-acting angiotensin and endothelin receptor antagonist (DARA) in development. It is an orally active compound combining endothelin type A (ET_A) receptor blockade with AT1 receptor antagonism in a single molecule. The US Food and Drug Administration (FDA) has granted sparsentan an orphan drug designation for the treatment of immunoglobulin A (IgA) nephropathy. Sparsentan has been used in trials of patients with focal segmental glomerulosclerosis (FSGS). In the DUET phase 2 randomised, double-blind, active-control, dose-escalation study patients with FSGS showed significantly greater reductions in proteinuria after 8 weeks' treatment with sparsentan versus irbesartan [10]. Sparsentan was safe and well tolerated in these patients. Mild-to-moderate oedema developed in a few patients, but they remained stable. The authors commented that the contribution of ET receptor inhibition to fluid retention was difficult to judge [10]. Based on these positive findings, the phase 3 randomised, double-blind, active-controlled DUPLEX trial has been initiated to assess antiproteinuric efficacy, nephroprotection, safety, and tolerability with combined inhibition of ETA and AT1 over a 2-year period [11].

23.3.3 ARBs that Inhibit Neprilysin Activity

Neprilysin is a membrane-bound zinc metalloproteinase that degrades endogenous natriuretic peptides involved in the regulation of sodium and water homeostasis.

Therefore, neprilysin inhibition is associated with increased levels of natriuretic peptides that have antihypertensive effects. However, neprilysin inhibition also increases levels of vasoconstrictor peptides, including angiotensin II [12]. Thus, dual inhibition of neprilysin and ACE has complementary effects and is more effective than inhibiting either one of these enzymes alone. However, combining neprilysin inhibition with an ACEI resulted in high bradykinin levels and a high incidence of angioedema [13]. Therefore, the combination of a neprilysin inhibitor with an ARB is preferable.

This new drug class is called angiotensin receptor-neprilysin inhibitor (ARNI), and sacubitril/valsartan is the first agent in this class to be approved. Sacubitril/valsartan is FDA approved for the treatment of patients with chronic heart failure with reduced ejection fraction (HFrEF) in New York Heart Association (NYHA) functional class II, III, or IV [14].

23.3.4 ARBs that Cross Blood–Brain Barrier

Earlier work has indicated that some, but not all, antihypertensive drugs may benefit cognition and reduce risk for Alzheimer's disease, independent of stroke [15]. One of these groups of antihypertensive agents is the ARBs.

Valsartan, telmisartan, and candesartan all cross the blood–brain barrier (BBB), whereas other ARBs do not (including irbesartan, olmesartan, losartan, and eprosartan) [16] (Fig. 23.4). In a study enrolling 1626 adults aged 55–91 years who did not have dementia, ARB users showed preserved memory over 3 years of follow-up compared with those taking other antihypertensive drugs [17]. Patients taking ARBs that crossed the BBB had better list-learning memory performance over time compared with other groups, including normotensives [17].

Older ARBs	Newer ARBs	BBB properties
Losartan Irbesartan Candesartan Valsartan Olmesartan Telmisartan Eprosartan	Azilsartan Fimasartan Sparsentan – dual acting angiotensin and endothelin receptor antagonist ARNI – neprilysin inhibitor (sacubitril) with valsartan	**BBB crossers** Telmisartan Valsartan Candesartan **BBB non-crossers** Losartan Irbesartan Olmesartan Eprosartan

Fig. 23.4 Angiotensin receptor blockers (ARBs) and their ability to cross the blood–brain barrier. *ARNI* angiotensin receptor-neprilysin inhibitor

23.4   Newer ARBs

23.4.1 Azilsartan

Azilsartan medoxomil was developed by Takeda Global Research USA and was the eighth ARB to be approved by FDA. It was approved in February 2011 for treatment in hypertension in adults in the USA and is now available worldwide either as the prodrug (azilsartan medoxomil) or the primary compound (azilsartan) and can be used alone or in combination with other antihypertensive drugs.

23.4.1.1   Pharmacokinetics

Azilsartan is formulated as its prodrug, azilsartan, medoxomil. It is absorbed rapidly in the gut and hydrolysed to the active metabolite azilsartan in the liver, primarily by the enzyme CYP2C9 [4]. Bioavailability after oral administration is 60%, with an elimination half-life of about 11 h. The time to peak plasma concentration is 1.5–3 h and steady-state concentrations are reached after 5 days. Azilsartan is excreted primarily in the faeces (55%) and 42% gets excreted in urine; 15% of azilsartan excreted in urine remains unchanged. Food intake has no impact on the bioavailability of azilsartan [18].

23.4.1.2   Dosage

The recommended starting dosage for azilsartan in patients with hypertension is 80 mg/day. However, therapy can be initiated at 40 mg/day in patients taking diuretics, and those volume-depleted or salt deprived [19]. No dose adjustment is required in the presence of mild-to-moderate hepatic or renal dysfunction, or in elderly patients. Azilsartan dosages studied in clinical trials compared with ACEI or ARBs is shown in Table 23.2.

Azilsartan Versus ACEIs

A double-blind, controlled, randomised trial compared the antihypertensive efficacy and safety of azilsartan versus ramipril in patients with clinic BP of 150–180 mmHg [23]. At doses of 40 and 80 mg/day azilsartan medoxomil reduced clinic and ambulatory systolic and diastolic BP (SBP and DBP) to a significantly greater extent than ramipril 10 mg/day. Azilsartan medoxomil had a similar safety profile to ramipril, but with a lower frequency of adverse effects and drug discontinuation [23]. Data from the EARLY register study showed that azilsartan lowered BP more effectively than ACEI; in addition, adherence to therapy was better in the azilsartan group [22]. Although ARBs are more effective in controlling BP than ACEIs, there does not

Table 23.2 Doses of azilsartan and comparator agents used in clinical studies

Study	Azilsartan dosage	Dosage of other ARBs
Sica et al. [20]	40 or 80 mg/day	Valsartan 320 mg/day
White et al. [21]	40 or 80 mg/day	Valsartan 320 mg/day
		Olmesartan 40 mg/day
Gitt et al. [22]	40 mg/day	Ramipril 10 mg/day

ARBs angiotensin receptor blockers

appear to be any differences in the effects of the two drug classes on mortality and cardiovascular outcomes [24].

Azilsartan Versus Other ARBs

Azilsartan has consistently been shown to be more effective than other ARBs at controlling systolic, diastolic, and 24-h ambulatory BP (Table 23.3) [4].

Table 23.3 Summary of clinical trial data comparing azilsartan with other angiotensin receptor blockers [4]

	Design	Number of patients	Inclusion criteria	Duration	Dose	Results
Sica et al. [20]	RCT, double blinded, placebo controlled	984	SBP 150–180 mmHg and 24-h mean SBP 130–170 mmHg	24 weeks	Azilsartan 40 or 80 mg OD vs. valsartan 320 mg OD	Azilsartan 40 mg (−14.9) and 80 mg (−15.3) significantly improved 24-h mean SBP (−11.3) $p < 0.0001$
Bakris et al. [25]	RCT, double blinded, placebo controlled	1275	SBP 150–180 mmHg or 24-h mean SBP 130–170 mmHg	6 weeks	Azilsartan 20, 40, 80 mmHg OD vs. olmesartan 40 mg OD vs. placebo	Azilsartan 80 mg (−14.6) significantly improved mean SBP vs. olmesartan (−12.6) ($p = 0.038$) 40 mg dose was noninferior to olmesartan
White et al. [21]	RCT, double blinded, placebo controlled	1291	SBP 150–180 mmHg and 24-h mean SBP 130–170 mmHg	6 weeks	Azilsartan 40, 80 mg OD vs. olmesartan 40 mg OD vs. valsartan 320 mg OD	Azilsartan 80 mg (−14.5) significantly improved mean SBP more than olmesartan (−11.7) and valsartan (−10.2). Azilsartan 40 mg (−13.4) noninferior to olmesartan
Rakugi et al. [26]	RCT, double blinded, placebo controlled	622	Grade I–II essential hypertension	16 weeks	Azilsartan 20–40 mg OD vs. candesartan 8–12 mg OD	Azilsartan significantly improved DBP (−12.4) vs. candesartan (−9.8) ($p = 0.0003$) and SBP azilsartan (−21.8) vs. candesartan (−17.5) ($p < 0.0001$)

DBP diastolic blood pressure, *OD* once daily, *RCT* randomised clinical trial, *SBP* systolic blood pressure

After 6 weeks' treatment, reductions in 24-h mean SBP in patients treated with azilsartan 40 mg/day, azilsartan 80 mg/day, valsartan 320 mg/day, or olmesartan 40 mg/day compared with placebo were −13.2, −14.4, −10.0, and −11.7 mmHg, respectively [21]. Statistical comparison showed that azilsartan 80 mg/day was superior to valsartan and olmesartan, while azilsartan 40 mg/day was noninferior to the other ARBs. In addition, azilsartan was not associated with an increase in adverse events [21]. In another study, reductions in 24-h mean SBP were significantly greater with azilsartan 80 mg/day versus olmesartan 40 mg/day (−14.6 vs. −12.6 mmHg; $p = 0.038$) [25]. Azilsartan dosages of 20 and 40 mg/day were noninferior to olmesartan 40 mg/day, and the tolerability of profile of both ARBs was like that of placebo [25]. Reductions in mean 24-h BP were also greater with azilsartan 40 and 80 mg/day compared with valsartan [20] and candesartan [26].

A meta-analysis of data from seven prospective randomised controlled trials of azilsartan or azilsartan medoxomil in 6152 patients with hypertension indicated a significant reduction in BP in patients treated with azilsartan 40 mg/day compared with control therapy [27].

In India, a prospective, multicentre, randomised study investigated the effects of azilsartan 40 mg/day, azilsartan 80 mg/day or telmisartan 40 mg/day in 303 adult patients with essential hypertension [28]. Based on clinic and 24-h BP measurements, it was concluded that azilsartan was an effective BP-lowering drug, was well tolerated, and was noninferior to telmisartan with respect to both efficacy and safety [28].

In comparison to other ARBs, the superior antihypertensive effects of azilsartan could be due to its unique binding properties and greater suppression of the sympathetic nervous system [29] (Fig. 23.5). It is also effective when combined with other antihypertensive drugs.

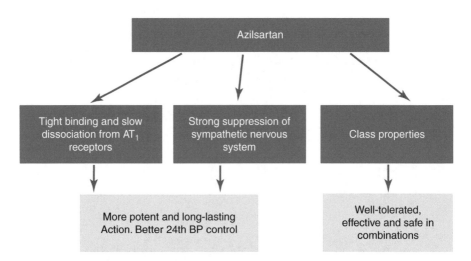

Fig. 23.5 Special features of antihypertensive effects of azilsartan

23.4.1.3 Effects Beyond BP Control

There is a growing body of data to support the "beyond BP" effects of azilsartan. It has inverse agonistic action resulting in amelioration of deleterious effects of angiotensin II such as cardiac hypertrophy, fibrosis, and insulin resistance [4]. In a small trial, 6 months' treatment with azilsartan was associated with improvements in left ventricular diastolic function in patients with hypertension and heart failure with preserved ejection fraction [30]. The mitral annular E/e' ratio on echocardiography decreased in the azilsartan group but remained unchanged in the candesartan group despite comparable BP reductions in the two groups [30].

Preclinical studies have shown that azilsartan medoxomil is a pleiotropic drug with favourable cardiometabolic effects [31] (Box 23.1). Azilsartan medoxomil showed antiproliferative effects in vascular cells in the absence of angiotensin II and in cells lacking AT1 receptors, suggesting involvement of mechanisms beyond AT1 receptor blockade [32]. In addition, azilsartan medoxomil was associated with favourable differentiation of adipocytes and stimulated the expression of genes for peroxisome proliferator-activated receptor (PPAR)-α, PPAR-δ, leptin, adipsin, and adiponectin to a greater extent than valsartan [32]. In addition, azilsartan modoxomil improved insulin sensitivity in animals by reducing the production of tumour necrosis factor-α and increasing expression of target genes for PPAR-γ, CCAAT/enhancer binding protein (C/EBP), and adiponectin (aP2) more effectively than candesartan [33, 34].

Box 23.1 Possible Pleiotropic Effects of Azilsartan
- Glucose metabolism: improved insulin sensitivity
- Adipocyte differentiation: better lipid profile
- Antioxidative and anti-inflammatory effects
- Antiproliferative and antifibrotic effects
- Renoprotection and cardioprotection

23.4.2 Fimasartan

Fimasartan is the latest selective AT1 receptor antagonist approved for clinical use. It was developed by a Korean company as an oral antihypertensive drug and was approved by regulatory authorities in that country in 2010. It is also approved in other countries, including China, Singapore, and Russia. Fimasartan was approved by the Central Drugs Standard Control Organization (CDSCO) in India in November 2018.

23.4.2.1 Chemistry

Fimasartan is a derivative of losartan in which the imidazole ring has been replaced (Fig. 23.6). This change results in a molecule with higher potency and a longer duration of action than losartan. Fimasartan provides a selective AT1 receptor antagonist effect with non-competitive binding.

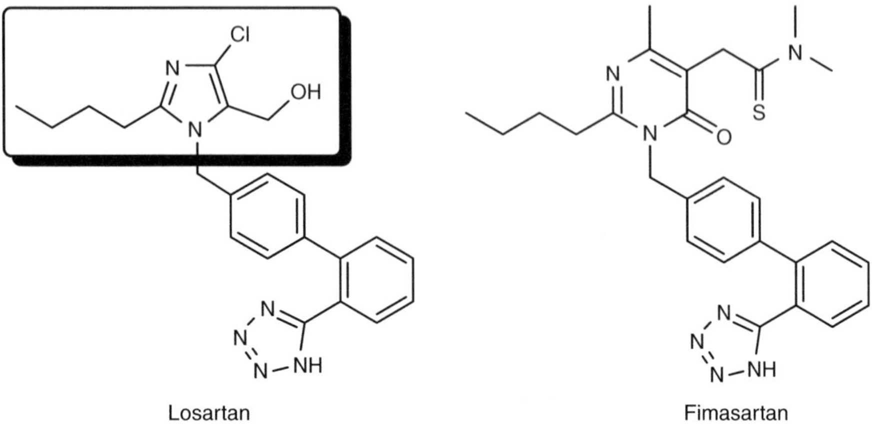

Losartan   Fimasartan

Fig. 23.6 Chemical structures of losartan and fimasartan

23.4.2.2 Pharmacokinetics

Fimasartan is a long-acting drug, allowing once daily dosing, usually at a dosage of 60–120 mg/day. Fimasartan is rapidly absorbed after oral intake, reaches maximum plasma concentrations at 3 h post-dose (range 0.5–5.0 h), and has an elimination half-life of 5.8 ± 1.6 h. Like other ARBs, the oral bioavailability of fimasartan is 18.6 ± 7.2% [5].

Fimasartan is relatively stable in terms of metabolism and circulates primarily in its parent form. It undergoes entero-hepatic circulation and is mainly eliminated in bile and faeces, with minimal excretion in urine. It is primarily catabolised by cytochrome P450 (CYP)3A.

It is well tolerated in patients with mild-to-moderate renal dysfunction, but dose modification is required in the presence of severe renal dysfunction (creatinine clearance <30 mL/min). Administration of fimasartan is safe in patients with mild hepatic dysfunction but should be avoided in those with severe hepatic dysfunction, patients on haemodialysis, pregnant or lactating mothers, and patients with hypersensitivity reaction.

23.4.2.3 Safety and Tolerability

Fimasartan was generally well tolerated in phase II and phase III clinical trials, with a similar rate of adverse events to placebo. In Safe-KanArb study of patients with arterial hypertension ($n > 14,000$), fimasartan was effective in controlling BP with good compliance and safety profile across all patient subgroups, including treatment-naïve patients and when added to current antihypertensive medication [35].

Data from the K-MetS study showed that fimasartan can safely be used in elderly patients [36]. Of the 6399 patients enrolled in the study, 2363 were elderly. Overall, fimasartan reduced clinic systolic and diastolic BP (SBP and DBP) from 144.1 ± 17.3 and 85.1 ± 10.4 mmHg at baseline to 127.7 ± 12.9 and 76.8 ± 8.4 mmHg at 1 year, respectively (all $p < 0.0001$). Results were similar for the effects of fimasartan on home BP, and all BP changes were similar in elderly and nonelderly patients.

However, pulse pressure, a better predictor of cardiovascular events in the elderly, decreased to a greater extent in elderly than in nonelderly patients (-8.2 ± 0.3 vs -7.0 ± 0.2 mmHg; $p < 0.0001$).

23.4.2.4   Evidence for Use

Currently available evidence for the clinical efficacy, safety, and tolerability of fimasartan is presented below. The efficacy and tolerability of fimasartan were compared with losartan in a 12-week, phase III, multicentre, prospective, randomised, double-blind clinical trial [37]. Patients with mild-to-moderate hypertension were randomly allocated to receive fimasartan 60 then 120 mg/day ($n = 256$) or losartan 50 then 100 mg/day ($n = 250$) for 12 weeks. Mean changes from baseline in office DBP at week 12 were -11.16 ± 7.53 mmHg in the fimasartan group and -8.56 ± 7.72 in the losartan group (both $p < 0.001$ vs baseline; between-group difference 2.70 mmHg; $p = 0.0002$). The proportion of responders (DBP <90 mmHg or change in DBP >10 mmHg) for treatment-naïve patients was similar in the fimasartan versus losartan group (73% vs. 69%) but was higher with fimasartan versus losartan in previously treated patients (62% vs. 48%; $p = 0.0143$) [37].

The FAST study, conducted by a Korean group, compared the BP-lowering effects of fimasartan with valsartan and olmesartan in patients with hypertension ($n = 365$) [38]. After a 2-week single-blind placebo run-in period, initial drug dosages were fimasartan 60 mg/day, valsartan 80 mg/day, and olmesartan 10 mg/day, given for 2 weeks, then these dosages were doubled for another 4 weeks of treatment. Sitting office SBP decreased to a significantly greater extent with fimasartan versus valsartan (change from baseline to week 6: -16.26 vs. -12.81 mmHg; $p = 0.0298$), but there was no significant difference in change from baseline in sitting SBP with fimasartan versus olmesartan (-16.25 vs. -14.78; $p = 0.4991$). There were no significant differences between treatment groups in the incidence of adverse events and adverse drug reactions [38].

Another study from Korea found that fimasartan 60 mg/day, fimasartan 120 mg/day, and candesartan 8 mg/day had similar BP-lowering effects in adults with mild-to-moderate hypertension [39]. Statistical analysis showed both dosages of fimasartan were noninferior to candesartan with respect to effect on DBP. The tolerability profiles of the two ARBs were generally similar, although there was a slightly higher rate of hepatic enzyme elevation in the fimasartan 120 mg/day group [39].

23.4.2.5   Combination with Other Drugs

Lee et al. evaluated the efficacy and safety of a fimasartan/amlodipine combination in patients with hypertension [40]. After a 2-week placebo run-in period, eligible patients were randomised to receive single or combined administration of fimasartan (0, 30, and 60 mg/day) and amlodipine (0, 5, and 10 mg/day) for 8 weeks. Fimasartan combined with amlodipine was associated with superior BP reductions and low levels of adverse events compared with either monotherapy, although all treatments reduced BP to a significantly greater extent than placebo. The greatest reduction was seen in the fimasartan 60 mg/amlodipine 10 mg combination group [40]. Fimasartan and amlodipine combination therapy has also been shown to be effective in patients with hypertension whose BP was not adequately controlled by fimasartan monotherapy [41].

A study by Rhee et al. randomised patients with a DBP ≥90 mmHg to receive either fimasartan/hydrochlorothiazide (HCTZ) 60 mg/12.5 mg or fimasartan 60 mg once daily for 4 weeks [42]. After 4 weeks, the drug dosages were increased to 120 mg/12.5 mg and 120 mg, respectively, if DBP was ≥90 mmHg. The combination therapy group showed a greater reduction in BP than the fimasartan monotherapy group, and the combination had a similar safety and tolerability profile to fimasartan monotherapy [42].

Taken together, currently available data show that fimasartan is an effective and safe antihypertensive agent in clinical trials, that may have anti-inflammatory and organ protection effects based on data from animal studies. These "beyond BP" effects appear to be like those of azilsartan. The effects of fimasartan in patients with chronic kidney disease are being investigated in the FANTASTIC trial [43], and the FABULOUS study is evaluating the BP-lowering effects of fimasartan in patients with acute stoke [44].

23.5 Nitrosamine Impurities in Some ARBs

Since 2018, specific ARB products have been recalled due to the presence of potentially carcinogenic nitrosamine impurities in at least some batches. Relevant companies have been voluntarily participating in addressing this issue, and most countries have acted in response. ARBs could be classified into three categories: (1) no impurities; (2) incompletely assessed; and (3) impurities above acceptable level [45–47].

It is estimated that if 8000 people took the highest daily valsartan dose (320 mg) of a formulation that contained NDMA for 4 years, there may be one additional case of cancer beyond the average cancer rate among Americans [48].

Not all ARBs are being recalled, and not all lots of valsartan, irbesartan, and losartan are affected. All healthcare professionals need to check the lists regularly and monitor updates from their regulatory agencies.

23.6 Conclusion

With the new generation of ARBs becoming available and new mechanisms and mode of action being identified for ARBs, clinicians may be able to better customise antihypertensive therapy for their patients. Some ARBs do more than just block AT1 receptors and target additional mechanisms of the disease process in hypertension, cardiovascular disease, and diabetes. This includes blockade of endothelin receptors, nitric oxide donor activity, inhibition of neprilysin activity, or stimulation of PPAR-γ. In addition, ARBs that cross the BBB may have beneficial effects on cognitive function in patients with hypertension. It is hoped that these new ARBs will have a beneficial impact on cardiovascular morbidity and mortality, but this remains to be determined.

References

1. Schmieder RE. Mechanisms for the clinical benefits of angiotensin II receptor blockers. Am J Hypertens. 2005;18(5):720–30.
2. Castellion AW, Fulton RW. Preclinical pharmacology of saralasin. Kidney Int Suppl. 1979;9:S11–9.
3. Carey RM, Wang ZQ, Siragy HM. Role of the angiotensin type 2 receptor in the regulation of blood pressure and renal function. Hypertension. 2000;35(1):155–63.
4. Pradhan A, Tiwari A, Sethi R. Azilsartan: current evidence and perspectives in management of hypertension. Int J Hypertens. 2019;2019:1824621.
5. Lee HY, Oh BH. Fimasartan: a new angiotensin receptor blocker. Drugs. 2016;76(10):1015–22.
6. Klahr S. The role of nitric oxide in hypertension and renal disease progression. Nephrol Dial Transplant. 2001;16(Suppl 1):60–2.
7. Breschi MC, Calderone V, Digiacomo M, Macchia M, Martelli A, Martinotti E, et al. New NO-releasing pharmacodynamic hybrids of losartan and its active metabolite: design, synthesis, and biopharmacological properties. J Med Chem. 2006;49(8):2628–39.
8. Breschi MC, Calderone V, Digiacomo M, Martelli A, Martinotti E, Minutolo F, et al. NO-sartans: a new class of pharmacodynamic hybrids as cardiovascular drugs. J Med Chem. 2004;47(23):5597–600.
9. Li YQ, Ji H, Zhang YH, Shi WB, Meng ZK, Chen XY, et al. WB1106, a novel nitric oxide-releasing derivative of telmisartan, inhibits hypertension and improves glucose metabolism in rats. Eur J Pharmacol. 2007;577(1):100–8.
10. Trachtman H, Nelson P, Adler S, Campbell KN, Chaudhuri A, Derebail VK, et al. DUET: a phase 2 study evaluating the efficacy and safety of sparsentan in patients with FSGS. J Am Soc Nephrol. 2018;29(11):2745–54.
11. Komers R, Diva U, Inrig JK, Loewen A, Trachtman H, Rote WE. Study design of the phase 3 sparsentan versus irbesartan (DUPLEX) study in patients with focal segmental glomerulosclerosis. Kidney Int Rep. 2020;5(4):494–502.
12. McMurray JJ. Neprilysin inhibition to treat heart failure: a tale of science, serendipity, and second chances. Eur J Heart Fail. 2015;17(3):242–7.
13. Jhund PS, McMurray JJ. The neprilysin pathway in heart failure: a review and guide on the use of sacubitril/valsartan. Heart. 2016;102(17):1342–7.
14. McMurray JJ, Packer M, Desai AS, Gong J, Lefkowitz MP, Rizkala AR, et al. Angiotensin-neprilysin inhibition versus enalapril in heart failure. N Engl J Med. 2014;371(11):993–1004.
15. Yasar S, Xia J, Yao W, Furberg CD, Xue QL, Mercado CI, et al. Antihypertensive drugs decrease risk of Alzheimer disease: ginkgo evaluation of memory study. Neurology. 2013;81(10):896–903.
16. Ho JK, Moriarty F, Manly JJ, Larson EB, Evans DA, Rajan KB, et al. Blood-brain barrier crossing renin-angiotensin drugs and cognition in the elderly: a meta-analysis. Hypertension. 2021;78(3):629–43.
17. Ho JK, Nation DA. Memory is preserved in older adults taking AT1 receptor blockers. Alzheimers Res Ther. 2017;9(1):1–14.
18. Angeli F, Verdecchia P, Pascucci C, Poltronieri C, Reboldi G. Pharmacokinetic evaluation and clinical utility of azilsartan medoxomil for the treatment of hypertension. Expert Opin Drug Metab Toxicol. 2013;9(3):379–85.
19. Kurtz TW, Klein U. Next generation multifunctional angiotensin receptor blockers. Hypertens Res. 2009;32(10):826–34.
20. Sica D, White WB, Weber MA, Bakris GL, Perez A, Cao C, et al. Comparison of the novel angiotensin II receptor blocker azilsartan medoxomil vs valsartan by ambulatory blood pressure monitoring. J Clin Hypertens. 2011;13(7):467–72.
21. White WB, Weber MA, Sica D, Bakris GL, Perez A, Cao C, et al. Effects of the angiotensin receptor blocker azilsartan medoxomil versus olmesartan and valsartan on ambulatory and clinic blood pressure in patients with stages 1 and 2 hypertension. Hypertension. 2011;57(3):413–20.

22. Gitt AK, Bramlage P, Potthoff SA, Baumgart P, Mahfoud F, Buhck H, et al. Azilsartan compared to ACE inhibitors in anti-hypertensive therapy: one-year outcomes of the observational EARLY registry. BMC Cardiovasc Disord. 2016;16:56.
23. Bönner G, Bakris GL, Sica D, Weber MA, White WB, Perez A, et al. Antihypertensive efficacy of the angiotensin receptor blocker azilsartan medoxomil compared with the angiotensin-converting enzyme inhibitor ramipril. J Hum Hypertens. 2013;27(8):479–86.
24. Chen R, Suchard MA, Krumholz HM, Schuemie MJ, Shea S, Duke J, et al. Comparative first-line effectiveness and safety of ACE (angiotensin-converting enzyme) inhibitors and angiotensin receptor blockers: a multinational cohort study. Hypertension. 2021;78(3):591–603.
25. Bakris GL, Sica D, Weber M, White WB, Roberts A, Perez A, et al. The comparative effects of azilsartan medoxomil and olmesartan on ambulatory and clinic blood pressure. J Clin Hypertens. 2011;13(2):81–8.
26. Rakugi H, Enya K, Sugiura K, Ikeda Y. Comparison of the efficacy and safety of azilsartan with that of candesartan cilexetil in Japanese patients with grade I-II essential hypertension: a randomized, double-blind clinical study. Hypertens Res. 2012;35(5):552–8.
27. Takagi H, Mizuno Y, Niwa M, Goto SN, Umemoto T. A meta-analysis of randomized controlled trials of azilsartan therapy for blood pressure reduction. Hypertens Res. 2014;37(5):432–7.
28. Sinha S, Chary S, Reddy Bandi M, Thakur P, Talluri L, Reddy VK, et al. Evaluation of the efficacy and safety of Azilsartan in adult patients with essential hypertension: a randomized, phase-III clinical study in India. J Assoc Phys. 2021;69(2):35–9.
29. Kusuyama T, Ogata H, Takeshita H, Kohno H, Shimodozono S, Iida H, et al. Effects of azilsartan compared to other angiotensin receptor blockers on left ventricular hypertrophy and the sympathetic nervous system in hemodialysis patients. Ther Apher Dial. 2014;18(5):398–403.
30. Sakamoto M, Asakura M, Nakano A, Kanzaki H, Sugano Y, Amaki M, et al. Azilsartan, but not candesartan improves left ventricular diastolic function in patients with hypertension and heart failure. Int J Gerontol. 2015;9(4):201–5.
31. Georgiopoulos G, Katsi V, Oikonomou D, Vamvakou G, Koutli E, Laina A, et al. Azilsartan as a potent antihypertensive drug with possible pleiotropic cardiometabolic effects: a review study. Front Pharmacol. 2016;7:235.
32. Kajiya T, Ho C, Wang J, Vilardi R, Kurtz TW. Molecular and cellular effects of azilsartan: a new generation angiotensin II receptor blocker. J Hypertens. 2011;29(12):2476–83.
33. Zhao M, Li Y, Wang J, Ebihara K, Rong X, Hosoda K, et al. Azilsartan treatment improves insulin sensitivity in obese spontaneously hypertensive Koletsky rats. Diabetes Obes Metab. 2011;13(12):1123–9.
34. Iwai M, Chen R, Imura Y, Horiuchi M. TAK-536, a new AT1 receptor blocker, improves glucose intolerance and adipocyte differentiation. Am J Hypertens. 2007;20(5):579–86.
35. Park JB, Sung KC, Kang SM, Cho EJ. Safety and efficacy of fimasartan in patients with arterial hypertension (Safe-KanArb study): an open-label observational study. Am J Cardiovasc Drugs. 2013;13(1):47–56.
36. Cho EJ, Sung KC, Kang SM, Shin MS, Joo SJ, Park JB. Fimasartan reduces clinic and home pulse pressure in elderly hypertensive patients: a K-MetS study. PLoS One. 2019;14(4):e0214293.
37. Lee SE, Kim YJ, Lee HY, Yang HM, Park CG, Kim JJ, et al. Efficacy and tolerability of fimasartan, a new angiotensin receptor blocker, compared with losartan (50/100 mg): a 12-week, phase III, multicenter, prospective, randomized, double-blind, parallel-group, dose escalation clinical trial with an optional 12-week extension phase in adult Korean patients with mild-to-moderate hypertension. Clin Ther. 2012;34(3):552–68.
38. Chung WB, Ihm SH, Jang SW, Her SH, Park CS, Lee JM, et al. Effect of Fimasartan versus valsartan and Olmesartan on office and ambulatory blood pressure in Korean patients with mild-to-moderate essential hypertension: a randomized, double-blind, active control, three-parallel group, forced titration, multicenter, phase IV Study (fimasartan achieving systolic blood pressure target (FAST) study). Drug Des Devel Ther. 2020;14:347–60.
39. Lee JH, Yang DH, Hwang JY, Hur SH, Cha TJ, Kim KS, et al. A Randomized, double-blind, candesartan-controlled, parallel group comparison clinical trial to evaluate the antihypertensive

efficacy and safety of fimasartan in patients with mild to moderate essential hypertension. Clin Ther. 2016;38(6):1485–97.

40. Lee HY, Kim YJ, Ahn T, Youn HJ, Chull Chae S, Seog Seo H, et al. A randomized, multicenter, double-blind, placebo-controlled, 3 × 3 factorial design, phase II study to evaluate the efficacy and safety of the combination of fimasartan/amlodipine in patients with essential hypertension. Clin Ther. 2015;37(11):2581–96.

41. Kim KI, Shin MS, Ihm SH, Youn HJ, Sung KC, Chae SC, et al. A randomized, double-blind, multicenter, phase iii study to evaluate the efficacy and safety of fimasartan/amlodipine combined therapy versus fimasartan monotherapy in patients with essential hypertension unresponsive to fimasartan monotherapy. Clin Ther. 2016;38(10):2159–70.

42. Rhee MY, Baek SH, Kim W, Park CG, Park SW, Oh BH, et al. Efficacy of fimasartan/hydrochlorothiazide combination in hypertensive patients inadequately controlled by fimasartan monotherapy. Drug Des Devel Ther. 2015;9:2847–54.

43. Kim JY, Son JW, Park S, Yoo TH, Kim YJ, Ryu DR, et al. Fimasartan proteinuria sustained reduction in comparison with losartan in diabetic chronic kidney disease (FANTASTIC): study protocol for randomized controlled trial. Trials. 2017;18(1):632.

44. https://clinicaltrials.gov/ct2/show/NCT03231293. Accessed 24 June 2021.

45. United States Food and Drug Administration. Search list of recalled angiotensin II receptor blockers (ARBs) including valsartan, losartan and irbesartan. https://www.fda.gov/drugs/drug-safety-and-availability/search-list-recalled-angiotensin-ii-receptor-blockers-arbs-including-valsartan-losartan-and. Accessed July 16, 2019.

46. United States Food and Drug Administration. Search list of recalled angiotensin II receptor blockers (ARBs) including valsartan, losartan and irbesartan. https://www.fda.gov/drugs/drug-safety-and-availability/search-list-recalled-angiotensin-ii-receptor-blockers-arbs-including-valsartan-losartan-and-FDA. Accessed 28 June 2021.

47. Health Sciences Authority. Singapore. 2019. https://www.hsa.gov.sg/sartanupdates. Accessed 28 June 2021.

48. Byrd JB, Chertow GM, Bhalla V. Hypertension hot potato - anatomy of the angiotensin-receptor blocker recalls. N Engl J Med. 2019;380(17):1589–91.

Newer Calcium Channel Blockers

24

Mangesh Tiwaskar

24.1 Introduction

In the 1960s, a drug discovery programme targeted molecules that acted as coronary dilators to provide relief from angina pectoris. This led to the development of calcium channel blockers (CCBs), which block voltage-operated calcium channels (VOCs). These channels facilitate the entry of calcium into cells, which results in contraction [1]. Conversely, blockade of these channels led to dilation and the benefits observed with these drugs. Since then, CCBs have been used for the treatment of hypertension, angina, peripheral vascular disease and some forms of arrhythmia [2].

24.1.1 Calcium Voltage-Operated Channels

Structurally, VOCs are composed of four subunits $\alpha1$, $\alpha2$-δ, β and γ. The $\alpha1$ subunit forms the core structure of the calcium ion channel. The $\alpha1$ subunits occur in ten different forms that are distributed in specific regions of the body and vary in their ion conductance [3]. Depending upon the types of $\alpha1$ subunit present, five different types of calcium channels have been identified: L-, N-, T-, P-, Q- and R-types [4]. Of these, only the T-type are low-voltage-operated channels, while the rest belong in the high-voltage category [5].

As shown in Table 24.1, neurons include several types of calcium channels whereas cardiac and vascular cells express only one or two types [6]. The subunits $\alpha1S$ ($Ca_v1.1$), $\alpha1C$ ($Ca_v1.2$), $\alpha1D$ ($Ca_v1.3$) or $\alpha1F$ ($Ca_v1.4$) expressed in L-type calcium channels [3] are the primary targets of commonly used CCBs like amlodipine. This is because these L-type calcium channels are crucial for the coupled processes

M. Tiwaskar (✉)
Karuna Hospital, Borivali West and Shilpa Medical Research Centre, Mumbai, India

© The Author(s), under exclusive license to Springer Nature Switzerland AG 2022
C. V. S. Ram et al. (eds.), *Hypertension and Cardiovascular Disease in Asia*,
Updates in Hypertension and Cardiovascular Protection,
https://doi.org/10.1007/978-3-030-95734-6_24

Table 24.1 Location of different types of voltage-operated calcium channels (VOCs) and associated indications for calcium channel blockers (CCBs)

Type of VOC	Systemic distribution	Intrarenal distribution	Indications for CCBs[a]
L	Skeletal/smooth muscle, pancreas, adrenal gland, brain and retina	Afferent arterioles, mesangial cells	Hypertension, angina, nerve damage from subarachnoid haemorrhage (Parkinson's disease, autism, schizophrenia, bipolar disorder)
T	Heart (sinoatrial and atrioventricular node), Purkinje fibre and along the nerve, brain, liver, adrenal gland	Afferent and efferent arterioles, mesangial cells, vasa recta, distal tubules, cortical collecting duct	Sleep disorders, pain, obesity, absence epilepsy, cancer
N	Nervous system and brain	Neuron on both afferent and efferent arterioles	Severe chronic refractory pain
P	Brain, neuron and pituitary gland		Migraine, ataxia, oncology
Q			
R			Seizure disorders, analgesia

[a]Possible new indications are given in brackets

of excitation and contraction in the cardiovascular system [6]. Thus, the clinical indications for different CCBs depend on the type of VOC being blocked (Table 24.1).

24.2 Types of Calcium Channel Blockers

Based on their chemical structure, commonly used CCBs were initially categorised into three distinct types (Fig. 24.1).

CCBs are thus referred to as dihydropyridines and non-dihydropyridines. All of the above agents block the entry of calcium into the cell by binding to the α1 subunit of the VOC. However, their differing chemical structures determine their binding site on the subunit. For example, benzothiazepines and dihydropyridines bind to a specific amino acid on the α1 subunit that is exposed to the cell surface, whereas phenylalkylamines bind to a region that is present on the inner surface of the cell membrane [3]. Dihydropyridines have greater affinity for calcium channels in vascular smooth muscle cells (VSMCs), whereas non-dihydropyridines such as verapamil act on the impulse conduction system to mediate negative inotropic, chronotropic and dromotropic effects. Diltiazem also acts on the impulse conduction system, particularly atrioventricular node conduction and the myocardium [9].

Dihydropyridines do not show negative chronotropic effects (due to reflex tachycardia that occurs secondary to peripheral vasodilation) [3]. Depending on the specific properties of each drug, its impact on the cardiovascular system is modified by the vasodilatory and baroceptor-mediated autonomic reflex [10]. The diarylaminopropylamine derivative bepridil has a negative inotropic effect [9].

Fig. 24.1 Initial classification of traditional CCBs (first reported by Fleckerstein et al. [7, 8])

Dihydrophyridinic agents :

Peripheral vessel – vasodilatation
Nifedipine

Phenilalchilaminic agents :

Cardiac - Negative inotropic effect
Verapamil

Benzothiazepinic agents :

Intermediate profile
Diltiazem

These are primarily L-type Ca2+ channel blockers

An overview of the cardiac and haemodynamic impact of representative drugs of dihydropyridine and non-dihydropyridine CCBs in adults with normal BP or hypertension is shown in Table 24.2.

Four generations of dihydropyridine CCBs have been developed with the goal of improving their efficacy and safety [4, 11]. Initially, CCBs blocked only the L-type VOC, but more recently developed agents block more than one type of calcium channel, improving their activity and benefits (Fig. 24.2) [3]. Some of the key differences between the different DHP CCBs are given in Table 24.3 [4, 12].

24.3 Mechanism of Action and Clinical Uses

All CCBs prevent the influx of calcium into various cells. Depending upon their affinity, CCBs can decrease VSMC contraction, reduce left ventricular (LV) contraction and heart rate or depress the excitement of the impulse conduction system of the heart. Their effects make them ideal candidates for the treatment of a variety of clinical conditions:

- CCBs prevent the entry of calcium into VSMC leading to dilation of peripheral vessels. This reduces peripheral resistance making CCBs effective antihypertensive agents [13].
- The dilatory effect of CCBs on the peripheral vasculature and coronary artery and their beneficial impact on LV contraction and heart rate decreases peripheral resistance, improves coronary blood flow and decreases myocardial oxygen consumption, which makes them useful as antianginal agents [14].
- CCBs have a positive influence on the impulse conduction system of the heart by decreasing the sinus node pacemaker rate and the conduction velocity of the atrioventricular node, leading to their antiarrhythmic effect [15, 16].
- CCBs also exhibit several pleiotropic effects, including anti-inflammatory and antioxidant effects, atherosclerotic plaque stabilisation, decreased VSMC migration/multiplication, inhibition of platelet aggregation and increased nitric oxide (NO) production [17].

Table 24.2 Cardiovascular and haemodynamic impact of dihydropyridine (DHP) and non-DHP CCBs

	Non-DHP (verapamil)		Non-DHP (diltiazem)		DHP (nifedipine)	
	Normotension	Hypertension	Normotension	Hypertension	Normotension	Hypertension
Effects on cardiovascular system						
Inotropic effect	↔,↓	→	↔,↓	↔	→	↔
Dromotropic effect	→	→	→	→	↔	↔
Chronotropic effect/heart rate	↔,↑	↔,↑	↔,↓	↑,↓	↔,↑	↔,↑
Myocardial oxygen consumption or RPP	↔	→	↔,↓	→	→	↔,↓

↑ increase, ↓ decrease, ↔ stable, *RPP* rate-pressure product

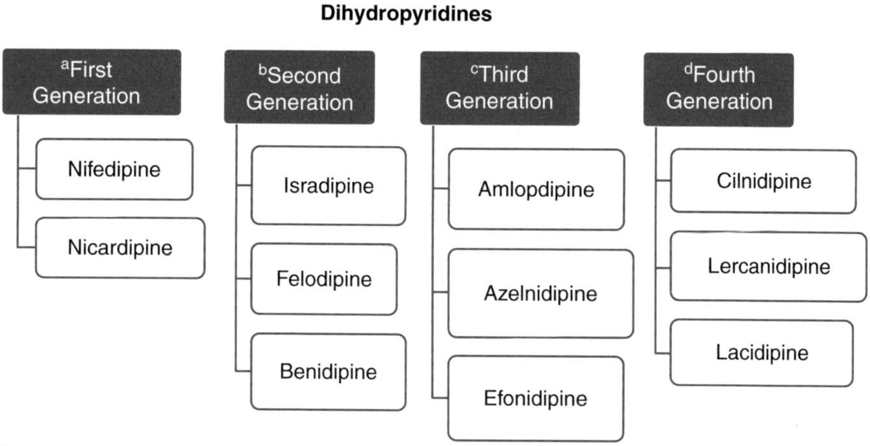

Dihydropyridines

ᵃFirst Generation	ᵇSecond Generation	ᶜThird Generation	ᵈFourth Generation
Nifedipine	Isradipine	Amlopdipine	Cilnidipine
Nicardipine	Felodipine	Azelnidipine	Lercanidipine
	Benidipine	Efonidipine	Lacidipine

ᵃ Short duration, more adverse effects
ᵇ Slow release formulations
ᶜ New formulations, more lipophilic, stable pharmacokinetics, well tolerated in HF
ᵈ (L/N-type) Stable activity, reduced adverse effects, broad spectrum

Fig. 24.2 Examples of different generations of dihydropyridine CCBs and their features

Table 24.3 Key differences between different generations of dihydropyridine CCBs

Parameters	Dihydropyridine CCBs			
	1st generation	2nd generation	3rd generation	4th generation
Calcium channel blocking activity	Primarily L-type	Primarily L-type or dual channel blockers	Primarily L-type or dual channel blockers	Dual channel blockers
Examples	Nifedipine	Nifedipine SR benidipine Efonidipine	Amlodipine, azelnidipine	Cilnidipine lercanidipine
Onset of action	Swift	Gradual	Slow	
Duration of action	Short-acting (half-life 2 h)	Moderate (half-life 7 h)	Long-acting (half-life 10–36 h)	Long-acting (half-life 7.5–10 h)
Lipophilicity	Less	Less	Moderately lipophilic	Highly lipophilic
Vasodilation	Equally effective			
Excitation of sympathetic nervous system	+++	++	+	-
Side effects				
Oedema	++	++	+	-
Tachycardia	+++	++		-
Headache	+++	++	+	
Flushing	+++	++	+	+

The most common clinical indications for CCBs include hypertension, angina pectoris, supraventricular dysrhythmias, coronary spasm [18], hypertrophic cardiomyopathy and pulmonary hypertension [15, 16]. In addition, CCBs are also prescribed for Raynaud's phenomenon, subarachnoid haemorrhage and migraine headaches [19]. Common adverse events with non-dihydropyridine CCBs include constipation, increased cardiac output and bradycardia, whereas the side effects of dihydropyridine CCBs include headaches, dizziness, flushing and peripheral oedema [20–22].

More details about some currently available CCBs are provided below and summarised in Table 24.4.

24.4   Amlodipine: The Classic CCB

Amlodipine has been used as one of the first line antihypertensives for several decades and is known as the gold standard CCB in terms of BP reduction [2, 27, 28]. It has various other effects, including antianginal and antiatherosclerotic properties [14]. Amlodipine is a long-acting dihydropyridine CCB that potently blocks the L-type calcium channels present in peripheral blood vessels and has N-type calcium channel blocking activity [3, 29]. Due to its slow onset of action, amlodipine does not cause reflex neuroendocrine activity, and its long half-life ensures 24-h BP control [29].

The beneficial effects of amlodipine on cardiovascular outcomes have been documented in several long-term landmark trials, including such as the Anglo-Scandinavian Cardiac Outcomes Trial-Blood Pressure Lowering Arm (ASCOT-BPLA) study [30] and the Valsartan Antihypertensive Long-term Use Evaluation (VALUE) trial [31]. In fact, the ASCOT study was stopped after 5.5 years because the amlodipine-based regimen showed clear mortality benefits compared with atenolol-based therapy [30]. In the VALUE trial, the BP-lowering effects of amlodipine-based regimen were evident early after treatment initiation, highlighting the rapid and sustained effect antihypertensive effects of amlodipine [31]. Given that combination antihypertensive therapy is often required to maintain sustained BP control, it is good that data from these trials also showed that amlodipine was highly effective when used in combination with other antihypertensive agents [32].

24.5   L-Type Channel Blockers

24.5.1 Clevidipine

Clevidipine is a CCB that can be administered intravenously (IV) and is available as a lipid emulsion. It is approved for use in settings where oral therapy is not possible. Clevidipine IV is an effective treatment for acute hypertension both pre- and postoperatively in adults undergoing cardiac surgery [23]. Clevidipine inhibits L-type calcium channels and in vitro studies confirm its high vascular selectivity. It has a

Table 24.4 Overview of dihydropyridine CCBs

Parameter	Amlodipine [2]	Clevidipine [23]	Efonidipine [24]	Azelnidipine [25]	Benidipine [26]	Cilnidipine [12]
Type of calcium channels blocked	L-type	L-type	L-/T-type	L-/T-type	L-/T-type; some effect on N-type	L-/N-type
Time to maximum plasma concentration	5.35 h	2–4 min	1.5–3.67 h	4.14 h	T_{max} <2 h	T_{max} 2 h
Duration of action	Prolonged (>36 h)	Short	Prolonged	Prolonged	Prolonged	Prolonged
Indication	Hypertension, coronary artery disease	For BP reduction when oral therapy is not possible; severe and perioperative hypertension	Essential hypertension, reno-parenchymal hypertension, angina pectoris	Hypertension	Hypertension, reno-parenchymal hypertension, angina pectoris	Hypertension, for end-organ protection
Dosage	5 mg/day (OD)	Initiate IV infusion at 1–2 mg/h	20–40 mg/day (OD) (40 mg/day BID for angina)	8–16 mg/day OD	Hypertension: 2–4 mg/day OD Angina: 8 mg/day BID	5–10 mg/day OD
Hepatic impairment	Start with lower dose (2.5 mg/day)	Not applicable	Use with caution	–	Liver function needs to be monitored	Use with caution; start at lower dose
Renal impairment	No dose adjustment needed	Can be used	–	Use with caution	–	No dose adjustment needed
Initial approval	US FDA (1987)	US FDA (2008)	Japan (1995)	Japan (2003)	Japan (1991)	Japan (1995)
CV outcome or mortality data	ALLHAT, ASCOT, VALUE, ACCOMPLISH	–	–	OSCAR trial	COPE trial	–

OD once daily, *BID* twice daily, *IV* intravenously, *US FDA* United State Food and Drug Administration, *ALLHAT* antihypertensive lipid-lowering treatment to prevent heart attack trial, *ASCOT* Anglo-Scandinavian cardiac outcomes trial, *VALUE* valsartan antihypertensive long-term use evaluation, *ACCOMPLISH* avoiding cardiovascular events through combination therapy to patients living with systolic hypertension, *OSCAR* OlmeSartan and calcium antagonists randomised, *COPE* combination therapy of hypertension to prevent cardiovascular events dose-dependent rapid onset and short duration of action. It reduces systemic vascular resistance without influencing preload and the effect is offset within 5–15 min of administration. In patients undergoing cardiac surgery, clevidipine is well-tolerated with a safety profile comparable to that of nicardipine, sodium nitroprusside or nitroglycerin. In patients with hypertension after coronary artery bypass grafting (CABG), clevidipine had better effects on arterial vasodilation and less effects on venodilatation compared with sodium nitroprusside [33].

24.6   L/T-Type Channel Blockers

L-type calcium channels are present in large vessels, whereas the T-type calcium channels are predominantly present in the microvessels. Compared to an L-type CCB, a dual L/T blocker can offer enhanced renal and cardiovascular benefits for a number of reasons [34]:

- T-type calcium channels are present on both the afferent and efferent arterioles of the glomerular apparatus, suggesting a positive impact on intraglomerular pressure [35].
- Reduced urinary protein excretion and glomerular filtration rate (GFR).
- Decreased aldosterone secretion, inflammation and oxidative stress [36].
- Improved endothelial function.
- Better overall antihypertensive and renoprotective effects.
- Lower risk of oedema because hydrostatic pressure across the capillary bed is equalised.

However, more studies on this type of CCBs are warranted because the promising T-type channel blocker mibefradil had to be withdrawn due to its interaction with the cytochrome P-450 3A4 enzyme [24].

24.6.1 Efonidipine

Efonidipine hydrochloride (efonidipine) is an L-/T-type DHP CCB with antihypertensive and antianginal properties with little/no reflex tachycardia [34]. It differs chemically from most other dihydropyridine CCBs because it has a phosphonate moiety at position 5 of the DHP ring. It is highly lipophilic and enters the cell membrane to reach its binding site. It has a slow onset and long duration of action due to its low dissociation constant [37].

Due to its dual blockade of calcium channels, efonidipine has antihypertensive, antianginal and cardio- and renoprotective properties. Efonidipine prolongs phase 4 or pacemaker depolarisation of the sinoatrial node action potential, an effect not found with several other CCBs, which contributes to its potent negative chronotropic effect and maintains cardiac output. In patients with hypertension and an elevated heart rate (>80 beats/min), efonidipine decreased BP to a similar extent as other dihydropyridine CCBs but maintained a lower heart rate, a property beneficial in patients with angina. Treatment with efonidipine also decreased myocardial oxygen demand and improved coronary blood flow, mitigating myocardial ischaemia and reduced the negative impact of ischaemia on myocardial adenosine triphosphate (ATP) level and energy charge potential [37].

Efonidipine has an equipotent dilatory impact on both afferent and efferent arterioles, resulting in increased GFR without elevated intraglomerular pressure, thereby preventing hypertension-induced renal damage. Efodipine decreased plasma aldosterone levels, even in patients undergoing maintenance dialysis, which may help prevent aldosterone-induced renal parenchymal fibrosis and offer additional cardiovascular protection. The renoprotective effects of efonidipine were found to be comparable to those of agents such as angiotensin-converting enzyme inhibitors (ACEIs). Side effects such as pedal oedema, frequent urination and enhanced triglycerides occurred in <0.1% of patients treated with efonidipine [38].

24.6.2 Azelnidipine

Azelnidipine is a lipophilic L-/T-type dihydropyridine CCB that has 17-fold higher lipophilicity than amlodipine [25] and a high affinity for vascular tissue [3]. It has a slow onset of action with a prolonged effect and is now being studied for its impact on post-ischaemic stroke management [39]. Azelnidipine has similar BP-lowering activity to amlodipine, does not elevate the pulse rate [40] and decreases cardiac hypertrophy in hypertension without influencing cardiac output while enhancing blood flow to the heart and the brain. Due to its diuretic and natriuretic properties, azelnidipine reduces renal injury and protein excretion caused by hypertension [41]. In patients with hypertension, azelnidipine decreases heart rate and proteinuria [39], controlled morning hypertension and lowered the pulse rate [25]. Overall, azelnidipine showed antiatherosclerotic, and cardio-, nephro- and neuroprotective effects and prevented insulin resistance [42]. The most common adverse events in clinical trials were light-headedness (0.5%), facial flushes (0.5%) and headache (1.1%), with no presence of oedema [39].

24.6.2.1 Clinical Study

In the OlmeSartan and Calcium Antagonists Randomized (OSCAR) study, high-risk elderly patients with uncontrolled hypertension ($n = 1165$; age >65 years) who had undergone prior treatment with olmesartan (20 mg/day) were treated with either high-dose olmesartan 40 mg/day or olmesartan 20 mg/day + amlodipine or azelnidipine [43, 44]. After 3 years' treatment, mean BP was lower in the combination group compared with the high-dose group (132.6/72.6 vs. 135.0/74.3 mmHg). The primary endpoint (a composite of cardiovascular events and non-cardiovascular death) occurred less often in the combination therapy versus high-dose olmesartan group (48 vs. 58 events; $p = 0.17$). The between-group difference in primary outcome events was statistically significant in the subgroup of patients who had existing cardiovascular disease at baseline.

24.6.3 Benidipine

Benidipine is referred to as an L-/T-type dihydropyridine CCB, although it has also been shown to inhibit N-type calcium channels [36]. Benidipine binds and dissociates slowly from its binding site and other tissues, resulting in a greater vasoselectivity and vasodilatory effects than other CCBs, including amlodipine, nifedipine and diltiazem [5]. Benidipine enters the cell membrane and then adheres to the dihydropyridine binding site. This unique mechanism of action is called the 'membrane approach' and is the key reason for its long-lasting activity [26].

Benidipine therapy has antihypertensive effects without affecting heart rate and improves endothelial function, pulse wave velocity and the augmentation index [26]. Benidipine has also been shown to have antioxidant effects, stimulate NO production, suppress expression of adhesion molecules, cause osteoblast differentiation, decrease proliferation of VSMC/mesangial cells and protect the myocardium [26, 45]. In patients with angina, benidipine improved prognosis to a greater extent than other CCBs [26]. In addition, the combination of benidipine with an angiotensin receptor blocker (ARB) showed additive benefits in patients with hypertension [26].

The renoprotective effects of benidipine are due to equivalent dilation of the glomerular afferent and efferent arterioles, which attenuates the decline of renal function compared with other L-type CCBs. Treatment with benidipine reduces aldosterone production, thus decreasing oxidative stress and reducing activation of the aldosterone-induced mineralocorticoid receptor, which contributes to its end-organ protective effects [46, 47]. Compared with amlodipine, benidipine also exerts a natriuretic effect, which could be related to enhanced renal blood flow [26].

24.6.3.1 Clinical Study

In the Combination Therapy of Hypertension to Prevent Cardiovascular Events (COPE) [48] trial, patients with uncontrolled hypertension (age 40–85 years) were randomised to treatment with benidipine/ARB ($n = 1166$, benidipine/β-blocker ($n = 1168$) or benidipine/thiazide diuretic ($n = 1168$); median follow-up was 3.61 years. The rate of the composite cardiovascular endpoint was lowest in the benidipine/thiazide diuretic group (Fig. 24.3). The proportion of patients who achieved target

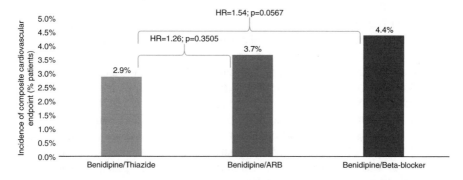

Fig. 24.3 Effect of benidipine-containing combination therapy regimens on cardiovascular events

BP was 64.1% with benidipine/ARB, 66.9% with benidipine/β-blocker and 66.0% with benidipine/thiazide diuretic. Compared with benidipine/β-blocker, there was a significantly lower incidence of fatal or nonfatal strokes in the benidipine/thiazide diuretic group ($p = 0.0109$), and a significantly lower rate of new-onset diabetes with benidipine/ARB ($p = 0.0240$). All trial treatments were well-tolerated. A sub-analysis of the COPE trial showed that the benidipine/thiazide diuretic combination reduced cardiovascular outcomes even in patients with poor BP control [49].

24.7 L-/N-Type Channel Blockers

24.7.1 Cilnidipine

Cilnidipine has L-/N-type calcium channel blocking activity. Due to the distribution of N-type channels along neurons and in the brain, cilnidipine prevents sympathetic nervous system activation in addition to its antihypertensive activity [3]. The dual impact of blocking L-type channels on VSMCs and N-type channels on neurons results in an antihypertensive effect with minimal reflex tachycardia and decreased heart rate variability [5].

L-/N-type CCBs have renoprotective effects based on decreased activation of the sympathetic nervous system and the renin–angiotensin–aldosterone system (RAAS), which causes arteriolar vasodilation and reduces proteinuria and podocyte injury [50]. Cilnidipine also has a beneficial effect on LV dysfunction, decreases LV mass, has antioxidant effects and is effective for white-coat and morning hypertension [51, 52]. Pedal oedema, a common side-effect with conventional dihydropyridine CCBs, to be much less frequent during treatment with cilnidipine [5, 12].

It has been suggested that the renoprotective effects of L-/N-type dihydropyridine CCBs in patients with diabetes may not be as marked as those of L-/T-type dihydropyridine CCBs due to the presence of diabetic neuropathy [53]. However, another study showed that switching to from L-type dihydropyridine CCBs to cilnidipine decreased urinary albumin excretion in normo-/microalbuminuric patients with diabetes [54]. Therefore, a greater understanding of the impact of cilnidipine in patients with diabetes is needed.

The renoprotective effects of dual-acting dihydropyridine CCBs (L-/N-type or L-/T-type) have been shown to be useful. A recent renal subanalysis of the JATOS trial demonstrated that a 2-year treatment with efonidipine elevated eGFR in elderly hypertensive patients, even when patients had reduced kidney function or diabetes at baseline [55].

24.7.1.1 Clinical Study

The impact of cilnidipine on morning hypertension, BP and pulse rate in patients with hypertension ($n = 2319$) was assessed over 12 weeks in the Ambulatory Blood Pressure Control and Home Blood Pressure (Morning and Evening) Lowering By N-Channel Blocker Cilnidipine (ACHIEVE-ONE) trial [56]. Both clinic and morning systolic BP were reduced during treatment with cilnidipine, by 19.6 and 17 mmHg, respectively. Decreases in morning systolic BP and pulse rate was also observed in the subgroup of patients with higher levels of these parameters at baseline [56]. In another analysis of data from the ACHIEVE-ONE study, patients ($n = $ 615) were divided into extreme dippers, dippers, nondippers or risers depending on their nocturnal BP. Cilnidipine significantly, but not completely, reversed abnormal nocturnal BP dipping status in these patients [57].

24.8 Conclusion

CCBs are an important part of a clinician's armamentarium for the treatment of hypertension and cardiovascular disease. L-type CCBs such as amlodipine have been in widespread clinical usage for many years due to their antihypertensive and antianginal efficacy and established cardiovascular benefits. Newer long-acting L-type, L-/T-type and L-/N-type CCBs that have additional activities need to be further evaluated in well-designed trials to facilitate their successful incorporation into clinical practice.

References

1. Godfraind T. Discovery and development of calcium channel blockers. Front Pharmacol. 2017;8:286.
2. Fares H, DiNicolantonio JJ, O'Keefe JH, Lavie CJ. Amlodipine in hypertension: a first-line agent with efficacy for improving blood pressure and patient outcomes. Open Heart. 2016;3(2):e000473.
3. Ozawa Y, Hayashi K, Kobori H. New generation calcium channel blockers in hypertensive treatment. Curr Hypertens Rev. 2006;2(2):103–11.
4. Wang AL, Iadecola C, Wang G. New generations of dihydropyridines for treatment of hypertension. J Geriatr Cardiol. 2017;14(1):67–72.
5. Tamargo J, Ruilope LM. Investigational calcium channel blockers for the treatment of hypertension. Expert Opin Investig Drugs. 2016;25(11):1295–309.
6. Richard S. Vascular effects of calcium channel antagonists: new evidence. Drugs. 2005;65(2):1–10.
7. Fleckenstein A, Tritthart H, Flackenstein B, Herbst A, Grün G. A new group of competitive divalent Ca-antagonists (iproveratril, D 600, prenylamine) with potent inhibitory effects on electromechanical coupling in mammalian myocardium. Pflugers Arch. 1969;307(2):25.
8. Fleckenstein A, Tritthart H, Döring HJ, Byon KY. B [BAY a 1040--a highly potent Ca ++ -antagonistic inhibitor of electro-mechanical coupling processes in mammalian myocardium]. Arzneimittelforschung. 1972;22(1):22–33.
9. Sueta D, Tabata N, Hokimoto S. Clinical roles of calcium channel blockers in ischemic heart diseases. Hypertens Res. 2017;40(5):423–8.
10. Wang J, McDonagh DL, Meng L. Calcium channel blockers in acute care: the links and missing links between hemodynamic effects and outcome evidence. Am J Cardiovasc Drugs. 2021;21(1):35–49.
11. Aouam K, Berdeaux A. Dihydropyridines from the first to the fourth generation: better effects and safety. Therapie. 2003;58(4):333–9.
12. Chandra KS, Ramesh G. The fourth-generation calcium channel blocker: cilnidipine. Indian Heart J. 2013;65(6):691–5.

13. Katz AM. Pharmacology and mechanisms of action of calcium-channel blockers. J Clin Hypertens. 1986;2(3):28–37.
14. Godfraind T. Calcium channel blockers in cardiovascular pharmacotherapy. J Cardiovasc Pharmacol Ther. 2014;19(6):501–15.
15. Singh BN, Hecht HS, Nademanee K, Chew CY. Electrophysiologic and hemodynamic effects of slow-channel blocking drugs. Prog Cardiovasc Dis. 1982;25(2):103–32.
16. Singh BN, Nademanee K, Baky SH. Calcium antagonists. Clinical use in the treatment of arrhythmias. Drugs. 1983;25(2):125–53.
17. Aftab K, Arain AA, Memon AAR, Memon AR, Phull QZ, Ansari MA. Evaluation of the anti platelet aggregation effects of diltiazem. J Pharmacol Clin Res. 2016;1(5):555–71.
18. Xie W, Zheng F, Evangelou E, Liu O, Yang Z, Chan Q, et al. Blood pressure-lowering drugs and secondary prevention of cardiovascular disease: systematic review and meta-analysis. J Hypertens. 2018;36(6):1256–65.
19. Solomon GD, Steel JG, Spaccavento LJ. Verapamil prophylaxis of migraine. A double-blind, placebo-controlled study. JAMA. 1983;250(18):2500–2.
20. Abernethy DR, Schwartz JB. Calcium-antagonist drugs. N Engl J Med. 1999;341(19):1447–57.
21. Makani H, Bangalore S, Romero J, Htyte N, Berrios RS, Makwana H, et al. Peripheral edema associated with calcium channel blockers: incidence and withdrawal rate--a meta-analysis of randomized trials. J Hypertens. 2011;29(7):1270–80.
22. Pedrinelli R, Dell'Omo G, Mariani M. Calcium channel blockers, postural vasoconstriction and dependent oedema in essential hypertension. J Hum Hypertens. 2001;15(7):455–61.
23. Keating GM. Clevidipine: a review of its use for managing blood pressure in perioperative and intensive care settings. Drugs. 2014;74(16):1947–60.
24. Tanaka H, Shigenobu K. Efonidipine hydrochloride: a dual blocker of L- and T-type ca(2+) channels. Cardiovasc Drug Rev. 2002;20(1):81–92.
25. Kario K, Sato Y, Shirayama M, Takahashi M, Shiosakai K, Hiramatsu K, Komiya M, Shimada K. Inhibitory effects of azelnidipine tablets on morning hypertension. Drugs. 2013;13(1):63–73.
26. Yao K, Nagashima K, Miki H. Pharmacological, pharmacokinetic, and clinical properties of benidipine hydrochloride, a novel, long-acting calcium channel blocker. J Pharmacol Sci. 2006;100(4):243–61.
27. Management of Hypertension. Indian guidelines on hypertension-IV. J Assoc Phys. 2019;67(Suppl):22–9.
28. Neutel J, Smith DH. Evaluation of angiotensin II receptor blockers for 24-hour blood pressure control: meta-analysis of a clinical database. J Clin Hypertens. 2003;5(1):58–63.
29. Tiwaskar M, Langote A, Kashyap R, Toppo A. Amlodipine in the era of new generation calcium channel blockers. J Assoc Phys. 2018;66(3):64–9.
30. Dahlöf B, Sever PS, Poulter NR, Wedel H, Beevers DG, Caulfield M, et al. Prevention of cardiovascular events with an antihypertensive regimen of amlodipine adding perindopril as required versus atenolol adding bendroflumethiazide as required, in the Anglo-Scandinavian Cardiac Outcomes Trial-Blood Pressure Lowering Arm (ASCOT-BPLA): a multicentre randomised controlled trial. Lancet. 2005;366(9489):895–906.
31. Julius S, Weber MA, Kjeldsen SE, McInnes GT, Zanchetti A, Brunner HR, et al. The valsartan antihypertensive long-term use evaluation (VALUE) trial: outcomes in patients receiving monotherapy. Hypertension. 2006;48(3):385–91.
32. Lacourcière Y, Poirier L, Lefebvre J, Archambault F, Cléroux J, Boileau G. Antihypertensive effects of amlodipine and hydrochlorothiazide in elderly patients with ambulatory hypertension. Am J Hypertens. 1995;8(12):1154–9.
33. Aronson S, Dyke CM, Stierer KA, Levy JH, Cheung AT, Lumb PD, et al. The ECLIPSE trials: comparative studies of clevidipine to nitroglycerin, sodium nitroprusside, and nicardipine for acute hypertension treatment in cardiac surgery patients. Anesth Analg. 2008;107(4):1110–21.
34. Ge W, Ren J. Combined L-/T-type calcium channel blockers: ready for prime time. Hypertension. 2009;53(4):592–4.
35. Weiss N, Zamponi GW. T-type calcium channels: from molecule to therapeutic opportunities. Int J Biochem Cell Biol. 2019;108:34–9.

36. Abe M, Soma M. Multifunctional L/N- and L/T-type calcium channel blockers for kidney protection. Hypertens Res. 2015;38(12):804–6.
37. Koh KK, Quon MJ, Lee SJ, Han SH, Ahn JY, Kim JA, et al. Efonidipine simultaneously improves blood pressure, endothelial function, and metabolic parameters in nondiabetic patients with hypertension. Diabetes Care. 2007 Jun;30(6):1605–7.
38. Efonidipine Pubchem. 2021. Available from https://pubchem.ncbi.nlm.nih.gov/compound/ Efonidipine.
39. Azelnidipine. Pubchem. 2021. Available from https://pubchem.ncbi.nlm.nih.gov/compound/ Azelnidipine
40. Kuramoto K, Ichikawa S, Hirai A, Kanada S, Nakachi T, Ogihara T. Azelnidipine and amlodipine: a comparison of their pharmacokinetics and effects on ambulatory blood pressure. Hypertens Res. 2003;26(3):201–8.
41. Yagil Y, Lustig A. Azelnidipine (CS-905), a novel dihydropyridine calcium channel blocker with gradual onset and prolonged duration of action. Cardiovasc Drug Rev. 1995;13(2):137–48.
42. Sada T, Saito H. Pharmacological profiles and clinical effects of azelnidipine, a long-acting calcium channel blocker. Nihon Yakurigaku Zasshi. 2003;122(6):539–47.
43. Ogawa H, Kim-Mitsuyama S, Matsui K, Jinnouchi T, Jinnouchi H, Arakawa K, et al. Angiotensin II receptor blocker-based therapy in Japanese elderly, high-risk, hypertensive patients. Am J Med. 2012;125(10):981–190.
44. Kim-Mitsuyama S, Ogawa H, Matsui K, Jinnouchi T, Jinnouchi H, Arakawa K. An angiotensin II receptor blocker-calcium channel blocker combination prevents cardiovascular events in elderly high-risk hypertensive patients with chronic kidney disease better than high-dose angiotensin II receptor blockade alone. Kidney Int. 2013;83(1):167–76.
45. Benidipine Pubchem. 2021. Available from https://pubchem.ncbi.nlm.nih.gov/compound/ Benidipine.
46. Tomino Y. Renoprotective effects of the L-/T-type calcium channel blocker benidipine in patients with hypertension. Curr Hypertens Rev. 2013;9(2):108–14.
47. Kosaka H, Hirayama K, Yoda N, Sasaki K, Kitayama T, Kusaka H, et al. The L-, N-, and T-type triple calcium channel blocker benidipine acts as an antagonist of mineralocorticoid receptor, a member of nuclear receptor family. Eur J Pharmacol. 2010;635(1-3):49–55.
48. Matsuzaki M, Ogihara T, Umemoto S, Rakugi H, Matsuoka H, Shimada K, et al. Prevention of cardiovascular events with calcium channel blocker-based combination therapies in patients with hypertension: a randomized controlled trial. J Hypertens. 2011;29(8):1649–59.
49. Umemoto S, Ogihara T, Matsuzaki M, Rakugi H, Ohashi Y, Saruta T. Effects of calcium channel blocker benidipine-based combination therapy on target blood pressure control and cardiovascular outcome: a sub-analysis of the COPE trial. Hypertens Res. 2017;40(4):376–84.
50. Thamcharoen N, Susantitaphong P, Wongrakpanich S, Chongsathidkiet P, Tantrachoti P, Pitukweerakul S, et al. Effect of N- and T-type calcium channel blocker on proteinuria, blood pressure and kidney function in hypertensive patients: a meta-analysis. Hypertens Res. 2015;38(12):847–55.
51. Tan HW, Li L, Zhang W, Ma ZY, Zhong XZ, Zhang Y. Effect of cilnidipine on left ventricular function in hypertensive patients as assessed by tissue Doppler Tei index. J Hum Hypertens. 2006;20(8):618–24.
52. Onose Y, Oki T, Yamada H, Manabe K, Kageji Y, Matsuoka M, et al. Effect of cilnidipine on left ventricular diastolic function in hypertensive patients as assessed by pulsed Doppler echocardiography and pulsed tissue Doppler imaging. Jpn Circ J. 2001;65(4):305–9.
53. Hayashi K. L-/T-type Ca channel blockers for kidney protection: ready for sophisticated use of Ca channel blockers. Hypertens Res. 2011;34(8):910–2.
54. Fukumoto S, Ishimura E, Motoyama K, Morioka T, Kimoto E, Wakikawa K, et al. Cilnidipine vs L-type calcium channel blockers evaluation of antihypertensive renoprotective effects in diabetic patients (CLEARED) study investigators. Antialbuminuric advantage of cilnidipine compared with L-type calcium channel blockers in type 2 diabetic patients with normoalbuminuria and microalbuminuria. Diabetes Res Clin Pract. 2012;97(1):91–8.

55. Hayashi K, Saruta T, Goto Y, Ishii M. Impact of renal function on cardiovascular events in elderly hypertensive patients treated with efonidipine. Hypertens Res. 2010;33:1211–20.
56. Kario K, Ando S, Kido H, Nariyama J, Takiuchi S, Yagi T, et al. The effects of the L/N-type calcium channel blocker (cilnidipine) on sympathetic hyperactive morning hypertension: results from ACHIEVE-ONE. J Clin Hypertens. 2013;15(2):133–42.
57. Kario K, Nariyama J, Kido H, Ando S, Takiuchi S, Eguchi K, et al. Effect of a novel calcium channel blocker on abnormal nocturnal blood pressure in hypertensive patients. J Clin Hypertens. 2013;15(7):465–72.

Resistant Hypertension: Recognition and Treatment

25

Raymond V. Oliva and George L. Bakris

25.1 Introduction

Hypertension is a very common chronic disease and suboptimal blood pressure (BP) control is the most attributable risk for death worldwide. The net and adjusted prevalence ratios of hypertension in the USA continue to increase, but due to increased awareness, there is slight improvement in the treatment and control of hypertension. Several large hypertension outcome trials, however, showed failure to achieve BP goals despite protocol defined treatment regimens, with 20–35% of the participants could not achieve BP control despite receiving more than three antihypertensive medications [1–4].

Resistant hypertension is defined in the 2018 American Heart Association [5], 2017 American College of Cardiology/American Heart Association [6], and the 2018 European Society of Hypertension/European Society of Cardiology's [7] scientific statements as blood pressure that remains above target despite the concurrent use of three antihypertensive agents of different classes. Ideally, one of the agents should be a diuretic, and all the drugs used are prescribed at maximum tolerated doses. Individuals whose BP is controlled but are requiring four or more antihypertensive medications should also be considered resistant to treatment. This definition does not apply to patients who are recently diagnosed with hypertension. Although used arbitrarily, resistant hypertension is defined to identify patients who are at high

R. V. Oliva (✉)
Division of Hypertension, Department of Medicine, University of the Philippines-Philippine General Hospital, Manila, Philippines
e-mail: rvoliva@up.edu.ph

G. L. Bakris
American Heart Association Comprehensive Hypertension Center, Section of Endocrinology, Diabetes and Metabolism, Department of Medicine, University of Chicago, Chicago, IL, USA
e-mail: gbakris@medicine.bsd.umedicine.edu

© The Author(s), under exclusive license to Springer Nature Switzerland AG 2022
C. V. S. Ram et al. (eds.), *Hypertension and Cardiovascular Disease in Asia*,
Updates in Hypertension and Cardiovascular Protection,
https://doi.org/10.1007/978-3-030-95734-6_25

risk of having reversible causes of hypertension, and who may benefit from special diagnostic and therapeutic considerations [8]. The abovementioned definition is for a patient with apparent resistant hypertension, there is a need to differentiate true resistant hypertension versus pseudoresistance by confirming above goal BP using proper measurement techniques in the office, or by confirming it by using out-of-office measurements such as ambulatory BP monitor or home BP.

25.2   Prevalence of Resistant Hypertension

To accurately determine the prevalence of resistant hypertension, a forced titration study of a large diverse hypertensive cohort should be required. However, there are hypertension outcome studies which offer an alternative look to the prevalence of this condition. The targets in the American guidelines have defined BP control as systolic BP (SBP) <130 mmHg and diastolic BP (DBP) <80 mmHg for all adults <65 years of age [5, 6]. Before this new target, when the goal was <140/90 mmHg, the prevalence of true resistant hypertension was 17.7% [8, 9]. With the latest definition, the prevalence of resistant hypertension in the USA is 19.7%, with ~3% of American adults taking a thiazide-like diuretic and 9% are taking a mineralocorticoid receptor blocker [9]. A major problem is that not all uncontrolled hypertension is resistant, the former includes patients who lack BP control secondary to poor adherence and/or inadequate treatment regimen [8]. In a single center in California with 200,000 newly diagnosed hypertensive individuals followed up for 5 years, approximately 21% were eventually prescribed with three or more antihypertensive medications, 1 in 50 was resistant to treatment with an incident rate of 1.9% [10].

In the Philippines, the Report of the Council of Hypertension of the Philippine Heart Association (PRESYON 3) showed that the prevalence of hypertension in the country is 28%, with about 73% of those treated with antihypertensive medications were uncontrolled [11]. Non-compliance to medications is high as almost half of those are not taking their medications [11]. In a retrospective cohort in a tertiary hospital in the country, 36.8% were identified to be on a three-drug regimen, while 42.1% are on four or more antihypertensive medications [12]. In a Spanish registry of over 68,000 patients, the prevalence of resistant hypertension was 14.8%, based on the AHA criteria. Since they utilized 24-h ambulatory BP monitoring (ABPM), they found out that some patients who were diagnosed resistant using office pressures were actually controlled when subjected to 24-h ABPM [13].

Clinical trials suggest that resistant hypertension may be considerably higher. Looking at the antihypertensive and lipid-lowering treatment to prevent heart attack trial (ALLHAT), after approximately 5 years of follow-up, 34% of the participants remained uncontrolled on an average of 2 medications. At the completion of the study, 27% of participants were on 3 or more medications. Overall, 49% were controlled on 1 or 2 medications, while the rest needed 3 or more BP medications indicating resistance. The ALLHAT results were significant as it included a large number of ethnically diverse participants [14]. In a recent analysis of the Anglo-Scandinavian Cardiac Outcome Trial (ASCOT) with a follow-up of 5 years, 35% of the individuals enrolled have been untreated prior to the conduct of the study, and 50% of the previously treated patients met the diagnostic criteria for resistant hypertension [15]. In a post-hoc analysis of the Systolic Blood Pressure Intervention Trial (SPRINT), 19.2% had baseline apparent treatment-resistant hypertension [16]. In a recent meta-analysis involving 91 studies with 3.2 million patients between 1991 and 2017, the prevalence of true resistant hypertension was 10.3%, apparent treatment resistant was 14.7%, and pseudoresistant hypertension was 10.3% [16].

25.3 Prognosis of Resistant Hypertension

No clinical trials have evaluated the prognosis of patients with resistant hypertension comparing it to easily controlled hypertensives. Presumably, prognosis is impaired among these types of patients due to long-standing elevated BP and its associated cardiovascular (CV), such as diabetes mellitus, left ventricular hypertrophy (LVH), obstructive sleep apnea (OSA), and/or CKD [17]. As shown in the Veterans Administration Cooperative Studies, there is a 96% reduction in the CV events over 18 months use of triple antihypertensive regimen compared to placebo, suggesting the substantial benefit of successful treatment [18]. How much of this benefit is still unknown.

25.4 Characteristics of Patients with Resistant Hypertension

In the analysis of the Framingham study, the strongest predictor of lack of BP control was older age. The data showed that patients aged 75 years and above have less than one-fourth as likely to have controlled systolic BP compared to younger participants younger than 60 years old [19]. This may be attributable to stiffening of the arteries causing isolated systolic hypertension in the elderly. Other strong predictors of lack of systolic BP control are the presence of LVH and obesity. In terms of controlling diastolic BP, the strongest negative predictor was obesity. The BP was controlled in one-third less as compared to lean patients [20]. In the ALLHAT study, older age, higher baseline systolic BP, LVH, and obesity all predicted treatment resistance, as was also seen in the Framingham study [14]. Overall, however, the strongest predictor of treatment resistance is chronic kidney disease (CKD), defined by serum creatinine >1.5 mg/dl [8]. Other predictors such as presence of diabetes, African American especially the women, and people living in Southeast US have also increased risk for resistant hypertension.

In a single center in the Philippines, individuals who are elderly, female, and with concomitant diabetes mellitus may contribute to the development of resistant hypertension. These patients may need a total of four to five drug combinations to control their blood pressure [12]. In South Asia, individuals with resistant hypertension were associated with obesity and diabetes mellitus [21].

In the Brazilian Longitudinal Study of Adult Health (ELSA-Brazil), individuals with resistant hypertension were older, black, less educated, and obese. They also have associated subclinical markers for end organ cardiovascular damage such as low glomerular filtration rate, increased proteinuria, left ventricular hypertrophy, and early atherosclerosis and arterial stiffness [22].

The above studies indicate that with progressive older and heavier population who has concomitant CKD and DM, prevalence of resistant hypertension can be anticipated to increase.

25.5 The Role of Genetics

Genetic assessments of patients with resistant hypertension are limited. A study in Finland screened 347 patients with resistant hypertension for mutations of the β and ϒ subunits of the epithelial sodium channel (ENaC), which can cause Liddle's syndrome. Results showed mutations in two β and ϒ ENaC gene variants were significantly more prevalent in resistant hypertensive patients [23]. The presence of these gene mutations was associated with increased urinary potassium excretions related to plasma renin levels. In another study, a particular allele coding for 11β-hydroxysteroid dehydrogenase type 2 (CYP3A5*1), which plays an important role in the metabolism of cortisol and corticosterone, has been associated with African Americans who have higher systolic BP levels and more resistant to treatment [24].

Although both studies are based on very small number of patients, these results support additional attempts to identify genotypes that may relate to treatment resistance. Identification of the specific genetic influence may lead to development of new therapeutic targets.

25.6 The Dilemma of Pseudoresistance

Pseudoresistance refers to poorly controlled hypertension that appears to be resistant but is actually attributable to other factors. Such factors include: (1) inaccurate measurement technique, (2) white coat hypertension, and (3) poor adherence to therapy [8, 25–27]. A careful evaluation to exclude these factors before labeling a patient with true resistant hypertension should be performed.

Although guidelines are available to properly assess office BP readings, several mistakes often produce falsely elevated readings. Two most common mistakes, measuring the BP before letting the patient sit quietly for at least 5 min and the use of small arm cuffs; may result in erroneous readings [8, 27]. Other mistakes identified include recent smoking, using only one reading instead of the average of three BP readings, and not fully supporting the arm at heart level [28]. In the elderly population, the presence of highly calcified or arteriosclerotic arteries results in the overestimation of intra-arterial BP [29].

White coat hypertension, also called isolated clinic/office hypertension, is defined as an elevation of BP during a clinic visit. About 25% of patients with

Table 25.1   Factors that may improve medication adherence of patients with resistant hypertension

Selection of agents with low side effect profiles such as RAAS blockers
Keep the regimen simple: once daily dosing with single pill combinations
Use of pill boxes
Encourage self BP monitoring
Family to help patients with memory deficits
Improve communication of doctor and patient
Implementing team approach in the management of resistant hypertension
Elevate medication adherence as a critical healthcare issue

elevated office BP readings achieve BP goal under treatment when ambulatory measurements are performed [30]. Patients with white coat hypertension have less target organ damage and less CV risks compared with true resistant hypertensives during ambulatory techniques [31].

Poor adherence to antihypertensive therapy is another cause of elevated BP readings. Studies suggest that up to 40% of patients with newly diagnosed hypertension will discontinue their medications during the first year of treatment [32, 33]. In post-marketing surveillance of about 4000 patients, there is a 50% discontinuation rate of using antihypertensives within the first year of treatment [34]. Factors that may improve medication adherence by patients are listed in Table 25.1.

A surprising contributing factor to pseudoresistance is the clinical inertia among physicians. Clinical inertia is defined as a conscious decision to not adequately treat a condition despite knowing that the disease is present [29]. It may be due to lack of training and the experience on the proper use of antihypertensive agents or an overestimation of the care already provided [35]. Unfortunately, poor adherence of patients to treatment is usually seen at the primary care level and is less common when patients are seen by hypertension specialists. In a retrospective study at a hypertension specialty clinic, poor adherence rate to medication is only seen in 16% of the evaluated patients [8] (Fig. 25.1).

In general, reversal of pseudoresistant hypertension adopts a two-step approach. First, there is a need to confirm true resistance, and second, identification of the factors that contribute to treatment resistance [36, 37]. The first step entails confirming the diagnosis with reliable office BP readings. Particular attention should be paid to the patient's posture, environment, and the arm used should be supported at heart level. Additionally, the use of appropriate sized cuffs and devices is mandatory. Physicians may get the average from three BP readings, instead of just one BP determination. Identification of white coat hypertension is very important. Either a qualified non-physician personnel performs office measurement or the use of an automated device with the patient alone in the room is useful in taking BP. The importance of home BP determination or the use of a 24 h ambulatory measurements may be useful in determining pseudoresistance. If the BP remains elevated after this evaluation, patient adherence to therapy should be evaluated.

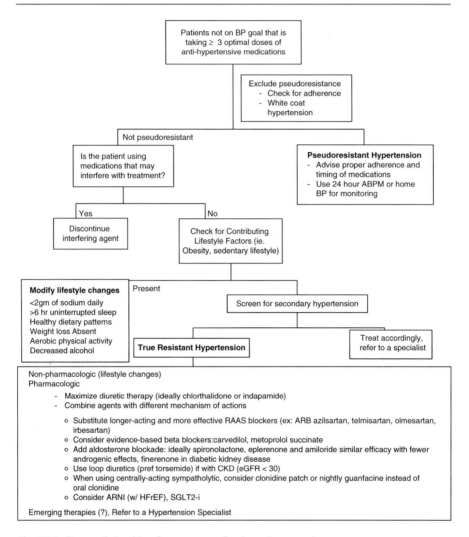

Fig. 25.1 Proposed algorithm for treatment of resistant hypertension

25.7   Factors Contributing to Resistant Hypertension

Several factors may contribute to the development of resistant hypertension. Classes of pharmacologic agents produced transient or even persistent elevations of BP. Non-steroidal anti-inflammatory drugs (NSAIDs) are common causes of worsening BP control [38]. Meta-analyses of their effects have indicated average increases in mean arterial pressure of approximately 5.0 mmHg [38]. This class of drugs can also blunt the BP lowering effects of several antihypertensive medications, with the exception of calcium channel blockers (CCBs) [39]. Similar effects have been seen with selective cyclooxygenase-2 (COX 2) inhibitors [40]. These effects presumably occur secondary to the inhibition of renal prostaglandin production, particularly prostaglandin E2 (PGE$_2$) and prostacyclin (PCI$_2$), with subsequent sodium and fluid retention. Sympathomimetics, such as nasal decongestants and anorexic pills, oral contraceptives, glucocorticoids, erythropoietin, and cyclosporine can also interfere with BP control. Black licorice found in oral tobacco products and herbal supplements, such as *ma huang*, increases BP by suppressing the metabolism of cortisol resulting in increased stimulation of mineralocorticoid receptors. Illicit drugs, such as cocaine and amphetamines, are also common causes of resistant hypertension [8, 29].

While modest alcohol consumption does not generally increase BP, heavy alcohol intake (>3 to 4 drinks/day) has a dose-related effect on BP, both on normal and hypertensive individuals [41]. In an analysis of Chinese adults ingesting ≥30 drinks per week, the risk of having hypertension increased from 12 to 14%. Cessation of heavy alcohol ingestion reduced the 24-h ambulatory BP measurements by 7.2 mmHg and diastolic BP by 6.6 mmHg [42].

Excess dietary salt is a key factor responsible for many causes of resistant hypertension by directly increasing BP through volume overload and by blunting the BP lowering effect of most classes of antihypertensive agents. The majority of resistant hypertensive patients have higher salt intake than the general population, exceeding more than 10 g/day [43–45]. The effects of increased dietary salt intake tend to be more pronounced in salt sensitive patients, including the elderly, African Americans, and in CKD patients [44].

Obesity is a very common feature of patients with resistant hypertension. Most obese patients tend to have an increased need of antihypertensive medications and increased likelihood of never achieving BP control. Mechanisms of obesity-induced hypertension are complex and not fully understood. These include insulin resistance, hyperinsulinemia, impaired sodium excretion, increased sympathetic nervous system activity, increase in aldosterone sensitivity, and presence of obstructive sleep apnea (OSA) have all been implicated as potential causes [46–48].

25.8   Secondary Causes of Hypertension

Secondary hypertension is common in patients with resistant hypertension. In studies, it has been shown that about 5–10% of patients have hypertension attributable to secondary causes [49]. Renal parenchymal disease is considered the most common medical cause of secondary hypertension. Renovascular hypertension, primary hyperaldosteronism, and OSA are other identifiable causes. Other less common forms of secondary hypertension are pheochromocytoma, Cushing's disease, coarctation of aorta, and others. The possibility of having a secondary cause increases with age as CKD, sleep apnea, and primary hyperaldosteronism are more prevalent in older patients [50]. Patients identified with secondary causes of hypertension should be treated accordingly specific to the disease.

25.9 Treatment of True Resistant Hypertension

The evaluation of hypertensive patients should be directed toward confirmation of true resistance. Physicians should be vigilant in identifying causes contributing to true resistance, exclude secondary causes, and to properly document target organ damage. The goal BP is ≤130/80 mmHg in average risk patients, according to the 2018 AHA guidelines [5]. There is evidence supporting a lower BP goal in patients with chronic kidney disease with associated proteinuria, with the latest Kidney Disease Improving Global Outcomes (KDIGO) recommendation [51], suggesting that BP targets be less than 120 mmHg systolic BP. The 2018 AHA guidelines recommend that both non-pharmacologic and pharmacologic therapies are warranted for treatment [8, 52].

25.9.1 Non-pharmacologic Interventions

Resistant hypertension is almost always multifactorial in etiology. Lifestyle changes, including weight loss, regular exercise, ingestion of high fiber, low fat, low salt diet, and moderate alcohol intake should be encouraged. A recent review of weight loss study showed that a 10 kg weight loss is associated with an average of 6.0 mmHg reduction of systolic BP [53]. Overweight and obese patients may find it difficult to achieve weight loss and it is even more difficult to maintain the weight loss, but these should be advised to patients with weight problems [53, 54].

The benefit of salt reduction is well documented in general hypertensive patients. Salt restriction to <3 g/day is associated with modest reductions in BP [55]. Current guidelines suggest that dietary sodium in hypertensive patients should be <100 mmol/day (2.0 g of sodium or 5 g sodium chloride) [55, 56]. Prolonged modest reduction in salt intake induces a relevant fall in both hypertensive and normotensive individuals, irrespective of sex and ethnic groups. Larger reduction in systolic blood pressure is attained with larger reduction in dietary salt [56]. Alcohol intake should be limited to no more than 1 ounce of ethanol/day in most men (~2 drinks) and 0.5 ounce of ethanol/day in women and lighter weight individuals [57]. Ingestion of a diet rich in fruits and vegetables, high in low fat dairy products, potassium, magnesium, calcium, and low in total saturated fats reduced BP by 11.4/5.5 mmHg [58–60].

In a meta-analysis, regular aerobic exercise produced an average reduction of 4 mmHg in systolic and 3 mmHg in diastolic BP [61]. In another study, reductions in diastolic BP were maintained after 32 weeks of exercise even when some of the medications were withdrawn. Based on the benefits of exercise, patients should be encouraged to exercise for a minimum of 30 min for most days of the week [62].

25.9.2 Pharmacologic Treatment

Pharmacologic treatment involves combination of three or more antihypertensive medications, including a diuretic. Recommendations in modification and intensification of treatment are based on pharmacologic principles in the context of underlying pathophysiology of hypertension, clinical experience of the physician, and treatment guidelines. Present rationale in treatment is to ensure that all possible mechanisms for BP elevations are blocked [63–65]. The regimen to be followed by the patient should be simplified and long-acting combination agents are used as much as possible to reduce the number of pills and to permit once daily dosing.

In general, most patients should be on a blocker of the renin–angiotensin–aldosterone system (RAAS) along with a calcium antagonist and an appropriately dosed diuretic. The physician must ensure that these agents are prescribed at full dosage, especially with obese individuals. This triple regimen of an angiotensin converting enzyme inhibitor (ACE-I) or angiotensin receptor blocker (ARB), a long-acting CCB (usually amlodipine) and a long-acting diuretic (preferably chlorthalidone), is often effective and well tolerated by patients. This triple regimen can be accomplished with two pills with the use of fixed dose combination drugs that are already available [29]. Combining an ACE-I with ARB was recently shown to be less effective in terms of BP reduction than adding a diuretic or a CCB and was not able to reduce cardiovascular or renal events [66, 67]. In a separate trial, the combination of the direct renin inhibitor, aliskiren, and an ARB was also associated with a small additional BP drop [68]. Thus, dual RAAS blockade is not recommended.

An appropriate diuretic remains the cornerstone of treatment, as persistent volume expansion contributes to resistant hypertension. This was seen in 279 patients taking thiazide diuretics which saw significant increase in levels of brain natriuretic peptide and atrial natriuretic peptide among resistant hypertensive patients, suggesting volume expansion [69]. Chlorthalidone is preferred over hydrochlorothiazide (HCTZ) for treatment, the former appearing to be more potent than the latter drug. Chlorthalidone 25 mg once daily demonstrated a greater ambulatory BP reduction with the largest difference occurring overnight, as compared to a 50-mg daily dose of the HCTZ [70]. Switching HCTZ users to chlorthalidone resulted in an additional 8-mmHg drop and increased number of patients reaching the BP goal [71].

It is recommended to start chlorthalidone at 12.5 mg once daily, with subsequent titration up to 25 mg daily. It is necessary to monitor serum electrolytes, with hypokalemia being a major problem when taking chlorthalidone. Among patients with an estimated glomerular filtration rate (eGFR) of less than 30 mL/min/1.72 m^2, thiazide diuretics may be less effective. Loop diuretics, such as furosemide or torsemide, may be necessary for effective volume control [8].

Adding aldosterone antagonists provides significant benefit to existing drug regimens in patients with resistant hypertension, as these patients were found to have higher levels of plasma aldosterone. In a study of patients on an average of four antihypertensive medications, addition of spironolactone resulted in an average of 25/12 mmHg reduction of BP after 6 months of treatment [72]. In the BP lowering arm of the Anglo-Scandinavian Cardiac Outcomes Trial (ASCOT), patients receiving spironolactone as a fourth line antihypertensive medication for uncontrolled BP resulted in a 21.9/9.5 mmHg drop in BP regardless of age, sex, smoking, and presence of diabetes [73, 74]. The advantage of using spironolactone is that it can lower BP even with normal or low aldosterone levels. In the PATHWAY-2 trial [75], a randomized, double-blind, cross-over study comparing spironolactone with placebo, doxazosin, or bisoprolol in 285 patients, the use of spironolactone reduced home systolic BP by nearly 9 mmHg compared with placebo, and by 4–5 mmHg compared with doxazosin and bisoprolol. Hyperkalemia developed in only 2% of the cohort. The most common adverse effect of spironolactone is breast tenderness, particularly in men, especially in doses above 25 mg/day. This can be avoided by using a more selective mineralocorticoid receptor antagonist, eplerenone, which has demonstrated BP lowering efficacy and can be beneficial on slowing kidney disease progression [76–78].

Amiloride is another potassium sparing diuretic associated with BP reduction. In a blinded comparison of amiloride 10 mg once daily, spironolactone 25 mg daily or a combination of both used in African American patients on two-drug regimen (diuretic and a CCB), the mean decreases in BP were 12.2/4.8 mmHg for amiloride, 7.3/3.3 mmHg for spironolactone and 14.1/5.1 mmHg for the combination of both drugs [79]. In a substudy of PATHWAY-2, amiloride was shown to be equally effective in lowering blood pressure compared to spironolactone 25 mg once daily [75]. Hyperkalemia may occur when prescribing these agents, especially in combination with an ACE-I or ARB, thus necessitating close monitoring if kidney function is not within the normal range. Physicians should educate patients to avoid food and supplements rich in potassium and to avoid medications such as NSAIDs [80].

If the patient is still hypertensive, additional medications are added sequentially. Possible agents that may be used include beta-blockers, centrally acting agents and vasodilators. Vasodilating beta-blockers, such as labetalol, carvedilol, or nebivolol, may provide more antihypertensive benefits with fewer side effects compared to traditional beta blockers [81]. If a centrally acting agent is required, long-acting drugs like guanfacine or clonidine transdermal patch are added to the regimen. Lastly, potent vasodilators, such as minoxidil or hydralazine, can be very effective, particularly at higher doses. Unfortunately, the use of these drugs, particularly in higher doses, has subsequent adverse events [57]. In particular, high dose of minoxidil can cause reflexive increases in heart rate and fluid retention.

Ultimately, the use of three or more combination drugs must be individualized taking into consideration prior benefit, risk of adverse events, and contributing factors including concomitant diseases such as CKD and diabetes. It must be noted that a referral to a hypertension specialist is warranted if therapy has progressed to adding a fourth agent [82].

25.9.3 Device-Based Experimental Therapies for Resistant Hypertension

New experimental techniques are being evaluated in the treatment of resistant hypertension. Catheter-based radiofrequency ablation of the renal sympathetic nerves was conducted in a randomized trial involving 106 patients on five antihypertensive medications including a diuretic and was observed for 6 months. The technique significantly decreased BP from 178/97 mmHg to 143/85 mmHg [82–84]. In another cohort study, blood pressure was reduced by 23/11 mmHg at 12 months and 32/14 mmHg at 24 months using the technique [84]. In the SYMPLICITY-HTN 3 study [85], 535 patients with treatment-resistant hypertension were assigned to either renal nerve denervation or a sham procedure. Blood pressure decreased to a similar degree in both groups at 6 months, with no difference of incidence in serious adverse events. In the SYMPATHY trial which randomized resistant hypertension patients to routine care versus renal nerve denervation plus routine care, 24-h ABPM decreased more in the routine care group (reduction of 6.6 mmHg) compared to the renal denervation group (decrease of 5.6 mmHg) [86]. Several unblinded studies showed renal nerve denervation could substantially lower BP in patients with resistant hypertension. The Renal Denervation for Hypertension (DENERHTN) showed that a stepped-wise hypertension therapy with renal denervation has more BP reduction compared to just adding oral medications (15.4 vs −9.5mmHg) in 24-h systolic BP. Complications related to radiofrequency ablation include femoral artery pseudoaneurysm and renal artery stenosis [87].

Another experimental therapy for resistant hypertension is the use of electrical stimulation of the carotid sinus baroreflex system. The surgical implantation of the RHEOS device was tested in 265 patients with resistant hypertension and results showed that there is no significant reduction in the systolic BP. In one year, the mean reduction of systolic BP was 25 mmHg but 35% of patients had serious procedure-related adverse events [88]. The device was unfortunately discontinued. A second baroreceptor activation therapy device, which was inserted in the carotid sinus by endovascular deployment lowered 24 h ambulatory BP in 6 months by 21/12 mmHg in 30 patients [89].

25.9.4 Experimental Therapy for Resistant Hypertension

Angiotensinogen is the unique substrate of all angiotensin peptides in the renin–angiotensin–aldosterone system. Therapeutic inhibition of this substrate using small interfering RNAs (siRNA) or antisense oligonucleotides (ASOs) has shown promise in reducing elevated BP in animal models [90]. These medications are given subcutaneously and at 4 weeks or longer intervals. This approach may improve better adherence for patients who are taking multiple oral medications [91].

25.10 Conclusions

The dilemma of resistant hypertension is growing and unfortunately is still understudied. Additional knowledge is needed to better identify, manage, and treat resistant hypertensives. Effective management should be based on pathophysiologic principles and clinical experience. It requires a careful examination and exclusion of factors associated with pseudoresistance. Modification of factors related to true resistance is needed to achieve better BP goals. An aggressive treatment should be designed to compensate for all mechanisms of BP elevation in a given patient. Proper combination of antihypertensive medications should be instituted immediately and is recommended to use triple regimen of an ACE-I or ARB, a long-acting CCB, and a diuretic. Diuretics, particularly chlorthalidone, are the cornerstone of treatment as these patients have volume overload. Aldosterone antagonists may be added as fourth line medication, and additional medication may be added if BP is still not at goal. Hopefully, experimental therapies may aid toward effective BP control in very difficult cases of resistant hypertension in the future.

Conflicts of Interest and Source of Support None.

References

1. Ong KL, Cheung BM, Man YB, Lau CP, Lam KS. Prevalence, awareness, treatment, and control of hypertension among United States adults 1999-2004. Hypertension. 2007;49:69–75.
2. Wolf-Maier K, Cooper RS, Kramer H, et al. Hypertension treatment and control in five European countries, Canada, and the United States. Hypertension. 2004;43:10–7.
3. Dahlof B, Devereux RB, Kjeldsen SE, et al. Cardiovascular morbidity and mortality in the losartan intervention for endpoint reduction in hypertension study (LIFE): a randomised trial against atenolol. Lancet. 2002;359:995–1003.
4. Pepine CJ, Handberg EM, Cooper-DeHoff RM, et al. A calcium antagonist vs a non-calcium antagonist hypertension treatment strategy for patients with coronary artery disease. The International Verapamil-Trandolapril Study (INVEST): a randomized controlled trial. JAMA. 2003;290:2805–16.
5. Carey RM, Calhoun DA, Bakris GL, et al. Resistant hypertension: detection, evaluation, and management from the American Heart Association. Hypertension. 2018;72:e53.
6. Whelton PK, Carey RM, Aronow WS, et al. 2017 ACC/AHA/AAPA/ABC/ACPM/AGS/APhA/ASH/ASPC/NMA/PCNA guideline for the prevention, detection, evaluation and management of high blood pressure in adults: a report of the American College of Cardiology/American Heart Association Task Force on Clinical Practice Guidelines. Hypertension. 2018;71:e13.
7. Williams B, Manicai G, Spiering W, et al. 2018 ESC/ESH Guidelines for the management of arterial hypertension. Eur Heart J. 2018;39:3021.
8. Calhoun DA, Jones D, Textor S, et al. Resistant hypertension: diagnosis, evaluation, and treatment: a scientific statement from the American Heart Association Professional Education Committee of the Council for High Blood Pressure Research. Circulation. 2008;117:e510–26.
9. Carey R, Sakhuja S, Calhoun DA, et al. Prevalence of apparent treatment-resistant hypertension in the United States. Hypertension. 2019;73(2):424–31.
10. Daugherty SL, Powers JD, Magid DJ, et al. Incidence and prognosis of resistant hypertension in hypertensive patients. Circulation. 2012;125(13):1635–42.
11. Sison J, Divinagracia R, Nailes J. Asian management of hypertension: current status, home blood pressure and specific concerns in Philippines (a country report). J Clin Hypertens. 2020;22:504–7.
12. Zabat GMA, Quiambao JL, Villaverde EC, et al. Clinical profile and management approach of diagnosed patients with resistant hypertension in a tertiary government hospital. Eur Heart J. 2013;34(supp 1):3226.
13. de la Sierra A, Segura J, Banegas JR, et al. Clinical features of 8295 patients with resistant hypertension classified on the basis of ambulatory blood pressure monitoring. Hypertension. 2011;57:898.
14. The Antihypertensive and Lipid-Lowering Treatment to Prevent Heart Attack Trial (ALLHAT). Major outcomes in high-risk hypertensive patients randomized to angiotensin-converting enzyme inhibitor or calcium channel blocker vs diuretic. JAMA. 2002;288:2981–97.

15. Gupta AK, Nasothimiou EG, Chang CL, Sever PS, Dahlof B, et al. Baseline predictors of resistant hypertension in the Anglo-Scandinavian cardiac outcome trial (ASCOT): a risk score to identify those at high risk. J Hypertens. 2011;29:2004–13.
16. Tsujimoto T, Kajio H. Intensive blood pressure treatment for hypertension. Hypertension. 2018;73(2):415–23.
17. Lewington S, MacMahon S. Blood pressure, cholesterol, and common causes of death: a review. Am J Hypertens. 1999;12:96–8.
18. Freis ED. Veterans administration cooperative study group on hypertensive agents: effects of age on treatment results. Am J Med. 1991;90:20S–3S.
19. Lloyd-Jones DM, Evans JC, Larson MG, O'Donnell CJ, Roccella EJ, Levy D. Differential control of systolic and diastolic blood pressure: factors associated with lack of blood pressure control in the community. Hypertension. 2000;36:594–9.
20. Lloyd-Jones DM, Evans JC, Larson MG, O'Donnell CJ, Levy D. Differential impact of systolic and diastolic blood pressure level on JNC-VI staging. Hypertension. 1999;34:381–5.
21. Naseem R, Adam AM, Khan F, et al. Prevalence and characteristics of resistant hypertensive patients in an Asian population. Indian Heart J. 2017;69:442–6.
22. Lotufo PA, Pereira AC, Vasconcellos PS, et al. Resistant hypertension: risk factors, subclinical atherosclerosis, and comorbidities among adults-Brazilian longitudinal study of adult health (ELSA-Brasil). J Clin Hypertens. 2015;17(1):74–80.
23. Hannila-Handelberg T, Kontula K, Tikkanen I, et al. Common variants of the beta and gamma subunits of the epithelial sodium channel and their relation to plasma renin and aldosterone levels in essential hypertension. BMC Med Genet. 2005;6:4.
24. Givens RC, Lin YS, Dowling AL, et al. CYP3A5 genotype predicts renal CYP3A activity and blood pressure in healthy adults. J Appl Physiol. 2003;95:1297–300.
25. Moser M, Cushman W, Handler J. Resistant or difficult-to-treat hypertension. J Clin Hypertens. 2006;8:434–40.
26. Kaplan NM. Resistant hypertension. J Hypertens. 2005;23:1441–4.
27. Pimenta E, Calhoun DA, Oparil S. Mechanisms and treatment of resistant hypertension. Arq Bras Cardiol. 2007;88:683–92.
28. Sarafidis PA, Bakris GL. State of hypertension management in the United States: confluence of risk factors and the prevalence of resistant hypertension. J Clin Hypertens. 2008;10:130–9.
29. Sarafidis PA, Bakris GL. Resistant hypertension: an overview of evaluation and treatment. J Am Coll Cardiol. 2008;52:1749–57.
30. Verdecchia P, Schillaci G, Borgioni C, et al. White coat hypertension and white coat effect. Similarities and differences. Am J Hypertens. 1995;8:790–8.
31. Redon J, Campos C, Narciso ML, Rodicio JL, Pascual JM, Ruilope LM. Prognostic value of ambulatory blood pressure monitoring in refractory hypertension: a prospective study. Hypertension. 1998;31:712–8.
32. Caro JJ, Salas M, Speckman JL, Raggio G, Jackson JD. Persistence with treatment for hypertension in actual practice. CMAJ. 1999;160:31–7.
33. Mazzaglia G, Mantovani LG, Sturkenboom MC, et al. Patterns of persistence with antihypertensive medications in newly diagnosed hypertensive patients in Italy: a retrospective cohort study in primary care. J Hypertens. 2005;23:2093–100.
34. Vrijens B, Vincze G, Kristanto P, Urquhart J, Burnier M. Adherence to prescribed antihypertensive drug treatments: longitudinal study of electronically compiled dosing histories. BMJ. 2008;336:1114–7.
35. Singer GM, Izhar M, Black HR. Goal-oriented hypertension management: translating clinical trials to practice. Hypertension. 2002;40:464–9.
36. Bakris G, Hill M, Mancia G, et al. Achieving blood pressure goals globally: five core actions for health-care professionals. A worldwide call to action. J Hum Hypertens. 2008;22:63–70.
37. Pickering TG. Measurement of blood pressure in and out of the office. J Clin Hypertens. 2005;7:123–9.
38. Johnson AG. NSAIDs and increased blood pressure. What is the clinical significance? Drug Saf. 1997;17:277–89.

39. Conlin PR, Moore TJ, Swartz SL, et al. Effect of indomethacin on blood pressure lowering by captopril and losartan in hypertensive patients. Hypertension. 2000;36:461–5.
40. Whelton A, White WB, Bello AE, Puma JA, Fort JG. Effects of celecoxib and rofecoxib on blood pressure and edema in patients > or =65 years of age with systemic hypertension and osteoarthritis. Am J Cardiol. 2002;90:959–63.
41. Aguilera MT, De la Sierra A, Coca A, Estruch R, Fernandez-Sola J, Urbano-Marquez A. Effect of alcohol abstinence on blood pressure: assessment by 24-hour ambulatory blood pressure monitoring. Hypertension. 1999;33:653–7.
42. Wildman RP, Gu D, Muntner P, et al. Alcohol intake and hypertension subtypes in Chinese men. J Hypertens. 2005;23:737–43.
43. He FJ, MacGregor GA. Effect of longer-term modest salt reduction on blood pressure. Cochrane Database Syst Rev. 2004;3:CD004937.
44. Luft FC, Weinberger MH. Review of salt restriction and the response to antihypertensive drugs. Satellite symposium on calcium antagonists. Hypertension. 1988;11:229–32.
45. Schafflhuber M, Volpi N, Dahlmann A, et al. Mobilization of osmotically inactive Na+ by growth and by dietary salt restriction in rats. Am J Physiol Renal Physiol. 2007;292:F1490–500.
46. Nishizaka MK, Pratt-Ubunama M, Zaman MA, Cofield S, Calhoun DA. Validity of plasma aldosterone-to-renin activity ratio in African American and white subjects with resistant hypertension. Am J Hypertens. 2005;18:805–12.
47. Hall JE. The kidney, hypertension, and obesity. Hypertension. 2003;41:625–33.
48. Hall JE, Kuo JJ, Da Silva AA, de Paula RB, Liu J, Tallam L. Obesity-associated hypertension and kidney disease. Curr Opin Nephrol Hypertens. 2003;12:195–200.
49. Singer GM, Setaro JF. Secondary hypertension: obesity and the metabolic syndrome. J Clin Hypertens. 2008;10:567–74.
50. Acelajado MC, Calhoun DA. Resistant hypertension, secondary hypertension, and hypertensive crises: diagnostic evaluation and treatment. Cardiol Clin. 2010;28:639–54.
51. Kidney Disease Improving Global Outcomes (KDIGO) Blood Pressure Working Group. KDIGO 2021 clinical practice guideline for the management of blood pressure in chronic kidney disease. Kidney Int. 2021;99(3):87.
52. Pisoni R, Ahmed MI, Calhoun DA. Characterization and treatment of resistant hypertension. Curr Cardiol Rep. 2009;11:407–13.
53. Aucott L, Rothnie H, McIntyre L, Thapa M, Waweru C, Gray D. Long-term weight loss from lifestyle intervention benefits blood pressure? A systematic review. Hypertension. 2009;54:756–62.
54. Poobalan AS, Aucott LS, Smith WC, Avenell A, Jung R, Broom J. Long-term weight loss effects on all cause mortality in overweight/obese populations. Obes Rev. 2007;8:503–13.
55. He FJ, Marciniak M, Visagie E, et al. Effect of modest salt reduction on blood pressure, urinary albumin, and pulse wave velocity in white, black, and Asian mild hypertensives. Hypertension. 2009;54:482–8.
56. Grillo A, Salvi L, Coruzzi P, et al. Sodium intake and hypertension. Nutrients. 2019;11:1–16.
57. Chobanian AV, Bakris GL, Black HR, et al. the seventh report of the joint national committee on prevention, detection, evaluation, and treatment of high blood pressure: the JNC 7 report. JAMA. 2003;289:2560–72.
58. Appel LJ, Miller ER, Jee SH, et al. Effect of dietary patterns on serum homocysteine: results of a randomized, controlled feeding study. Circulation. 2000;102:852–7.
59. Moore TJ, Vollmer WM, Appel LJ, et al. Effect of dietary patterns on ambulatory blood pressure: results from the dietary approaches to stop hypertension (DASH) trial. Hypertension. 1999;34:472–7.
60. Svetkey LP, Simons-Morton D, Vollmer WM, et al. Effects of dietary patterns on blood pressure: subgroup analysis of the dietary approaches to stop hypertension (DASH) randomized clinical trial. Arch Intern Med. 1999;159:285–93.
61. Kokkinos PF, Narayan P, Colleran JA, et al. Effects of regular exercise on blood pressure and left ventricular hypertrophy in African-American men with severe hypertension. N Engl J Med. 1995;333:1462–7.

62. Whelton SP, Chin A, Xin X, He J. Effect of aerobic exercise on blood pressure: a meta-analysis of randomized, controlled trials. Ann Intern Med. 2002;136:493–503.
63. Czarina AM, Calhoun DA. Treatment of resistant hypertension. Minerva Cardioangiol. 2009;57:787–812.
64. Graves JW, Bloomfield RL, Buckalew VM Jr. Plasma volume in resistant hypertension: guide to pathophysiology and therapy. Am J Med Sci. 1989;298:361–5.
65. Taler SJ. Treatment of resistant hypertension. Curr Hypertens Rep. 2005;7:323–9.
66. Mann JF, Schmieder RE, McQueen M, et al. Renal outcomes with telmisartan, ramipril, or both, in people at high vascular risk (the ONTARGET study): a multicentre, randomised, double-blind, controlled trial. Lancet. 2008;372:547–53.
67. Yusuf S, Teo KK, Pogue J, et al. Telmisartan, ramipril, or both in patients at high risk for vascular events. N Engl J Med. 2008;358:1547–59.
68. Oparil S, Yarows SA, Patel S, Fang H, Zhang J, Satlin A. Efficacy and safety of combined use of aliskiren and valsartan in patients with hypertension: a randomised, double-blind trial. Lancet. 2007;370:221–9.
69. Gaddam KK, Nishizaka MK, Pratt-Ubunama MN, et al. Characterization of resistant hypertension: association between resistant hypertension, aldosterone, and persistent intravascular volume expansion. Arch Intern Med. 2008;168:1159–64.
70. Ernst ME, Carter BL, Goerdt CJ, et al. Comparative antihypertensive effects of hydrochlorothiazide and chlorthalidone on ambulatory and office blood pressure. Hypertension. 2006;47:352–8.
71. Khosla N, Chua DY, Elliott WJ, Bakris GL. Are chlorthalidone and hydrochlorothiazide equivalent blood-pressure-lowering medications? J Clin Hypertens. 2005;7:354–6.
72. Nishizaka MK, Zaman MA, Calhoun DA. Efficacy of low-dose spironolactone in subjects with resistant hypertension. Am J Hypertens. 2003;16:925–30.
73. Oparil S. The ASCOT blood pressure lowering trial. Curr Hypertens Rep. 2006;8:229–31.
74. Chapman JN, Kirby P, Caulfield MC, Poulter NR. Cardiovascular risk factors in a cohort of 30,000 high-risk men and women in the UK: cross-sectional, retrospective and prospective studies of screenees for the Anglo-Scandinavian Cardiac Outcomes Trial (ASCOT). J Hum Hypertens. 2001;15(1):23–6.
75. Williams B, MacDonald TM, Morant S, et al. Spironolactone versus placebo, bisoprolol and doxazosin to determine the treatment for drug-resistant hypertension (PATHWAY-2): a randomised, double-blind, crossover trial. Lancet. 2015;386:2059.
76. Gheorghiade M, Khan S, Blair JE, et al. The effects of eplerenone on length of stay and total days of heart failure hospitalization after myocardial infarction in patients with left ventricular systolic dysfunction. Am Heart J. 2009;158:437–43.
77. Pitt B, Reichek N, Willenbrock R, et al. Effects of eplerenone, enalapril, and eplerenone/enalapril in patients with essential hypertension and left ventricular hypertrophy: the 4E-left ventricular hypertrophy study. Circulation. 2003;108:1831–8.
78. Pitt B, Williams G, Remme W, et al. The EPHESUS trial: eplerenone in patients with heart failure due to systolic dysfunction complicating acute myocardial infarction. Cardiovasc Drugs Ther. 2001;15:79–87.
79. Eide IK, Torjesen PA, Drolsum A, Babovic A, Lilledahl NP. Low-renin status in therapy-resistant hypertension: a clue to efficient treatment. J Hypertens. 2004;22:2217–26.
80. Calhoun DA, Zaman MA, Nishizaka MK. Resistant hypertension. Curr Hypertens Rep. 2002;4:221–8.
81. Townsend RR, DiPette DJ, Goodman R, et al. Combined alpha/beta-blockade versus beta 1-selective blockade in essential hypertension in black and white patients. Clin Pharmacol Ther. 1990;48:665–75.
82. Mahfoud F, Schlaich M, Kindermann I, et al. Effect of renal sympathetic denervation on glucose metabolism in patients with resistant hypertension: a pilot study. Circulation. 2011;123:1940–6.

83. Esler MD, Krum H, Sobotka PA, Schlaich MP, Schmieder RE, Bohm M. Renal sympathetic denervation in patients with treatment-resistant hypertension (the symplicity HTN-2 trial): a randomised controlled trial. Lancet. 2010;376:1903–9.
84. Krum H, Schlaich M, Whitbourn R, et al. Catheter-based renal sympathetic denervation for resistant hypertension: a multicentre safety and proof-of-principle cohort study. Lancet. 2009;373:1275–81.
85. Bhatt DL, Kandzri DE, O'Neill WW, et al. A controlled trial of renal denervation for resistant hypertension. N Engl J Med. 2014;370:1393.
86. de Jager RL, de Beus E, Beeftink MM, et al. Impact of medical adherence on the effect of renal denervation: the SYMPATHY trial. Hypertension. 2017;69:678.
87. Azizi M, Sapoval M, Gosse P, et al. Optimum and stepped care standardised antihypertensive with or without denervation for resistant hypertension (DENERHTN): a multicentre, open-label, randomised controlled trial. Lancet. 2015;385:1957.
88. Bisognano JD, Bakris G, Nadim MK, et al. Baroreflex activation therapy lowers blood pressure in patients with resistant hypertension: results from the double blind, randomized, placebo-controlled rheos pivotal trial. J Am Coll Cardiol. 2011;58:765.
89. Spiering W, Williams B, der Heyden V, et al. Endovascular baroreflex amplification for resistant hypertension: a safety and proof-of-principle clinical study. Lancet. 2017;390:2655.
90. Wu CH, Wang Y, Murong M, et al. Antisense oligonucleotides targeting angiotensinogen: insights from animal studies. Biosci Rep. 2019;39(1):201.
91. Ren L, Colafella KMM, Bovee DM, et al. Targeting angiotensinogen with RNA-based therapeutics. Curr Opin Nephrol Hypertens. 2020;29(2):180–9.

Hypertensive Urgency and Emergency: Diagnostic and Therapeutic Considerations

26

Tiny Nair

26.1 Terminology and Definitions

Patients with known or undiagnosed hypertension can present with acute elevations of blood pressure (BP) to >220/120 mmHg. These individuals can present in outpatient or emergency department settings. There is a spectrum of presentation, and the term 'accelerated hypertension' is used to describe an acute rise in BP. This rise in BP can accompany a spectrum of presentations, including hypertensive urgency [1, 2] and hypertensive emergency, the latter term is used when acute organ damage is imminent. The challenge is to differentiate a 'hypertensive emergency' from 'hypertensive urgency' because the former requires BP to be controlled within minutes to hours, while BP can be controlled over days and weeks in the latter and more rapid lowering of BP may result in a poor outcome [3].

The term 'malignant hypertension' was initially used when acute elevations of BP were accompanied by papilledema [4]. More recently, however, this term is used when there is evidence of damage to three different target organs [5]. However, use of the terms hypertensive urgency and hypertensive emergency is considered better and more clinically relevant because these include acute damage to other target organs.

26.1.1 Hypertensive Emergency

In a hypertensive emergency, BP is extremely high (often exceeding 220/120 mmHg), accompanied by signs and symptoms of acute, progressive, hypertension-mediated organ damage (HMOD) [6, 7]. Patients may also present with a variety of symptoms based on the impact of extremely high BP on different target organs [1,

T. Nair (✉)
Department of Cardiology, PRS Hospital, Trivandrum, Kerala, India

8]. These include cerebral symptoms (e.g., headache, altered consciousness, confusion, or frank neurological deficit), cardiopulmonary symptoms (e.g., chest heaviness, chest pain, or breathlessness), ophthalmic symptoms (e.g., blurred vision), and/or renal symptoms (e.g., reduced urine output or pedal oedema) [9, 10]. Symptoms may be mild in some patients, but clinical examination or investigations might reveal features of acute and ongoing organ damage. This is important to understand because HMOD could also be present as a chronic state. The presence of target organ damage alone does not necessarily imply a diagnosis of hypertensive emergency, mandate the urgent initiation of parenteral antihypertensive therapy, or necessitate hospital admission. Acute and ongoing organ damage need to be present to make a diagnosis of hypertensive emergency.

Sometimes the clinical features of hypertensive emergency might match one of the several clinical syndromes, such as cerebrovascular accident, acute coronary syndrome, acute left ventricular failure, aortic dissection, or eclampsia. Urgent and rapid control of BP, within minutes or hours, using parenteral antihypertensive therapy is generally indicated in hypertensive emergency [8, 11].

26.1.2 Hypertensive Urgency

In contrast to hypertensive emergency, the course of hypertensive urgency is much less risky. Patients usually have a long-standing history of hypertension, often with irregular treatment or poor compliance [12]. BP is generally very high, usually >180/120 mmHg and there are no signs of acute or ongoing HMOD, although signs of chronic target organ damage may be present. The presence of left ventricular hypertrophy (LVH), grade 2 hypertensive retinopathy or urinary microalbuminuria may just indicate poor long-term BP control. Unlike hypertensive emergency, patients with hypertensive emergency do not show any acute, progressive, or ongoing organ damage [1, 2, 11]. In addition there is no need for hospital admission or use of parenteral drugs to urgently lower BP. In fact, urgent lowering of BP in stable patients can cause more complications [13, 14].

Available data suggest that the risk of short-term complications in those presenting with high BP alone and no acute organ damage is the same when less aggressive BP control is implemented compared with use of parenteral agents [15, 16]. The rates of major adverse cardiovascular event endpoint (including death, myocardial infarction [MI], and stroke) have been shown to be similar in patients managed aggressively or non-aggressively [15]. In 2013, Levy et al. assessed outcomes in 435 patients who presented with markedly elevated BP but no acute HMOD. Again, short-term prognosis did not differ between patients treated aggressively or otherwise [14].

Aggressive BP lowering can cause postural hypotension, decrease cerebral hypoperfusion resulting in cerebral ischaemia, cause acute coronary syndrome, and reduce renal perfusion causing acute renal failure [15]. In the setting of a chronic hypertension, many of these target organs are constantly perfused at higher mean perfusion pressures, resetting their autoregulation. Acute reduction of BP in such cases can be harmful [15, 17].

26.2   Clinical Features

26.2.1   Symptoms

Common signs and symptoms of hypertensive emergency based on involvement of the cerebrovascular and cardiovascular systems, eyes, and kidneys are summarised in Table 26.1 and Box 26.1. Clinical syndromes in which patients might present with hypertensive emergency are shown in Box 26.2. Patients presenting with severe hypertension in the ER, especially those with hypertensive urgency, have some common risk factors. These include poor socioeconomic status, medication non-compliance, irregular follow-up, and alcohol and drug abuse (Box 26.3) [6, 17–19].

Table 26.1   Symptoms indicating hypertensive emergency in the presence of increasingly high BP

Symptoms			
Neurologic	Ophthalmic	Cardiopulmonary	Renal
Headache	Blurred vision	Chest heaviness, pain	Reduced urine output
Confusion		Breathlessness	Pedal oedema
Altered consciousness		Palpitation	
Seizure		Pink frothy sputum	

Box 26.1 Clinical Signs that Indicate Hypertensive Emergency
- Focal neurologic deficit
- Tachycardia
- Sweating
- Abdominal tenderness
- Absent lower limb pulses
- Acute pulmonary oedema

Box 26.2 Clinical Syndromes in Which Patients Present with Hypertensive Emergency
- Stroke: Atherothrombotic, haemorrhagic
- Acute coronary syndrome: non-ST segment elevation myocardial infarction, ST-elevation myocardial infarction
- Acute left ventricular failure
- Aortic dissection
- Acute Renal Failure
- Retinal haemorrhage, papilledema
- Preeclampsia/eclampsia

Box 26.3 Risk Factors for Hypertensive Emergency and Urgency
- Poor socioeconomic status
- Poor healthcare access
- Medication noncompliance and nonadherence
- Smoking, alcohol use
- Oral contraceptive pill use
- Drug abuse

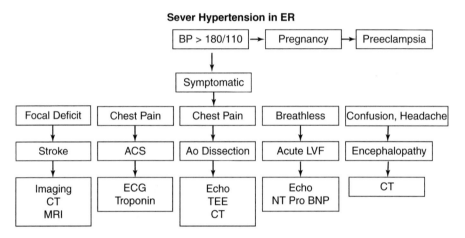

Fig. 26.1 Initial approach to evaluation of patients with hypertensive emergency depending on initial symptoms. *ACS* acute coronary syndrome, *CT* computed tomography, *ECG* electrocardiogram, *Echo* echocardiography, *ER* emergency room, *MRI* magnetic resonance imaging, *NT-proBNP* N-terminal pro B-type natriuretic peptide, *TEE* trans-oesophageal echocardiography

26.3   Investigations

The approach to management and initial evaluation depends on patient presentation. Suspicion of stroke initiates a stroke protocol involving brain imaging (computed tomographic or magnetic resonance imaging scan) to determine the need for thrombolysis or interventional therapy. Patients with acute coronary syndrome, especially ST-elevation MI should be referred to the catheterisation laboratory for an angiogram and, if indicated, angioplasty to open the occluded coronary artery [6, 18, 20]. A simple algorithm showing the initial approach to evaluation of patients with hypertensive emergency is provided in Fig. 26.1.

Investigations are both general and specific (Table 26.2). They help to determine the degree of target organ damage and rule out secondary causes of hypertension [18–20].

Table 26.2 Common investigations performed in patients with hypertensive emergencies and their interpretation

Investigations	Findings	Interpretation
Blood count	Peripheral smear (schistocytes)	Microangiopathic haemolytic anaemia
Urine examination	Proteinuria, RBCs, RBC cast	Acute glomerular and tubular injury
CT scan	Cerebral oedema, infarction, haemorrhage	Acute stroke
ECG	ST elevation, ST depression, LVH strain pattern	Acute coronary syndrome
Chest X-ray	Bilateral central pulmonary shadows (bat wing appearance)	Acute pulmonary oedema
Echocardiography	LVH, LV diastolic and/or systolic dysfunction	Long-standing uncontrolled hypertension
Chest CT, abdominal ultrasound	Flap of dissection visible	Aortic dissection

CT computed tomography, *ECG* electrocardiography, *LV* left ventricular, *LVH* left ventricular hypertrophy, *RBC* red blood cell

26.4 Pathophysiology

Overall, it is estimated that 1% of patients with hypertension develop true hypertensive emergency. The pathology of ongoing organ damage presents with specific pathologic, haemodynamic, and rheologic patterns.

26.4.1 Haemodynamic Changes

As BP rises, the cerebral arteries constrict to prevent the brain from being exposed to higher pressures, a mechanism known as cerebral autoregulation [21]. With further increases in BP, this autoregulation tends to fail. Finally, the vessel gives way to the dilating pressure, and exposure of the cerebral parenchyma to high pressure results in cerebral oedema [22] (Fig. 26.2).

26.4.2 Pathologic Changes

Damage to the small arterioles, known as fibrinoid necrosis, can occur in multiple organs resulting in multi-organ dysfunction. In the brain, this is seen in the penetrating and subcortical arteries. Chronic changes associated with hypertension are characterised by thickening of the subintima and medial wall. Hypertensive emergency tends to create an ongoing degenerative pathology, especially in the small arterioles of brain and kidneys with deposition of fibrin-like material in the subintima [10, 23]. The pathology is very similar to autoimmune vasculitis [24].

26.4.3 Rheologic Changes

Small arteriolar changes result in altered flow characteristics that cause damage to blood particles like red blood cells (RBCs). This damage may manifest as microangiopathic haemolytic anaemia, which can be a complication of a full-blown hypertensive crisis [15, 24].

26.4.4 Secondary Cases of High BP in Hypertensive Crisis

The most missed causes of markedly elevated BP in patients with hypertensive urgency or emergency are shown in Box 26.4 [18, 25, 26].

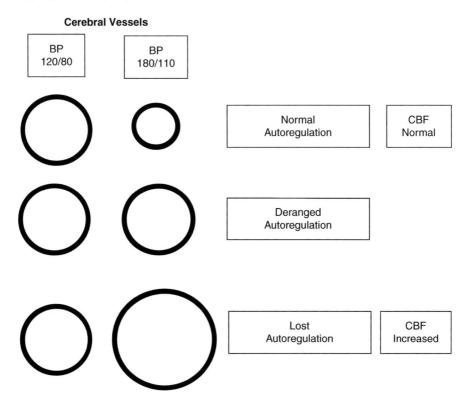

Fig. 26.2 Changes in cerebral vessels under increasing blood pressure (BP). *CBF* cerebral blood flow

Box 26.4 Commonly Missed Secondary Causes of Hypertension in Hypertensive Crisis
- Pheochromocytoma
- Renal artery stenosis
- Alcohol, cocaine, treatment with monoamine oxidase inhibitors
- Clonidine withdrawal
- Treatment with oral contraceptives or steroids

Box 26.5 Relative Incidence of Organ Involvement in Hypertensive Emergencies [16, 28]
- Cerebral infarction: 20–25%
- Pulmonary oedema: 15–30%
- Acute coronary syndrome: 15–25%
- Intracerebral or subarachnoid haemorrhage: 5–15%
- Hypertensive encephalopathy: 0–15%
- Eclampsia: 0–4%
- Aortic dissection: 0–2%

26.4.5 Frequency of Target Organ Damage

The relative frequency of organ damage in acute severe hypertension varies. The most common sites of target organ damage are the central nervous system (CNS) and the cardiovascular system (CVS) (Box 26.5) [27].

26.5 Management

26.5.1 Goal BP

26.5.1.1 Hypertensive Emergency
In the first hour, the general goal is to reduce mean arterial pressure (MAP) by 25%. Over the next 2–6 h, a BP goal of 160 mmHg for systolic BP and 100–110 mmHg for diastolic BP would be realistic. This BP level needs to be maintained for the first 24 h by strict titration of therapy. Over the next 48 h, the goal should be to reach the guideline-defined office BP goal in the absence of specific comorbidity [2, 6, 13].

26.5.1.2 Hypertensive Urgency
There is no hurry to reduce BP in these patients. If the patient is known to have hypertension and had not been taking prescribed medication, these treatments should be restarted. For patients with high BP despite good compliance with therapy, additional oral medications should be added [18, 20].

26.5.2 Non-pharmacologic Measures

Patients whose BP falls by 20/10 mmHg after reassurance, rest, and pain relief may not actually have hypertensive urgency and may not need emergent drug therapy [29].

Table 26.3 Parenteral antihypertensive agents that can be used in hypertensive emergency

Group	Agents
Vasodilators	Nitroglycerin, sodium nitroprusside, hydralazine
Calcium channel blockers	Clevidipine, nicardipine
Beta-blockers	Esmolol, labetalol, metoprolol
Angiotensin-converting enzyme inhibitors	Enalaprilat
Alpha-blockers	Phentolamine

26.5.3 Pharmacological Therapy for Hypertensive Emergencies

There are many fast acting IV drugs available for the management of hypertensive emergency (Table 26.3) [8, 27].

Sodium nitroprusside reduces BP very rapidly and usually requires invasive BP monitoring. It has a rapid onset of action (in seconds) and short duration (1–2 min), making it easy to titrate, especially in labile situations. The usual dosage ranges from 0.25 to 8 µg/kg/min. Common side effects are nausea, vomiting, and muscle cramps. Cyanide toxicity is rare but can occur during prolonged therapy [30]. The drug degrades on exposure to light, making it challenging to infuse [8, 31]. It should be used in an intensive care unit setting for short periods and with close monitoring only.

Nitroglycerin is a commonly used anti-ischaemic drug that is less potent than sodium nitroprusside. Its effect on reducing preload and relieving coronary ischaemia makes it a good choice in the presence of ischaemic syndromes with ongoing cardiac pain, and in cardiac failure. The infusion dosage ranges from 5 to 100 µg/min depending on BP. Headache and vomiting are known side effects [32, 33]. Development of methemoglobinemia and tolerance to prolonged infusion is a challenge [31].

Esmolol is an ultra-short-acting beta-blocker that can be easily titrated due to a 1–2 min onset of action and a 10–30 min duration of action. In addition to beta-blockade, this agent has negative ionotropic and chronotropic effects that can prolonged the PR interval and worsen heart failure. Bronchospasm is a relative contraindication. The dose is 250–500 µg/kg IV bolus followed by 50–100 µg/kg/min IV infusion. It is the drug of choice in hypertensive emergencies associated with dissection of aorta because it reduces the force of myocardial contraction, slowing the progression of dissection [34]. It is also indicated in acute coronary syndrome, thyrotoxicosis and after coronary artery bypass grafting-related severe hypertension [31, 35].

Labetalol is a combined alpha- and beta-blocker. It is a gentle beta-blocker that is safe in eclampsia. Onset of clinical effect occurs in 5–10 min and lasts for 3–6 h. The dosage is 20–80 mg IV bolus. Common side effects are nausea, vomiting, flushing, bradycardia, and orthostatic hypotension [36]. Heart block and acute heart failure are contraindications to its use [31, 35].

Enalaprilat is the intravenous formulation of the popular angiotensin-converting enzyme inhibitor (ACEI), enalapril. The BP-lowering effects of this agent are difficult to reverse, and therefore acute hypotension is a major problem with IV enalaprilat. However, it may be used in select cases of severe hypertension with heart failure and scleroderma-related hypertensive crisis. The dose is 1.25–5 mg every 6 h [37]. Onset of action occurs in 15–30 min and lasts for 6–12 h [31, 35].

The IV CCB nicardipine has the advantage of reducing cerebral ischaemia and therefore has an advantage in cerebrovascular accident. Usual side effects are reflex tachycardia, headache, and a sensation of flushing. The dose is 5–15 mg/h IV. Onset of action occurs within 5–10 min of treatment initiation and lasts for 1–4 h [30]. Dosage is not based on body weight meaning that ease of administration is an added advantage for nicardipine [35].

Hydralazine can be administered via the intramuscular (IM) or IV route. It is especially useful in eclampsia. BP-lowering effects are seen within 10–20 min of starting an IV infusion and duration is action of 1–4 h. For IM injection, the onset of action is seen within 20–30 min and the duration of action is 4–6 h. Common vasodilatory side effects, such as flushing and reflex tachycardia, are common. Aortic dissection and coronary ischaemia are contraindications to use of hydralazine [38].

26.5.4 Transition to Oral Therapy

This is planned when BP is under control. Given that most IV agents are vasodilators, tachyphylaxis is common. The addition of low-dose non-loop diuretics (typically hydrochlorothiazide or chlorthalidone) tends to be associated with good reductions in BP. Once a patient is out of crisis, a routine work-up to rule out secondary causes (pheochromocytoma, aldosteronism, renal artery stenosis, renal parenchymal disease, or obstructive sleep apnoea) can help in targeting subsequent therapy [16].

26.6 Special Situations

26.6.1 Aortic Dissection

Progression of aortic dissection depends on the magnitude of raised systolic BP and associated shear stress on the damaged aorta. The goal is to rapidly reduce systolic BP to 120 mmHg within 20–30 min. This can be achieved using short-acting IV beta-blockers such as esmolol and labetalol. Monotherapy with direct-acting vasodilators is contraindicated because this might increase stroke volume and worsen the dissection [39]. Confirmation of diagnosis and timely surgical procedure is the key to good outcomes [27, 40, 41].

26.6.2 Acute Coronary Syndrome/ST-Elevation MI

The goal in ACS is to control BP and relieve ischaemic pain. This is usually achieved using a combination of nitroglycerin and beta-blockers. Ongoing ischaemia requires rapid reopening of the culprit artery, especially in ST-elevation MI [42]. Cases of ACS presenting with non-ST-elevation MI without haemodynamic derangement, ongoing ischaemia, major arrhythmias, or other life-threatening features can be stabilised on medical management and BP controlled prior to angiography and necessary intervention [31, 35].

26.6.3 Acute Pulmonary Oedema

Intravenous nitroglycerin is the drug of choice because most cases of acute pulmonary oedema result from ischaemic syndromes. IV loop diuretics given to relieve congestion can also reduce BP. Careful initiation of therapy with RAAS blockers (including ACEIs, angiotensin receptor blockers [ARBs] and angiotensin receptor-neprilysin inhibitors [ARNIs]) should be performed. Recent data show that ARNIs are preferred over ARBs and ACEIs even in acute hospital settings. These can be combined with a loop diuretic and a mineralocorticoid receptor antagonist (MRA). (Tiny 19 b) Ischaemia as the underlying cause for heart failure needs to be actively investigated and revascularisation therapy offered if required [35].

26.6.4 Ischaemic Stroke

Severely elevated BP in the setting of an acute ischaemic stroke is often a compensatory mechanism to increase blood flow to the ischaemic area and reduce the potentially ischaemic 'penumbra'. Reducing BP in the acute phase can result in progression of the area of ischaemia and worsening of neurologic status. The cerebral perfusion pressure (CPP) depends on the difference between MAP and intracranial pressure (ICP). ICP increases during the acute phase of stroke means that it is mandatory for the MAP to increase to maintain perfusion. It is important not to reduce BP abruptly [21]. The general guideline is not to lower BP unless this is above 220/120 mmHg. Gradual reduction in BP by 15% compared with baseline is recommended in the first 24 h to avoid excessive lowering of MAP. The goal BP during and after thrombolysis is 185/110 mmHg, with a slightly lower goal of 180/105 mmHg after 24 h. Labetalol and nicardipine are preferred agents due to their ability to maintain cerebral perfusion [43]. IV sodium nitroprusside or esmolol can also be used because dosages are easily titrated [43, 44].

26.6.5 Haemorrhagic Stroke

Reductions in BP also need to be gradual in patients with haemorrhagic stroke, but BP-lowering therapy is needed when systolic BP is >220 mmHg and there is also benefit to gradually lowering systolic BP from 150–220 mmHg. A goal BP of <160 mmHg provides a good trade-off between maintaining perfusion and increasing haematoma. In a trial of aggressive BP lowering to 140 mmHg resulted in similar neurologic outcomes but worse renal results [45, 46]. Nimodipine is a weak dihydropyridine CCB that reduces cerebral artery vasospasm in cerebral haemorrhage and is known to reduce the risk of re-bleeding and improve outcomes [46, 47].

26.6.6 Preeclampsia

Pregnant patients at >20 weeks' gestation who present with hypertension constitute a special risk category, especially when this is accompanied by a specific risk factor such as proteinuria >300 mg/day, hepatic enzyme elevation, creatinine >1.1 mg%, and platelet count <100,000. Untreated preeclampsia can progress to seizures and eclampsia with grave maternal and foetal outcomes. A full-fledged haemolysis, elevated liver enzymes and low platelets (HELLP) syndrome is particularly dangerous. Drugs that are safe in pregnancy-induced hypertension include magnesium sulphate, hydralazine, labetalol, alpha methyldopa, and nifedipine. ACEIs, ARBs, and sodium nitroprusside are contraindicated [48].

26.6.7 Perioperative Hypertension

Catecholamine release during surgery, change of volume status, post-operative pain, and anxiety are often associated with high BP. It is generally recommended to continue parenteral antihypertensive medications until oral intake is permitted and then restart oral agents as soon as possible [20]. Intraoperative hypertensions can be controlled using IV agents if BP is very high (>180/110 mmHg) [6, 20].

26.7 Conclusion

Acute severe hypertension in the emergency room is a challenge to diagnose and treat. The disease spectrum ranges from a true emergency needing hospital admission, close monitoring, and IV drugs to marked BP elevation that does not need to be managed using such drastic measures. Early diagnosis, proper triage, and careful therapy titration can provide a favourable outcome in these conditions.

References

1. Muiesan ML, Salvetti M, Amadoro V, di Somma S, Perlini S, Semplicini A, et al. An update on hypertensive emergencies and urgencies. J Cardiovasc Med. 2015;16(5):372–82.
2. Wolf SJ, Lo B, Shih RD, Smith MD, Fesmire FM. Clinical policy: critical issues in the evaluation and management of adult patients in the emergency department with asymptomatic elevated blood pressure. Ann Emerg Med. 2013;62(1):59–68.
3. Astarita A, Covella M, Vallelonga F, Cesareo M, Totaro S, Ventre L, et al. Hypertensive emergencies and urgencies in emergency departments: a systematic review and meta-analysis. J Hypertens. 2020;38(7):1203–10.
4. Rubin S, Cremer A, Boulestreau R, Rigothier C, Kuntz S, Gosse P. Malignant hypertension: diagnosis, treatment and prognosis with experience from the Bordeaux cohort. J Hypertens. 2019;37(2):316–24.
5. Domek M, Gumprecht J, Lip GYH, Shantsila A. Malignant hypertension: does this still exist? J Hum Hypertens. 2020;34(1):1–4.
6. Elliott WJ. Clinical features in the management of selected hypertensive emergencies. Prog Cardiovasc Dis. 2006;48(5):316–25.
7. Zampaglione B, Pascale C, Marchisio M, Cavallo-Perin P. Hypertensive urgencies and emergencies. Prevalence and clinical presentation. Hypertension. 1996;27(1):144–7.
8. Rosei EA, Salvetti M, Farsang C. European Society of Hypertension Scientific Newsletter: treatment of hypertensive urgencies and emergencies. J Hypertens. 2006;24(12):2482–5.
9. Varounis C, Katsi V, Nihoyannopoulos P, Lekakis J, Tousoulis D. Cardiovascular hypertensive crisis: recent evidence and review of the literature. Front Cardiovasc Med. 2017;3:51.
10. Ault MJ, Ellrodt AG. Pathophysiological events leading to the end-organ effects of acute hypertension. Am J Emerg Med. 1985;3(6):10–5.
11. Adebayo O, Rogers RL. Hypertensive emergencies in the emergency department. Emerg Med Clin North Am. 2015;33(3):539–51.
12. Paini A, Tarozzi L, Bertacchini F, Aggiusti C, Rosei CA, De Ciuceis C, et al. Cardiovascular prognosis in patients admitted to an emergency department with hypertensive emergencies and urgencies. J Hypertens. 2021;39(12):2514–20.
13. Freis ED, Arias LA, Armstrong ML, Blount AW, Calabresi M, Castle CH, et al. Effects of treatment on morbidity in hypertension. Results in patients with diastolic blood pressures averaging 115 through 129 mm Hg. JAMA. 1967;202(11):1028–34.
14. Levy PD, Mahn JJ, Miller J, Shelby A, Brody A, Davidson R, et al. Blood pressure treatment and outcomes in hypertensive patients without acute target organ damage: a retrospective cohort. Am J Emerg Med. 2015;33(9):1219–24.
15. Patel KK, Young L, Howell EH, Hu B, Rutecki G, Thomas G, et al. Characteristics and outcomes of patients presenting with hypertensive urgency in the office setting. JAMA Intern Med. 2016;176(7):981–8.
16. Kaplan NM. Chapter 8: hypertensive emergencies. In: Kaplan NM, Victor RG, editors. Kaplan's clinical hypertension. 11th ed. Philadelphia: Wolters Kluwer Health/Lippincott Williams & Wilkins; 2015. p. 263.
17. DeFelice A, Willard J, Lawrence J, Hung J, Gordon MA, Karkowsky A, et al. The risks associated with short-term placebo-controlled antihypertensive clinical trials: a descriptive meta-analysis. J Hum Hypertens. 2008;22(10):659–68.
18. Rodriguez MA, Kumar SK, De Caro M. Hypertensive crisis. Cardiol Rev. 2010;18(2):102–7.
19. Johnson W, Nguyen ML, Patel R. Hypertension crisis in the emergency department. Cardiol Clin. 2012;30(4):533–43.
20. Gifford RW. Management of hypertensive crises. JAMA. 1991;266(6):829–35.
21. Ruland S, Aiyagari V. Cerebral autoregulation and blood pressure lowering. Hypertension. 2007;49(5):977–8.
22. Kaplan NM. Management of hypertensive emergencies. Lancet. 1994;344(8933):1335–8.

23. Cremer A, Amraoui F, Lip GY, Morales E, Rubin S, Segura J, Van den Born BJ, Gosse P. From malignant hypertension to hypertension-MOD: a modern definition for an old but still dangerous emergency. J Hum Hypertens. 2016;30(8):463–6.
24. Shantsila A, Dwivedi G, Shantsila E, Butt M, Beevers DG, Lip GY. Persistent macrovascular and microvascular dysfunction in patients with malignant hypertension. Hypertension. 2011;57(3):490–6.
25. Marik PE, Varon J. Hypertensive crises: challenges and management. Chest. 2007;131(6):1949–62.
26. Hollander JE. Cocaine intoxication and hypertension. Ann Emerg Med. 2008;51(3):18–20.
27. Ram CV, Silverstein RL. Treatment of hypertensive urgencies and emergencies. Curr Hypertens Rep. 2009;11(5):307–14.
28. van der Veen PH, Geerlings MI, Visseren FL, Nathoe HM, Mali WP, van der Graaf Y, et al. Hypertensive target organ damage and longitudinal changes in brain structure and function: the second manifestations of arterial disease-magnetic resonance study. Hypertension. 2015;66(6):1152–8.
29. Grassi D, O'Flaherty M, Pellizzari M, Bendersky M, Rodriguez P, Turri D, et al. Hypertensive urgencies in the emergency department: evaluating blood pressure response to rest and to antihypertensive drugs with different profiles. J Clin Hypertens. 2008;10(9):662–7.
30. Neutel JM, Smith DH, Wallin D, Cook E, Ram CV, Fletcher E, Maher KE, Turlepaty P, Grandy S, Lee R, et al. A comparison of intravenous nicardipine and sodium nitroprusside in the immediate treatment of severe hypertension. Am J Hypertens. 1994;7(7):623–8.
31. Perez MI, Musini VM. Pharmacological interventions for hypertensive emergencies: a Cochrane systematic review. J Hum Hypertens. 2008;22(9):596–607.
32. Abdelwahab W, Frishman W, Landau A. Management of hypertensive urgencies and emergencies. J Clin Pharmacol. 1995;35(8):747–62.
33. Hsieh YT, Lee TY, Kao JS, Hsu HL, Chong CF. Treating acute hypertensive cardiogenic pulmonary edema with high-dose nitroglycerin. Turk J Emerg Med. 2018;18(1):34–6.
34. Varon J. Treatment of acute severe hypertension: current and newer agents. Drugs. 2008;68(3):283–97.
35. Padilla Ramos A, Varon J. Current and newer agents for hypertensive emergencies. Curr Hypertens Rep. 2014;16(7):450.
36. Peacock WF, Varon J, Baumann BM, Borczuk P, Cannon CM, Chandra A, et al. CLUE: a randomized comparative effectiveness trial of IV nicardipine versus labetalol use in the emergency department. Crit Care. 2011;15(3):157.
37. Hirschl MM, Binder M, Bur A, Herkner H, Brunner M, Müllner M, et al. Clinical evaluation of different doses of intravenous enalaprilat in patients with hypertensive crises. Arch Intern Med. 1995;155(20):2217–23.
38. Campbell P, Baker WL, Bendel SD, White WB. Intravenous hydralazine for blood pressure management in the hospitalized patient: its use is often unjustified. J Am Soc Hypertens. 2011;5(6):473–7.
39. Gupta PK, Gupta H, Khoynezhad A. Hypertensive emergency in aortic dissection and thoracic aortic aneurysm-a review of management. Pharmaceuticals. 2009;2(3):66–76.
40. Hiratzka LF, Bakris GL, Beckman JA, Bersin RM, Carr VF, Casey DE, et al. 2010 ACCF/AHA/AATS/ACR/ASA/SCA/SCAI/SIR/STS/SVM guidelines for the diagnosis and management of patients with Thoracic Aortic Disease: a report of the American College of Cardiology Foundation/American Heart Association Task Force on Practice Guidelines. Circulation. 2010;121(13):266–369.
41. Papaioannou TG, Stefanadis C. Vascular wall shear stress: basic principles and methods. Hell J Cardiol. 2005;46(1):9–15.
42. Picariello C, Lazzeri C, Attanà P, Chiostri M, Gensini GF, Valente S. The impact of hypertension on patients with acute coronary syndromes. Int J Hypertens. 2011;2011:563657.
43. Ahmed N, Wahlgren N, Brainin M, Castillo J, Ford GA, Kaste M, et al. Relationship of blood pressure, antihypertensive therapy, and outcome in ischemic stroke treated with intravenous

thrombolysis: retrospective analysis from safe implementation of thrombolysis in stroke-international stroke thrombolysis register (SITS-ISTR). Stroke. 2009;40(7):2442–9.

44. Figueroa SA, Zhao W, Aiyagari V. Emergency and critical care management of acute ischaemic stroke. CNS Drugs. 2015;29(1):17–28.
45. Anderson CS, Heeley E, Huang Y, Wang J, Stapf C, Delcourt C, Lindley R, Robinson T, Lavados P, Neal B, Hata J, Arima H, Parsons M, Li Y, Wang J, Heritier S, Li Q, Woodward M, Simes RJ, Davis SM, Chalmers J. Rapid blood-pressure lowering in patients with acute intracerebral hemorrhage. N Engl J Med. 2013;368(25):2355–65.
46. Qureshi AI, Palesch YY, Barsan WG, Hanley DF, Hsu CY, Martin RL, et al. Intensive blood-pressure lowering in patients with acute cerebral hemorrhage. N Engl J Med. 2016;375(11):1033–43.
47. Reed WG, Anderson RJ. Effects of rapid blood pressure reduction on cerebral blood flow. Am Heart J. 1986;111(1):226–8.
48. Sibai BM. Diagnosis and management of gestational hypertension and preeclampsia. Obstet Gynecol. 2003;102(1):181–92.

Mechanical Interventional Therapies for Hypertension: Present Status and Future Prospects

27

Emmett Tsz Yeung Wong ⓘ and Adrian Fatt Hoe Low

27.1 Introduction

Hypertension affects more than 1 billion people globally and is the number one risk factor for cardiovascular mortality and morbidity [1]. It increases the risk of acute coronary events, stroke, heart failure, atrial fibrillation, and kidney failure [2]. The World Health Organization (WHO) estimates that between 2000 and 2025, the number of people diagnosed with hypertension will increase by 60% to up to 1.56 billion, and Asia will contribute to a large proportion of this population. In China and India, the total number of hypertensive patients is expected to increase to more than 500 million by 2025 [3, 4]. Hypertension is also very common among the elderly in many Asian countries, with a prevalence of more than 60% among patients above the age of 60–65 in several countries [5–8]. With the rapid increase in the aging population in these Asian countries, it is expected that the prevalence of hypertension and related cardiovascular morbidity will

E. T. Y. Wong (✉)
Division of Nephrology, Department of Medicine, National University Hospital, Singapore, Singapore

National University Centre for Organ Transplantation, National University Hospital, Singapore, Singapore

Department of Medicine, Yong Loo Lin School of Medicine, National University of Singapore, Singapore, Singapore
e-mail: emmett_ty_wong@nuhs.edu.sg

A. F. H. Low
Department of Medicine, Yong Loo Lin School of Medicine, National University of Singapore, Singapore, Singapore

Department of Cardiology, National University Heart Centre, Singapore, National University Hospital, Singapore, Singapore
e-mail: adrian_low@nuhs.edu.sg

© The Author(s), under exclusive license to Springer Nature Switzerland AG 2022
C. V. S. Ram et al. (eds.), *Hypertension and Cardiovascular Disease in Asia*,
Updates in Hypertension and Cardiovascular Protection,
https://doi.org/10.1007/978-3-030-95734-6_27

continue to rise, resulting in an increasing socioeconomic burden [9]. Lifestyle modifications and medications have been the cornerstone of therapy for the treatment of hypertension. Although drug treatment strategies to lower blood pressure (BP) are well described, poor control rates of hypertension remain a significant challenge to clinicians. A major problem is non-adherence to anti-hypertensive medications, which is caused in part by intolerable or unfavorable side effects or adverse drug reactions. In addition, there is a subgroup of patients with treatment-resistant hypertension, when both lifestyle and pharmacological interventions fail to control blood pressure. As recent guidelines have also recommended more stringent BP goals, achieving these targets often requires multiple medications, which result in high dosing complexity, pill burden, and cost. Mechanical interventional therapies are thus an emerging and important field for the treatment of hypertension.

27.2   Premise Behind Mechanical Intervention Therapies for Hypertension

Renal sympathetic nerves are important in the initiation of hypertension and maintenance of the hypertensive state. Increased renal sympathetic activity results in (1) increased renin secretion mediated by direct adrenergic innervation, (2) increased renal tubular sodium reabsorption and sodium retention mediated by direct contact between nerve endings and basolateral membranes of renal tubular epithelial cells throughout the nephron, and (3) renal vasoconstriction resulting in decreased glomerular filtration rate and renal blood flow. The impact of renal innervation can thus occur long before changes in renal hemodynamics are evident. It was found that renal sympathetic nerves are activated in patients with essential hypertension, and the increase in renal sympathetic activity is most apparent in hypertensive patients younger than 40 years of age [10–13].

Sympathetic drive is elevated in multiple types of hypertension [14]. There is an interplay between renal nerves and the sympathetic nervous system. Efferent sympathetic signals from the central nervous system (CNS) modulate the physiology of the kidneys, which are a source of afferent central sympathetic activity, sending signals to the CNS [15]. The integral role of the sympathetic nervous system in the pathogenesis of hypertension has made it an attractive target for both pharmacological and mechanical interventions for its treatment. Interventions in animal models to abrogate renal sympathetic signaling prevent both the development of hypertension and lower BP [11, 12, 16].

This chapter will briefly examine the history and origins of mechanical intervention therapies, study its present state, and look at some of its future directions. While an increasing number of technologies are work in progress at various stages of development, the review will limit itself to therapies for which human feasibility studies have been published.

27.3 Mechanical Intervention Therapies for Hypertension

27.3.1 Surgical Sympathectomies

In the 1930s–1950s, prior to the advent of effective anti-hypertensive drugs, non-selective surgical sympathectomies, often called "splanchnicectomies," were performed in patients with uncontrolled hypertension [17, 18]. While medically effective, successful in reducing BP, and even improving survival in a population where the mortality rate was almost 100% in 5 years [19], the broad nature of these radical approaches led to significant morbidity, many perioperative adverse effects, and long-term complications, including bowel, bladder, and erectile dysfunction, as well as the dreaded side effects of palpitations, postural hypotension, and tachycardia due to non-specific sympathetic denervation of the viscera and lower extremity vasculature [17, 20–22]. It is also worth noting that in addition to improving BP in about half of the patients, surgical sympathectomies also rendered BP more sensitive to anti-hypertensive medications [19].

The use of radical surgical sympathectomies and therapeutic splanchnicectomies was eventually abandoned in the 1960s, not only because of the severe adverse effects and complications but also due to the development of effective anti-hypertensive medications [23].

27.3.2 Endovascular, Catheter-Based Renal Sympathetic Denervation

Increased renal sympathetic outflow, demonstrated in human hypertension, suggests that renal sympathetic nerves, conveniently located in a peri-arterial distribution, might be an attractive target for the treatment of hypertension [24]. The use of newly developed endovascular catheter technology thus takes advantage of the benefit of selective renal denervation while avoiding the downfalls of non-selective sympathectomies [13]. The first catheter developed for this purpose was the Symplicity™ catheter system (Ardian Inc., Palo Alto, CA, USA; now Medtronic, Minneapolis, MN, USA) in 2005. The original thoughts were to utilize a catheter to basically do what Smithvick and Thompson had done in a slightly different approach without creating a stepwise function in changes in sympathetic tone. This involves ablation of both afferent and efferent renal sympathetic nerve fibers located in the adventitia of the arteries with a radiofrequency (RF)-emitting catheter inserted percutaneously into the renal artery via a transfemoral approach [13, 23, 25]. The device consists of a 5-F blind-ending catheter that houses a flexible RF wire. The back-end portion of the catheter handle connects to an RF generator that supplies power. The procedure is performed under local anesthesia and sedation. Catheter control by the handle allows bringing this device into contact with the endothelial surface of the artery. Four to six 2-minute treatments are delivered at different locations longitudinally and rotationally in order to achieve a helical pattern of ablation within each renal artery. The energy delivered is 5 –8 W. Both renal arteries are treated on the same day. Before the procedure, the patient receives anticoagulation and intravenous administration of 200 ug of nitroglycerin [25, 26]. Preclinical studies performed by Ardian Inc. in juvenile swine then reported the effectiveness of this technique at achieving renal denervation without causing severe vascular or renal injury up to 6 months following the procedure [26].

27.3.2.1 The Symplicity-HTN Studies

In April 2007, Dr. Henry Krum and colleagues successfully performed the first percutaneous renal sympathetic denervation (RDN) in a patient with severe treatment-resistant hypertension [27]. Two years later, the Symplicity HTN-1 feasibility study was published, for which the Symplicity Flex™ (Medtronic, Minneapolis, MN, USA) catheter was used for RF ablation. Aiming primarily to assess safety and proof-of-principle, Krum's study [26] had a relatively small sample size of 50 patients, was unrandomized and not placebo-controlled, did not restrict medication adjustments, or maintain strict recruitment criteria, and failed to exclude secondary forms of hypertension. Nevertheless, it ignited the interest of the nephrology, hypertension, and interventional communities in the field after demonstrating substantial and safe office BP reduction of 27/17 mmHg in patients with resistant hypertension at 12 months. The subsequent open-label, unblinded, randomized, non-sham-controlled Symplicity HTN-2 study [28] provided additional insights into the effectiveness and safety of RDN in patients with resistant hypertension. This study of 106 patients found a 33/11 mmHg reduction in office-based BP in the renal denervation group compared to the control group at six-month follow-up ($p < 0.0001$ for systolic and diastolic BP). These changes were also paralleled by similar reductions in home-based BP measurements, average BP measurements derived from 24 h-hour ambulatory BP recordings (although performed in only half the patients, and less impressive reduction of 11/7 mmHg), and the need for anti-hypertensive medications. Less than 5% of patients who underwent renal denervation experienced procedural adverse events, portraying it as a promising and relatively safe procedure.

Heterogeneity of response to RDN was beginning to emerge in early studies [29]. Frequent criticisms of these small, uncontrolled studies surround common themes of sub-optimal evaluation for secondary hypertension, lack of blinded BP endpoints, lack of sham-controlled procedures, and inadequacy of follow-up [30]. To address these issues, the Symplicity HTN-3 study [31] was undertaken and reported in 2014. It is the largest study of RDN to date, enrolled more than 500 participants with resistant hypertension and was sham (renal-angiogram)-controlled. To the surprise of many clinicians, the study demonstrated no difference in both office and ambulatory BP reduction with renal denervation compared with sham control procedures, although crucially, it proved the safety of RDN. Substantial limitations have been subsequently identified by the investigators and denervation enthusiasts and have been the subject of extensive commentary—including variable number of ablations, enrolment of a heterogeneous patient population with different baseline medication usage, unstable medications at baseline, frequent drug changes and variable adherence to medications, and reliance on a large number of centers with little practical experience in renal denervation [32–36]. In particular, only 19 of 364 patients (5%) treated with RDN actually received bilateral ablation in all four quadrants of the renal artery. Not surprisingly, those that did receive per-protocol ablation therapy demonstrated the greatest reductions in office, home, and ambulatory systolic BP [32].

Prior to Symplicity-HTN 3, several thousand patients had been treated worldwide, mostly using the first-generation single-electrode Simplicity catheter. Most of these patients were treated as a standard of care rather than in a clinical trial, although data was captured in the Global Symplicity Registry, which reports that RDN is a safe and effective treatment for resistant hypertension [24, 37]. The UK Renal Denervation Affiliation has also reported large reductions in office and ambulatory BP in 253 patients with severe hypertension treated according to strict criteria with different RDN catheters and suggests that the real-world application of RDN is successful when done per-protocol [38].

27.3.2.2 Beyond Symplicity-HTN

Since Symplicity-HTN 3, RDN technologies have advanced. Current approaches increasingly make use of multi-electrode catheters for RF ablation, irrigated balloon catheters for ultrasound ablation, chemical ablation, or cryoablation [24, 39–42]. More recent trials also addressed the shortcomings and limitations of earlier studies by introducing standardized study design features, such as only including participants with combined systolic and diastolic hypertension, using standardized protocols for anti-hypertensive medications with washout and standardized titration, monitoring for anti-hypertensive medication adherence, and requiring that procedures be performed by a single, experienced interventionist at each center [43–47].

To date, four (often-termed second generation) randomized, multi-center, single-blinded, sham-controlled trials have reported results using these improved approaches—SPYRAL HTN-OFF MED, SPYRAL HTN-ON MED, RADIANCE-HTN SOLO, and RADIANCE-HTN TRIO, with further active studies now in the analysis stage, such as REQUIRE (ClinicalTrials.gov Identifier: NCT02918305). Both SYPRAL trials used the Symplicity Spyral™ (Medtronic, Minneapolis, MN, USA) device, a flexible 4-electrode catheter design used to perform four simultaneous ablations with a helical pattern, which can perform comprehensive RDN including ablation in the distal main renal artery and arterial branches. The SPYRAL HTN-OFF MED pivotal trial enrolled 311 participants with office systolic BP 150 to <180 mmHg off all anti-hypertensive medications and used circumferential RF ablation small enough to reach the distal renal artery and accessory vessels [43]. The SYPRAL HTN-ON MED trial enrolled 467 participants patients with office systolic BP 150–180 mmHg, office diastolic BP ≥90 mmHg, and 24-h ambulatory systolic BP 140–170 mmHg on 1–3 anti-hypertensive medications [44].

The RADIANCE trials used an alternative technology of balloon-type ultrasonic PARADISE™ (ReCor Medical, Palo Alto, CA, USA) RDN catheters to allow complete circumferential denervation in a more reliable fashion than RF ablation by generating frictional heating through the interaction of high-frequency sound waves that pass through the surrounding fluids. The RADIANCE-HTN SOLO proof-of-concept trial enrolled 146 participants with daytime ambulatory BP ≥135/85 mmHg and <170/105 mmHg after 1 month of discontinuing up to two anti-hypertensive medications [45]. The RADIANCE-HTN TRIO randomized 136 patients with office BP ≥140/90 mmHg despite three or more anti-hypertensive medications including a diuretic, to ultrasound ablation or sham procedure after a 4-week standardized switch to a once-daily fixed-dose single-pill combination of calcium-channel blocker, angiotensin-receptor blocker, and a thiazide diuretic [47]. All four trials reported consistent reductions in ambulatory and/or office BP in the short-term (2–3 terms) and mid-term (6 months) post-procedure with RF (SPYRAL trials) or ultrasound-based (RADIANCE trials) denervation [43–45, 47].

27.3.2.3   RDN: Not Ready for Prime Time Yet?

A survey of consensus statements by hypertension and cardiology societies internationally suggests that the jury is still out on the role of RDN as the standard of care in clinical practice. However, this may soon change and evolve as new evidence support the effectiveness and safety of RDN. The 2013 International Expert Consensus Statement on Percutaneous Transluminal Renal Denervation for the Treatment of Resistant Hypertension concluded that RDN should be considered only in patients whose BP cannot be controlled with a combination of modification and pharmacologic therapy that is tailored according to current guidelines, and while renal nerve ablation could have beneficial effects in other conditions characterized by elevated renal sympathetic nerve activity, its potential use for such indications should currently be limited to formal research studies of its safety and efficacy [48]. The 2018 European Society of Hypertension (ESH) position paper on renal denervation [49] also supports the 2018 European Society of Cardiology/ESH Guidelines for the management of hypertension in not recommending device-based therapies for routine use in the treatment of hypertension at least at the current moment [50]. The more recent Joint UK Societies' 2019 Consensus Statement on Renal Denervation also note that subsequent to Symplicity HTN-3, clinical trial design and rigor of execution has been greatly improved resulting in sham-controlled randomized controlled trials (RCTs) that demonstrate a short-term reduction in ambulatory BP without any significant safety concerns in both medication-naïve and medication-treated hypertensive patients, but despite this, still feel that further evaluation of this therapy is warranted and that RDN should not be offered to patients outside of the context of clinical trials [51]. The main reservations arise from the short duration of follow-up, lack of a physiologic test to confirm achievement of successful and complete renal ablation, and substantial heterogeneity of BP-lowering responses to RDN as well as to the sham procedure even in the second-generation randomized trials.

27.3.2.4   A View from Asia

There were case reports of successful treatment of resistant hypertension with renal denervation by Ong et al. [52] and Ho et al. [53] in Singapore as early as 2012 and Yang et al. [54] in Korea in 2013. There was also an early prospective study carried out in China in 2012 with the aim of evaluating the safety and short-term efficacy of

RDN in eight patients, with significant decreases in mean systolic and diastolic BP compared to baseline and no significant complications or changes in kidney function [55]. Symplicity HTN-Japan was the first RCT to evaluate RDN in an Asian population. Twenty-two patients were randomized to RDN and 19 to control, and 11 patients were crossed over and received RDN at 6 months post-randomization. There was a significant and sustained reduction in office systolic (-32.8 ± 20.1 mmHg) and diastolic (-15.8 ± 12.6 mmHg) BP in the RDN group at 36 months post-procedure ($p < 0.001$), and no procedural-, device- or treatment-related safety events throughout the 36 months. The study concluded that despite the small sample size, results were encouraging that RDN maintained a reduction in BP up to 36 months [56]. A sub-analysis of the Global Symplicity Registry evaluated the outcomes of RDN among patients from Korea. Compared with Caucasians with uncontrolled hypertension, the Korean cohort of patients was younger, had a lower body mass index and baseline systolic BP, and a higher prevalence of cardiovascular disease (CVD). After propensity score matching, the Koreans had a systolic BP change similar to that of Caucasians at 6 months (-19.4 ± 17.2 mmHg vs. -20.9 ± 21.4 mmHg; adjusted $p = 0.998$), whereas the change in BP reduction was significantly higher in Koreans than in Caucasians (-27.2 ± 18.1 mmHg vs. -20.1 ± 23.9 mmHg; adjusted $p = 0.002$) at 12 months [57].

The 2019 Consensus Statement of the Taiwan Hypertension Society and the Taiwan Society of Cardiology on Renal Denervation for the Management of Arterial Hypertension considers RDN as a legitimate alternative anti-hypertensive strategy and recommends that RDN be performed in the context of registry and clinical studies (Class I Recommendation, Level C evidence) and that it should not be performed routinely, without detailed evaluation of various causes of secondary hypertension and renal artery anatomy (Class III Recommendation, Level C evidence). In particular, RDN should not be viewed as an anti-hypertensive regimen strategy only for patients with resistant hypertension, and the presence of secondary causes of hypertension should not be viewed as a contraindication for RDN [58]. The 2020 Consensus Statement of the Asia Renal Denervation Consortium (ARDEC), in considering the ethnic differences of the hypertension profile and demographics of CVD demonstrated in the Symplicity HTN-Japan and Global Symplicity Registry data from Korea and Taiwan, notes that RDN might be an effective hypertension management strategy in Asia, and that based on available evidence, recommends that RDN not be considered a therapy of last resort, but as an initial therapy option that may be applied alone or as a complementary therapy to anti-hypertensive medication [59].

It is well-known that there are significant ethnic differences in the determinants of hypertension and the risk of hypertension-related CVD [60–62]. The association between BP and CVD is stronger in Asian patients than in Caucasian patients [62–64]. Masked hypertension, which is associated with increased sympathetic activity, is more prevalent in Asians [65–67], and East Asian populations have at least a two-fold higher sensitivity to β-blockade effects of propranolol, compared to white populations. These ethnic differences suggest that Asian populations might be particularly sensitive to sympathetic modulation, thereby making RDN an attractive strategy for the treatment of hypertension for Asians [59, 68].

27.3.3 Baroreflex Activation Therapy

Arterial baroreceptors located along the carotid sinus and aortic arch are stimulated in response to arterial BP elevation and reflexively send afferent nerve impulses into the nucleus tractus solitarius in the CNS. This in turn downregulates sympathetic efferent output and BP-lowering as well as increased parasympathetic outflow resulting in bradycardia [69, 70]. This can be achieved by delivering electrical field stimulation at the carotid sinus.

The first-generation Rheos™ (CVRx. Maple Grove, MN, USA) system utilized an implantable pulse generator and bipolar electrodes which were surgically attached to the carotid sinus under general anesthesia. It was initially evaluated in the prospective, non-randomized US Rheos Feasibility Trial in which ten patients underwent successful implantation with a consistent reduction in systolic BP of 41 mmHg [71], and subsequently in the DEBuT-HT trial, another prospective, non-randomized study which enrolled 45 patients with resistant hypertension and showed an average office BP reduction of 21/12 mmHg at 3 months and 33/22 mmHg at 2 years [72]. Finally, in the randomized, sham-controlled Rheos Pivotal Trial, 265 patients were randomized to early (1 month post-implantation) or delayed (6 months post-implantation) device activation. There was no significant difference between groups in the primary efficacy endpoint of at least 10 mmHg reduction in systolic BP. The study also failed to meet its early safety endpoint with 9% of patients developing transient or permanent facial nerve injury [73]. A 6-year open follow-up of patients who had been included in one of the three above trials showed that baroreflex activation therapy (BAT) maintained its efficacy for persistent reduction of office BP (systolic BP fell from 179 ± 24 mmHg to 144 ± 28 mmHg, $p < 0.0001$; diastolic BP fell from 103 ± 16 mmHg to 85 ± 18 mmHg, $p < 0.001$) in patients with resistant hypertension without major safety issues, and that the effect of BAT is greater than average in patients with signs of heart failure [74].

A second-generation device, Barostim Neo™ (CVRx, Maple Grove, MN, USA) has replaced Rheos™ and encompasses a single unipolar electrode and miniaturized generator with improved battery life [75]. The procedure can now be done under conscious sedation with a much-improved safety profile. A preliminary study in 30 patients with resistant hypertension demonstrated office BP reduction of 26.0/12.4 mmHg at 6 months, with shorter implantation and hospitalization times, less immediate procedure-related complications, and no reports of either temporary or permanent facial injury, compared with the Rheos™ system [76]. CVRx was previously undertaking the Barostim Hypertension Pivotal Trial (ClinicalTrials.gov Identifier: NCT01679132), which was eventually suspended as its resources only allow adequate oversight for one pivotal trial at a time, which was the BeAT-HF Trial. The BeAT-HF Trial was a multi-center, prospective, RCT in which subjects who had heart failure with reduced ejection fraction were randomized to BAT with optimal medical management or optimal medical management alone. Following the announcement of results that BAT resulted in significantly improved quality of life, exercise capacity, and NT-proBNP [77] at the Heart Rhythm Society 2019 Scientific Sessions, Barostim Neo™ obtained US FDA approval for BAT in heart failure later that year. The BAROSTIM THERAPY™ Registry (ClinicalTrials.gov Identifier: NCT02880631) is underway.

27.4 Future Prospects

27.4.1 Renal Denervation Therapies

With the results of the second-generation RDN trials, there is renewed interest in RDN as a therapy for resistant hypertension and perhaps beyond. One key unanswered question and future research direction revolve around physiological and biochemical markers to evaluate the extent of RDN at the time of the procedure to provide feedback of successful and complete RDN [78, 79]. Several markers have been proposed, including renal blood flow parameters [80], blood levels of brain-derived neurotrophic factor [81], and the BP response to catheter-based renal nerve stimulation [82, 83]. Another important question is whether there is a potential for functional reinnervation after RDN, or that of renal nerve fibers regaining their patency, with the implication of the durability of effects of RDN—this will require studies with a longer duration of post-procedural follow-up to ascertain clinically meaningful success. Finally, more studies are required to identify the characteristics of participants who will respond most to RDN, as well as to determine the effect of RDN on cardiovascular outcomes [24, 78, 79].

There are also newer and upcoming technologies for RDN, including a separate class of catheters that make use of microinjection of a neurotoxin (e.g. alcohol) to chemically ablate renal nerves to facilitate deeper nerve injury while avoiding endothelial damage [84] and non-vascular catheter-based technology that deploys a transurethral approach to ablate the renal pelvis [85].

27.4.2 Other Mechanical Interventional Therapies

In addition to RDN and BAT, there are many novel device technologies in the process of undergoing evaluation. Central iliac arteriovenous anastomosis creation using a nitinol stent-like device (ROX AV coupler™, ROX Medical, San Clemente, CA, USA) is thought to principally address mechanical aspects of the circulation by creating a fixed caliber conduit between the proximal arterial and low-resistance venous circulation, which helps to restore the Windkessel function of the central circulation and improve proximal vascular compliance [86, 87]. This device has received CE Mark approval for hypertension.

A proof-of-concept study has also been conducted for carotid body ablation using the Cibiem Carotid Body Modulation System™ in patients with resistant hypertension, demonstrating significant, durable office BP reduction of 23/12 mmHg at 6 months [88], with a subsequent feasibility study showing that unilateral carotid body resection lowered BP by 26 mmHg in 57% of patients with drug-resistant hypertension [89].

Other device technologies that are under investigation include baroreceptor amplification therapy, deep brain stimulation, vagal nerve stimulation, and median nerve stimulation [24]. There is also a renewed interest in the potential of RDN. These render mechanical intervention therapies for hypertension an exciting field ahead and an important space to watch.

References

1. Forouzanfar MH, Liu P, Roth GA, Ng M, Biryukov S, Marczak L, et al. Global burden of hypertension and systolic blood pressure of at least 110 to 115 mm Hg, 1990–2015. JAMA. 2017;317(2):165–82.
2. Messerli FH, Williams B, Ritz E. Essential hypertension. Lancet. 2007;370(9587):591–603.
3. Kearney PM, Whelton M, Reynolds K, Muntner P, Whelton PK, He J. Global burden of hypertension: analysis of worldwide data. Lancet. 2005;365(9455):217–23.
4. Jin CN, Yu CM, Sun JP, Fang F, Wen YN, Liu M, et al. The healthcare burden of hypertension in asia. Heart Asia. 2013;5(1):238–43.
5. Malhotra R, Chan A, Malhotra C, Østbye T. Prevalence, awareness, treatment and control of hypertension in the elderly population of Singapore. Hypertens Res. 2010;33(12):1223–31.
6. Kim KI, Chang HJ, Cho YS, Youn TJ, Chung WY, Chae IH, et al. Current status and characteristics of hypertension control in community resident elderly Korean people: data from a Korean longitudinal study on health and aging (Klosha study). Hypertens Res. 2008;31(1):97–105.
7. Group HS. Prevalence, awareness, treatment and control of hypertension among the elderly in Bangladesh and India: a multicentre study. Bull World Health Organ. 2001;79(6):490–500.
8. Lu FH, Tang SJ, Wu JS, Yang YC, Chang CJ. Hypertension in elderly persons: its prevalence and associated cardiovascular risk factors in Tainan City, southern Taiwan. J Gerontol A Biol Sci Med Sci. 2000;55(8):M463–8.
9. Park JB, Kario K, Wang JG. Systolic hypertension: an increasing clinical challenge in Asia. Hypertens Res. 2015;38(4):227–36.
10. Esler M, Jennings G, Korner P, Willett I, Dudley F, Hasking G, et al. Assessment of human sympathetic nervous system activity from measurements of norepinephrine turnover. Hypertension. 1988;11(1):3–20.
11. Gf D. Neural control of the kidney: past, present, and future. Hypertension. 2003;41(3 Pt 2):621–4.
12. Dibona GF, Esler M. Translational medicine: the antihypertensive effect of renal denervation. Am J Physiol Regul Integr Comp Physiol. 2010;298(2):R245–53.
13. Gewirtz JR, Bisognano JD. Catheter-based renal sympathetic denervation: a targeted approach to resistant hypertension. Cardiol J. 2011;18(1):97–102.
14. Smith PA, Graham LN, Mackintosh AF, Stoker JB, Mary DA. Relationship between central sympathetic activity and stages of human hypertension. Am J Hypertens. 2004;17(3):217–22.
15. Schlaich MP, Sobotka PA, Krum H, Whitbourn R, Walton A, Esler MD. Renal denervation As a therapeutic approach for hypertension: novel implications for an old concept. Hypertension. 2009;54(6):1195–201.
16. Dibona GF, Kopp UC. Neural control of renal function. Physiol Rev. 1997;77(1):75–197.
17. Smithvick RH, Thompson JE. Splanchnicectomy for essential hypertension; results in 1,266 cases. J Am Med Assoc. 1953;152(16):1501–4.
18. Longland CJ, Gibb WE. Sympathectomy in the treatment of benign and malignant hypertension; a review of 76 patients. Br J Surg. 1954;41(168):382–92.
19. Smithvick RH. Hypertensive vascular disease; results of and indications for Splanchnicectomy. J Chronic Dis. 1955;1(5):477–96.
20. Doumas M, Faselis C, Papademetriou V. Renal sympathetic denervation and systemic hypertension. Am J Cardiol. 2010;105(4):570–6.

21. Morrissey DM, Brookes VS, Cooke WT. Sympathectomy in the treatment of hypertension; review of 122 cases. Lancet. 1953;1(6757):403–8.
22. Evelyn KA, Singh MM, Chapman WP, Perera GA, Thaler H. Effect of thoracolumbar sympathectomy on the clinical course of primary (essential) hypertension. A ten-year study of 100 sympathectomized patients compared with individually matched, symptomatically treated control subjects. Am J Med. 1960;28:188–221.
23. Castro Torres Y, Katholi RE. Renal denervation for treating resistant hypertension: current evidence and future insights from a global perspective. Int J Hypertens. 2013;2013:513214.
24. Lobo MD, Sobotka PA, Pathak A. Interventional procedures and future drug therapy for hypertension. Eur Heart J. 2017;38(15):1101–11.
25. Schlaich MP, Sobotka PA, Krum H, Lambert E, Esler MD. Renal sympathetic-nerve ablation for uncontrolled hypertension. N Engl J Med. 2009;361(9):932–4.
26. Krum H, Schlaich M, Whitbourn R, Sobotka PA, Sadowski J, Bartus K, et al. Catheter-based renal sympathetic denervation for resistant hypertension: a multicentre safety and proof-of-principle cohort study. Lancet. 2009;373(9671):1275–81.
27. Rocha-Singh KJ. Medtronic Ardian Simplicity™ renal denervation devices. In: Heuser R, Schlaich M, Sievert H, Editors. Renal Denervation. London: Springer; 2015.
28. Esler MD, Krum H, Sobotka PA, Schlaich MP, Schmieder RE, Böhm M, et al. Renal sympathetic denervation in patients with treatment-resistant hypertension (the Symplicity Htn-2 trial): a randomised controlled trial. Lancet. 2010;376(9756):1903–9.
29. Persu A, Jin Y, Azizi M, Baelen M, Völz S, Elvan A, et al. Blood pressure changes after renal denervation at 10 European expert centers. J Hum Hypertens. 2014;28(3):150–6.
30. Persu A, Renkin J, Thijs L, Staessen JA. Renal denervation: ultima ratio or standard in treatment-resistant hypertension. Hypertension. 2012;60(3):596–606.
31. Bhatt DL, Kandzari DE, O'neill WW, D'agostino R, Flack JM, Katzen BT, et al. A controlled trial of renal denervation for resistant hypertension. N Engl J Med. 2014;370(15):1393–401.
32. Kandzari DE, Bhatt DL, Brar S, Devireddy CM, Esler M, Fahy M, et al. Predictors of blood pressure response in the Symplicity Htn-3 trial. Eur Heart J. 2015;36(4):219–27.
33. Kandzari DE, Kario K, Mahfoud F, Cohen SA, Pilcher G, Pocock S, et al. The Spyral Htn global clinical trial program: rationale and design for studies of renal denervation in the absence (Spyral Htn off-med) and presence (Spyral Htn on-med) of antihypertensive medications. Am Heart J. 2016;171(1):82–91.
34. Esler M. Renal denervation for hypertension: observations and predictions of a founder. Eur Heart J. 2014;35(18):1178–85.
35. Lobo MD, De Belder MA, Cleveland T, Collier D, Dasgupta I, Deanfield J, et al. Joint UK Societies' 2014 consensus statement on renal denervation for resistant hypertension. Heart. 2015;101(1):10–6.
36. Mahfoud F, Böhm M, Azizi M, Pathak A, Durand Zaleski I, Ewen S, et al. Proceedings from the European clinical consensus conference for renal denervation: considerations on future clinical trial design. Eur Heart J. 2015;36(33):2219–27.
37. Böhm M, Mahfoud F, Ukena C, Hoppe UC, Narkiewicz K, Negoita M, et al. First report of the global symplicity registry on the effect of renal artery denervation in patients with uncontrolled hypertension. Hypertension. 2015;65(4):766–74.
38. Sharp AS, Davies JE, Lobo MD, Bent CL, Mark PB, Burchell AE, et al. Renal artery sympathetic denervation: observations from the UK experience. Clin Res Cardiol. 2016;105(6):544–52.
39. Kapil V, Jain AK, Lobo MD. Renal sympathetic denervation - a review of applications in current practice. Interv Cardiol. 2014;9(1):54–61.
40. Kiuchi MG, Maia GL, De Queiroz Carreira MA, Kiuchi T, Chen S, Andrea BR, et al. Effects of renal denervation with a standard irrigated cardiac ablation catheter on blood pressure and renal function in patients with chronic kidney disease and resistant hypertension. Eur Heart J. 2013;34(28):2114–21.
41. Worthley SG, Tsioufis CP, Worthley MI, Sinhal A, Chew DP, Meredith IT, et al. Safety and efficacy of a multi-electrode renal sympathetic denervation system in resistant hypertension: the Enlightn I trial. Eur Heart J. 2013;34(28):2132–40.

42. Pathak A, Coleman L, Roth A, Stanley J, Bailey L, Markham P, et al. Renal sympathetic nerve denervation using intraluminal ultrasound within a cooling balloon preserves the Arterial Wall and reduces sympathetic nerve activity. EuroIntervention. 2015;11(4):477–84.
43. Böhm M, Kario K, Kandzari DE, Mahfoud F, Weber MA, Schmieder RE, et al. Efficacy of catheter-based renal denervation in the absence of antihypertensive medications (Spyral Htn-off med pivotal): a multicentre, randomised, sham-controlled trial. Lancet. 2020;395(10234):1444–51.
44. Kandzari DE, Böhm M, Mahfoud F, Townsend RR, Weber MA, Pocock S, et al. Effect of renal denervation on blood pressure in the presence of antihypertensive drugs: 6-month efficacy and safety results from the Spyral Htn-on med proof-of-concept randomised trial. Lancet. 2018;391(10137):2346–55.
45. Azizi M, Schmieder RE, Mahfoud F, Weber MA, Daemen J, Lobo MD, et al. Six-month results of treatment-blinded medication titration for hypertension control following randomization to endovascular ultrasound renal denervation or a sham procedure in the Radiance-Htn solo trial. Circulation. 2019;139(22):2542–53.
46. Mauri L, Kario K, Basile J, Daemen J, Davies J, Kirtane AJ, et al. A multinational clinical approach to assessing the effectiveness of catheter-based ultrasound renal denervation: the radiance-Htn and require clinical study designs. Am Heart J. 2018;195:115–29.
47. Azizi M, Sanghvi K, Saxena M, Gosse P, Reilly JP, Levy T, et al. Ultrasound renal denervation for hypertension resistant to a triple medication pill (Radiance-Htn Trio): a randomised, multicentre, single-blind, sham-controlled trial. Lancet. 2021;397(10293):2476–86.
48. Schlaich MP, Schmieder RE, Bakris G, Blankestijn PJ, Böhm M, Campese VM, et al. International expert consensus statement: percutaneous transluminal renal denervation for the treatment of resistant hypertension. J Am Coll Cardiol. 2013;62(22):2031–45.
49. Schmieder RE, Mahfoud F, Azizi M, Pathak A, Dimitriadis K, Kroon AA, et al. European society of hypertension position paper on renal denervation 2018. J Hypertens. 2018;36(10):2042–8.
50. Williams B, Mancia G, Spiering W, Agabiti Rosei E, Azizi M, Burnier M, et al. 2018 esc/ Esh guidelines for the management of arterial hypertension: the task force for the management of arterial hypertension of the European Society of Cardiology and the European Society of Hypertension: the task force for the management of arterial hypertension of the European Society of Cardiology and the European Society of Hypertension. J Hypertens. 2018;36(10):1953–2041.
51. Lobo MD, Sharp ASP, Kapil V, Davies J, De Belder MA, Cleveland T, et al. Joint Uk Societies' 2019 consensus statement on renal denervation. Heart. 2019;105(19):1456–63.
52. Ong PJ, Foo D, Ho HH. Successful treatment of resistant hypertension with percutaneous renal denervation therapy. Heart. 2012;98(23):1754–5.
53. Ho HH, Foo D, Ong PJ. Successful preoperative treatment of a patient with resistant hypertension who had percutaneous renal denervation therapy before bariatric surgery. J Clin Hypertens (Greenwich). 2012;14(8):569–70.
54. Yang JH, Choi SH, Gwon HC. Percutaneous renal sympathetic denervation for the treatment of resistant hypertension with heart failure: first experience in Korea. J Korean Med Sci. 2013;28(6):951–4.
55. Jiang XJ, Liang T, Dong H, Peng M, Ma WJ, Guan T, et al. Safety and short-term efficacy of renal sympathetic denervation in the treatment of resistant hypertension. Zhonghua Yi Xue Za Zhi. 2012;92(46):3265–8.
56. Kario K, Yamamoto E, Tomita H, Okura T, Saito S, Ueno T, et al. Sufficient and persistent blood pressure reduction in the final long-term results from Symplicity Htn-Japan - safety and efficacy of renal denervation at 3 years. Circ J. 2019;83(3):622–9.
57. Kim BK, Böhm M, Mahfoud F, Mancia G, Park S, Hong MK, et al. Renal denervation for treatment of uncontrolled hypertension in an Asian population: results from the global Symplicity registry in South Korea (Gsr Korea). J Hum Hypertens. 2016;30(5):315–21.
58. Wang TD, Lee YH, Chang SS, Tung YC, Yeh CF, Lin YH, et al. 2019 consensus statement of the Taiwan hypertension society and the Taiwan society of cardiology on renal denervation for the management of arterial hypertension. Acta Cardiol Sin. 2019;35(3):199–230.

59. Kario K, Kim BK, Aoki J, Wong AY, Lee YH, Wongpraparut N, et al. Renal denervation in Asia: consensus statement of the Asia renal denervation consortium. Hypertension. 2020;75(3):590–602.

60. Ueshima H, Sekikawa A, Miura K, Turin TC, Takashima N, Kita Y, et al. Cardiovascular disease and risk factors in Asia: a selected review. Circulation. 2008;118(25):2702–9.

61. Kario K, Chen CH, Park S, Park CG, Hoshide S, Cheng HM, et al. Consensus document on improving hypertension management in Asian patients, taking into account Asian characteristics. Hypertension. 2018;71(3):375–82.

62. Wang TD, Goto S, Bhatt DL, Steg PG, Chan JC, Richard AJ, et al. Ethnic differences in the relationships of anthropometric measures to metabolic risk factors in Asian patients at risk of Atherothrombosis: results from the reduction of Atherothrombosis for continued health (reach) registry. Metabolism. 2010;59(3):400–8.

63. Hoshide S, Wang JG, Park S, Chen CH, Cheng HM, Huang QF, et al. Treatment considerations of clinical physician on hypertension management in Asia. Curr Hypertens Rev. 2016;12(2):164–8.

64. Arima H, Murakami Y, Lam TH, Kim HC, Ueshima H, Woo J, et al. Effects of prehypertension and hypertension subtype on cardiovascular disease in the Asia-Pacific region. Hypertension. 2012;59(6):1118–23.

65. Kario K, Bhatt DL, Brar S, Bakris GL. Differences in dynamic diurnal blood pressure variability between Japanese and American treatment-resistant hypertensive populations. Circ J. 2017;81(9):1337–45.

66. Uzu T, Ishikawa K, Fujii T, Nakamura S, Inenaga T, Kimura G. Sodium restriction shifts circadian rhythm of blood pressure from nondipper to dipper in essential hypertension. Circulation. 1997;96(6):1859–62.

67. Omboni S, Aristizabal D, De La Sierra A, Dolan E, Head G, Kahan T, et al. Hypertension types defined by clinic and ambulatory blood pressure in 14143 patients referred to hypertension clinics worldwide. Data from the Artemis study. J Hypertens. 2016;34(11):2187–98.

68. Choi K, Choi S. Current status and future perspectives of renal denervation. Korean Circ J. 2021;51(9):717–32.

69. Victor RG. Carotid Baroreflex activation therapy for resistant hypertension. Nat Rev Cardiol. 2015;12(8):451–63.

70. Mancia G, Grassi G. The autonomic nervous system and hypertension. Circ Res. 2014;114(11):1804–14.

71. Illig KA, Levy M, Sanchez L, Trachiotis GD, Shanley C, Irwin E, et al. An implantable carotid sinus stimulator for drug-resistant hypertension: surgical technique and short-term outcome from the multicenter phase ii Rheos feasibility trial. J Vasc Surg. 2006;44(6):1213–8.

72. Scheffers IJ, Kroon AA, Schmidli J, Jordan J, Tordoir JJ, Mohaupt MG, et al. Novel Baroreflex activation therapy in resistant hypertension: results of a European multi-center feasibility study. J Am Coll Cardiol. 2010;56(15):1254–8.

73. Bisognano JD, Bakris G, Nadim MK, Sanchez L, Kroon AA, Schafer J, et al. Baroreflex activation therapy lowers blood pressure in patients with resistant hypertension: results from the double-blind, randomized, placebo-controlled Rheos pivotal trial. J Am Coll Cardiol. 2011;58(7):765–73.

74. De Leeuw PW, Bisognano JD, Bakris GL, Nadim MK, Haller H, Kroon AA, et al. Sustained reduction of blood pressure with baroreceptor activation therapy: results of the 6-year open follow-up. Hypertension. 2017;69(5):836–43.

75. Alnima T, De Leeuw PW, Kroon AA. Baropacing As a new option for treatment of resistant hypertension. Eur J Pharmacol. 2015;763(Pt A):23–7.

76. Hoppe UC, Brandt MC, Wachter R, Beige J, Rump LC, Kroon AA, et al. Minimally invasive system for Baroreflex activation therapy chronically lowers blood pressure with pacemaker-like safety profile: results from the Barostim neo trial. J Am Soc Hypertens. 2012;6(4):270–6.

77. Zile MR, Lindenfeld J, Weaver FA, Zannad F, Galle E, Rogers T, et al. Baroreflex activation therapy in patients with heart failure with reduced ejection fraction. J Am Coll Cardiol. 2020;76(1):1–13.

78. Weber MA, Mahfoud F, Schmieder RE, Kandzari DE, Tsioufis KP, Townsend RR, et al. Renal denervation for treating hypertension: current scientific and clinical evidence. Jacc Cardiovasc Interv. 2019;12(12):1095–105.
79. Sarathy H, Cohen J. Renal denervation for the treatment of hypertension. Clin J Am Soc Nephrol. 2021;16(9):1426–8.
80. Tsioufis C, Papademetriou V, Dimitriadis K, Tsiachris D, Thomopoulos C, Park E, et al. Catheter-based renal sympathetic denervation exerts acute and chronic effects on renal hemodynamics in swine. Int J Cardiol. 2013;168(2):987–92.
81. Dörr O, Liebetrau C, Möllmann H, Gaede L, Troidl C, Haidner V, et al. Brain-derived neurotrophic factor as a marker for immediate assessment of the success of renal sympathetic denervation. J Am Coll Cardiol. 2015;65(11):1151–3.
82. De Jong MR, Hoogerwaard AF, Adiyaman A, Smit JJJ, Heeg JE, Van Hasselt BAAM, et al. Renal nerve stimulation identifies Aorticorenal innervation and prevents inadvertent ablation of vagal nerves during renal denervation. Blood Press. 2018;27(5):271–9.
83. De Jong MR, Adiyaman A, Gal P, Smit JJ, Delnoy PP, Heeg JE, et al. Renal nerve stimulation-induced blood pressure changes predict ambulatory blood pressure response after renal denervation. Hypertension. 2016;68(3):707–14.
84. Fischell TA, Vega F, Raju N, Johnson ET, Kent DJ, Ragland RR, et al. Ethanol-mediated perivascular renal sympathetic denervation: preclinical validation of safety and efficacy in a porcine model. EuroIntervention. 2013;9(1):140–7.
85. Heuser RR, Mhatre AU, Buelna TJ, Berci WL, Hubbard BS. A novel non-vascular system to treat resistant hypertension. EuroIntervention. 2013;9(1):135–9.
86. Burchell AE, Lobo MD, Sulke N, Sobotka PA, Paton JF. Arteriovenous anastomosis: is this the way to control hypertension? Hypertension. 2014;64(1):6–12.
87. Kapil V, Sobotka PA, Saxena M, Mathur A, Knight C, Dolan E, et al. Central iliac arteriovenous anastomosis for hypertension: targeting mechanical aspects of the circulation. Curr Hypertens Rep. 2015;17(9):585.
88. Ratcliffe L, Hart E, Patel N, Szydler A, Chrostowska N, Wolf J. Unilateral carotid body resection As an anti-hypertensive strategy: a proof of principle study in resistant hypertensive patients. J Hum Hypertens. 2015;29(10):625.
89. Narkiewicz K, Ratcliffe LE, Hart EC, Briant LJ, Chrostowska M, Wolf J, et al. Unilateral carotid body resection in resistant hypertension: a safety and feasibility trial. Jacc Basic Transl Sci. 2016;1(5):313–24.

Printed in the United States
by Baker & Taylor Publisher Services